KITCHEN SCIENCE FRACTALS

A Lab Manual for Fractal Geometry

Other Related Titles from World Scientific

Benoit Mandelbrot: A Life in Many Dimensions
edited by Michael Frame and Nathan Cohen
ISBN: 978-981-4366-06-9

Analysis, Probability and Mathematical Physics on Fractals
edited by Patricia Alonso Ruiz, Joe P Chen, Luke G Rogers,
Robert S Strichartz and Alexander Teplyaev
ISBN: 978-981-121-552-0

Elegant Fractals: Automated Generation of Computer Art
by Julien Clinton Sprott
ISBN: 978-981-3237-13-1

From Fractals and Cellular Automata to Biology: Information as Order Hidden Within Chance
by Alberto Strumia
ISBN: 978-981-121-715-9

Clouds Are Not Spheres: A Portrait of Benoît Mandelbrot, The Founding Father of Fractal Geometry
by Nigel Lesmoir-Gordon
ISBN: 978-1-78634-474-8

KITCHEN SCIENCE FRACTALS

A Lab Manual for Fractal Geometry

Michael Frame
Yale University
New Haven, Connecticut, USA

Nial Neger
Trumbull High School
Trumbull, Connecticut, USA

World Scientific

NEW JERSEY • LONDON • SINGAPORE • BEIJING • SHANGHAI • HONG KONG • TAIPEI • CHENNAI • TOKYO

Published by

World Scientific Publishing Co. Pte. Ltd.
5 Toh Tuck Link, Singapore 596224
USA office: 27 Warren Street, Suite 401-402, Hackensack, NJ 07601
UK office: 57 Shelton Street, Covent Garden, London WC2H 9HE

Library of Congress Control Number: 2021946322

British Library Cataloguing-in-Publication Data
A catalogue record for this book is available from the British Library.

KITCHEN SCIENCE FRACTALS
A Lab Manual for Fractal Geometry

ISBN 978-981-121-845-3 (hardcover)
ISBN 978-981-121-892-7 (paperback)
ISBN 978-981-121-846-0 (ebook for institutions)
ISBN 978-981-121-847-7 (ebook for individuals)

For any available supplementary material, please visit
https://www.worldscientific.com/worldscibooks/10.1142/11774#t=suppl

Typeset by Stallion Press
Email: enquiries@stallionpress.com

Printed in Singapore

***Left to right*: Michael Frame, Benoit Mandelbrot, Nial Neger**

Michael Frame taught math for 43 years, the last 20 at Yale, working with Benoit Mandelbrot. At Yale he won three teaching awards, including the Devane Medal. His previous books include *Chaos Under Control* with David Peak, *Fractals, Graphics, and Mathematics Education* with Benoit Mandelbrot, *Fractal Worlds* with Amelia Urry, *Benoit Mandelbrot: A Life in many Dimensions* with Nathan Cohen, *Mathematical Models in the Biosciences* 1 and 2, and *Geometry of Grief*. With Nial Neger and Benoit Mandelbrot he spent six delightful summers running the workshops that are the basis for this book.

Nial Neger taught math at Trumbull High School in Trumbull Connecticut for 33 years. While in Trumbull he held the positions of the k-12 district program leader in mathematics and the chairman of the high school mathematics department. In 1999 he retired and took a course in Fractal Geometry taught by Michael Frame at Yale University. There he met Benoit Mandelbrot and worked with Michael and Benoit researching fractal geometry and running summer workshops on fractals until 2005. He has co-authored several articles on fractals with Michael Frame and has contributed to the books *Fractals, Graphics, and Mathematics Education*, by Benoit Mandelbrot and Michael Frame, *Content Area Mathematics for Secondary Teachers*, by Allen Cook and Natalia Romalis, and *Benoit Mandelbrot A Life in Many Dimensions* by Michael Frame and Nathan Cohen.

Contents

Prologue

What do the shapes of Fig. 1 have in common? Perhaps first we should ask what they are. We'll argue that in one way or another, all are fractals, shapes made of pieces each of which looks like the whole shape. As an important organizing principle of geometry this was first postulated by Benoit Mandelbrot, whose meditations and manifesto on this geometry is *The Fractal Geometry of Nature* [1].

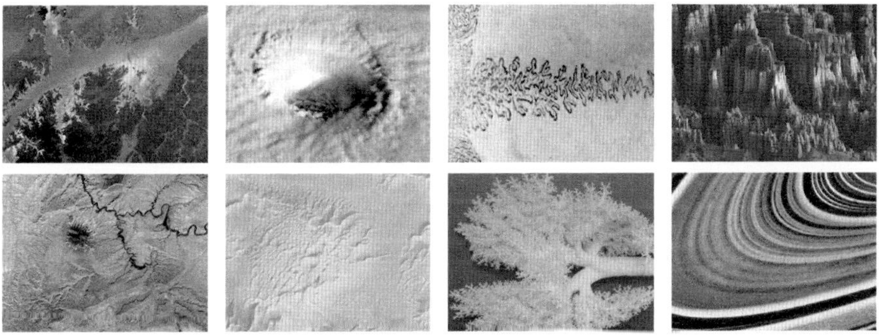

Figure 1: Some natural fractals.

The first picture of the top row is a coastline, an appropriate image to begin our gallery of fractals because Benoit Mandelbrot's 1967 paper "How long is the coast of Britain?" [2] was the first widely-available result to present a fractal aspect of nature, though Mandelbrot had not yet introduced the word "fractal". This paper does contain the term "self-similarity" which means that the shape, here a coastline, consists of bits that are similar to the whole. Bays and inlets are decorated with with smaller bays and inlets, those in turn by still smaller bays and inlets, for several more levels. With Google Maps, zooming in on a coastline, especially of Norway or Wales, shows similar structures over several levels.

The second picture of the top row is a view of a hurricane from space. The hurricane's eye, itself a spiral of clouds, appears to be surrounded by a collection of smaller spirals, some of which reveal their own haloes of still smaller spirals. The first time we showed this image in a summer workshop, the initial guess was the top of a cup of latte from which someone has spooned out some foam. After a bit of laughter, we saw this picture introduced another feature of fractals, *scale ambiguity*, by which we mean it is difficult to distinguish a nearby small object from a distant large object. That is, the object has no natural length scale. Water waves do, which is why in 1950s monster movies the 300 foot behemoth stomping through a harbor clearly is a guy in a rubber suit splashing through a swimming pool.

The third picture is the suture in a deer skull. The wiggles and meanders of the suture are modified by wiggles and meanders of smaller and smaller sizes. These lock together the pieces of the deer skull, accommodating growth by expanding skull plates and sutures alike. The clarity of this image emphasizes another point about physical fractals. Whereas mathematical fractals exhibit copies on arbitrarily small scales, for physical fractals we see similar structures only down to a certain size. The *scaling range* is the extent of sizes over which similar structures exist. It represents the range over which the same physical forces or biological processes dominate the development of the object. For example, the hydrodynamic forces that sculpt a coastline hold over a range from kilometers to centimeters. The appropriateness of a fractal description of an object depends on the scaling range. A common rule of thumb is that scaling over two orders of magnitude is needed to support a claim of fractality. Sadly, this rule was violated in hundreds of papers in the early days of fractals.

The fourth picture is mountains of course. Sculpted by erosion, similar forces at work over many scales. Spires consist of smaller spires, surrounded by ever larger populations of ever smaller spires. A consequence of self-similarity is that the number of small spires exceeds the number of large. But this vague relation is not sufficient to establish fractality. The number of pieces should be (approximately) a power of the size of the pieces. This is called a *power law relation*. Mathematical fractals exhibit an exact power law relation; physical fractals an approximate relation.

The fifth picture is a river network: rivers with meanders and tributaries, each with its own meanders and tributaries. Zooming in, we would find additional levels of structure below the resolution of this image. These networks are determined in part by the local topography, but the weathering that sculpts topography is driven in part by the flow of water. Moreover, the path of a river changes with time. This emphasizes that while fractals may look complicated, the processes that grow them can be simple. The old notion that simple effects have simple causes and complex effects have complex causes is broken by fractals.

The sixth picture is a network of sand dunes. This multi-level branching is formed by the interaction of wind and the flow of sand, not a liquid, not a solid, but a collection of solids interacting by gravity, friction, and collision. Aggregates of solid particles—sand in a desert, dust clumps under your bed, iron pellets in an iron mine pelletizing factory, grain in a grain elevator—exhibit dynamical behaviors that differ from those of liquids and those of solids. An interesting demonstration is to hold a broomstick vertically into an empty bucket, then fill the bucket with (uncooked) rice. Grasp the top of the broomstick and lift. The bucket rises with the broomstick. The structures the rice grains assume to allow this are worth a look. Hit the side of the bucket with a hammer and you can lift the broomstick out of the rice. Fractals can be found in states of matter between the familiar solid, liquid, and gas.

The seventh picture is of a cast of a dog's lung. On average, human lungs have 23 successive levels of branching and are a fine example of a biological fractal. Iterated branching is evolution's way to fold a large surface area into a small volume, with similar airway length from the trachea to each of the branch tips. Rather than a detailed structural map, the genetic instructions that govern lung growth likely are simple rules—grow a multiple of the branch circumference and then split—that are repeated over many levels. Typical lung area is about 130 m^2, contained in a volume of 4 to 6 liters. Starting with level 16, these branches are covered with alveoli, about half a billion in total. Over many millions of years, evolution has discovered efficiencies. These include using the laws of physics and of geometry, including fractal geometry. Familiarity with fractal geometry can help us recognize patterns in biology.

The eighth picture is of the rings of Saturn. Many years ago Saturn's rings were thought to consist of a small number of pieces, but spacecraft photos have shown them to be many many individual thin rings, grouped together with a hierarchy of gaps. Rings and gaps, larger gaps surrounded by a cascade of smaller gaps, interwoven with still smaller gaps, and so on. The gravitational pull of Saturn's many moons sweep ring particles from the gaps. Now we know that Saturn is not the only ringed planet: Jupiter, Uranus, and Neptune have rings, though more modest than those of Saturn. Where's the fractal? The cross-section of the rings is an example of a fractal called a fat Cantor set, which we'll discuss in Lab 2.3.

(Not long after the first Voyager images of Saturn's rings were obtained, one of us (MF) team-taught an astronomy course with a physicist. In class the physicist showed a Voyager picture of Saturn's rings and said that Saturn had thousands of rings, "like the groves in a phonograph record." I pointed out that a phonograph record has a single spiral grove, not a bunch of concentric circular groves. The students laughed, and I am sorry to say this was neither the first time nor the last time I was called a rude name by another teacher.)

As a graduate student in the University of Paris, Benoit Mandelbrot studied probability theory. Benoit's memoirs, *The Fractalist* [3], is a good sketch of his life. But despite his work in pure mathematics, Benoit's work never was far from applications. Often Benoit recounted the story of Antaeus, a Gigantes of Greek mythology who could not be defeated when he kept contact with the Earth. He likened mathematics to Antaeus and the applied sciences to the Earth. Math is strongest when it remembers its connections, its source. And from childhood he was interested in maps, in complicated shapes, in geometry. Early studies of errors in telephone circuits, part of IBM's efforts to let computers talk with one another over phone lines, led him to discover that noise occurs in clumps and that the distribution of clump sizes follows a power law.

We mentioned power laws a few paragraphs earlier. Now let's look in a little more detail. The variable y exhibits a power law dependence on the variable x if there are constants a and b for which

$$y = ax^b \tag{1}$$

This relationship may be difficult to recognize from a plot of y versus x, but take the logarithm of both sides of Eq. (1) and use these properties of logarithms,

$\log(A \cdot B) = \log(A) + \log(B)$ and $\log(A^B) = B\log(A)$, to find a relation with a clearer graphical signature.

$$\log(y) = b\log(x) + \log(a) \qquad (2)$$

That is, a plot of $\log(y)$ versus $\log(x)$ gives points that lie on a straight line with slope b and intercept $\log(a)$.

Power law relations are common in science. For example,

• Hooke's law for springs, $F(x) = -kx$, the restoring force of a spring is proportional to the displacement, either compression or extension, of the spring.

• Newton's law of gravitation, $F(r) = -GMm/r^2$, the gravitational force between points of mass M and m is proportional to the reciprocal of the square of the distance between the points.

• In 1932 Max Kleiber [4] analyzed data over an immense range of scales (21 orders of magnitude, from microbes to whales) and found that the metabolic rate per unit mass is proportional to $M^{-1/4}$. This was a surprise, because in 1883 Max Rubner [5] (yes, another Max) argued that the metabolic rate per unit mass is proportional to $M^{-1/3}$. Rubner's argument is roughly this: the surface area of an animal is proportional to L^2, where L is the animal's length. The animal's mass M is proportional to its volume, hence to L^3. The "proportional to" is to accommodate different geometries, among other things. Heat is dissipated across the surface of an animal, so the metabolic rate is proportional to L^2 and the metabolic rate per unit mass is proportional to $L^2/L^3 = 1/L = 1/M^{1/3} = M^{-1/3}$. The difference between Rubner's theory and Kleiner's empirical data still is not resolved to everyone's satisfaction.

• Stellpflug's pumpkin law, $W(r) \approx kr^{2.74}$, the weight of a pumpkin is proportional to its radius to the 2.74 power. Why not the power 3? Because larger pumpkins have more than proportionally larger hollow centers.

Benoit found other examples of scaling in rainfall patterns [6], in fluctuations of the annual flooding of the Nile [7], in the distribution of incomes [8], in stock price fluctuations [9], in turbulence [10], in the numbers of mutants in old bacterial cultures [11], among many others. Gradually Benoit saw a common pattern below the surface in these examples. All were described by a geometry based on scaling relations. This is fractal geometry. Although first glimpsed through instances in the physical world, fractal geometry is easier to grasp in the mathematical world. In later chapters we'll investigate physical fractals; mathematical fractals are the subjects of the labs in Chapter 1.

Among the geometric fractals, the most familiar examples are self-similar. These are easy to recognize: a *self-similar set* is composed of smaller exact copies of itself. This decomposition sets in motion a zoom through ever decreasing scales: if a set is made of smaller copies of the whole, then those pieces must themselves be made of still smaller copies of the whole, and so on. For mathematical fractals this cascade continues forever, past arbitrarily small copies. Earlier we mentioned that for physical fractals this iterated decomposition extends only over a limited range, which as we mentioned earlier is called the scaling range. A fractal description of an object is appropriate only if it gives a simplified way to view the object over a significant range of scales. Often Benoit said that no one benefits from the attempted application of fractality to inappropriate settings. "Fractals everywhere" means that many places we look, our gaze includes objects that can be described as fractal. It does not mean that every object we see is a fractal.

One of the most famous self-similar sets is the Sierpinski gasket, the left image shown here. It is easy to see that this fractal is made up of copies of itself: it has three large copies, one of which is shown magnified on the right. These in turn contain copies of copies, and so on.

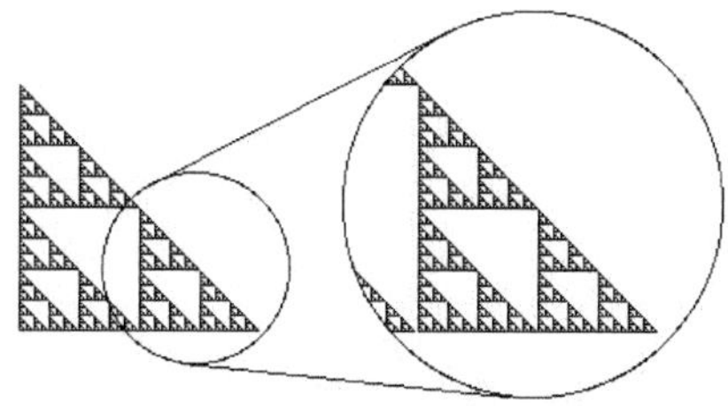

Note that in theory the complete gasket is the limit set of an infinite process, and as a result, any picture we can see is not the actual gasket but a stage in the process of generating the gasket. No one has ever seen a complete gasket, or any other self-similar fractal, except with the imagery of our minds. To be fair, no one has seen an exact circle, either.

The first book on fractals is Benoit Mandelbrot's *The Fractal Geometry of Nature*[1]. Since then books have been published on fractals for little kids [12], for general readers [13, 14], for teachers [15], for undergraduates [47, 17, 18, 19], for graduate students [20], and probably hundreds of volumes of conference proceedings.

Kitchen science generally refers to experiments that can be performed with a minimum amount of equipment. No particle accelerators, no autoclaves, no gene sequencers, no X-ray spectroscopes, no CRISPR technology. Just things you'd find in or near a kitchen. When we began this project in 2000, we added the subtitle "if there's a computer in your kitchen." Nowadays this addition does not seem so relevant. We must mention that a few of the experiments in Chapter 5 do use some proper lab equipment. Some chemicals, a DC power supply (or batteries and a rheostat), breadboards, resistors, capacitors, inductors. One uses a signal generator and an oscilloscope. Probably your kitchen doesn't have an oscilloscope, and probably your kitchen breadboard won't be the kind where you assemble circuits. But most of the physical experiments are pretty simple.

From 2000 to 2006 with Benoit Mandelbrot we ran a series of summer workshops on fractal geometry for middle school, high school, and college teachers. Initially we ran a five-day program on iterated function systems (IFS), dimension and measurement, Julia sets and the Mandelbrot set, cellular automata, and random fractals and the stock market. Morning sessions were some approximation of lectures by one of us (usually MF) based on his fractal geometry course at Yale. The 20 or 30 interested teachers in the room guaranteed the mornings involved a lot of interaction. Afternoon sessions were lab exercises lead by the other of us (usually NN). A discussion of the main points ended the day, with emphasis on how parts of the material could be incorporated into lessons in math and science classes. While some colleges do offer fractal geometry courses, most high school curricula don't have room for a course devoted to fractals. Senior math classes after the AP exams were the exception: with no more national test preparation, teachers had to fill in the rest of the year with something, so it may as well be something interesting. But because our goal was to fold fractal geometry in the high school math curriculum, the only approach that could work was to use fractals as tools to engage students in the study of established topics. In this the teachers were particularly enthusiastic and creative.

• Do you want to get your students interested in the geometry of transformations of the plane? Teach them about IFS and have a contest for finding the transformations that generate a given fractal. This approach is especially effective for middle school students, reinforcing the notion that visualization is a good way to make geometry interesting even for very young students.

• Logarithms may be the least interesting topic in high school math, in part because they appear to have no use other than to solve the problems of the logarithm chapter of the math text. But after students have seen a selection of fractal images and are asked to rank the fractals by degree of visual complexity, dimension as a measure of complexity makes a lot of sense. And logarithms are the tool to compute dimensions.

• Often complex numbers are introduced to make sense of some applications of the quadratic formula. After years of hearing that "you can't take the square root of a negative number," what do you do with

$$x = \frac{-b \pm \sqrt{b^2 - 4ac}}{2a}$$

when $4ac > b^2$ and so $b^2 - 4ac$ is negative? This does necessitate the introduction of complex numbers, but isn't such an effective motivation if students aren't interested in quadratic equations; it's just another complication in algebra. But show a few pictures of the Mandelbrot set, a couple of animated zooms to find a small copy of the Mandelbrot set around the border of the Mandelbrot set, then show the formula $z_{n+1} = z_n^2 + c$. *That* tiny formula produces *all those* pictures? Do you want to see how? Oh, yes. This motivates the use of complex numbers.

Every summer that we ran the workshops, Benoit Mandelbrot arrived mid morning one day to talk with the teachers. The effect was remarkable. Even though we'd told the teachers that Benoit would be there, many weren't prepared to see the person who had discovered all this math. Benoit talked for a while, then asked for questions. The teachers were silent. So Benoit said something funny. For example, "People ask what is the hardest theorem I've proved? I've proved only simple theorems. But I've asked some very very good questions." Silence for a beat, then laughter, then lots of questions. Despite his stratospheric reputation, Benoit was down-to-earth, funny, and genuinely interested in what the teachers do. Certainly, Benoit's visit was the highlight of the workshops.

At the end of the first summer workshops, we thanked the teachers for their participation but were unprepared for their response, "So what are we doing next summer?" This was a surprise. We'd planned to run the same sequence of five workshops, with some tweaks to be sure, for a different group of teachers. And we did that every summer till the workshops ended. But the second year we added a part 2 workshop, three days on new topics: fractals generated by Pascal's triangle, circle inversions, and nonlinear tilings. At the end of these, again the question, "So what are we doing next summer?" We settled into a pattern: roughly the same five workshops for new participants, then each year three new topics for the part 2 workshops. We had to continue to add new topics because some people came to the part 2 workshops year after year. From the total of 225 teachers who attended the workshops, 14 attended two of the part 2 workshops, 12 attended three, 8 attended four, and 7 attended every one of the five part 2 workshops. We are not that entertaining, so these people must have found the topics useful and interesting.

With these new topics in the part 2 workshops each year, we could explore additional labs: IFS with memory, fractals in music, fractal video feedback, multifractals, algebra of dimensions, negative dimensions, fractals and finance, fractals in four dimensions, the Mandelbrot set for cubic and quartic polynomials, fractal painting, and variations on these. Some of these worked reasonably well, others not so much. At any level beyond middle school arts and crafts, fractal tilings require math more sophisticated than that familiar to high school, and many undergraduate, students. While the music and finance labs were popular, both music and finance are inherently high-dimensional so in the short exploration of a lab we are unlikely to uncover interesting patterns. Another problem is that if music and finance admit fractal patterns, they will be fractals in time, not space. We can't perceive the whole pattern all at once. Certainly, we can look at the score of a composition or at market price charts, but any scaling structures in these are subtle and a challenge to find. For lack of good, accessible results we have dropped these labs. However we must mention that the finance lab did lead to the biggest laugh of the entire six-year program. On one visit to the workshops, Benoit talked about fractals in finance. One of the teachers asked Benoit if he'd applied his techniques to make any money. Benoit replied, "There are four topics I never discuss in public: politics, religion, my portfolio, and sex." Looks of great puzzlement from the teachers. Did he really say what they thought he said? No one made a remark; Benoit waited. Eventually one of us (MF) said, "I understand why you don't discuss politics or religion, and I get the point about not discussing your portfolio. If you answer 'yes', then everyone starts to use the same tools and you lose your edge. If you answer 'no', the question will be 'Why not?' But really, Benoit, has anyone ever asked you about sex?" The immediate reply, "No, but I keep hoping." Silence for a few seconds, then the room exploded. One person laughed so hard he fell out of his chair. Really. And it wasn't either of us. Benoit did not often deploy his sense of humor, but when he did it hit like a cruise missile.

Until the proliferation of computers, lab manuals were written only for the experimental sciences: mostly physics, chemistry, biology, and electronics. Then when computers became household appliances, the list of experimental sciences expanded to include math. This expansion was not embraced joyfully by all mathematicians, but some visionaries, including Benoit, agitated tirelessly for this advance. So far as we know, the first lab manual for fractal geometry experiments was *Chaos, Fractals, and Dynamics: Computer Experiments in Mathematics* [21] by Robert Devaney. Others followed, including the three volumes of *Fractals for the Classroom: Strategic Activities* [22], and the three volumes *Iteration* [23], *Fractals* [24], and *Chaos* [25] of *A Tool Kit of*

Dynamics Activities. Our book includes both computer and physical experiments, as well as some chemistry, art experiments and work in the kitchen and in the garden.

While most of the labs involved computer explorations, we did some physical projects: Labs 1.4, 1.5, 1.6, 2.2, 2.4, 2.5, 2.6, and 4.1. This last, a painting lab, generated an art show that demonstrated some teachers have considerable artistic flair. Over the years after the workshops ended, we found more ways that fractals arise in the sciences and arts, and in cooking (Labs 4.2—4.9 and 5.1—5.9). These physical experiments were especially enjoyable because they provided motivation for one of us (MF) to set up a little laboratory in the attic of his house, a temporal bookend for the lab he had as a child in a corner of his father's workshop.

That Benoit died before we got to these physical experiments is a great pity. In the twenty years we worked together, the happiest times were when we went exploring. Which idea led to what, how these notions are related, on and on. The image—so clear then, a treasured memory now—was that of little kids, running around in a big field under a sunny sky, discovering all sorts of interesting things and happily sharing these discoveries. The design of some of the experiments in this book has brought an echo of that state of mind. We hope that you will have similar fun, or better yet, design and carry out your own fractal laboratory experiments.

Michael Frame
Hamden, CT

Nial Neger
Trumbull, CT

Software and solutions

In order to benefit fully from this book, readers should have access to either Mathematica or Python and be able to run programs in one of those languages. No programming skill is needed: the programs can be copied from the supplementary material to support this book, available at

https://www.worldscientific.com/worldscibooks/10.1142/11774#t=suppl

There you will find four files

KSFMma.html, which contains Mathematica code used for the computer exercises. Open the html file in your web browser, copy the program you want to use, and paste it into your own copy of Mathematica.

KSFPython.html, which contains Python code used for the computer exercises. Open the html file in your web browser, copy the program you want to use, and paste it into your own copy of Python.

amylase.html, which contains the DNA sequence for amylase. This is used with the symbol-driven IFS software.

KSFSol.pdf, which contains solutions to selected exercises from the labs.

Chapter 1

IFS Labs

A way to generate any self-similar set is to start with some shape. (Technically, a closed and bounded figure. These are called *compact*. See Appendix B.1.1.) Then perform a family of scalings(contractions), reflections, rotations, and translations on the figure. These transformations, called an *iterated function system* (IFS), generate a collection of smaller copies of the original figure. Group these together and apply the IFS to the grouped image. Repeat again and again, until no change is visible from one generation to the next. This is an approximation of the fractal produced by that family of transformations.

For the gasket, all we need are scalings and translations. Start with a filled-in isosceles right triangle. Scale this triangle by 1/2 in both directions, make three copies of this smaller triangle. Then translate one copy over and one copy up, by the same distances, to obtain the configuration on the right of Fig. 1.1. If we suppose the original triangle has base and altitude 1 and its lower left vertex is the origin, then the transformations become

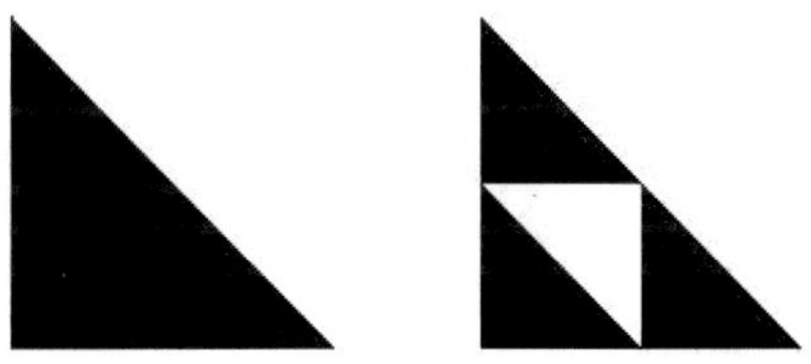

Figure 1.1: First step to build a gasket.

1. Scale by 1/2 horizontally and vertically. That is, multiply the x- and y-coordinates by 1/2. Note that scaling does not move the origin.
2. Scale by 1/2 horizontally and vertically, and translate by 1/2 horizontally.
3. Scale by 1/2 horizontally and vertically, and translate by 1/2 vertically.

After we apply these three transformations to the triangle to obtain three smaller triangles, we apply the three transformations to the union of these three triangles to obtain nine still smaller triangles, and so on. See Fig. 1.2.

Figure 1.2: The first three steps to build a gasket.

In principle, we continue this process forever. In practice after a few iterations the changes are smaller than pixels on the computer screen, so we stop. But if we could continue this process forever, this sequence of shapes converges to a gasket, a mathematical object, not physical or even graphical.

Note that applying these three transformations to the gasket leaves the gasket unchanged. The gasket is the only (compact) shape for which this is true. In that sense, these three transformations determine the gasket.

This is a bit of magic: among all the infinitely many shapes you can draw in the plane, the gasket is the *only* shape left unchanged by these three transformations. Here's a larger piece of magic: no matter what picture you use—a sketch of a cat, for example—applying these three transformations again and again gives a sequence of pictures that converges to the gasket.

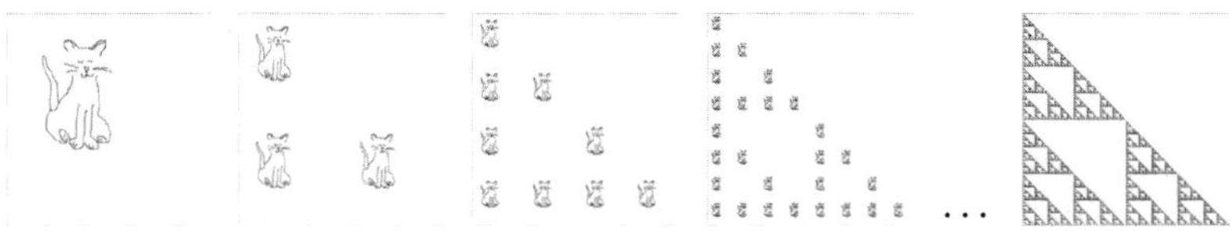

Figure 1.3: How to turn a cat into a gasket.

This method to construct fractals by families of contractions (transformations that decrease the distance between every pair of points) was pioneered by John Hutchinson [26]. The name "iterated function system" was proposed by Michael Barnsley and Stephen Demko [27] and popularized as a method of compressing images by Barnsley and Alan Sloan [28]. Good references on IFS are [16, 20, 29, 31, 32].

We can build fractals with transformations more general than scaling and translation. The pieces also can be rotated and reflected. This we'll study systematically in Lab 1.1. For now, we'll mention an interesting problem. Consider fractals made of three pieces, each scaled by a factor of $1/2$, possibly reflected across the x-axis or across the y-axis, or rotated $90°$, $180°$, or $270°$, and translated so one piece occupies the lower left corner, one the lower right corner, and one the upper left corner, the relative positions of the pieces of the gasket. These are called *Sierpinski gasket relatives.*

The Sierpinski gasket has many relatives, but how many? Each piece can be rotated $0°$, $90°$, $180°$, or $270°$. In addition, each piece can be reflected across the x-axis, across the y-axis, across both, or across neither. This appears to be 16 combinations, each of which can be applied independently to each of the three pieces, giving $16^3 = 4096$ Sierpinski relatives. But really there are not that many, because some combinations of reflection and rotation have the same effect. For instance, reflection across the x- axis and across the y-axis is equivalent to rotation by $180°$, and reflection across the y-axis followed by a $270°$ rotation is equivalent to reflection across the x-axis and rotation by $90°$. When we take into account these equivalences, we find 8 distinct transformations for each piece, hence $8^3 = 512$ Sierpinski relatives. Eight of these are symmetric across the diagonal line $y = x$. One of these is the Sierpinski gasket; see page 236 of [18] for the other seven. Reflection across the diagonal is equivalent to reflection across the x-axis followed by a $90°$ rotation. If the fractal is symmetric across the diagonal, then two transformations—one that involves a reflection across $y = x$, one that doesn't—generate each piece and so eight combinations of transformations produce each symmetric fractal. Consequently we have a total of 64 Sierpinski relatives that are symmetric across the diagonal. This leaves 448 Sierpinski relatives, but because these are not symmetric across $y = x$ these 448 relatives can be viewed as 224 pairs, where reflection across $y = x$ takes one fractal of each pair to the other. On pages 232–234 of [18] we see pictures of these 224 gasket relatives.

While these fractals exhibit a variety of visual characteristics, they can be organized into four groups according to their topology, that is, the relations between the pieces of the fractals. We'll need four concepts from topology. Topology involves properties of a space that are preserved under continuous deformation. For example, if every pair of points in a space can be joined by a path that lies in the space, then this property holds if the space is stretched or twisted, but not broken. Breaking is a discontinuous transformation. We'll use these four topological properties.

- *Connected*: every pair of points in the space lies on a path in the space. (This is *path-connected*, a notion that is generally more restrictive than "connected," but the two are equivalent for subsets of Euclidean space. Path-connectivity is more easily visualized.) That is, the space is all one piece.

- *Disconnected*: the space contains at least one pair of points that do not lie on any path in the space. That is, the space consists of at least two pieces. (Does self-similarity then imply the disconnected fractal must consist of infinitely many pieces?)

- *totally disconnected*: single points are the only connected subsets of the space. *Dust* is an evocative name for a totally disconnected space.

- *simply-connected*: path-connected and every closed loop in the space can be shrunk to a point without leaving the space.

Each gasket relative falls into one of four classes, illustrated in Fig. 1.4.
(a) connected but not simply-connected, (In fact, we see infinitely many loops.)
(b) connected and simply-connected,
(c) disconnected, and
(d) totally disconnected.

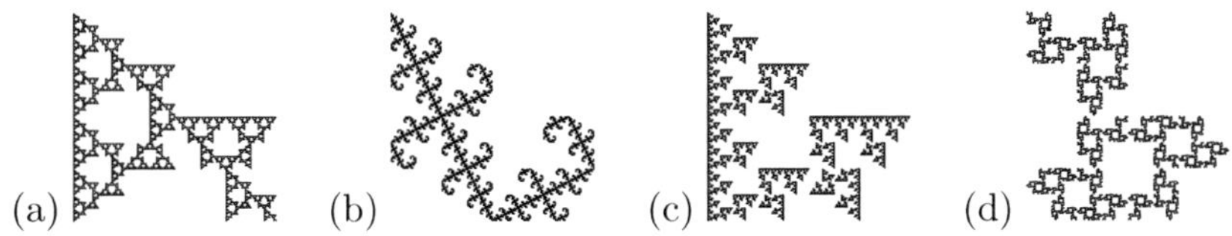

Figure 1.4: Examples of gasket relatives.

Tara Taylor has studied gasket relatives [33, 34] and has found conditions that give rise to each type (a), (b), (c), and (d), though a simple overall criterion that distinguishes these types still is undiscovered. But of course there may be no simple criterion. To illustrate the difficulty of this problem, the fractal pictured in Fig. 1.5—[18] page 233, row 2, number 4—is totally disconnected. Can you see a disconnection? When we wrote this book, the general problem remained unsolved. Perhaps you'll have an idea if you experiment with these fractals.

The labs in this chapter build familiarity with many aspects of the IFS construction. In Labs 1.1–1.4 we'll find ways to approach the inverse problem, that is, find the transformations that generate a given a fractal. This includes two methods, the deterministic algorithm and the random algorithm, for rendering fractals from IFS transformations. In Lab 1.5 we show how to construct a fractal by accretion rather than by refinement.

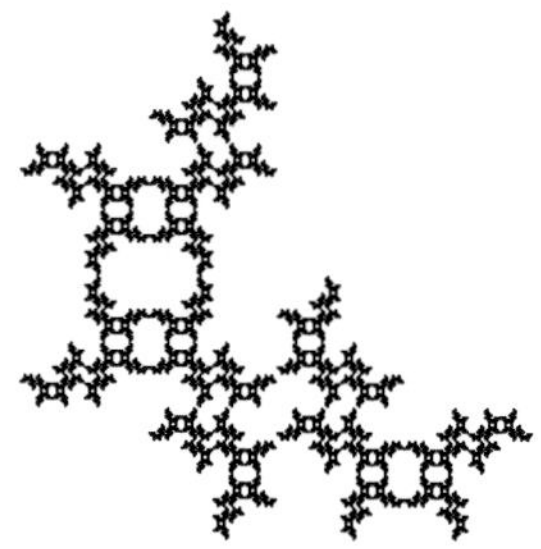

Figure 1.5: Totally disconnected.

The random IFS algorithm, also called the chaos game, sometimes was introduced manually: each student did 20 or 30 iterations with a ruler and a die and the plots they produced were (unconvincingly) proclaimed to look like a gasket. In Lab 1.6 we show that the superposition of several short runs of the random IFS algorithm is similar to a single long run, which does look like a gasket.

The notion of the address of a portion of a fractal is central to much of the study of IFS. We introduce addresses in Lab 1.7; in Lab 1.8 we show that decimals are addresses of an IFS that generates the unit interval. In Labs 1.9 and 1.10 we apply addresses in an extension of IFS to limit which combinations of transformations can be produced. With this we can generate new fractals that cannot be produced by the familiar IFS algorithm. Finally, in Lab 1.11 we modify the random IFS algorithm to give a graphical representation of data sequences.

1.1 Finding IFS for fractal images

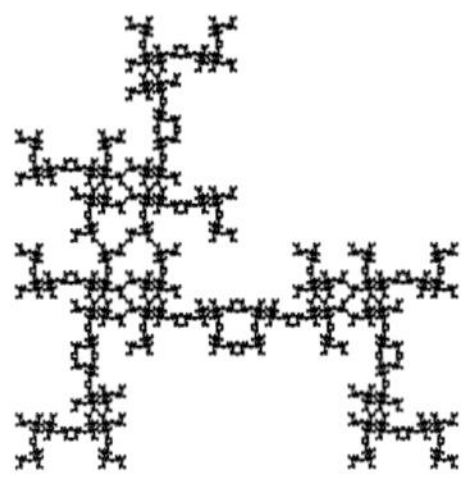

Iteration is a powerful tool to develop intuition about self-similarity. Familiarity with a large collection of examples is another powerful tool. This lab gives some practice in both.

1.1.1 Purpose

We'll learn how to find an iterated function system (IFS) that generates a given fractal. This involves a geometric interpretation of the parameters of a linear transformation of the plane, as well as the iteration of images.

1.1.2 Materials

We'll use IFS software (1.1 of the Mathematica or Python codes), a ruler, and small pieces of thin paper, tracing paper works well. If you have a lot of patience, scissors, tape, and a photocopier can replace the software, if none of the transformations involve reflections. If reflections are involved, overhead transparency films and a marker pen can replace tracing paper.

1.1.3 Background

In this lab we'll learn the basics of generating simple fractals by iterated transformations of the plane. For a given fractal, the collection of these transformations is called the *iterated function system* (IFS) of the fractal. We are interested in the *inverse problem*: given a fractal, find transformations of an IFS that produces the fractal [35]. In this lab IFS generate fractals by a procedure called the *deterministic algorithm*. In the examples we explore it is easy to see that each IFS generates a sequence of shapes that converges to a limit shape, the fractal generated by the IFS, the proof that this always works is a bit involved. We sketch the proof in Appendix B.1.

To produce interesting fractals we need a richer variety of transformations than those used to produce the gasket. We begin by describing these transformations.

1.1.3.1 Geometry of plane transformations

Here is the fundamental geometry of the transformations we'll use.

• *Scaling* measures by how much an image is shrunk. Scaling always means shrink toward the origin. To accommodate general transformations, we allow scaling by different amounts horizontally and vertically. The amount of horizontal scaling is denoted by r; the amount of vertical scaling by s.

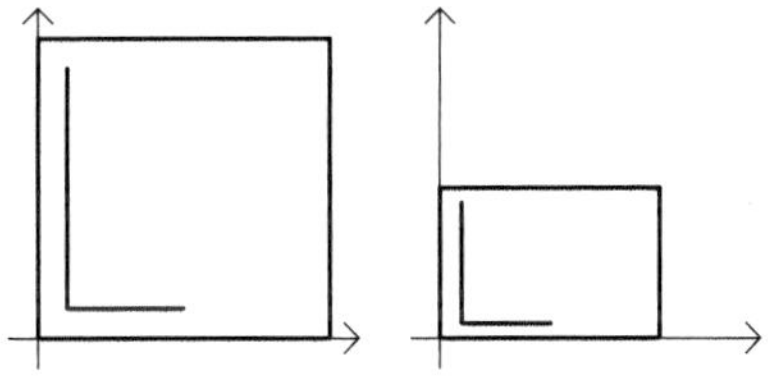

Figure 1.6: Scaling.

In Fig. 1.6 the left image has $r = s = 1$ (no scaling) and the right has $r = 0.75$ and $s = 0.5$. Note that the lower left corner of the contracted figure is still at the origin.

• *Reflection*: Negative r values reflect the shape across the y-axis. Negative s values reflect the shape across the x-axis. In Fig. 1.7 suppose we start with the left image of Fig. 1.6, not shown here to reduce crowding in the image. So the initial shape is a square, in the first quadrant, with an L near the lower left corner of the square. The image in the second quadrant of Fig. 1.7 has $r = -1$ and is a reflection left-right of the initial shape across the y-axis. The image in the fourth quadrant has $s = -1$ and is a reflection up-down of the initial shape across the x-axis.

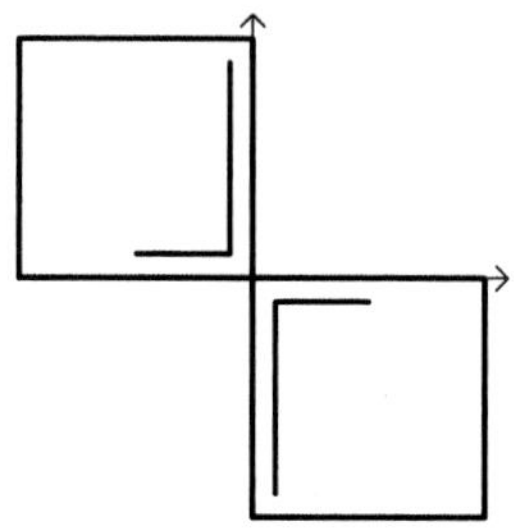

Figure 1.7: Reflections.

• *Rotation*: Counter-clockwise rotation is done by positive angles. Clockwise rotation is done by negaive angles. Rotations are about the origin, the lower left corner of the figure in this example. General linear transformations use two angles, θ for rotations of horizontal lines, φ for rotations of vertical lines In Fig. 1.8 we see $\theta = 30°, \varphi = 0$ (left), $\theta = 0, \varphi = 60°$ (center), and $\theta = \varphi = 45°$ (right). In general, rotations with $\theta = \varphi$ are rigid rotations.

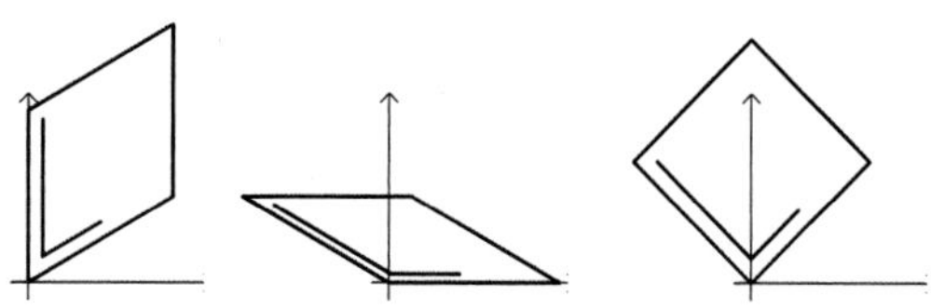

Figure 1.8: Rotations.

Do you see that reflection across the x-axis followed by reflection across the y-axis is equivalent to rotation by 180°?

• *Translation*: The magnitude of horizontal translation is denoted by e; the amount of vertical translation is denoted by f. The left image of Fig. 1.9 has $e = 0.25$, the right has $f = 0.25$. In both cases the transformation begins from the square with lower left corner at the origin.

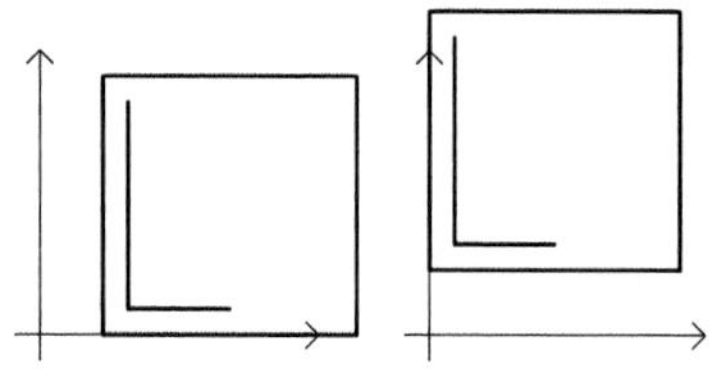

Figure 1.9: Translations.

The mathematical formulation of the most general transformation is

$$\begin{bmatrix} x \\ y \end{bmatrix} \rightarrow \begin{bmatrix} r\cos(\theta) & -s\sin(\varphi) \\ r\sin(\theta) & s\cos(\varphi) \end{bmatrix} \begin{bmatrix} x \\ y \end{bmatrix} + \begin{bmatrix} e \\ f \end{bmatrix} \tag{1.1}$$

Order matters. Note that a reflection followed by rotation does not necessarily produce the same image as this rotation followed by this reflection. To understand the geometry of these transformations, the formulation of Eq. (1.1), and the code presented in 1.1 and 1.2 of the Mathematica and Python codes, perform the operations in this way: scalings first, then reflections, then rotations, and finally translations.

We encode a transformation by listing these parameters as a row in a table. The first two entries involve scaling and reflection, the second two rotation, the final two translation.

r	s	θ	φ	e	f

1.1.3.2 IFS formalism

The only requirement for IFS transformations is that they are *contractions*. That is, for any pair of points (w, x) and (y, z), we have

$$d(T(w,x), T(y,z)) \leq r \cdot d((w,x),(y,z))$$

where d is the Euclidean distance, for instance, $d((w,x),(y,z)) = \sqrt{(w-y)^2 + (x-z)^2}$, and r is a constant, $0 < r < 1$. See Fig. 1.10. The smallest r making this inequality valid for all pairs of points is the *contraction factor* of T.

An *iterated function system* is a collection of contractions $T_1, \ldots, T_n$. In Sects. 1.1.3.3, B.1.2, and B.1.3, we see that given this set of contractions, there is a unique set A satisfying

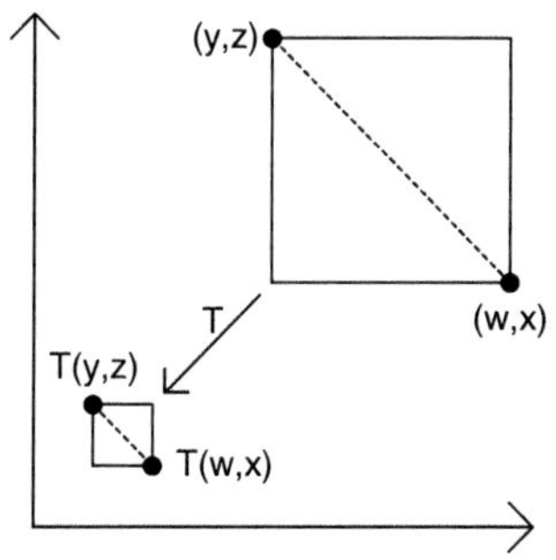

Figure 1.10: A contraction.

$$\mathcal{T}(A) = T_1(A) \cup \cdots \cup T_n(A) = A \tag{1.2}$$

The gasket IFS is shown in Table 1.1.

	r	s	θ	φ	e	f
T_1	0.5	0.5	0	0	0	0
T_2	0.5	0.5	0	0	0.5	0
T_3	0.5	0.5	0	0	0	0.5

Table 1.1: IFS rules for the gasket.

1.1.3.3 The deterministic algorithm

Suppose the three gasket transformations are applied to a filled-in unit square. The result is three smaller filled-in squares, shown in Fig. 1.11(a). We call this the first iteration of the gasket transformations to the unit square.

Applying the three gasket transformations to Fig. 1.11(a) gives Fig. 1.11(b), the second iteration. Fig. 1.11(c) and (d) are the third and eighth.

Given contractions $T_1, \ldots, T_n$ and an initial spape B_0, the *deterministic IFS algorithm* is the process of repeatedly applying all the transformations, obtaining a sequence of shapes $B_1, B_2, \ldots$. That is,

$$B_1 = T_1(B_0) \cup \cdots \cup T_n(B_0) = \mathcal{T}(B_0)$$
$$B_2 = T_1(B_1) \cup \cdots \cup T_n(B_1) = \mathcal{T}(B_1)$$
$$B_3 = T_1(B_2) \cup \cdots \cup T_n(B_2) = \mathcal{T}(B_2)$$

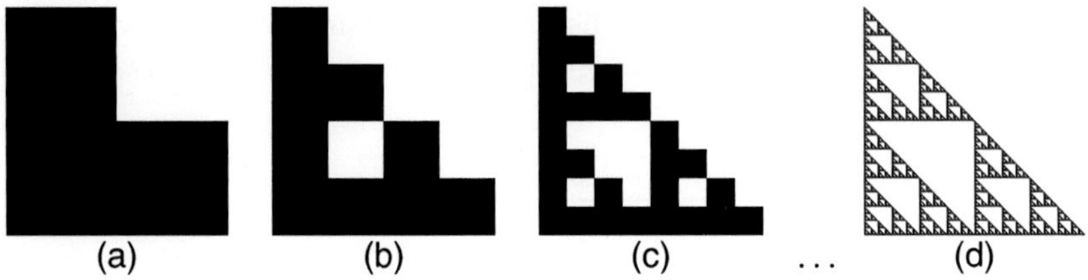

Figure 1.11: The first, second, third, and eighth iterations of applying the gasket rules to the filled-in unit square.

and so on. In Appendix B.1.3 we illustrate the sense in which this sequence $A_0, A_1, A_2 \ldots$ converges.

1.1.3.4 The inverse problem

The solution of the inverse problem involves two steps. The self-similarity of a fractal A guarantees that it can be decomposed into pieces $A_1, A_2, \ldots, A_n$ each of which is similar to A. The first step is to find this decomposition. The second step is this: for each piece A_i find the transformation that turns the whole fractal A into the piece A_i. By "find the transformation" we mean find the r, s, θ, φ, e, and f values defined in Sect. 1.1.3.1.

1.1.4 Procedure

For each fractal image, follow these steps.

• Subdivide the image into smaller copies of itself. To visualize the subdivision it may be useful to enclose the fractal in a square and subdivide the square. See Fig. 1.12.

• For each small copy, find the transformation—that is, find the r, s, θ, φ, e, and f values—that turns the whole into that piece. If the scaling factors aren't apparent, divide the side length of small copy by the length of the corresponding side of the whole fractal.

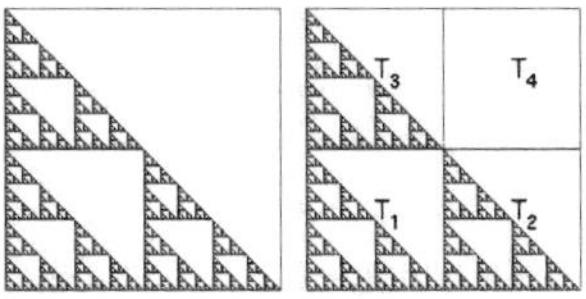

Figure 1.12: The Sierpinski gasket enclosed in a subdivided square.

After determining the values of r, s, θ, φ, e, and f for each part of a fractal, run the IFS program with those values. If the image generated does not match the fractal, use the differences to find appropriate modifications of the parameters. For more complicated images, this process may take several tries.

If finding values for a picture is difficult, here is an approach to build intuition.

• Trace the main features of the fractal and cut out smaller copies of the tracing.

• To allow for reflections, flip the small copies and on the back trace over the lines on the front.

• Place the small copies, perhaps rotating or reflecting them, to make a copy of the original fractal.

1.1.5 Sample A

Note the top and bottom left pieces have the same orientation as the entire fractal, while the bottom right piece is reflected left to right, that is across a vertical line. Because our transformation rules allow only reflections across the x- and y-axes, some care must be taken with the translation after the reflection.

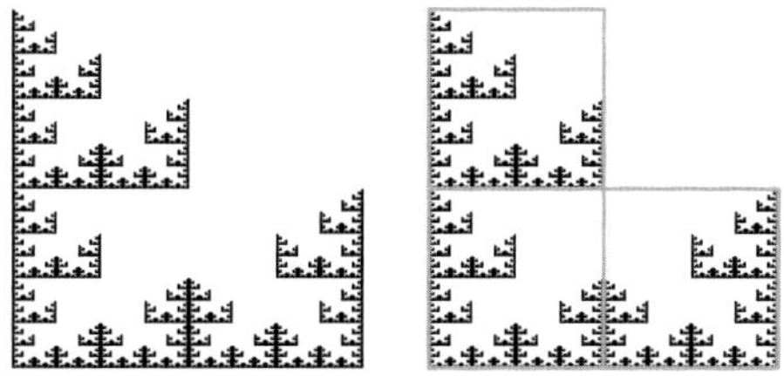

Figure 1.13: A fractal (left) decomposed into three pieces (right).

First note all copies have been scaled by 1/2. The copy in the T_1 position involves no translation. The copy in the T_3 position has been scaled and translated vertically by 1/2.

The copy in the T_2 position has been scaled, reflected over the y-axis and translated to the right. After reflecting over the y-axis, the scaled copy will be on the left side of the y-axis. To get it in the correct final position, it must be translated to the right by 1.

Table 1.2 shows rules that generate the fractal of this example.

	r	s	θ	φ	e	f
T_1	0.5	0.5	0	0	0	0
T_2	-0.5	0.5	0	0	1.0	0
T_3	0.5	0.5	0	0	0	0.5

Table 1.2: IFS rules for the Sample A fractal.

1.1.6 Sample B

Note the top and bottom left pieces have the same orientation as the entire fractal, while the bottom right piece is rotated. Because our transformation rules allow only rotations fixing the origin, some care must be taken with the translation after the rotation.

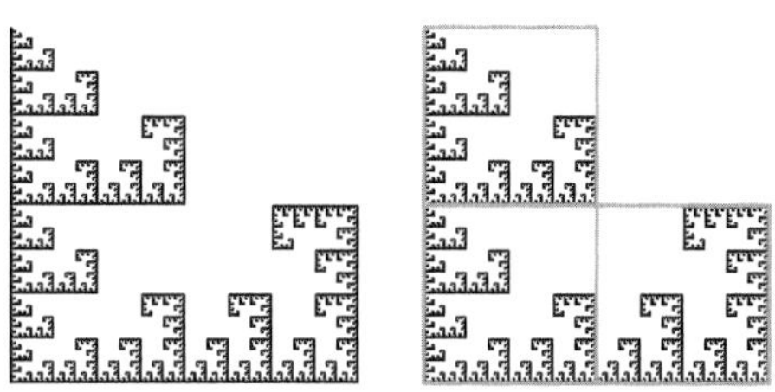

Figure 1.14: A fractal (left) decomposed into three pieces (right).

All copies have been scaled by 1/2. The copy in the T_1 position involves no translation. The copy in the T_3 position has been scaled and translated vertically by 1/2.

	r	s	θ	φ	e	f
T_1	0.5	0.5	0	0	0	0
T_2	0.5	0.5	90	90	1.0	0
T_3	0.5	0.5	0	0	0	0.5

Table 1.3: IFS rules for the Sample B fractal.

The copy in the T_2 position has been scaled, rotated 90° counterclockwise and translated. After rotation, the scaled copy will be on the left side of the y-axis. To get it in the correct final position, it must be translated to the right by 1.

Table 1.3 shows rules that generate the fractal of this example.

1.1.7 Conclusion

Finding IFS rules for fractals is very good practice for working with scaling, reflection, rotation, and translation. Some of the more complicated images are difficult to decompose, especially when combinations of reflections and rotations are involved. Scalings of different values in the same diagram also present challenges to finding rules. Aids such as the paper copies and using software can help, but be careful about where the copies end up after reflecting or rotating. Mistakes here can cause mistakes in translation.

Notice that there is very little difference between the rules for the two samples. Often a small change in the rules can produce very different images. This can make for some interesting experimentation but be careful, not every set of rules can be used. For these exercises the final image must be within the unit square and many sets of rules will take the image outside this area.

1.1.8 Exercises

Find IFS rules to generate each of these fractals. Some of these may have more than one correct answer.

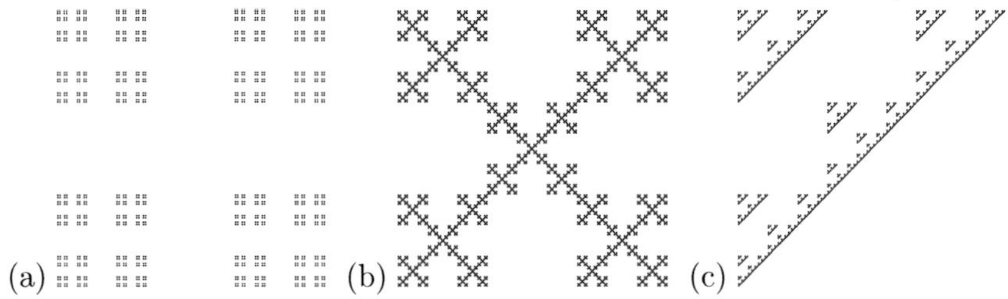

(d) 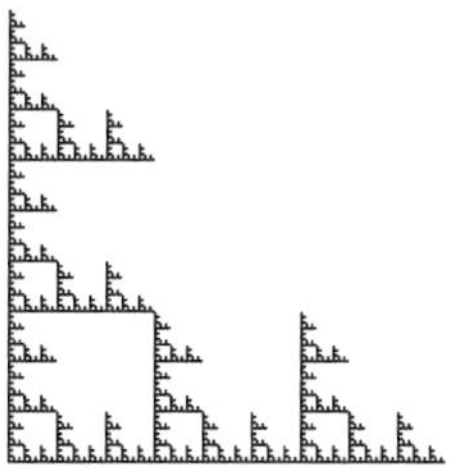 (e) 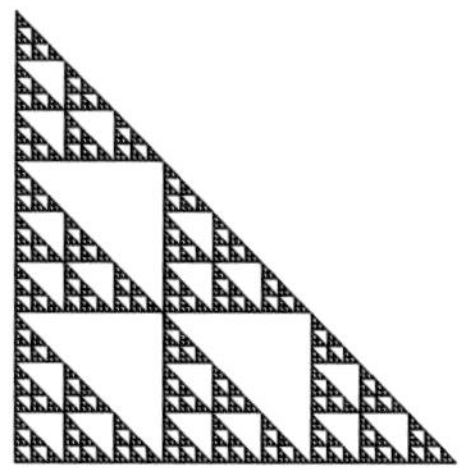 (f) 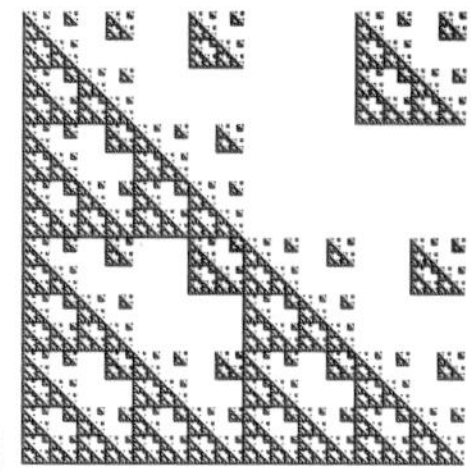

(g)  (h) 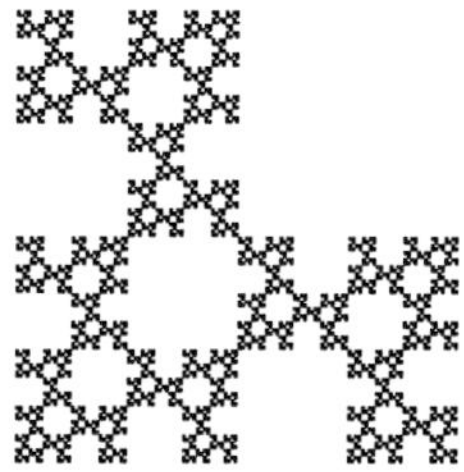 (i) 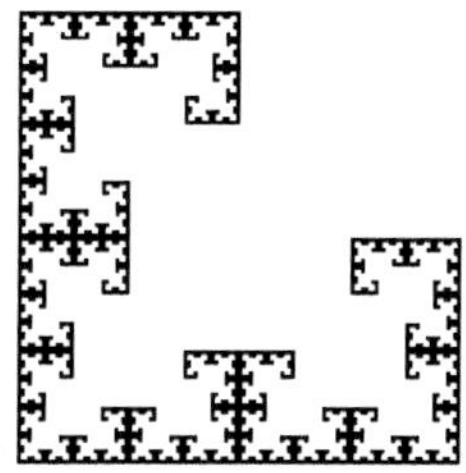

(j) 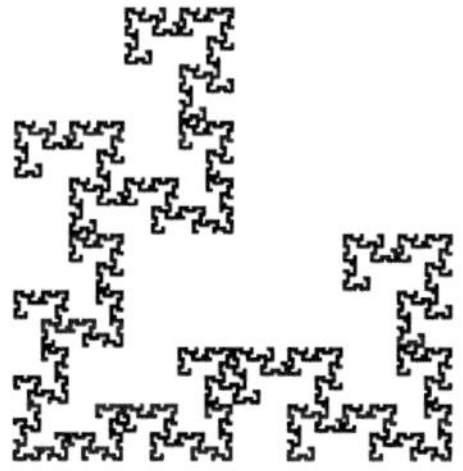 (k) 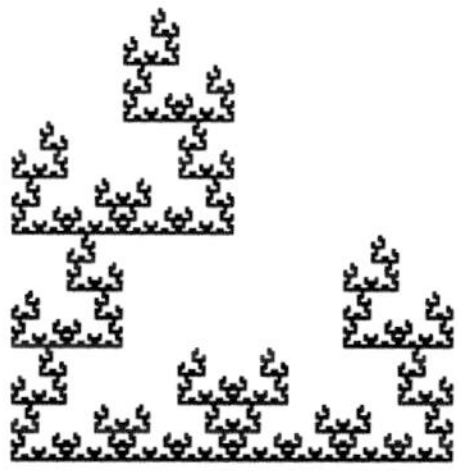 (l) 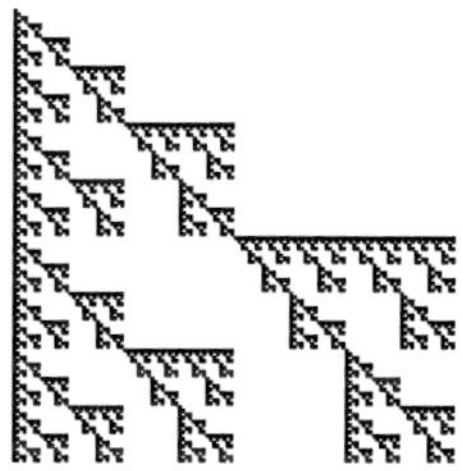

(m) 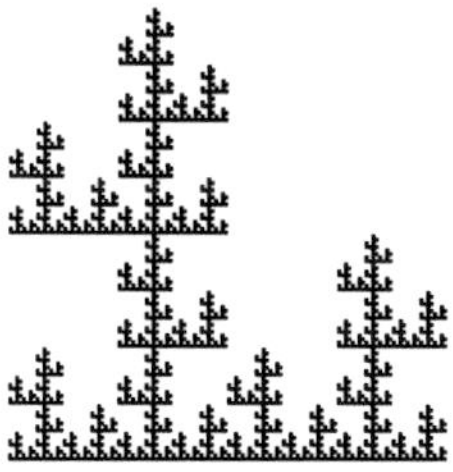 (n) 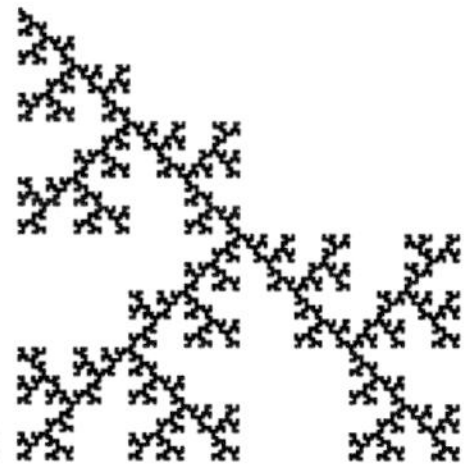 (o) 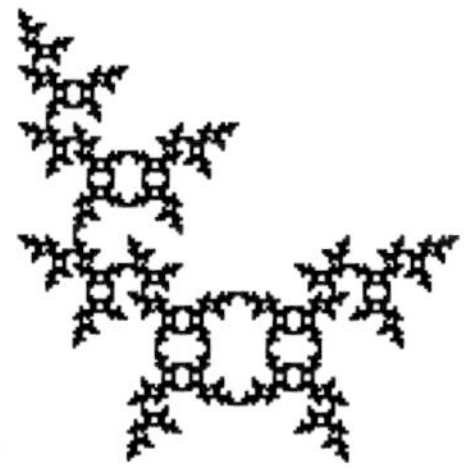

(p)  (q)  (r) 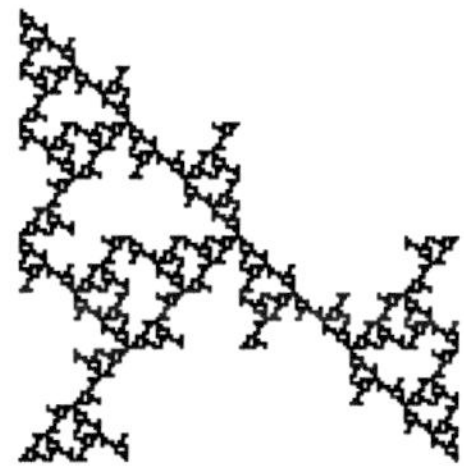

(s)

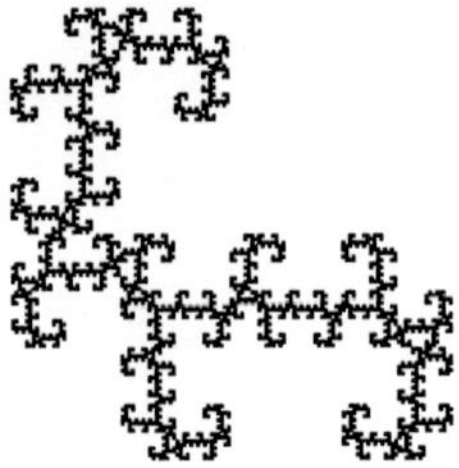

(t)

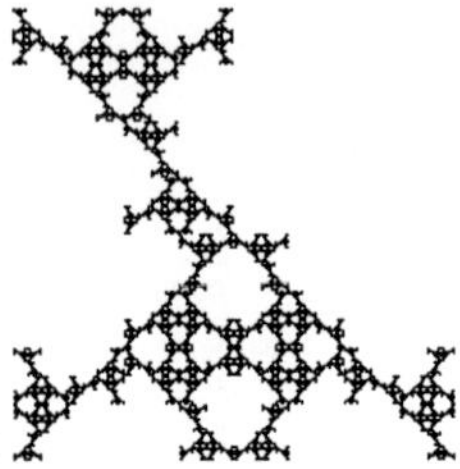

(u)

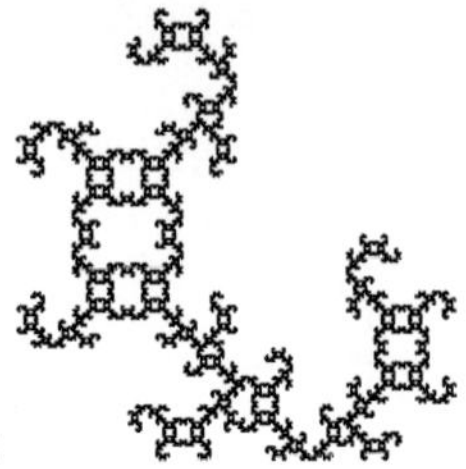

1.2 Spiral fractals from IFS

Here we extend the deconstruction method of Lab 1.1 to study spiral fractals. Informed identification of the pieces is the new trick.

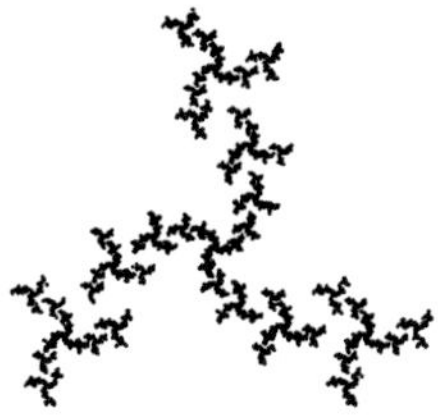

1.2.1 Purpose

We'll find IFS rules that generate spiral fractals. This includes a new way, the random algorithm, to render IFS images. Also we'll see a different image decomposition scheme: "a few small bits and everything else." This is effective for some natural fractals as well.

1.2.2 Materials

We'll use IFS software (1.2 of the Mathematica or Python codes), a ruler calibrated in mm, a protractor, and a calculator.

1.2.3 Background

Here we present a simple method of decomposing spiral fractals into a small number of copies. Because one of the copies often is scaled by a factor only slightly less than 1, the deterministic IFS algorithm converges very slowly. To overcome this problem, here we present another way to generate fractals, the random IFS algorithm.

1.2.3.1 Decomposition of a spiral

From the Background for Lab 1.1, recall the affine transformations of the plane are characterized by six parameters, r, s, θ, φ, e, and f. Recall also the two steps for finding IFS rules for a given fractal A.

- Decompose A into some number N of scaled copies of itself: $A = A_1 \cup \cdots \cup A_N$.
- For each A_i, find an affine transformation T_i satisfying $T_i(A) = A_i$.

Figure 1.15: A spiral fractal.

For this discussion, we use the spiral fractal of Fig. 1.15. With some practice, the same general idea can be applied to many fractals. In particular, all spiral fractals work this way.

A first guess at decomposing the spiral might be discouraging. The largest (right-most, in this case) subspiral is scaled copy of the whole spiral. The largest subspiral of what remains is a scaled copy of the whole spiral, and so on. The first 300 of these are outlined by boxes in Fig. 1.16. Viewed this way, the mathematical fractal consists of infinitely many scaled copies of itself.

We labeled this a "useless decomposition," but maybe a better description is "inelegant decomposition." Certainly, a decomposition that consists of infinitely many pieces is not of practical use. But from the point of view of generating an image on a computer screen, once the boxes are so small that all remaining boxes are contained in the central pixel, we needn't continue the decomposition. Fig. 1.16 shows that this decomposition needs about 300 transformations. Then the infinitely many remaining transformations are clumped into a single box, hence the description "inelegant."

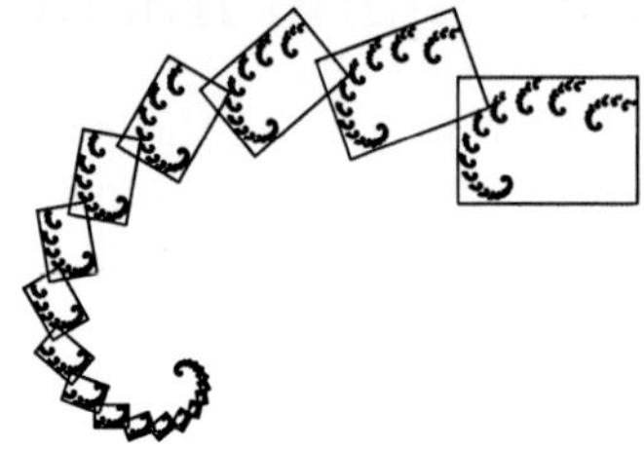

Figure 1.16: A useless decomposition.

However, the inelegance of this decomposition, clumping infinitely many copies into a single box, can give the idea for a very elegant, clever decomposition. The idea of separating off one (in this case) or several small copies of the fractal and noting that what remains is itself a scaled copy of the fractal is an approach useful for many fractals in addition to spiral fractals.

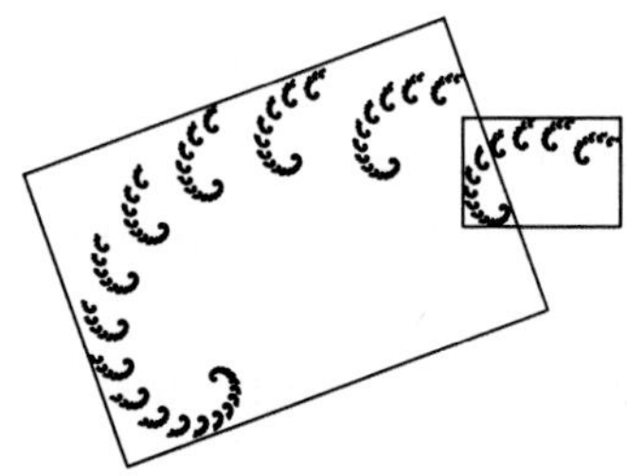

Figure 1.17: A clever decomposition.

What makes this work is that when you cover the right-most subspiral, what remains is a slightly smaller copy of the whole spiral, rotated a bit in the counterclockwise direction. Once we understand this approach, applying it to different spiral fractals is simple.

1.2.3.2 The random algorithm

In Sample 1.2.5 we see that this spiral fractal can be generated by this IFS Table

	r	s	θ	φ	e	f	prob
T_1	0.29	0.29	0	0	0.71	0.41	0.11
T_2	0.83	0.83	20	20	0	0	0.89

For later reference we have labeled the transformations T_1 and T_2. Applied to this IFS, the deterministic algorithm, described in the Background of Lab 1.1, yields a surprise. Suppose we start with the unit sqaure, outlined in grey in the left of Fig. 1.18. One application of T_1 and T_2 give the two filled-in squares on the left. Two applications give the four filled-in squares left of center. Thirteen applications gives the $2^{13} = 8192$ squares shown right of center. The most apparent problem is that the square near the center of the spiral remains quite large. This is a consequence of T_2 having a contraction factor r that is not much less than to 1. The side length of the largest square is $.83^{13} = 0.0887$. In order for the largest square to have side length 0.01—still visible as a square—the deterministic algorithm would need to be run for

24 generations, yielding $2^{24} = 16,777,216$ squares, a nontrivial demand on a computer's memory and processor time.

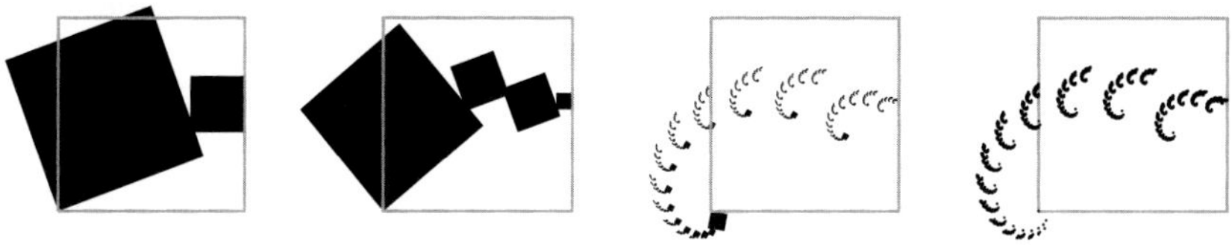

Figure 1.18: Some applications of the deterministic algorithm to a spiral IFS.

This problem results from taking the unit square as a starting shape. A smaller square would shrink down to pixel size more quickly, but why not remove the shrinking problem completely by taking a point as the starting shape? The right picture above shows the 15th iteration of the deterministic algorithm, taking the point $(0,0)$ as starting shape. Here we have $2^{15} = 32,768$ points. While this is better than the picture to its left, still this is not as good as Fig. 1.15, and that contains only $30,000$ points. Note especially how badly the right side of Fig. 1.18 fills in around the center of the spiral. How was the image in Fig. 1.15 generated?

Rendering IFS images by the deterministic algorithm provides a simple explanation of why this process converges to a shape left invariant by the simultaneous application of all the T_i. Another method for rendering the images is the random algorithm. Properly applied, this gives a good approximation of the fractal with a more modest number of points. For the random algorithm an additional column must be added to the IFS table. This column contains the probability, p_i, of applying each T_i. Note the p_i must sum to 1. The random algorithm consists of these steps.

- (1) Select an initial point (x_0, y_0).
- (2) Given a point (x_j, y_j), select a transformation T_i according to the probabilities p_i.
- (3) To the list of points making up the image, add $(x_{j+1}, y_{j+1}) = T_i(x_j, y_j)$.
- (4) Repeat steps (2) and (3), stopping when a preset number of points has been reached.

That the random and the deterministic algorithms produce the same images (in the limit of many, many applications) is not obvious, and is the subject of Appendix B.2.2. How the probabilities may be chosen is the subject of Sect. 1.2.3.3. We end this section with a discussion of ways to select the initial point (x_0, y_0).

Figure 1.19: An effect of the choice of initial point.

The left image of Fig. 1.19 is a picture generated by the random algorithm, 30,000 points with $(x_0, y_0) = (0,0)$; on the right $(x_0, y_0) = (0.9, 0)$. Visible on the right are some points that do not belong to the spiral. Because all the

T_i are contractions, these points get ever nearer to the spiral. In the world of pixels the problem of these points is solved by not displaying the first few (or few hundred) points. Beyond those, the mathematical points generated by the random algorithm lie so close to the spiral that each point occupies a pixel also occupied by points of the spiral. In the mathematical world, the fractal is the limit set of the sequence of points generated by the random algorithm, so isolated points do not appear.

The left image of Fig. 1.19 was generated by taking $(x_0, y_0) = (0, 0)$, the fixed point of T_2. If we start with a point (x_0, y_0) on the fractal, all sequences of points

$$(x_1, y_1) = T_{i_1}(x_0, y_0), \quad (x_2, y_2) = T_{i_2}(x_1, y_1), \quad (x_3, y_3) = T_{i_3}(x_2, y_2), \ldots$$

lie on the fractal. This is a consequence of Eq. (1.2), $\mathcal{T}(A) = A$. In Appendix B.2.1 we show that the fixed point of each T_i belongs to A. So if we begin the random algorithm with a fixed point of one of the IFS transformations, all points generated by the random algorithm will lie on the fractal.

1.2.3.3 Computing the probabilities

The random IFS algorithm generates an image by applying the IFS transformations one at a time in random order, with frequency determined by the probability assigned to each transformation. Say p_i is the probability of applying T_i.

Figure 1.20: Left: $p_1 = p_2 = 0.5$. Right: $p_1 = 0.11$, $p_2 = 0.89$.

In the spiral IFS if $p_1 = p_2 = 0.5$, then T_1 and T_2 are applied about equally often. Because every application fo T_1 produces a point in A_1, the right-most subspiral, and every application of T_2 produces a point in A_2, everything else in the spiral, on the left of Fig. 1.20 about 15,000 points are in the right-most subspiral and about 15,000 in everything else. The result is that the right-most subspiral is filled in fairly densely, and the rest is filled much less well, especially near the center of the spiral.

To achieve a more uniform fill, the probability of applying each transformation T_i should equal the fraction of the total shape S occupied by $T_i(S)$.

If all rotations are rigid ($\theta_i = \varphi_i$ for each i), then if we start with the unit square S, the area of each piece $T_i(S)$ is $|r_i s_i|$. (Absolute value because r_i and s_i can be negative if T_i involves reflections.) Many fractals in the plane have zero area. For example, in Lab 2.3 we'll see that if the dimension of a shape is less than 2, that shape has zero area. So we can't base calculations on relative area, but we can use this estimate. If the overlaps of the pieces are not substantial—just along edges, say—then we'll take

$$p_i = \frac{|r_i s_i|}{|r_1 s_1| + |r_2 s_2| + \cdots + |r_n s_n|}$$

For the spiral example

$$p_1 = \frac{0.29^2}{0.29^2 + 0.83^2} \approx 0.11 \quad \text{and} \quad p_2 = \frac{0.83^2}{0.29^2 + 0.83^2} \approx 0.89$$

With these probabilities, the random algorithm fills in 30,000 points for the right image of Fig. 1.20. Note that all the regions are filled in approximately evenly.

1.2.3.4 The effect of scaling the translation terms

A common misconception about IFS involves scaling the translation terms, e and f. If all the e_i and f_i are replaced by $0.5 \cdot e_i$ and $0.5 \cdot f_i$, then surely this will move the pieces of the fractal closer together. If the pieces were just touching initially, then after scaling the translations, the pieces should overlap. Or so some people think, before careful reflection.

In fact, this is not true. The r_i and s_i determine what portion of the whole shape is contained in each piece. So if the r_i and s_i are unchanged, the overlap status of the pieces is unchanged. On the other hand, the locations of the pieces will change. On the left of Fig. 1.21 is the spiral we have been studying, on the right is the spiral in the same box, with all the translations scaled by 1/2.

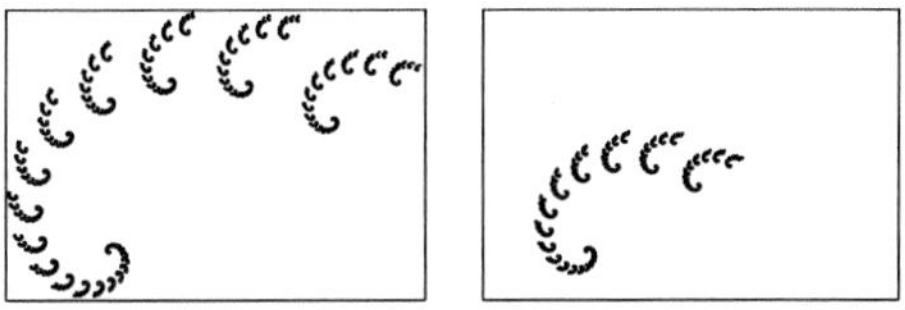

Figure 1.21: The effect of scaling all the translation parameters.

So we see that multiplying all the translations by the same factor c produces a fractal similar to the original, but scaled by c.

1.2.4 Procedure

To illustrate the procedure we use a spiral that can be decomposed into two pieces: the right-most subspiral and everything else.

In general, each transformation is determined by values for r, s, θ, φ, e, and f. Our spirals are self-similar, so for each piece $r = s$ and $\theta = \varphi$.

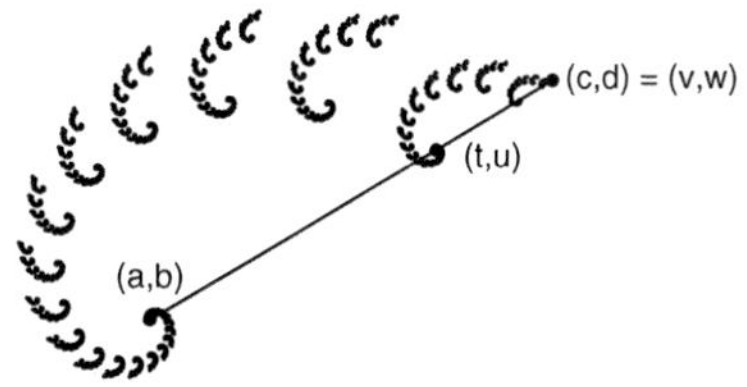

Figure 1.22: Corresponding points in the right-most supspiral.

To determine these values, select two points (a, b) and (c, d) in the fractal. To find the parameters for the transformation giving the smaller of the self-similar pieces of the fractal, in this piece locate the points (t, u) and (v, w) that correspond to (a, b) and (c, d). In Fig. 1.22 we see the points (a, b) and (c, d) and the points (t, u) and (v, w) for the right-most subspiral.

To find r, divide a distance in the whole spiral into the corresponding distance in the subspiral. For example,

$$r = \frac{\text{dist}((\text{t,u}),(\text{v,w}))}{\text{dist}((a,b),(c,d))}$$

For self-similar fractals $s = r$.

To find θ, measure the angle between the lines (a, b) to (c, d) and (t, u) to (v, w). (In this example, the angle is 0.) Again, for self-similar fractals $\varphi = \theta$.

To find e, measure the horizontal distance between (a, b) and (t, u). That is, $e = t - a$.

To find f, measure the vertical distance between (a, b) and (t, u). That is, $f = u - b$.

From Sect. 1.2.3.4, recall that multiplying all e and f values by the same constant simply changes the scale of the picture, not the relative sizes or positions of the pieces within the picture. As a first attempt at guaranteeing the image approximately fills the window, divide all e and f values by the maximum of the vertical and horizontal extents of the fractal. With additional, similar multiplications of all e and f terms, an appropriate size image can be obtained.

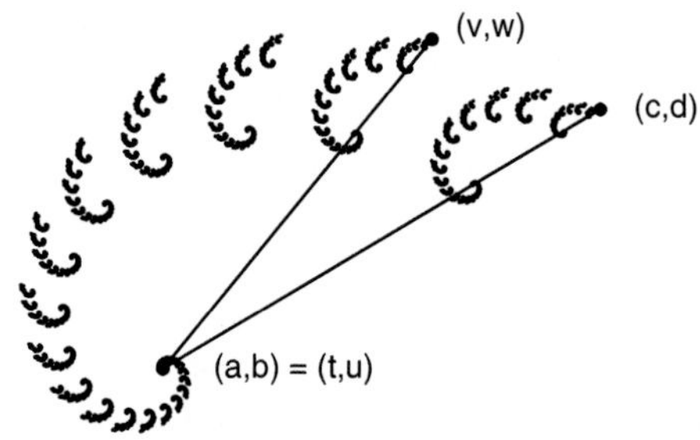

Figure 1.23: Corresponding points in the whole spiral and everything else.

Fig. 1.23 shows the location of (t, u) and (v, w) for the everything else piece of this spiral. The angle between the lines (a, b) to (c, d) and (t, u) to (v, w) gives the angle $\theta = \varphi$ for this piece of the spiral.

1.2.5 Sample

Note the measurements reported here refer to a particular printing of the picture. Measuring a diffrent size image will not change r and s, as these are ratios. The angles θ and φ certainly are not altered by the size of the image. The *values* of e and f do depend on the size of the image, but as mentioned above, these can be multiplied by a common factor to adjust the size of the IFS image. Consequently, the size of the image is not a significant factor in these calculations.

Right-most subspiral. Select points (a, b) and (c, d) on the whole spiral. The corresponding points on the right-most spiral are (t, u) and (v, w). See Fig. 1.24.

To find r and s, we measure 55mm for dist$((a, b), (c, d))$ and 16mm for dist$((t, u), (v, w))$. Consequently, $r = 16/55 = 0.29 = s$.

To find θ and φ the line (t, u) to (v, w) is parallel to the line (a, b) to (c, d). So $\theta = 0 = \varphi$.

To find e, the horizontal distance from (a, b) to (t, u) is 35mm. As a scale we take the horizontal distance from (a, b) to (c, d), measured as 49mm. So $e = 35/49 = 0.71$.

To find f, the vertical distance from (a, b) to (t, u) is 20mm. We must use the same scale for all vertical and horizontal translations, so $f = 20/49 = 0.41$.

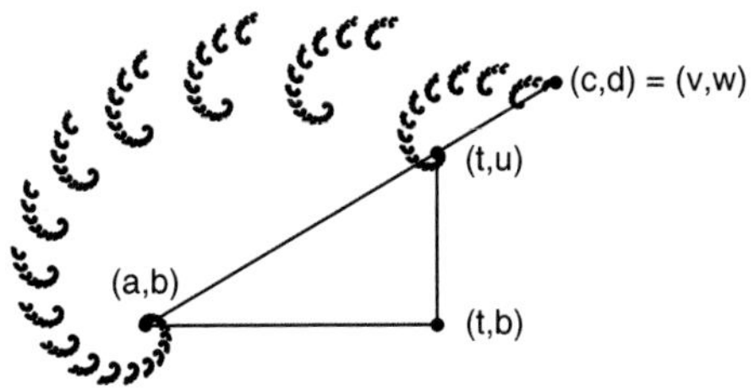

Figure 1.24: IFS parameters for the right-most subspiral.

Everything else. Here are the points (t, u) and (v, w) for this piece. See Fig. 1.25.

To find r and s, measure 55mm for dist$((a, b), (c, d))$ and 46mm for dist$((t, u), (v, w))$. Consequently, $r = 46/55 = 0.84 = s$.

To find θ and φ, the line (t, u) to (v, w) makes an angle of about 20° with the line (a, b) to (c, d). So $\theta = 20° = \varphi$.

To find e and f, the horizontal distance from (a, b) to (t, u) is 0mm, so $e = 0$; the vertical distance from (a, b) to (t, u) is 0mm, so $f = 0$.

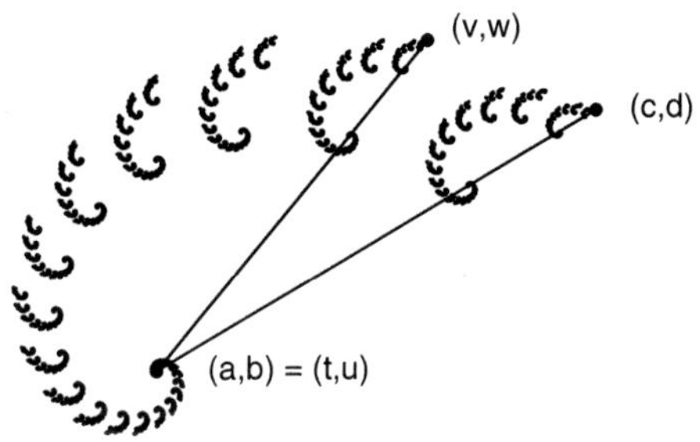

Figure 1.25: IFS parameters for everything else.

Probabilities. Calculate the probabilities by the method described in Sect. 1.2.3.3. The probabilities of the smaller and larger subspirals are

$$\frac{0.29^2}{0.29^2 + 0.83^2} \approx 0.11 \qquad \frac{0.83^2}{0.29^2 + 0.83^2} \approx 0.89$$

Then the IFS table for this spiral is

r	s	θ	φ	e	f	prob
0.29	0.29	0	0	0.71	0.41	0.11
0.83	0.83	20	20	0	0	0.89

1.2.6 Conclusion

Trying to get the correct rules for spiral fractals is not as easy as for relatives of the gasket in Lab 1.1. Even after finding an efficient decomposition into pieces, determining the r, s, θ, φ, e, and f values for each piece may require measuring lengths and angles, with some attention to rescaling.

The idea of finding IFS parameters by measuring distances and angles on an image of the fractal can be applied to many other shapes. This is one of the methods used in Lab 1.4, finding IFS rules for a leaf.

1.2.7 Exercises

Prob 1.2.1 In Fig. 1.24, suppose $(a, b) = (0, 0)$. With the parameters given in the IFS table of Sample 1.2.5, compute the coordinates of the point (c, d).

Prob 1.2.2 Use the IFS table of Sample 1.2.5 and the answer of Exercise 1.2.1 to compute the coordinates of the point (v, w) in Fig. 1.23.

Prob 1.2.3 Find IFS tables to generate these spiral fractals.

(a)

(b)

(c)

(d)

(e)

(f)

Prob 1.2.4 Find a natural fractal that comes close to one of these spiral fractals. A Nautilus shell is *not* an example.

1.3 Finding IFS rules from images of points

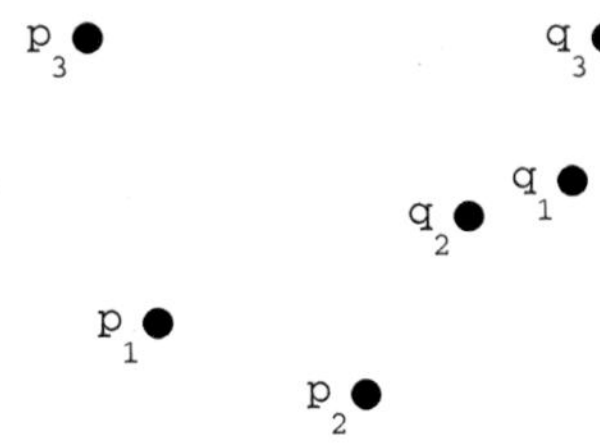

Lab 1.1 involved simple geometric fractals and we found the IFS parameters by inspection. Spiral fractals, the subject of Lab 1.2, are a bit less straightforward. We had to measure the lengths and angles, still relatively straightforward because the rotations are rigid and the pieces are similar ($r_i = s_i$) to the whole. Shapes found in nature—leaves and ferns, for example—rarely are so simple. We need a way to find IFS parameters that works for more general shapes. That's the subject of this lab.

1.3.1 Purpose

We'll learn how to find the r, s, θ, φ, e, and f values of an IFS transformation, based on the selection of three (noncollinear) points in the fractal A and their images on each copy $T_i(A)$.

1.3.2 Materials

We'll use a ruler calibrated in mm, graph paper, affine transformation code (1.3 of the Mathematica or Python codes).

1.3.3 Background

Here we derive the formula for the parameters—r, s, θ, φ, e, and f—of the transformation that takes p_1, p_2, and p_3 to q_1, q_2, and q_3.

1.3.3.1 General formulation

Given three non-collinear points, the *initial points*

$$p_1 = (x_1, y_1),\ p_2 = (x_2, y_2),\ \text{and } p_3 = (x_3, y_3)$$

and three *image points*

$$q_1 = (u_1, v_1),\ q_2 = (u_2, v_2),\ \text{and } q_3 = (u_3, v_3)$$

we seek an affine transformation T satisfying $T(p_1) = q_1$, $T(p_2) = q_2$, and $T(p_3) = q_3$. In general, such a transformation T is determined by six parameters, a, b, c, d, e, and f, in this fashion: $T(x, y) = (u, v)$ means

$$\begin{aligned} ax + by + e &= u \\ cx + dy + f &= v \end{aligned}$$

Then the three equations $T(p_1) = q_1$, $T(p_2) = q_2$, and $T(p_3) = q_3$ can be written as

$$\begin{aligned} &ax_1 + by_1 + e = u_1 \qquad && ax_2 + by_2 + e = u_2 \qquad && ax_3 + by_3 + e = u_3 \\ &cx_1 + dy_1 + f = v_1 && cx_2 + dy_2 + f = v_2 && cx_3 + dy_3 + f = v_3 \end{aligned}$$

We are accustomed to thinking of x_1, y_1, u_1, v_1, etc., as variables, but here they are the coordinates of the points, hence known. The variables are a, b, c, d, e, and f, the parameters of the transformation T. Group together the equations containing a, b, and e, and group together the equations containing c, d, and f:

$$\begin{aligned} ax_1 + by_1 + e &= u_1 & \qquad cx_1 + dy_1 + f &= v_1 \\ ax_2 + by_2 + e &= u_2 & \qquad cx_2 + dy_2 + f &= v_2 \\ ax_3 + by_3 + e &= u_3 & \qquad cx_3 + dy_3 + f &= v_3 \end{aligned} \tag{1.3}$$

The left system of equations can be solved for a, b, amd e, the right system for c, d, and f.

Four steps remain:
(1) give the matrix formulation of this problem,
(2) prove there is a unique solution if and only if the points p_1, p_2, and p_3 are non-collinear,
(3) give the solution, and
(4) convert a, b, c, and d into r, s, θ, and φ.

1.3.3.2 Matrix formulation

The equations (1.3) an be written in matrix form as

$$\begin{bmatrix} x_1 & y_1 & 1 \\ x_2 & y_2 & 1 \\ x_3 & y_3 & 1 \end{bmatrix} \begin{bmatrix} a \\ b \\ e \end{bmatrix} = \begin{bmatrix} u_1 \\ u_2 \\ u_3 \end{bmatrix} \quad \text{and} \quad \begin{bmatrix} x_1 & y_1 & 1 \\ x_2 & y_2 & 1 \\ x_3 & y_3 & 1 \end{bmatrix} \begin{bmatrix} c \\ d \\ f \end{bmatrix} = \begin{bmatrix} v_1 \\ v_2 \\ v_3 \end{bmatrix} \tag{1.4}$$

Note both equations have the same coefficient matrix, so if that matrix is invertible, the solutions are easy.

$$\begin{bmatrix} a \\ b \\ e \end{bmatrix} = \begin{bmatrix} x_1 & y_1 & 1 \\ x_2 & y_2 & 1 \\ x_3 & y_3 & 1 \end{bmatrix}^{-1} \begin{bmatrix} u_1 \\ u_2 \\ u_3 \end{bmatrix} \quad \text{and} \quad \begin{bmatrix} c \\ d \\ f \end{bmatrix} = \begin{bmatrix} x_1 & y_1 & 1 \\ x_2 & y_2 & 1 \\ x_3 & y_3 & 1 \end{bmatrix}^{-1} \begin{bmatrix} v_1 \\ v_2 \\ v_3 \end{bmatrix} \tag{1.5}$$

1.3.3.3 Proof of a unique solution

Write $p_1 = (x_1, y_1)$, $p_2 = (x_2, y_2)$, and $p_3 = (x_3, y_3)$, and form the vectors $\vec{S} = \langle x_2 - x_1, y_2 - y_1 \rangle$ and $\vec{T} = \langle x_3 - x_1, y_3 - y_1 \rangle$. The angle θ between $\vec{S}$ and $\vec{T}$ is 0° or 180° if and only if the points p_1, p_2, and p_3 are collinear. So the condition that these points are not collinear—which we always can arrange because we choose these points—guarantees that $\theta \neq 0°$ and $\theta \neq 180°$

The vector cross-product measures the angle between vectors, but is defined only for vectors in 3-dimensional space (and, it turns out, also in 7-dimensional space, but we won't use that construction). This is resolved easily by defining auxiliary vectors $\vec{S}' = \langle x_2 - x_1, y_2 - y_1, 0 \rangle$ and $\vec{T}' = \langle x_3 - x_1, y_3 - y_1, 0 \rangle$. Then the length $|\vec{S}' \times \vec{T}'|$ of the cross-product satisfies $|\vec{S}' \times \vec{T}'| = |\vec{S}'||\vec{T}'|\sin(\theta)$.

As long as p_1, p_2, and p_3 are distinct points, the lengths $|\vec{S}'|$ and $|\vec{T}'|$ are nonzero. Consequently, $|\vec{S}' \times \vec{T}'| = 0$ if and only if $\sin(\theta) = 0$. That is, if and only if the points p_1, p_2, and p_3 are collinear. Because the points are not collinear, $|\vec{S}' \times \vec{T}'| \neq 0$.

Recall the matrix definition of the cross-product:

$$\begin{aligned}\vec{S}' \times \vec{T}' &= \det \begin{bmatrix} \vec{i} & \vec{j} & \vec{k} \\ x_2 - x_1 & y_2 - y_1 & 0 \\ x_3 - x_1 & y_3 - y_1 & 0 \end{bmatrix} \\ &= 0\vec{i} - 0\vec{j} + ((x_2 - x_1)(y_3 - y_1) - (x_3 - x_1)(y_2 - y_1))\vec{k}\end{aligned}$$

Consequently,

$$|\vec{S}' \times \vec{T}'| = x_2y_3 - x_2y_1 - x_1y_3 - x_3y_2 + x_3y_1 + x_1y_2 = \det \begin{bmatrix} x_1 & y_1 & 1 \\ x_2 & y_2 & 1 \\ x_3 & y_3 & 1 \end{bmatrix}$$

A square matrix is invertible if and only if its determinant is nonzero, so the coefficient matrix M of Eqs. (1.4) is invertible and the solutions for the parameters a, b, c, d, e, and f is given by Eqs. (1.5).

1.3.3.4 And the solution is

The standard inversion formula applied to the coefficient matrix M of Eqs. (1.4) gives

$$\begin{bmatrix} x_1 & y_1 & 1 \\ x_2 & y_2 & 1 \\ x_3 & y_3 & 1 \end{bmatrix}^{-1} = \frac{1}{\det(M)} \begin{bmatrix} y_2 - y_3 & y_3 - y_1 & y_1 - y_2 \\ x_3 - x_2 & x_1 - x_3 & x_2 - x_1 \\ x_2y_3 - y_2x_3 & x_3y_1 - y_3x_1 & x_1y_2 - y_1x_2 \end{bmatrix}$$

Then Eqs. (1.5) give values for a, b, c, d, e, and f.

1.3.3.5 Conversion to r, S, θ, and φ

The last step is to convert the values for a, b, c, and d into values for r, s, θ and φ. (The e and f values are the same in both representations.) In order to build an organized way to make this conversion, first note that we do not need to allow reflection across both axes. In Fig. 1.26 we see that reflection across the x-axis (top images) is equivalent to reflection across the y-axis followed by a 180° rotation (bottom images). Consequently, both reflections can be achieved by reflection across the y-axis. That is, we never need to take s to be negative.

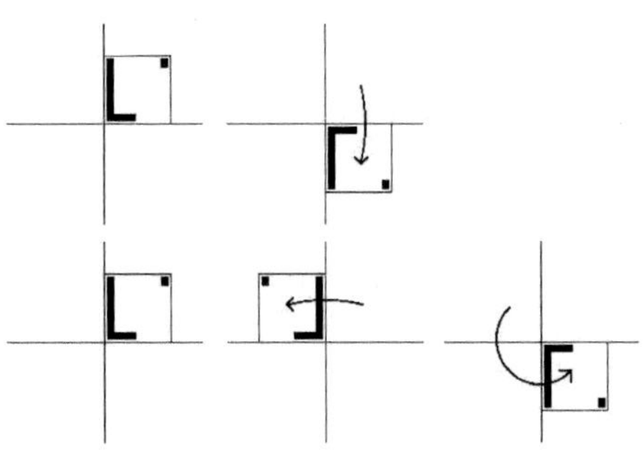

Figure 1.26: Comparing a reflection to a rotation and a reflection.

To convert a, b, c, and d to r, s, θ and φ write

$$a = r\cos(\theta), \quad b = -s\sin(\varphi), \quad c = r\sin(\theta), \quad d = s\cos(\varphi)$$

and refer to Fig. 1.27. The magnitudes of r and s are easy to find.

$$a^2 + c^2 = r^2 \cos^2(\theta) + r^2 \sin^2(\theta) = r^2 \quad \text{so} \quad r = \pm\sqrt{a^2 + c^2}$$
$$b^2 + d^2 = s^2 \cos^2(\varphi) + s^2 \sin^2(\varphi) = s^2 \quad \text{so} \quad s = \sqrt{b^2 + d^2}$$

The sign of r is + if the transformation does not involve a reflection, − if it does. Here is how the initial and image points determine the presence of a reflection.

Reflections

To determine if the transformation T involves a reflection, consider the initial points

$$p_1 = (x_1, y_1), \quad p_2 = (x_2, y_2), \quad \text{and} \quad p_3 = (x_3, y_3)$$

and their images

$$q_1 = T(p_1) = (u_1, v_1), \quad q_2 = T(p_2) = (u_2, v_2), \quad \text{and} \quad q_3 = T(p_3) = (u_3, v_3)$$

View these as points in the xy-plane in 3-dimensional space and form the cross-products

$$(p_2 - p_1) \times (p_3 - p_1) = 0\vec{i} - 0\vec{j} + ((x_2 - x_1)(y_3 - y_1) - (x_3 - x_1)(y_2 - y_1))\vec{k}$$
$$(q_2 - q_1) \times (q_3 - q_1) = 0\vec{i} - 0\vec{j} + ((u_2 - u_1)(v_3 - v_1) - (u_3 - u_1)(v_2 - v_1))\vec{k}$$

If both vectors point in the same direction, the orientation of the triple of image points is the same as that of the triple of initial points, so T does not involve a reflection. If the vectors point in opposite directions, T does inolve a reflection.

This can be coded with the *sign function*, $sgn(x)$, which takes the values +1, 0, and −1 when $x > 0$, $x = 0$, and $x < 0$. Then the transformation involves a reflection across across the y-axis if $z =$

$$sgn(((x_2-x_1)(y_3-y_1)-(x_3-x_1)(y_2-y_1))\cdot((u_2-u_1)(v_3-v_1)-(u_3-u_1)(v_2-v_1)))$$

is negative. Then we account for reflection by taking $r = z \cdot \sqrt{a^2 + c^2}$.

Rotations

Now to find the angles, from the left circle of Fig. 1.27 we see $a = r\cos(\theta)$ and $c = r\sin(\theta)$ so

$$\frac{c}{a} = \frac{r\sin(\theta)}{r\cos(\theta)} = \tan(\theta)$$

From the right circle of Fig. 1.27 we see $b = -s\sin(\varphi)$ and $d = s\cos(\varphi)$ so

$$\frac{b}{d} = \frac{-s\sin(\varphi)}{s\cos(\varphi)} = -\tan(\varphi)$$

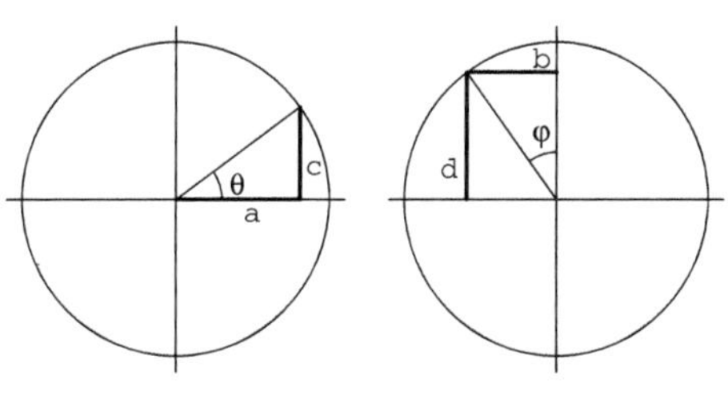

Figure 1.27: The angles θ and φ.

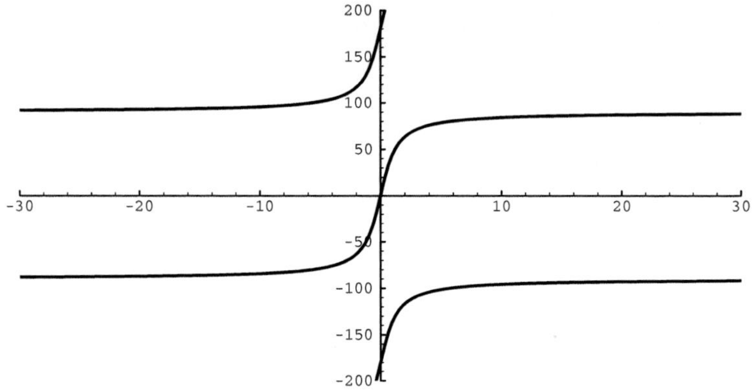

Figure 1.28: θ on the y-axis, $\arctan(c/a)$ on the x-axis.

So $\theta = \arctan(c/a)$ and $\varphi = \arctan(-b/d)$. The mathematical function $\arctan(y/x)$ is single-valued only between $-90°$ and $90°$ (See Fig. 1.28), but both Mathematica and Python have versions of arctan that depend on x and y, not just on their ratio. That is, ArcTan[x,y] and math.atan2(x,y) account for the signs of the x and y and return values between $-180°$ and $180°$.

If the transformation involves a reflection, that is, if $z < 0$, then add $180°$ to θ. If this sum exceeds $180°$, subtract $360°$ to return the angle to the range between $-180°$ and $180°$; if the sum is less than $-180°$, add $360°$.

1.3.4 Sample

Find the transformation taking p_1 to q_1, p_2 to q_2, and p_3 to q_3 in Fig. 1.29.

The choice of axes and scale does not affect the transformation parameters, except that multiplying e and f by the same factor multiplies the measured distance between points by the same factor. So take p_1 to be the origin, and p_2 along the x-axis. The initial points can be selected by the user, so p_2 and p_3 can be taken to be unit distances along the x- and y-axes. This is only a simplification and in any case, we must select the points p_1, p_2 and p_3 so their images q_1, q_2, and q_3 can be identified clearly in each copy $T_i(A)$ of the fractal. Our simple choice may not always be appropriate. The code will work for all (non-collinear) choices of initial points.

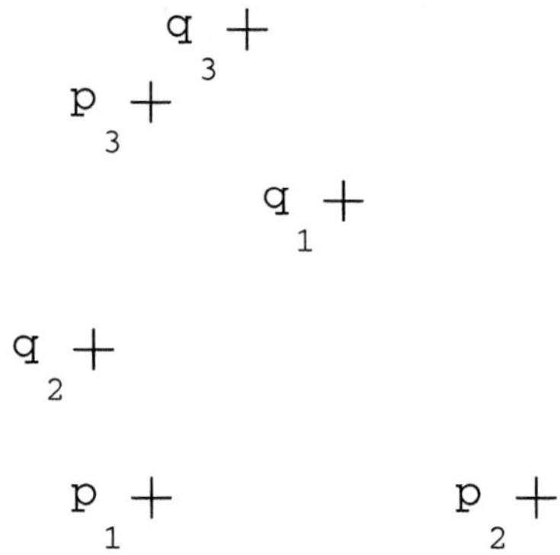

Figure 1.29: Points and images.

So we'll take $p_1 = (x_1, y_1) = (0, 0)$, $p_2 = (1, 0) = (x_2, y_2)$, and $p_3 = (0, 1) = (x_3, y_3)$. Then measure $q_1 = (u_1, v_1) = (0.5, 0.75)$, $q_2 = (u_2, v_2) = (-0.15, 0.375)$, and $q_3 = (u_3, v_3) = (0.25, 1.18)$.

We'll work through the calculation by hand, then present the results of the Mathematica and Python code.

The coefficient matrix M from Eq. (1.4) is

$$M = \begin{bmatrix} 0 & 0 & 1 \\ 1 & 0 & 1 \\ 0 & 1 & 1 \end{bmatrix} \text{ and } M^{-1} = \begin{bmatrix} -1 & 1 & 0 \\ -1 & 0 & 1 \\ 1 & 0 & 0 \end{bmatrix}$$

Then

$$\begin{bmatrix} a \\ b \\ e \end{bmatrix} = M^{-1} \begin{bmatrix} u_1 \\ u_2 \\ u_3 \end{bmatrix} = M^{-1} \begin{bmatrix} -0.15 \\ 0.5 \\ 0.25 \end{bmatrix} = \begin{bmatrix} -0.65 \\ -0.25 \\ 0.5 \end{bmatrix}$$

and

$$\begin{bmatrix} c \\ d \\ f \end{bmatrix} = M^{-1} \begin{bmatrix} v_1 \\ v_2 \\ v_3 \end{bmatrix} = M^{-1} \begin{bmatrix} 0.375 \\ 0.75 \\ 1.18 \end{bmatrix} = \begin{bmatrix} -0.375 \\ 0.43 \\ 0.75 \end{bmatrix}$$

Next, $z = sgn(-0.373) = -1$ so $r = -\sqrt{a^2 + c^2} \approx -0.7504$. The coordinates of the points are measured, so we shouldn't trust more than two digits to the right of the decimal. Consequently, we take $r = -0.75$. Similarly, $s = \sqrt{b^2 + d^2} \approx 0.50$. Now $z < 0$ so $\theta = \arctan(c/a) + 180° = \arctan(-0.375/-0.65) + 180° \approx -150° + 180° = 30°$. Finally, $\varphi = \arctan(-b/d) = \arctan(0.25/0.43) \approx 30°$.

If we enter the coordinates of the initial points p_1, p_2, and p_3, and those of the image points q_1, q_2, and q_3 into 1.3 of the Mathematica or Python codes, it returns

$$r = -0.75,\ s = 0.50,\ \theta = 30°,\ \varphi = 30°,\ e = 0.5,\ f = 0.75$$

1.3.5 Conclusion

In Labs 1.1 and 1.2 we relied on visual inspection, sometimes with a ruler and a protractor to measure lengths and angles. But mostly we relied on inspection to identify the transformation of the whole fractal to each of its pieces. In order to make IFS a tool useful for the synthesis of natural images, we must be able to identify transformation parameters in the presence of visual noise. Every fern, every tree, every leaf, every coastline, every lung, every cloud, every mountain range, every circulatory system, every nervous system ..., all grow in environments filled with random perturbations. This leads to the first of two essential points in the search for natural fractals.

- Natural objects exhibit no exact physical symmetry. This is as true for fractals as it is for Euclidean shapes: Nature contains neither exact gaskets nor exact circles. Squint a bit when you look for natural fractals. The general pattern, not the precise details, are your guide.
- While geometrical fractals can be magnified without end, natural fractals cannot. Sufficient magnification reveals molecules, and the molecules of a sheep do not look like tiny sheep. Natural fractals exhibit self-similarity over only a limited range of scales. This is called the *scaling range*. The utility of a fractal description is related directly to the extent of its scaling range.

For those fractals that exhibit only approximate self-similarity over a limited range of scales, the technique of this lab can be of use in the search for transformation parameters.

1.3.6 Exercises

In Exercises 1.3.1–1.3.6 take $p_1 = (1, 0)$, $p_2 = (0, 0)$, and $p_3 = (0, 1)$. Find the r, s, θ, φ, e, and f values of the transformation that takes p_1 to q_1, p_2 to q_2, and p_3 to q_3 for the image points q_1, q_2, and q_3 given in the problem.

Prob 1.3.1 $q_1 = (0.25, 0.5)$, $q_2 = (0.5, 0.75)$, and $q_3 = (0.25, 1.0)$.

Prob 1.3.2 $q_1 = (0.67, 0.72)$, $q_2 = (0.5, 0.25)$, and $q_3 = (-0.20, 0.51)$.

Prob 1.3.3 $q_1 = (0.75, 0.68)$, $q_2 = (0.5, 0.25)$, and $q_3 = (0.25, 0.68)$.

Prob 1.3.4 $q_1 = (0.93, 0.5)$, $q_2 = (0.5, 0.25)$, and $q_3 = (0.875, -0.40)$.

Prob 1.3.5 $q_1 = (0.72, 0.375)$, $q_2 = (0.5, 0.25)$, and $q_3 = (0.125, 0.90)$.

Prob 1.3.6 $q_1 = (0.72, 0.375)$, $q_2 = (0.5, 0.25)$, and $q_3 = (0.63, 1.0)$.

Prob 1.3.7 For each of (a), (b), (c), and (d), find the r, s, θ, φ, e, and f parameters that take p_1 to q_1, p_2 to q_2, and p_3 to q_3. Use $p_1 = (1, 0)$, $p_2 = (0, 0)$, and $p_3 = (0, 1)$.

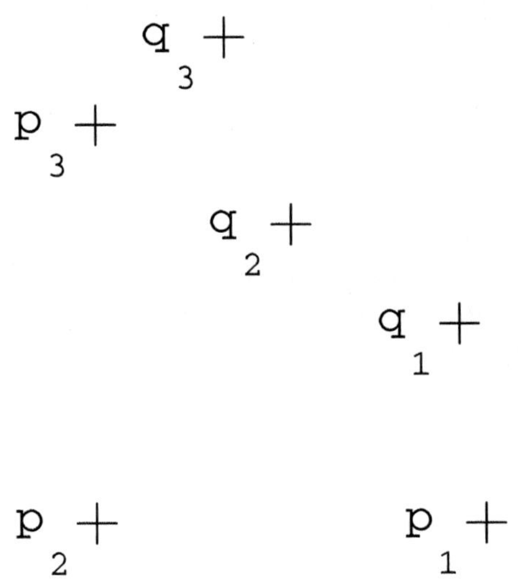

(a)

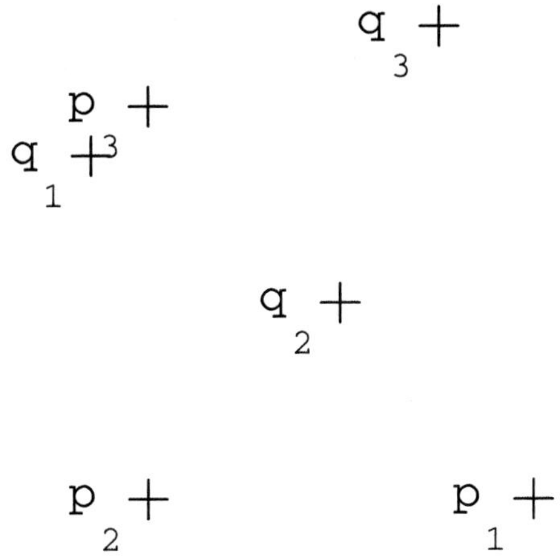

(b)

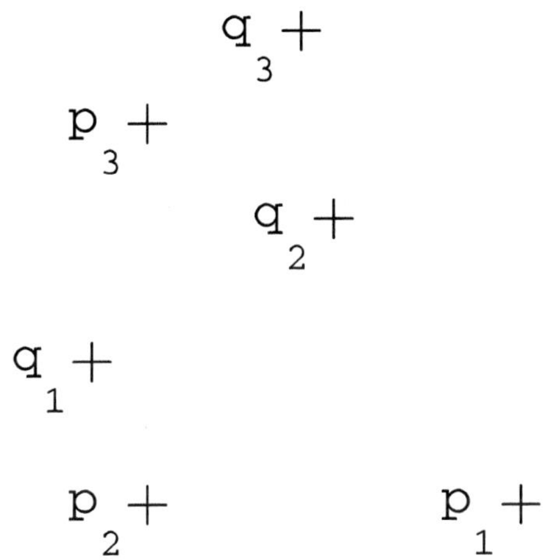

(c)

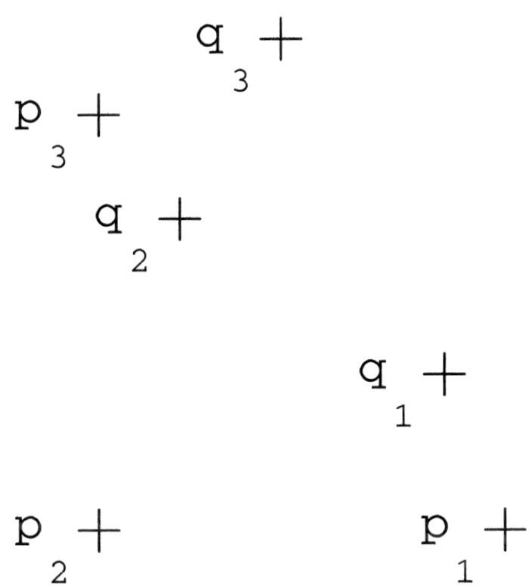

(d)

1.4 A fractal leaf by IFS

Here we apply the method of Lab 1.1 and software of Labs 1.2 (1.2 of the Mathematica or Python codes) and 1.3 (1.3 of the Mathematica or Python codes) to build fractal images of leaves.

1.4.1 Purpose

We'll learn how to find IFS rules that generate a fractal which approximates a leaf. Along the way we'll see that the self-similarity of IFS images allows small changes in IFS parameters to accumulate as considerable changes in the final image.

1.4.2 Materials

We'll use a leaf with a complex outline—a maple leaf is a good choice—tracing paper or an overhead transparency, a ruler, a protractor, three-point transformation calculator (1.3 of the Mathematica or Python codes), and random IFS software (1.2 of the Mathematica or Python codes).

1.4.3 Background

The geometry of plane transformations is introduced in Sect. 1.1.3.1, the deterministic IFS algorithm in Lab 1.1, the random IFS algorithm in Lab 1.2, and the three-point method of finding IFS parameters in Lab 1.3.

In Lab 1.2 we described the visual method of estimating IFS parameters. Here we'll review this approach.

In the original shape select two points, W and X that lie on a horizontal line, and two points Y and Z that lie on a vertical line. In each small copy of the shape given by its self-similar geometry, locate the corresponding points W', X', Y', and Z'. Then $|r| = |W'X'|/|WX|$ and $s = |Y'Z'|/|YZ|$; the orientation of the piece relative to the whole determines the sign of r. Next, θ is the angle between WX and $W'X'$, and φ is the angle between YZ and $Y'Z'$. Finally, e is the difference of the x-coordinates of (say) W' and W, f is the difference of the y-coordinates of W' and W.

1.4.4 Procedure

Trace the outline of the leaf on tracing paper, or on an overhead transparency. Sketch approximate smaller copies of the leaf that together cover the leaf. You must guess at the appropriate sizes. Find the corresponding IFS rules by either the visual approach or the three-point approach.

Visual approach. Rotate and translate the copies to cover (approximately) the original outline. The size reductions, rotations, and translations of each copy give the r, s, θ, φ, e, and f values for the corresponding transformations.

Three-point approach. Select three non-collinear points A, B, and C in readily identifiable positions on the leaf. Identify the corresponding points A', B', and

C' in each copy and use the three-point software in 1.3 of the Mathematica or Python codes to find the parameters r, s, θ, φ, e, and f for each transformation.

Once the IFS rules are found, run the random IFS software in 1.2 of the Mathematica or Python codes to generate the simulated leaf image. Compare the result with the original outline. If the match is not good, isolate the features causing the most trouble, correct the rules, and run the IFS program again. Repeat until the image is adequately close to the original.

1.4.5 Sample

In Fig. 1.30 we see a picture of a maple leaf and a tracing of its outline.

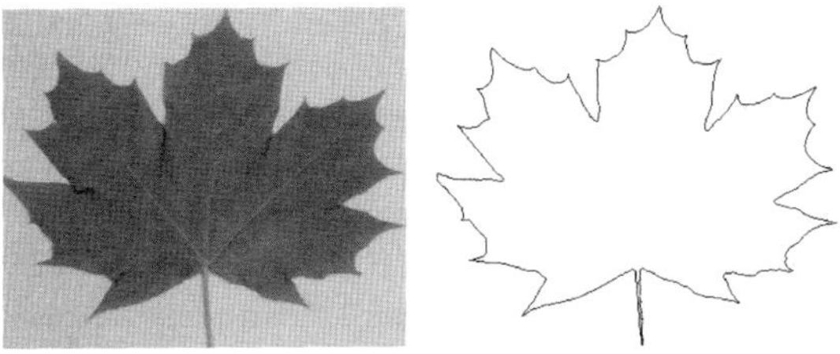

Figure 1.30: A leaf and its outline.

All of these steps can be done electronically with a scanner or digital camera and a graphics program that has scale, rotate, reflect, and place (geometrically equivalent to translate) functions. We expect that now most people have access to equipment and software with these capabilities—who knows, maybe the latest iPhone can do this—but the variety of such programs is so great that we'll leave the details to people who own the equipment. We'll stick with a combination of manual and visual processing.

We begin with scaling by 0.75 in the x- and y-directions, rotating 30° clockwise, and translating by $e = 0.7$ and $f = 0.3$. See the left side of Fig. 1.31.

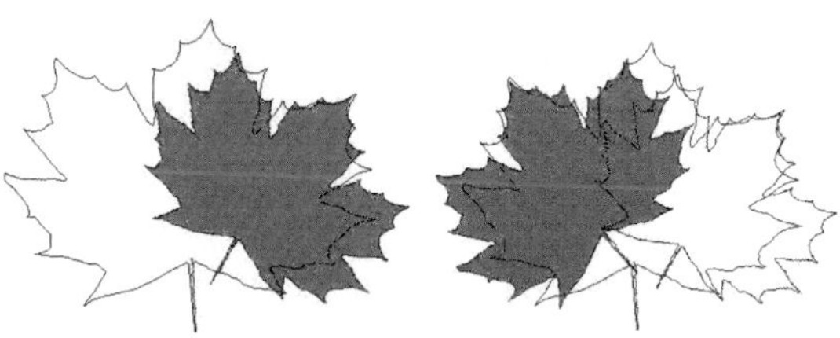

Figure 1.31: Covering the right and left sides.

Next scale by 0.80 in the x- and y-directions, rotate 45° counterclockwise, and translate by $e = -0.4$ and $f = 0.4$. See the right side of Fig. 1.31.

To cover the bottom part of the leaf, trying to match the fairly long curves on both sides of the stem, scale by 0.70 in the x- and y-directions, do not rotate, and do not translate. See the left side of Fig. 1.32.

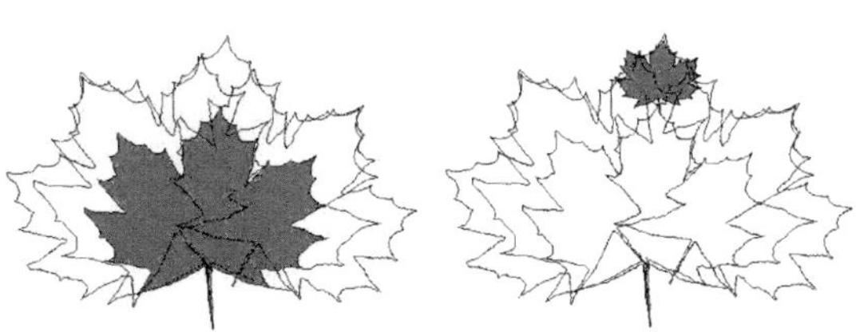

Figure 1.32: Covering the bottom and top.

To cover the top part of the leaf, scale by 0.25 in the x- and y-directions, do not rotate, and translate by $e = 0.2$ and $f = 2.2$. See the right side of Fig. 1.32.

Putting together all these pieces, we have this IFS

r	s	θ	φ	e	f	prob
0.75	0.75	-30	-30	0.7	0.3	0.32
0.8	0.8	45	45	-0.4	0.4	0.36
0.7	0.7	0	0	0	0	0.28
0.25	0.25	0	0	0.2	2.2	0.04

Table 1.4: A first IFS table to generate the leaf image.

where the probabilities were estimated using the method of Sect. 1.2.3.3. For example, $p_1 = r_1^2/(r_1^2 + r_2^2 + r_3^2 + r_4^2)$.

Before we run the random IFS program with the parameters of Table 1.4, let's think about just how closely we expect the image generated by the IFS to match the actual leaf. A first step is to inspect the Figs. 1.31 and 1.32, and compare the aggregate of these four pieces with the outline of the leaf, the right side of Fig. 1.30. The match along the top and the bottom of the leaf is pretty good, but the left and right pieces extend beyond the sides of the original image. So our first guess is that the IFS image should agree with the shape of the leaf along the top and bottom, but that the sides should be too wide. Is this what we see?

The IFS of Table 1.4 produces the picture on the left side of Fig. 1.33. As we expected, the IFS side lobes are too wide, but also the indentations between the side lobes and the top lobe are much smaller. This can be a consequence of the side lobe size (Do you see why?), so we change r_1 to 0.65 and r_2 to 0.7. This gives the picture on the right side of Fig. 1.33.

Figure 1.33: First (left) and second (right) leaf forgeries.

The right side of Fig. 1.33 looks a bit better, but we need to fill in the gap at 10 o'clock. So we change s_2 to 0.6, e_2 to -0.5, and f_2 to 0.5. This gives the left picture in Fig. 1.34.

This introduces another gap at the top. One approach to this is to stretch the fourth piece vertically. We do this by changing s_4 to 0.4. See the right side of Fig. 1.34. To assess the success of this IFS, we compare the picture it generates to the outline of the leaf.

Figure 1.34: Third (left) and fourth (right) leaf forgeries.

This is not a very good match, so we make a few additional refinements.

To get rid of the lump on the left side of the central part of the leaf, we widen the second copy and move it back to the right: change r_2 to 0.8 and e_2 to -0.3. To soften the jagged top of the central part, we move the fourth part down by changing f_4 to 1.75. In Fig. 1.35 we see the picture, together with the IFS producing it. This is a bit better match, but additional refinements still could improve the image. In particular, the leaf has two large lobes, one to the left of the central lobe, one to the right, both are approximately rotated copies of the central lobe. So far our simulation does not capture that feature.

Note, however, that the outline of the IFS-generated image will be a fractal on all levels, while a real leaf has this property over only a few levels.

Collecting together all the changes we've made, the IFS that generates the left image of Fig. 1.35 is given in Table 1.5. The probability p_1 is given by

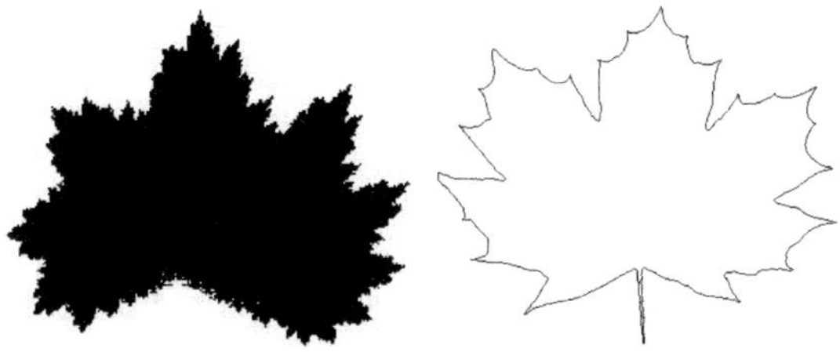

Figure 1.35: Last leaf forgery, and the original outline.

$$p_1 = \frac{0.65 \cdot 0.75}{0.65 \cdot 0.75 + 0.8 \cdot 0.6 + 0.7 \cdot 0.7 + 0.25 \cdot 0.4} \approx 0.31$$

and the others are computed similarly.

r	s	θ	φ	e	f	prob
0.65	0.75	-30	-30	0.7	0.3	0.31
0.8	0.6	45	45	-0.3	0.5	0.31
0.7	0.7	0	0	0	0	0.31
0.25	0.4	0	0	0.2	1.75	0.07

Table 1.5: A IFS table for the left image of Fig. 1.35.

Now we'll illustrate the use of the three-point transformation method to find the first IFS transformation for the leaf image. Take the point A as the origin, we measure these coordinates. $A = (0, 0)$, $B = (-3.9, 5.3)$, and $C = (4.5, 4.6)$. We approximate the corresponding points $A' = (1.2, 0.7)$, $B' = (1.2, 5.7)$, and $C' = (4.7, 1.6)$.

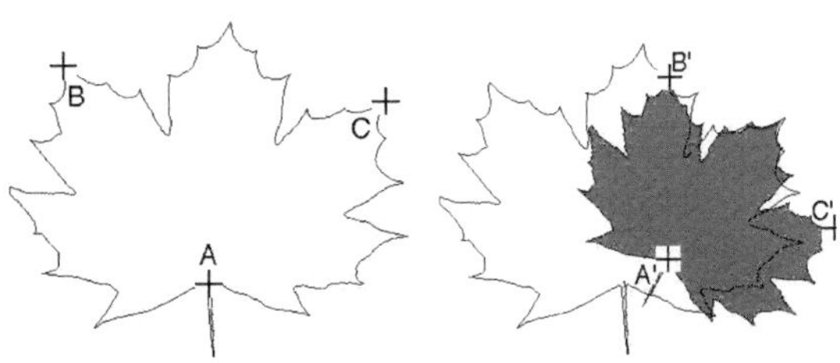

Figure 1.36: An illustration of the three-point method.

For these points the software in 1.3 of the Mathematica or Python codes gives

$$r = 0.62,\ s = 0.70,\ \theta = -45,\ \varphi = -28,\ e = 1.2,\ f = 0.7$$

a bit different from the first row of Table 1.5. However, we note (1) the measurements of the coordinates of A, B, C, A', B', and C' of necessity involve some uncertainty, and (2) small changes in IFS parameters produce small—though still sometimes quite noticeable—changes in the image.

1.4.6 Conclusion

Covers of a leaf with reduced copies of the leaf image can lead to IFS tables that produce reasonable forgeries of the leaf. When the reduced copies do not cover the leaf image exactly—and this will almost always happen—the first set of IFS rules may give a less than convincing image. The self-similarity of IFS images allows errors to propagate across levels of the image, sometimes accumulating in ways more noticeable than expected. A sequence of successive modifications can produce better results.

1.4.7 Exercises

Prob 1.4.1 Collect some leaves and find IFS to synthesize images like that reproduce those leaves.

Prob 1.4.2 If you can, get leaves from different kinds of trees. Are maple leaves easier to synthesize than oak or elm?

Prob 1.4.3 If you can make an image of the vein system of a leaf, can you find IFS rules for that?

1.5 Fractal wallpaper

Here we grow fractals by accretion. Fractality is revealed by zooming out rather than in.

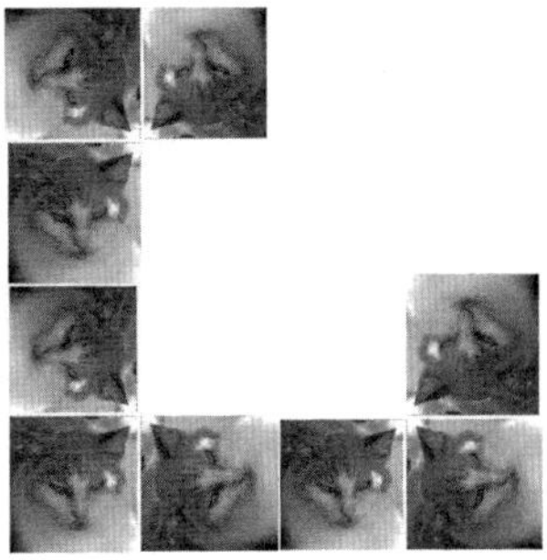

1.5.1 Purpose

We'll learn how to create a sequence of wall hangings representing stages of a fractal pattern. The hangings will increase in size as the number of stages increase. The deterministic algorithm introduced in Lab 1.1 can be thought of as generating a fractal by erosion: in Fig. 1.11 we began with a filled in unit square, removed the upper right corner, removed the upper right corners of what remains, and so on. It is sensible to say that the fractal (the Sierpinski gasket) is produced by erosion. Many natural fractals, coastlines are an obvious example, are produced by erosion. The same patterns occur on different length scales because the same forces work across different scales.

In this lab we grow fractals by accretion: we begin with a small collection of objects and add more and more of the same size objects in the same pattern over larger scales. Corals grow this way. What about the pulmonary and circulatory systems? Do you think they grow by a process closer to erosion or accretion?

1.5.2 Materials

For a physical realization of this lab, 1,084 one inch square pictures (the same or different, perhaps pictures of friends or pets), scissors, glue sticks, clear tape, 361 pieces of white 2×2 inch paper, 120 pieces of white 4×4 inch paper, 40 pieces of white 8×8 inch paper.

For a digital realization of this lab, a digital picture, graphics software for cropping and resizing pictures, and for cutting and pasting images, and a printer.

1.5.3 Background

We'll need the geometry of plane transformations of Sect. 1.1.3.1 and the deterministic algorithm of Sect. 1.1.3.3. Mostly, though, we need to compare producing fractals by erosion, which we call "smaller and smaller," and by accretion, which we call "larger and larger."

Often self-similarity is expressed as scaling under magnification: zooming in on any part of the figure reveals a sequence of shapes that eventually includes the original shape. An apparently unending zoom animation can be made from a simple loop of a few images. We showed this animation on the first day of the workshops. After it had run for a few seconds, one of us said, "This animation took a very long time to make." Then the "Oh, I get it" light went on in the eyes of the audience, a wave of grins swept through the room, and we knew that one of the main implications of self-similarity was clear.

Natural fractals depart from mathematical fractals in several ways, one of which is that for natural fractals this scaling under magnification holds over only a limited range. Years ago one of our students observed, "In this class I thought I'd learn the atoms of a sheep look like little sheep. Of course, this is wrong, and I learned the truth is much more interesting."

This observation leads to another way to build fractals: start with an *atom*, the smallest unit from which the fractal is made, and combine these in some pattern. Group together the atoms of that pattern and make copies of it. Arrange the copies in the same way the atoms were arranged to form the first group. Continue. In this way we build a shape that exhibits self-similarity over a larger and larger range as it grows. See Sects. 1.5.4 and 1.5.5 for an illustration.

1.5.4 Procedure

To create this display we use one inch square pictures (of NN's cat Chandon) and secure them to a blank wall about ten feet long and at least six feet high. Begin with 1,084 one inch square pictures. Start at the left end of the wall and attach a single picture a little more than three feet off the floor. Select a fractal pattern; one of the four examples in the Exercises section, for example. Using a glue stick, attach three pictures, in the correct orientations and positions for the selected pattern, to a 2×2 square of paper. Fig. 1.37 illustrates the idea, with a cat image. Secure this piece of paper to the wall, at the same height as the first picture and several inches to the right of it. This is the first stage of the display.

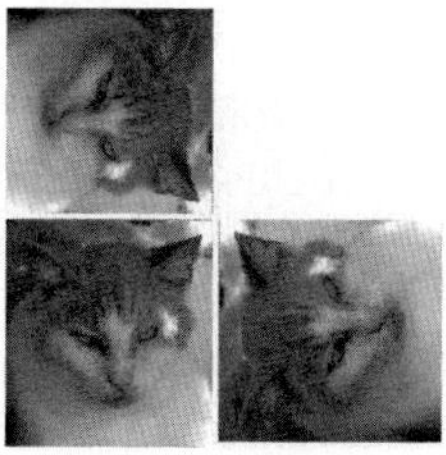

Figure 1.37: The first stage.

The next five stages use the 2×2 squares as building blocks. To complete the display, a total of 363 more 2×2 squares are needed. Make all these before assembling any later stages of the display.

Use the 2×2 squares, three at a time, to make the second stage. Orient and position the 2×2 squares to fit the fractal pattern. This is the second stage, shown in Fig. 1.38. Attach them by taping between the squares on their blank sides. Fill in the empty quadrant with blank paper to make a complete 4×4 square. You should have 121 of these 4×4 squares. Attach one to the wall to the right of the first stage.

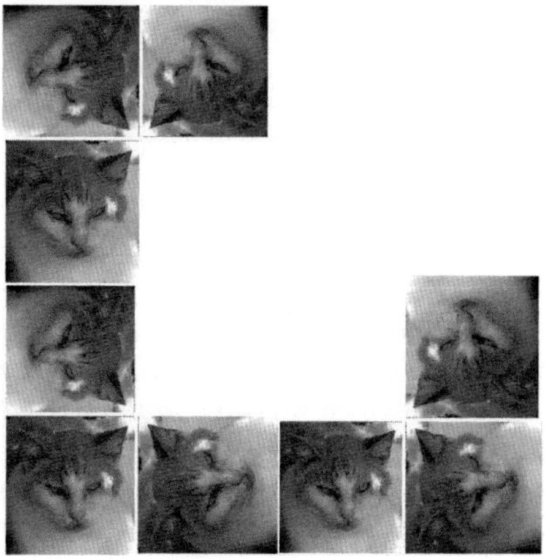

Figure 1.38: The second stage.

Now combine the remaining 120 copies of the 4×4 squares three at a time following the fractal pattern, taping them on their backs to makle 40 copies of an 8 × 8 square, filling in the empty quadrants with blank paper. Attach one of these 8 × 8 squares to the right of the second stage. This is stage three.

Take the 39 remaining 8 × 8 squares that remain and attach them three at a time in the orientations and positions following the fractal pattern, making 13 copies of a 16 × 16 square after filling in the empty quadrant with blank paper. This is stage four. Attach a copy to the right of stage three on the wall.

Use the 12 remaining 16 × 16 squares following the fractal pattern to make four 32 × 32 squares, filling in the empty quadrant with blank paper. These are stage five. Attach one to the right of stage four.

Finally, assemble the three remaining 32×32 squares into one 64×64 square, filling in the empty quadrant with blank paper. This is stage six. Attach it to the wall, to the right of stage five.

1.5.5 Sample

Here are some photos of the construction and display of the sample image from one of the workshops.

Note that the left picture of the second row demonstrates Benoit's second career as a photography model. The right picture of the second row shows a *very loose* connection between the products of this lab and the post 1988 paintings of Chuck Close. If you are unfamiliar with the work of Chuck Close, Google is your friend. In 1988 Close suffered a spinal arterial collapse that left him largely paralyzed. His motor control no longer adequate to continue his earlier immense hyperrealist paintings, Close composed paintings that when viewed from a distance were impressive pixellated portraits, but when viewed from nearby the pixels are seen as small abstract paintings, often consisting of a few simple curves. We turn faces into fractals. Not nearly as interesting, but still, it is a bit of a surprise to see faces appear as you zoom in (that is, step closer to) a fractal wallhanging.

1.5.6 Conclusion

This method of growing fractals by accretion gives a physical sense of how the same process operates on different scales. By taking photos as the atoms of the fractal, we have a clear example of how different features are visible at different scales. If photos of students are used, mixing the pictures before beginning the assembly process can generate interesting discussions.

The basic idea of this construction was presented by Vicki Fegers and Mary Beth Johnson in "Fractals—energizing the mathematics classroom," [37], Chapter 9 of [15].

1.5.7 Exercises

Carry out the fractal wallpaper construction for each of these schematic first stages. Recall no reflections are used for the physical realizations of the wallpaper. Computer realizations do not have this restriction. The capital T in each square indicates the relative orientation of each square. The initial shape is a square that contains a T in its usual orientation.

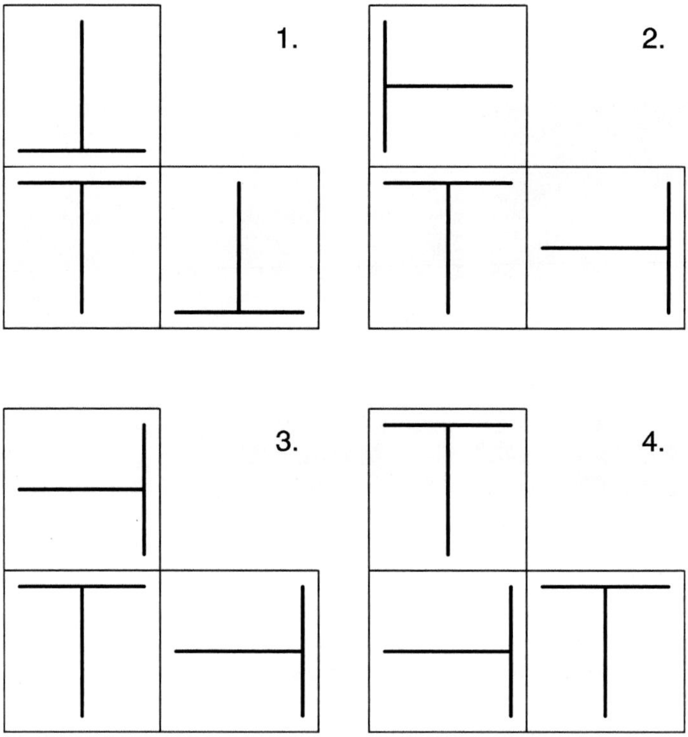

1.6 Cumulative gasket pictures

Here we show that a single long run of the random IFS algorithm is can be approximated very closely by the superposition of several shorter runs.

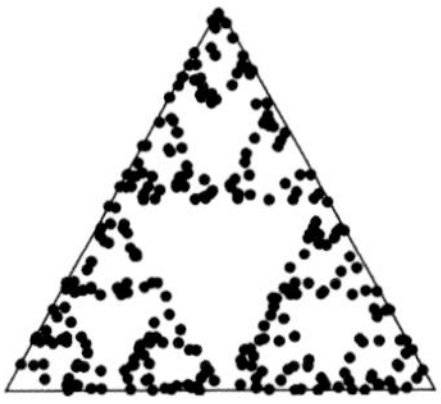

1.6.1 Purpose

Some texts [31] refer to special cases of the random IFS algorithm as the "chaos game." The manual implementation of the chaos game has a familiar shortcoming: patience limits images to a few dozen points, not enough to reveal any underlying pattern. In this lab we'll show a way to overcome this problem: the superposition of several moderate-length chaos game sequences can be quite close to a single longer sequence of the chaos game. Also, we'll use the chaos game to illustrate an interesting property of random sequences of numbers.

1.6.2 Materials

We'll use a triangle template and a gasket template, about ten overhead transparencies, a die (singluar of dice), adhesive tape, a ruler calibrated in mm, a permanent marking pen, and an overhead projector to display the data.

1.6.3 Background

We'll define the chaos game and compare it to the random IFS algorithm. Also, we'll illustrate the shortcoming of the manual chaos game.

1.6.3.1 The chaos game

One of the simplest and most commonly used introductions to fractals comes by a process that Barnsley named the *chaos game*.

• First select three points, V_1, V_2, and V_3 in the plane; vertices of an equilateral triangle are a popular choice.

• Select a starting point P_0.

• Given P_i, produce P_{i+1} by selecting randomly one of the V_j, all with equal probability. Then P_{i+1} is the midpoint of the line segment between P_i and V_j.

The points P_0, P_1, P_2, and P_3 are illustrated in on the left of Fig. 1.39. Here the sequence of vertex choices is 1 then 3 then 2.

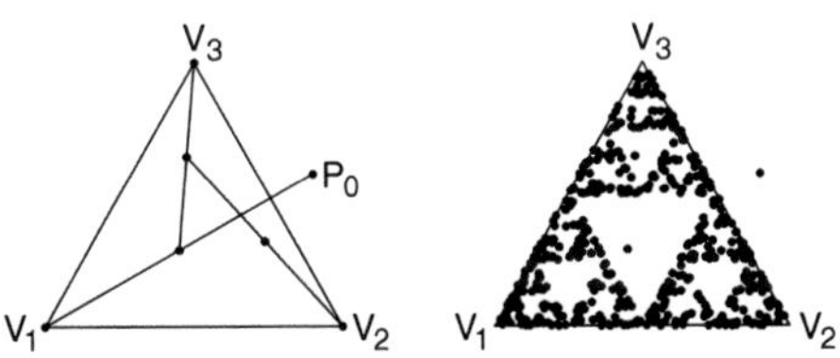

Figure 1.39: Left: a few chaos game points. Right: a lot of points.

Some of the earliest illustrations of the chaos game, used before IFS became widely known, involved a surprise. First, the chaos game was described, as above. Next, a few points were generated by hand, often using students as random vertex selectors. The game was played until a few of the points were within the triangle. Then the teacher observed that because the next point lies half-way between the current point and one of the vertices of the triangle, and the triangle is a convex figure, once a point enters the triangle, all subsequent points will lie within the triangle. Nodding heads, murmurs of assent from the class. Next, the teacher remarked that because the vertices are selected uniformly randomly, the points will fill the triangle uniformly. More nodding heads, more agreement, and now a few bored looks from some students. Then a computer projection system, usually an LCD plate placed on an overhead projector (remember, this was in the mid 1980s), was set up, the button was pushed, and something like the right side of Fig. 1.39 appeared, usually followed by gasps from the class, sometimes an expletive or two, a "whiskey tango foxtrot" immediately followed by a blush. Those were more innocent times. How did this elementary game produce a Sierpinski gasket?

Here's a sketch of the main step. Suppose one of the early iterates (x_i, y_i) lies in the largest empty triangle in the gasket. Draw a few iterates by hand and you'll see that (x_{i+1}, y_{i+1}) leaves this triangle and no later iterates can re-enter that empty triangle. Then (x_{i+1}, y_{i+1}) lies in one of the three next-largest empty triangles; later iterates exit these three empty triangles and never return. And so on. A more complete argument can be built around the method presented in the next section.

1.6.3.2 The random IFS algorithm

In Sect. 1.6.3.3 we'll see that the chaos game is a special case of the random IFS algorithm, described in Sect. 1.2.3.2. To prepare for the next section, here we'll review the main points of the random IFS algorithm.

An IFS is determined by a collection of transformations $T_1, \dots, T_n$, that are contractions. By this we mean that for any pair of points (x, y) and (x', y'), the distance between $T_j(x, y)$ and $T_j(x', y')$ is less than the distance between (x, y) and (x', y'). Then there is a unique compact set A left invariant by the simultaneous application of all the T_j. That is, $A = T_1(A) \cup \cdots \cup T_n(A)$.

The random IFS algorithm is a way of generating a quick, good, approximation of A. For this we associate to each T_j a probability, p_j, of applying that transformation. (In Sect. 1.2.3.3 we presented an assignment of probabilities that fills A approximately uniformly. For the chaos game transformations, taking all the p_j equal is the corresponding assignment.) The random IFS algorithm begins by selecting a point $P_0 = (x_0, y_0)$. Then given the point $P_i = (x_i, y_i)$, the next point P_{i+1} is generated by selecting a T_j randomly according to the probabilities p_j and taking $P_{i+1} = T_j(x_i, y_i)$.

On the computer, one continues to generate points until no appreciable change appears. The assignment of probabilities determines how quickly this occurs. So long as none of the $p_j = 0$, *any* assignent of probabilities will lead to a picture of A, though the wait may be substantial.

Mathematically we collect the infinite sequence of points generated and take the limit set, those points approximated arbitrarily closely by points of the sequence. We mention this in case you wondered how a countable collection—the set of points generated by the random IFS algorithm—can fill an uncountable set such as the gasket. (One of us has gotten this question in class more than once.) The answer is that it can't, but its limit set can. This isn't a surprise: the rational numbers are countable, and their limit set is the real numbers, which are uncountable, as Cantor's diagonal argument shows.

1.6.3.3 The chaos game as random IFS

Write the vertex $V_j = (a_j, b_j)$. Then the chaos game rule, "take the midpoint of the segment between (x_i, y_i) and (a_j, b_j)" can be written as

$$(x_{i+1}, y_{i+1}) = \left(\frac{x_i + a_j}{2}, \frac{y_i + b_j}{2}\right) = \left(\frac{x_i}{2}, \frac{y_i}{2}\right) + \left(\frac{a_j}{2}, \frac{b_j}{2}\right)$$

In terms of the familiar IFS parameters introduced in Sect. 1.1.3.1 r, s, θ, φ, e, and f, this chaos game transformation has $r = s = 1/2$, $\theta = \varphi = 0$, $e = a_j/2$, and $f = b_j/2$. All transformations have the same scaling factor, 1/2, no transformation involves a rotation or a reflection, and the translations are determined by the chaos game vertices.

Let's focus on the equilateral triangle chaos game. These three chaos game rules are equivalent to the IFS rules T_1, T_2, and T_3 that generate the equilateral Sierpinski gasket A, and we know $A = T_1(A) \cup T_2(A) \cup T_3(A)$. The sides of the equilateral triangle belong to the gasket, so if we start the chaos game with a point on one of these sides, all the points the chaos game generates will lie on the gasket.

1.6.3.4 A shortcoming of the chaos game played manually.

So far, so good. The chaos game is an easy introduction to generating (to be sure, a limited class of) fractals, demonstrating the power of iteration without concern for the mechanics of general affine transformations, or for probabilities. It provides a simple introduction to the more general iterated function system (IFS) appraoch.

A problem arose when the chaos game was introduced to wider audiences, sometimes without much thought. Playing the game by hand, using a die for a vertex selector, is very attractive. The rules are simple, the process repetitive, easily learned and implemented. Some teachers began to emphasize the manual chaos game and omit the computer-generated concluding image. This resulted in some confusion, because students lose patience with manual chaos game constructions after only about 30 iterations. Fig. 1.40 shows two 30 point runs of the chaos game. Neither is a convincing approximation of a gasket, especially because both begin outside the triangle.

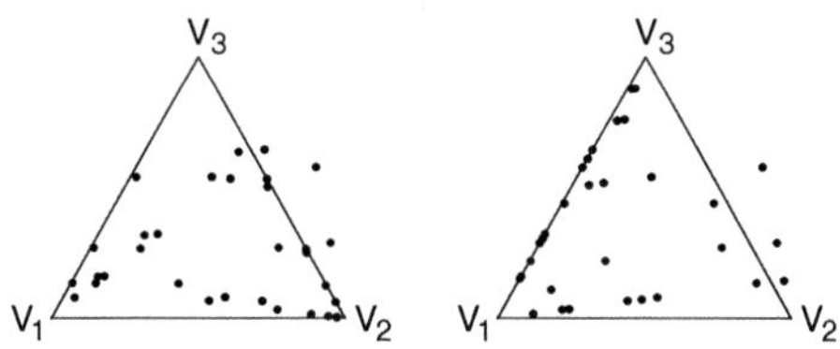

Figure 1.40: Two 30 point chaos game runs, starting outside the triangle.

In Appendix B.3 we develop a method to estimate the probability of encountering specific strings in a long sequence of trials. Think of successive runs of heads in a sequence of coin tosses. We interpret this in Exercise 1.6.6.

1.6.4 Procedure

1. Place an overhead transparency over the triangle template. Secure the transparency to the template with tape.
2. Select one of the three sides of the triangle and select a starting point P_0 anywhere on that side.
3. Roll the die.
• If 1 or 2 comes up, use the ruler to locate the point P_1 half-way between P_0 and the lower left corner, call this the $1, 2$ corner, of the triangle template.
• If 3 or 4 comes up, use the ruler to locate the point P_1 half-way between P_0 and the lower right corner, call this the $3, 4$ corner, of the triangle template.
• If 5 or 6 comes up, use the ruler to locate the point P_1 half-way between P_0 and the upper corner, call this the $5, 6$ corner of the triangle template.
4. Repeat step 3, using P_1 to generate P_2, P_2 to generate P_3, and so on, until P_{30} is plotted.
5. Repeat steps 1–4, plotting 30 points on all ten transparencies, or until patience is exhausted. With ten people or groups of people, each group need do steps 1–4 only once.
6. Overlay the transparencies. Interpret the result.

1.6.5 Sample

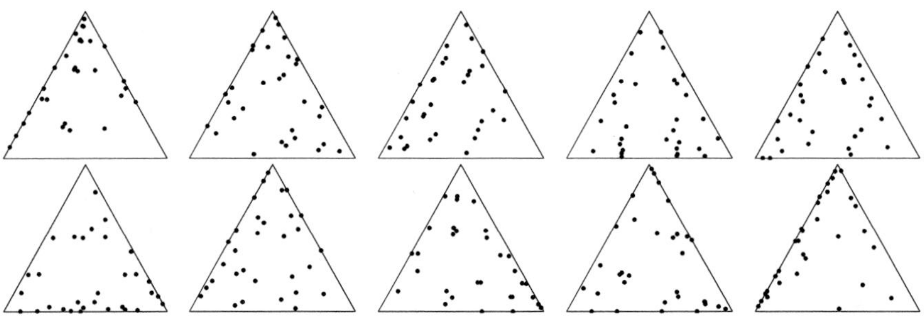

Figure 1.41: Ten 30 point runs of the equilateral triangle chaos game.

In Fig. 1.41 we see ten samples of 30 points each. For manual implementation, ten is about the maximum number of transparencies, because overhead transparencies are not really transparent. Also, for experiments done by hand, 30 points are about the most that anyone will generate before their patience is exhausted. Does any one of these ten images look much like a gasket? Maybe the second from the left of the second line, which shows the beginning of the empty central triangle. But in general, none of these is even vaguely suggestive of the gasket. If someone had tenacity sufficient to generate a plot of 300 points, a picture more suggestive of a gasket would appear.

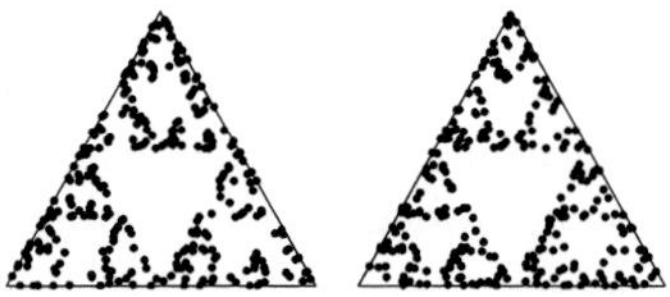

Figure 1.42: Left: the aggregate of the pictures of Fig. 1.41. Right: a run of 300 points.

Now for the goal of this lab. The left of Fig. 1.42 is the aggregate of these ten pictures; the right is the picture of 300 iterates of one point. Both are reasonable representations of the gasket.

The 30 point pictures illustrate two important aspects of the chaos game.

- Manual chaos game experiments require a great deal of patience to produce a recognizable image.
- The evident variability between short chaos game runs is considerable.

1.6.6 Conclusion

The superposition of several short runs of the chaos game is at least visually indistinguishable from a longer run of the same total number of points. Both reveal sequences of points marching toward a vertex for more consecutive steps than some intuition suggests.

1.6.7 Exercises

Prob 1.6.1 Perform the experiment described in the Procedure. Use the triangle template of Fig. 1.43. For use in later problems record the sequence of transformations applied, that is, the sequence of vertices selected.

Recall that in Sect. 1.6.3.3 we saw that by starting the chaos game with a point on a side of the triangle, we are guaranteed that all the points generated lie on the gasket. In exercises 1.6.2–1.6.5 we investigate some of the consequences of taking a starting point not on the gasket.

Prob 1.6.2 For this problem, start with a point P_0 near the middle of the triangle and generate points P_1 through P_{30} with the sequence of Exercise 1.6.1. After the points are generated, place the transparency over the gasket template of Fig. 1.44. Observe some of the points lie in the triangles removed in forming the gasket.

Prob 1.6.3 Continuing Exercise 1.6.2, on closer examination of the gasket overlay note that exactly one point lies in the largest removed triangle, exactly one point lies in the three next-largest removed triangles, exactly one point lies in the nine next-largest removed triangles, and so on. Why is this?

Prob 1.6.4 Suppose the base of the gasket has length 512 pixels, and suppose P_0 lies inside the largest removed triangle. What is the smallest integer N for which we can we be sure P_N is indistinguishable from a point of the gasket?

Prob 1.6.5 Repeat Exercise 1.6.2 with P_0 lying outside the triangle and with a different sequence of random numbers. Must P_1 lie in the largest removed triangle of the gasket?

Prob 1.6.6 (a) Without using any physical implements including die tossing or computer random number generators, write a random sequence of 300 1s, 2s, and 3s. This will take a while, but we'll use your sequence to make an interesting point.
(b) How long is the longest string of consecutive 1s in this sequence?

(c) Compute the probability of obtaining five consecutive 1s, assuming each additional digit in the sequence has equal probability of being 1, 2, or 3. (Hint: This is tricky. Our solution uses a Markov chain. It may be simpler to write a program to generate strings of 300 1s, 2s, and 3s, and test the string for the presence of 5 consecutive 1s. Run the program a lot of times and use the ratio of the number of runs with five consecutive 1s to the total number of runs as an estimate of this probability. Or peek at the solution. But before you do that, make a guess.)
(d) Apply this result to interpreting the equilateral triangle chaos game.
(e) Compute the probability of obtaining five consecutive 1s in the superposition of ten 30 point runs of the chaos game. Compare this with (c).
The argument in Appendix B.3 gives some guidance.

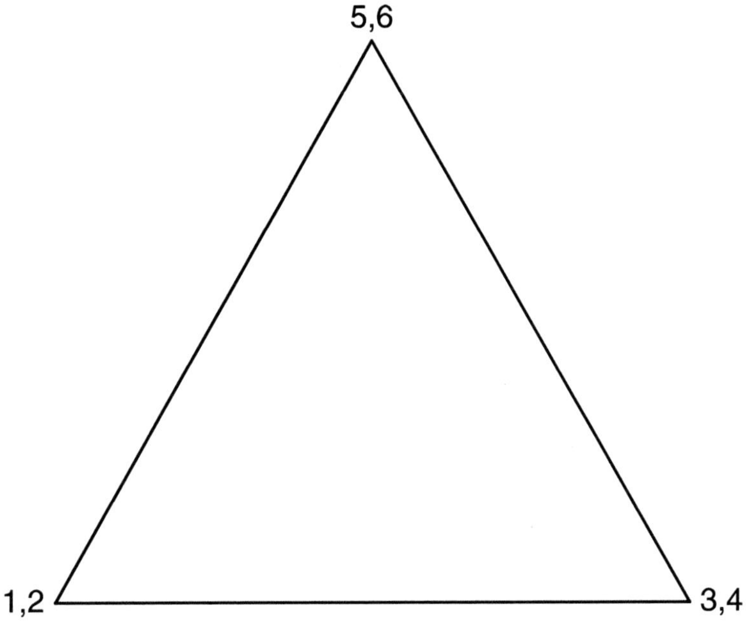

Figure 1.43: Triangle template

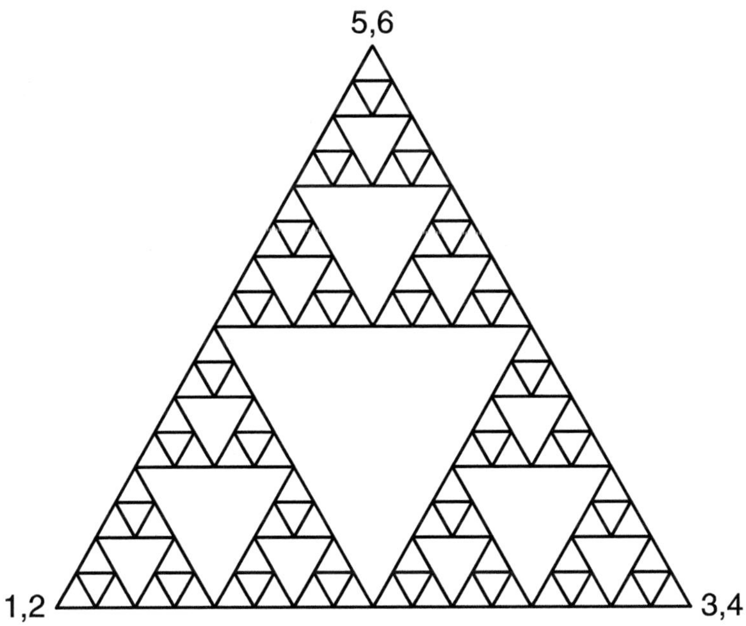

Figure 1.44: Gasket template

1.7 IFS and addresses

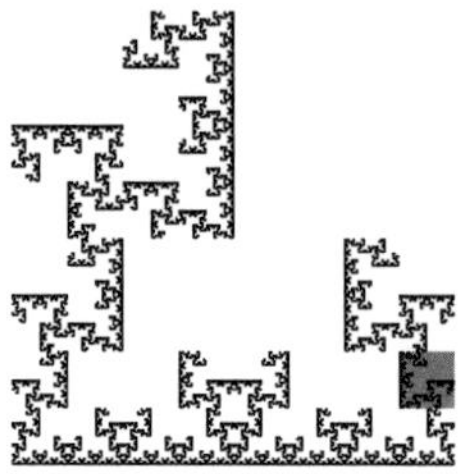

Here we'll use compositions of IFS transformations to assign labels, called addresses, to regions of the attractor. The address of a region is the sequence of transformations needed to enter that region.

1.7.1 Purpose

Here we'll become familiar with the addresses of regions in a fractal. We'll use this in Labs 1.8, 1.9, 1.10, and 1.11.

1.7.2 Materials

We'll use paper and pencil. A version of random IFS software that can highlight addresses (1.7 of the Mathematica or Python codes) is useful, but not necessary.

1.7.3 Background

Here we have only one topic to explore: the definition of addresses by compositions of transformations. Because IFS express fractals as sets left invariant by collections of transformations, each region of a fractal can be described by the sequence of transformations giving rise to that region. That sequence is the address of the region. Here we work with a specific example, these transformations

$$\begin{aligned} T_1(x,y) &= (x/2, y/2) & T_2(x,y) &= (x/2, y/2) + (1/2, 0) \\ T_3(x,y) &= (x/2, y/2) + (0, 1/2) & T_4(x,y) &= (x/2, y/2) + (1/2, 1/2) \end{aligned} \tag{1.6}$$

For the filled-in unit square S, we see $S = T_1(S) \cup T_2(S) \cup T_3(S) \cup T_4(S)$ with overlaps only along edges.

The square S has corners $(0,0)$, $(1,0)$, $(0,1)$, and $(1,1)$, so the square $T_1(S)$ has corners

$$T_1(0,0) = (0,0),\ T_1(1,0) = (1/2,0),\ T_1(0,1) = (0,1/2),\ T_1(1,1) = (1/2,1/2)$$

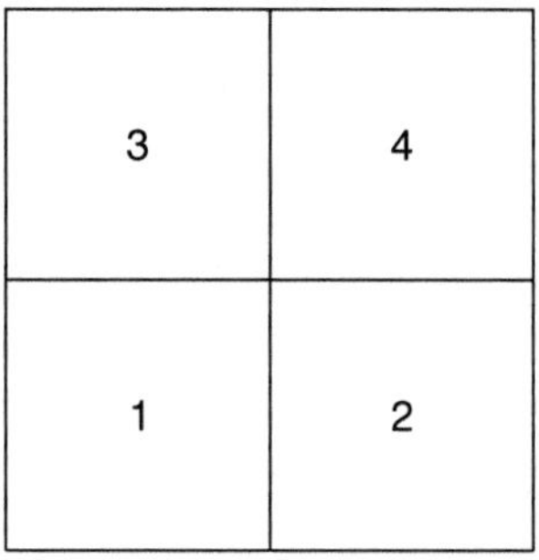

The *address* of a region is the transformation or combination of transformations needed to specify the region. For example, the four quarters of the unit square have the addresses shown on the right. Because they are specified by one digit, we call these *length-1 addresses*. Longer addresses specify smaller regions, that is, determine the position more precisely.

We specify addresses by the order of the composition of transformations. For example, the square $T_1(T_2S))$ has address 12. Here's a way to see this.

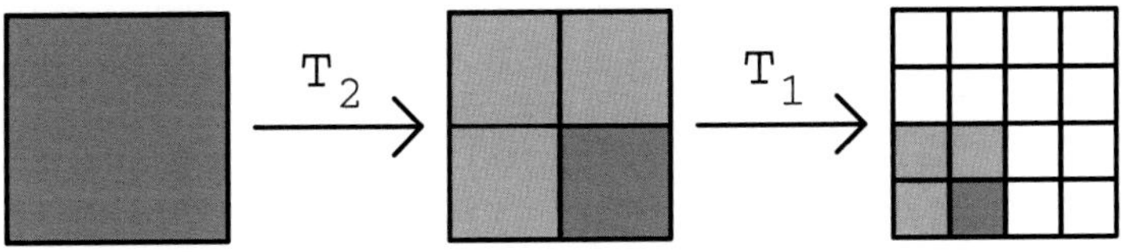

Figure 1.45: The composition $T_1(T_2)$ and the region it determines.

The full square S, shaded dark on the left, is mapped by T_2 to address 2, the dark square on the lower right in the middle. The whole square S is mapped to the lower left subsquare by T_1, so the dark region on the right image is $T_1(T_2(S))$. That is, this dark square has address 12. Another way to think of this address is that it is the 2 part of square 1.

If we continue in this fashion, we obtain the *length-2 addresses*, the *length-3 addresses*, and so on.

33	34	43	44
31	32	41	42
13	14	23	24
11	12	21	22

The notion of address is central to the proof that the determinisitc and random IFS algorithms produce the same images. See Appendix B.2.2. We'll see applications of addresses in other labs as well.

Addresses of fractals work like geographic addresses in that longer addresses specify locations more precisely. With the transformations of this section, the address 143 describes all points in a $1/8 \times 1/8$ square; the address 143221 describes all points in a $1/64 \times 1/64$ square. The address New Haven, Connecticut, USA describes all points in an 20 square mile area; Room 446 of 10 Hillhouse Ave., New Haven, Connecticut, USA describes all points in a 200 square foot office.

1.7.4 Sample

With the transformations of Eq. (1.6) the shaded square has address 123. We can see this by "relative coordinates". The square with address 123 is in the square with address 1. Within that square, it is in the 2 part, that is, in the square with address 12. Within that square, it is the 3 part. That is, the shaded regions is the 3 part of the 2 part of the 1 square.

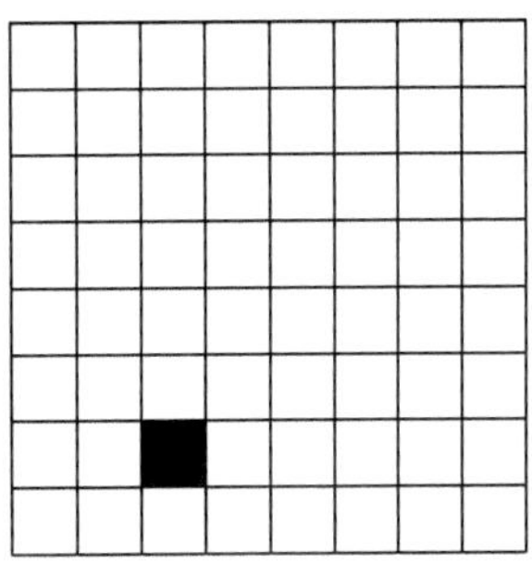

We can find the coordinates of the corners of the square with address 123 by inspection. They are $(2/8, 1/8)$, $(3/8, 1/8)$, $(3/8, 2/8)$, and $(2/8, 2/8)$.

Another way to find these coordinates of the corners of this square is to find the formula for the composition $T_1(T_2(T_3(x, y)))$. We start with T_3 and work outward in this composition.

$$T_1(T_2(T_3(x,y))) = T_1\left(T_2\left(\frac{x}{2}, \frac{y+1}{2})\right)\right) = T_1\left(\frac{1}{2}\frac{x}{2} + \frac{1}{2}, \frac{1}{2}\frac{y+1}{2}\right)$$
$$= T_1\left(\frac{x+2}{4}, \frac{y+1}{4}\right) = \left(\frac{1}{2}\frac{x+2}{4}, \frac{1}{2}\frac{y+1}{4}\right) = \left(\frac{x+2}{8}, \frac{y+1}{8}\right)$$

That is, $T_1(T_2(T_3(x,y))) = (x/8, y/8) + (1/4, 1/8)$. Evaluating this function at the four corners $(0,0)$, $(1,0)$, $(1,1)$, and $(0,1)$ of the unit square gives the coordinates of the four corners found by inspection.

1.7.5 Conclusion

Addresses of regions in a fractal specify the order of the transformations producing that region. Longer addresses specify smaller regions. Because different collections of transformations can generate the same fractal, we see that the adddress of a region depends on the choice of transformations.

1.7.6 Exercises

Prob 1.7.1 With the transformations of Eq. (1.6), shade the region with address 312 and the region with address 124.

Prob 1.7.2 With T_1, T_2, and T_4 of Eq. (1.6), and with $T_3(x,y) = (-x/2, y/2) + (1/2, 1/2)$, shade the regions with addresses 123, 312, and 124.

Prob 1.7.3 Now change T_3 so its IFS parameters are $r = s = 1/2$, $\theta = \varphi = 90°$, and $e = f = 1/2$. Shade the region with address 312. Compare that region with the region 312 of Exercise 1.7.2. Comment on the differences.

Prob 1.7.4 The standard IFS rules for the equilateral gasket are
$T_1(x,y) = (x/2, y/2)$
$T_2(x,y) = (x/2, y/2) + (1/2, 0)$
$T_3(x,y) = (x/2, y/2) + (1/4, \sqrt{3}/4)$
Find changes to T_3 making each of the shaded regions have address 321. Explain how you arrived at your answer

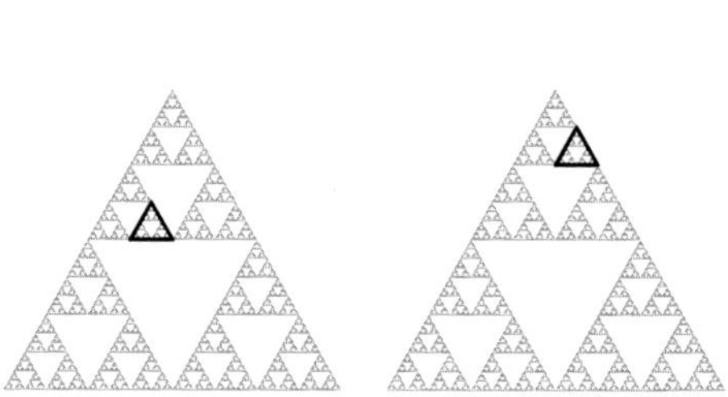

1.8 Decimals as addresses

Decimals are addresses. Perhaps this approach would make the introduction of decimals more intuitive for young students.

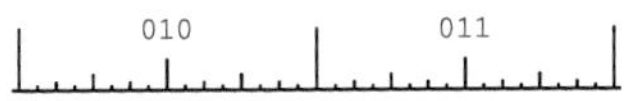

1.8.1 Purpose

We'll show that the notion of addresses, explored in Lab 1.7, can be adapted to provide a visual interpretation of decimals that illuminates some properties of decimals. For example, we'll see why $0.199999\cdots = 0.200000\ldots$, without summing geometric series.

1.8.2 Materials

For this lab we'll use just paper, pencil, and a ruler with an inch and a mm scale.

1.8.3 Background

Labs 1.1 and 1.7 are all you need.

1.8.4 Procedure

Use a ruler to subdivide a line segment into pieces of a given length. Attach an address to each piece. Subdivide each of these pieces in the same fashion. Attach an address to each of these smaller pieces. This seems simple enough, but we may find a surprise when we generalize these ideas to addresses on a Koch curve of Fig. 1.46.

1.8.5 Sample A

Here we use the centimeter scale; for the exercise we use the inch scale because this allows a greater number of levels of subdivision without eye strain.
1. Draw a horizontal line segment of length 10 cm. Mark the endpoints with vertical segments of length 1 cm.

2. Divide the horizontal segment into pieces of length 1 cm. Mark these segments with vertical segments of length 0.5 cm.

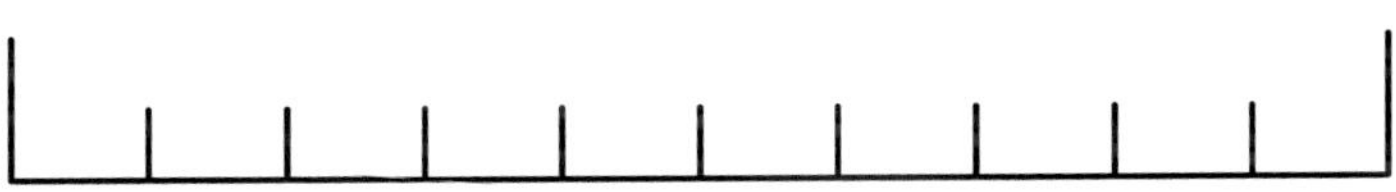

Assign an address 0 to the leftmost tenth, 1 to the next tenth, 2 to the next tenth, ..., 9 to the rightmost tenth. Note the address of each interval is the number of tenths—the first digit in the decimal expansion—of all points in that interval.

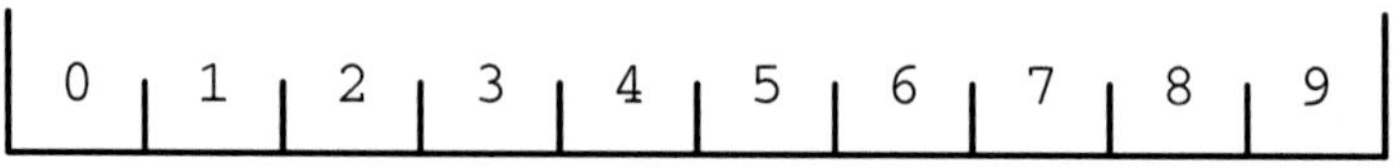

3. Divide the horizontal segment into pieces of length 0.1 cm. Mark these segments with vertical segments of length 0.25 cm.

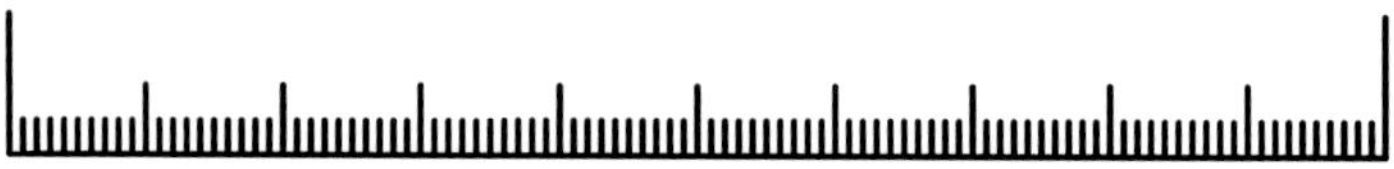

To the subdivisions of the interval with address 0 assign addresses $00, 01, \dots, 09$; to the subdivisions of the interval with address 1 assign addresses $10, 11, \dots, 19$; and so on. Note the address of each interval is the number of hundredths of all points in that interval.
4. To relate this to the more familiar notion of addresses through IFS, the subdivision of this Sample can be achieved through these transformations,

$$f_i(x) = \frac{x}{10} + \frac{i}{10} \qquad \text{for } i = 0, 1, \dots 9$$

If we think of the original interval of this Sample as $[0, 1]$, the region with address 37 is $f_3(f_7([0, 1]))$. Note $f_7([0, 1]) = [f_7(0), f_7(1)] = [7/10, 8/10]$ so $f_3(f_7([0, 1])) = [3/10 + 7/100, 3/10 + 8/100] = [37/100, 38/100]$.

1.8.6 Sample B

Here we'll develop some techniques useful for the solution of Exercises 1.8.7 and 1.8.8.
(a) Suppose a point x in $[0, 1]$ has binary address that consists of a block $a_1 \dots a_n$ of 0s and 1s that repeats forever. What can we say about x? The coordinate of the point x is

$$\begin{aligned}
x &= \frac{a_1}{2^1} + \dots + \frac{a_n}{2^n} + \frac{a_1}{2^{n+1}} + \dots + \frac{a_n}{2^{2n}} + \frac{a_1}{2^{2n+1}} + \dots + \frac{a_n}{2^{3n}} + \dots \\
&= \left(\frac{a_1}{2^1} + \dots + \frac{a_n}{2^n}\right) \cdot \left(1 + \frac{1}{2^n} + \frac{1}{2^{2n}} + \dots +\right) \\
&= \left(\frac{a_1}{2^1} + \dots + \frac{a_n}{2^n}\right) \cdot \frac{2^n}{2^n - 1} \\
&= \left(\frac{2^{n-1}a_1 + \dots + a_n}{2^n}\right) \cdot \frac{2^n}{2^n - 1} \\
&= \frac{2^{n-1}a_1 + \dots + a_n}{2^n - 1}
\end{aligned}$$

Right off we see that x is a rational number. Can we say more? Yes: the denominator is an odd number, so every factor of the denominator is odd. If we cancel every common factor of the numerator and the denominator, the denominator still is odd. That is, if a point x has (binary) address that is periodic, then x is a rational number that, in lowest form, has odd denominator.

(b) What about the converse? Suppose x in $[0,1]$ is a rational in lowest form and has an odd denominator, $x = p/q$ with p and q relatively prime and q odd. Is the binary expansion of x periodic? We'll illustrate this with an example, and let you think about how to extend this to a general argument.

Take $x = 1/5$. Its binary expansion is

$$x = \frac{a_1}{2} + \frac{a_2}{2^2} + \frac{a_3}{2^3} + \frac{a_4}{2^4} + \cdots$$

• Because $x = 1/5 < 1/2$, we see that $a_1 = 0$.

To determine if $a_2 = 0$ or 1, if $a_1 = 0$ we need to find if x lies in $[0, 1/4)$ or $[1/4, 1/2)$, and if $a_1 = 1$ we determine if x lies in $[1/2, 3/4)$ or $[3/4, 1)$. Can we find a_2 without specifying these two cases? It turns out there is a way, and it's cute. Not as cute as a 3 month old kitten, but it is math cute. Here's how. When $a_1 = 0$, if $2x < 1/2$ then $a_2 = 0$ and if $2x \geq 1/2$ then $a_2 = 1$. What if $a_1 = 1$? Then $2x \geq 1$, but $2x - 1$ lies between 0 and 1, and (here's the main point) if $2x - 1 < 1/2$ then $x < 3/4$, while if $2x - 1 \geq 1/2$ then $x \geq 3/4$. To combine both cases into one statement we use the mod function,

$$x \pmod 1 = x - [x] \quad \text{where } [x] \text{ is the integer part of } x$$

For example, $1.7 \pmod 1 = 1.7 - [1.7] = 1.7 - 1 = 0.7$. Then the general condition to find the coefficient a_n in the expansion of x is this:

$$a_n = 0 \text{ if } 2^n x \pmod 1 < 1/2 \quad \text{and} \quad a_n = 1 \text{ if } 2^n x \pmod 1 \geq 1/2$$

So let's continue with the expansion of $x = 1/5$.
• Because $2x = 2/5 < 1/2$, we see that $a_2 = 0$.
• Because $2^2 x = 4/5 \geq 1/2$, we see that $a_3 = 1$.
• Because $2^3 x \pmod 1 = 8/5 \pmod 1 = 3/5 \geq 1/2$, we see that $a_4 = 1$.
• Because $2^4 x \pmod 1 = 6/5 \pmod 1 = 1/5 < 1/2$, we see that $a_5 = 0$.
But notice something else: $2^4 x \pmod 1 = 1/5$ and $x = 1/5$ so all the additional a_n will repeat this pattern of 0011. That is, the address of 1/5 is $(0011)^\infty$. Verify that this geometric series

$$\frac{0}{2^1} + \frac{0}{2^2} + \frac{1}{2^3} + \frac{1}{2^4} + \frac{0}{2^5} + \frac{0}{2^6} + \frac{1}{2^7} + \frac{1}{2^8} + \frac{0}{2^9} + \frac{0}{2^{11}} + \frac{1}{2^{11}} + \frac{1}{2^{12}} + \cdots$$

sums to 1/5.

(c) Finally, we mention that this problem can be approached in another way, using the fact that periodic addresses determine fixed points. Suppose the address of x is $(a_1 a_2 \ldots a_n)^\infty$. Then $f_{a_1} f_{a_2} \cdots f_{a_n}(x)$ has address

$$(a_1 a_2 \ldots a_n)(a_1 a_2 \ldots a_n)^\infty = (a_1 a_2 \ldots a_n)^\infty,$$

that is, $f_{a_1} f_{a_2} \cdots f_{a_n}(x) = x$.

Each $f_{a_i}(x)$ divides x by 2 and adds 0 or 1/2, depending on whether a_i is 0 or 1. So

$$f_{a_1} f_{a_2} \cdots f_{a_n}(x) = \frac{x}{2^n} + \frac{a_1}{2} + \frac{a_2}{2^2} + \cdots + \frac{a_n}{2^n} = \frac{x}{2^n} + \frac{k}{2^n}$$

where $k = a_1 2^{n-1} + a_2 2^{n-2} + \cdots + a_n$. Then $x = f_{a_1} f_{a_2} \cdots f_{a_n}(x)$ becomes

$$x = \frac{x}{2^n} + \frac{k}{2^n}$$

which gives $x = k/(2^n - 1)$. That is, any number whose address is a repeating block can be written as a rational number with denominator of the form $2^n - 1$ for some n. Even if this fraction is not in lowest form, odd numbers have only odd factors and so in lowest form the denominator still is odd.

1.8.7 Conclusion

Address provide another way to interpret successive binary (halves, quarters, eighths, and so on) and decimal digits. This approach gives a geometric interpretation of the limits of geometric series. We can apply addresses to any fractal generated by IFS, but the ideas particular to decimal addresses can give some insights about fractal curves. For example, we show that the Koch curve has no tangent lines. Perhaps surprisingly, this occurs in two different ways, depending on the address of the point on the curve.

1.8.8 Exercises

Prob 1.8.1 *Subdividing the inch* Consider an inch. We'll successively subdivide it into halves, and the halves into quarters, and so on. Seeking the scaled patterns within patterns is a natural approach for people accustomed to thinking about fractals.
(a) First, divide the inch into halves, left and right. To each point in this inch we assign an address, 0 to the points on the left half, 1 to the points on the right half. (In Exercise 1.8.5 we'll see that the point $x = 1/2$ has two addresses, one that ends in an infinite string of 1s and one that ends in an infinite string of 0s. We'll always take the address that ends in 0s.) Then because an address that begins with 0 means the point is less than 1/2 and an address that begins with 1 means the point is greater than or equal to 1/2, we can think of this address as giving the number of 1/2s in the coordinate of the point.
(b) Next divide each half into quarters. The second address digit, 0 or 1, indicates whether the point is in the left or right side of the half to which it belongs. Write the address digits of these four intervals and interpret the address of each interval in terms of the number of 1/4s of each point of the interval. (Extend the argument of (a) about infinite strings of 0s and infinite strings of 1s, to select the addresses of 1/4 and 3/4.)
(c) Now divide the quarters into eighths. Write the address digits of these eight intervals.
(d) Note the order of the address digits. The first (left) digit tells us to which half of the inch the point belongs, the next digit to which quarter, and so on. Compare this with the usual order we write addresses on letters (physical mail, not email). Give a concrete example to illustrate your interpretation.
(e) To strengthen the tie to fractals, find two transformations f_0 and f_1 that generate $[0, 1]$ as an IFS. Show that the order of the address digits is identical to the order of the composition of these functions.

Prob 1.8.2 *Locating numbers* With the address assignment of Exercise 1.8.1 and our convention about infinite strings of 0s or of 1s, if x lies in the interval with address 0, we know $0 \leq x < 1/2$. If x lies in the interval with address 1, we know $1/2 \leq x \leq 1$.
(a) What can be said about a point x in the interval with address 0110?
(b) Find the intervals of address length-1, -2, -3, and -4 that contain the point $x = 2/7$.

Prob 1.8.3 *Place values* In decimal expansions, the first digit to the right of the decimal is the tenths, the second is the hundredths, and so on.
(a) In binary expansions, find the first place with value smaller than one-thousandth.
(b) Solve problem (a) using base 2 logarithms, using base 10 logarithms.
(c) Decimal expansions correspond to subdividing each interval into ten pieces, binary expansions into two. Into how many pieces must each interval be subdivided so the fourth place corresponds to about 0.0004?

Prob 1.8.4 *Comparing numbers* In decimal, changing the 5th digit moves a point by $0.00001 = 10^{-5}$. In binary, changing the 5th digit moves a point by $2^{-5} = 0.03125$
(a) Which binary digits must be changed to move a point by $3/16$? By $11/16$?
(b) Which binary digits must be changed to move a point by $1/3$? By $1/7$? By $1/15$? For which fractions $1/n$ can this pattern of changes be continued?
(c) Which binary digits must be changed to move a point by $3/7$? By $7/15$? By $15/31$? For which fractions m/n can this pattern of changes be continued?

Prob 1.8.5 *Non-unique addresses* Using binary subdivision show that every fraction of the form $m/2^n$ for $1 \leq m \leq 2^n - 1$ has two addresses, one that ends in an infinite string of 0s, the other that ends in an infinite string of 1s.
(a) On successive graphs indicate the intervals with these addresses: 01 and 10, 011 and 100, 0111 and 1000.
(b) Find the address of the right endpoint of the interval with one-digit address 0; find the address of the left endpoint of the interval with one-digit address 1.
(c) Noting both addresses of (b) correspond to the point $x = 1/2$, interpret this equality in terms of infinite series.
(d) Find the pair of addresses of $1/4$, $3/4$, $1/8$, and $5/16$.

Prob 1.8.6 *Addresses and functions* Here we investigate how functions alter the addresses of points.
(a) What is the effect of the functions $f_0(x) = x/2$ and $f_1(x) = x/2 + 1/2$ of the address of a point x in $[0, 1]$?
(b) Find the fixed points of f_0 and f_1, both by addresses and algebraically.
(c) Find the fixed points of $f_0(f_1)$ and of $f_1(f_0)$, both by addresses and algebraically.

Prob 1.8.7 *Fractions and addresses* Here we use the binary subdivision.
(a) Suppose the address of x terminates in an infinite string of 0s or in an infinite string of 1s. Show x is a rational number with denominator a power of 2. (This can be viewed as the converse of Exercise 1.8.5.)

(b) Suppose the address of x *eventually* consists of a block $a_1a_2 \ldots a_n$ of 0s and 1s that repeats forever. Show that x is a rational number with even denominator Hint: Recall that in Sample 1.8.6 we analyzed the case where the address of x consists of a block that repeats forever.
(c) Adopt the method of Sample 1.8.6 (b) to show that for every rational number p/q in $[0, 1]$, the address of p/q is either periodic or eventually periodic.
(d) Suppose the address of x in $[0, 1]$ is not periodic or eventually periodic. What can we say about x?

Prob 1.8.8 Adapt the method of Sample 1.8.6 (c) to provide a fixed point argument that to show that every address which is eventually periodic, but not periodic, is the address of a rational number with even denominator.

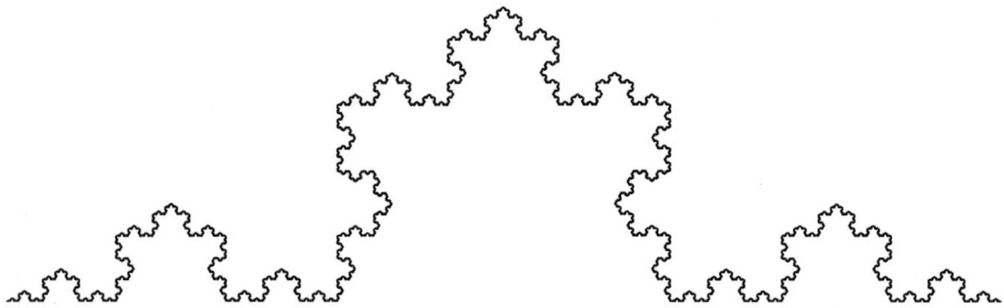

Figure 1.46: A Koch curve.

Prob 1.8.9 *Addresses of points on the Koch curve* A similar addressing scheme can be applied to the Koch curve shown in Fig. 1.46. The Koch curve is a fractal that consists of four pieces scaled by 1/3, with appropriate translations and rotations. By using addresses we'll show no point of the Koch curve has a well-defined tangent. With the Koch curve IFS table here, we can assign addresses 0, 1, 2, and 3 to parts of the Koch curve.

	r	s	θ	φ	e	f
T_0	1/3	1/3	0	0	0	0
T_1	1/3	1/3	60	60	1/3	0
T_2	1/3	1/3	−60	−60	1/2	$\sqrt{3}/6$
T_3	1/3	1/3	0	0	2/3	0

(a) On a copy of Fig. 1.46, indicate the regions with addresses 00, 01, 02, 03, 10, 11, 12, and 13.
(b) *Multiple Addresses* On a copy of Fig. 1.46, indicate the points with addresses $0(3^\infty)$ and $1(0^\infty)$. This can be interpreted as an extension of the ideas introduced in Exercise 1.8.5.

As an application of this addressing notion, we show that at no point of the Koch curve can a tangent be defined.
(c) Consider the point $1(3^\infty) = 2(0^\infty)$, the apex of the Koch curve. To show this point has no tangent, we produce a sequence of points converging to the apex, with the chords from these points to the apex alternating over a 30°

range. Locate the points 0^∞, $1(0^\infty)$, $12(0^\infty)$, $13(0^\infty)$, $132(0^\infty)$, and $133(0^\infty)$. Compute the angles between successive chords from these points to the apex.
(d) Next, we consider points with addresses ending in an infinite sequence of 1s or an infinite sequence of 2s. For example, take the point 1^∞ and consider the sequence of chords to this point from $1(0^\infty)$, $11(0^\infty)$, $111(0^\infty)$, and so on. What is the general pattern?
(e) Addresses that terminate in a constant are treated in parts (c) and (d). Given any address $a_1a_2a_3\ldots$ that does not terminate in a constant string, what can be said about the angles of the chords to $a_1a_2a_3\ldots$ from $a_1a_2a_3\ldots a_k(0^\infty)$ for $k = 1, 2, 3, \ldots$?

Prob 1.8.10 Describe two distinct methods to specify the location of your house. Don't use GPS coordinates.

1.9 IFS with memory

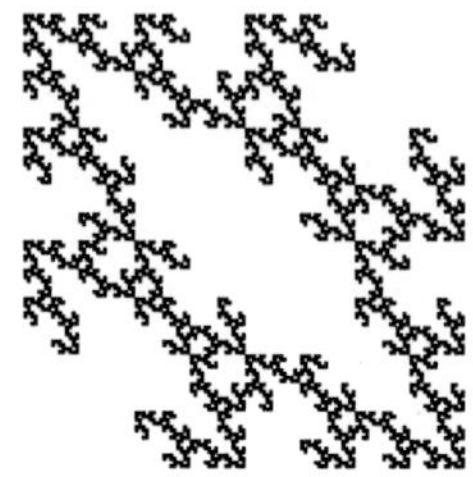

Here we apply the notion of addresses to define a new type of IFS: keep the transformations fixed and vary the allowed compositions of these transformations.

In Lab 1.10 we explore IFS with 2-step memory. In Lab 1.11 we use some examples from the current lab to assess how many consecutive data points of a time series are needed to predict the next entry.

1.9.1 Purpose

We'll learn how images can be generated with simple IFS rules with 1-step memory, that is, IFS rules for which we prescribe which pairs of transformations can be composed. We'll learn how to recognize if particular geometrical patterns can be attributed to patterns of allowed pairs of transformations. And we'll learn to determine when the attractor of an IFS with 1-step memory also is the attractor of an IFS with 0-step memory, that is, an IFS generated by the deterministic algorithm of Lab 1.1.

1.9.2 Materials

We'll use IFS with memory software (1.9 of the Mathematica or Python codes), and deterministic IFS software (1.1 of the Mathematica or Python codes).

1.9.3 Background

Here we consider only one IFS,

$$\begin{array}{ll} T_1(x,y) = (x/2, y/2) & T_2(x,y) = (x/2, y/2) + (1/2, 0) \\ T_3(x,y) = (x/2, y/2) + (0, 1/2) & T_4(x,y) = (x/2, y/2) + (1/2, 1/2) \end{array} \quad (1.7)$$

With these transformations, the deterministic IFS algorithm generates the filled-in unit square S.

Because this background section is fairly long, we'll begin with a map of what we'll do.

1. We introduce the notion of IFS with 1-step memory through the example of *allowed pairs*, illustrating this with a graphical representation of allowed pairs.

2. Next we give the software representation of allowed pairs.

3. As a simple example, we find how to determine which allowed pairs generate lines.

4. Then we introduce *romes*, central to understanding which 1-step memory IFS attractors also are the attractors of IFS without memory, but with more transformations.

1.9.3.1 Allowed pairs

In the standard IFS formalism, all combinations of transformations are allowed. By allowing only certain combinations, we open the possibility of generating many, many more images. Every combination that is not *allowed* is *forbidden*, so it suffices to specify either the allowed or the forbidden combinations.

Note that if $T_{i_1} \dots T_{i_N}$ is forbidden, then so is $T_{i_a} \dots T_{i_b} T_{i_1} \dots T_{i_N} T_{i_c} \dots T_{i_d}$ for all sequences $i_a \dots i_b$ and $i_c \dots i_d$, including empty sequences.

Consequently, the simplest IFS with memory is specified by forbidden pairs, equivalently, by allowed pairs.

For our transformations T_1, T_2, T_3, and T_4, this information can be represented visually by a *transition graph.* The graph has

- four vertices, one for each T_i, and
- an edge $i \to j$ from vertex i to vertex j if the composition $T_j(T_i)$ is allowed.

For example, the graph in Fig. 1.47 specifies that T_4 cannot follow T_1, because the graph has no arrow from vertex 1 to vertex 4.

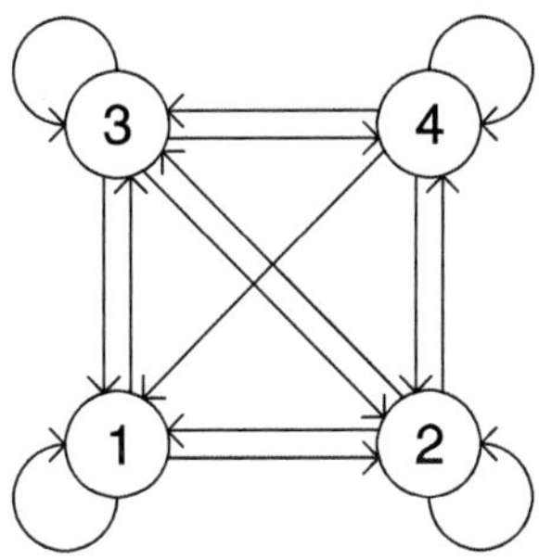

Figure 1.47: Transition graph with $1 \to 4$ forbidden.

For the graph of Fig. 1.47, the only forbidden transition is $1 \to 4$, indicated by the 0 in column 1 ("from") and in row 4 ("to"). This means that address 41 is empty. This is the largest empty square in Fig. 1.48. The other empty squares all are consequences of 41 being empty. The four next-largest empty squares are have addresses 141, 241, 341, and 441. To see why these must be empty, recall S is the filled-in unit square and denote $T_j(T_i(S))$ by S_{ji}, $T_k(T_j(T_i(S)))$ by S_{kji}, and so on. We know S_{41} is empty, and because $S_{141} = T_1(S_{41})$, we see that S_{141} is empty. The other empty squares are images, perhaps under the composition of many T_i, of the single empty square S_{41}.

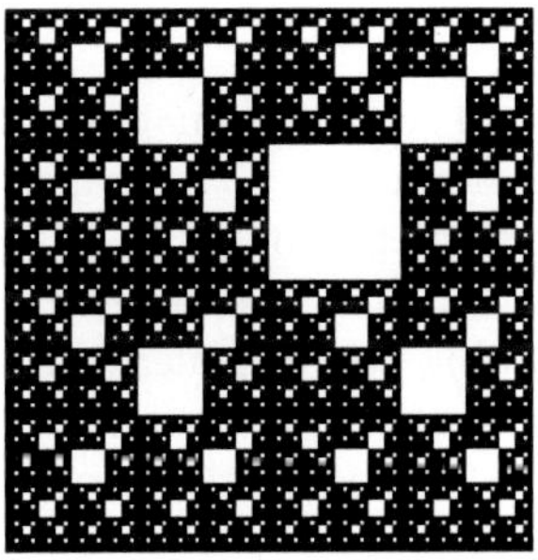

Figure 1.48: IFS of the transition graph of Fig. 1.47.

Suppose you add up the areas of the empty squares. This is a bit tricky: we see 1 empty length-2 address square, 4 empty length-3 address squares, but only 15 empty length-4 address squares. Do you see the general pattern? When you find the pattern and account for all the empty squares, their areas sum to 1. Carried to its mathematical limit, the shape of Fig. 1.48 has zero area. In Chapter 2 we'll have much more to say about this.

1.9.3.2 Lines

For each pair of transformations from among T_1, T_2, T_3, and T_4, an IFS that allows all combinations of that pair generates a line between the fixed points

$p_1 = (0,0)$, $p_2 = (1,0)$, $p_3 = (0,1)$, and $p_4 = (1,1)$ of the transformations. With this we can generate six lines. The left images of Fig. 1.49 shows two such lines, transition graph above, IFS plot below. This example shows that some of these lines can be generated together. Some other combinations, for example, the lines that join p_1 to p_2 and p_1 to p_3 cannot be generated together without producing many other shorter lines.

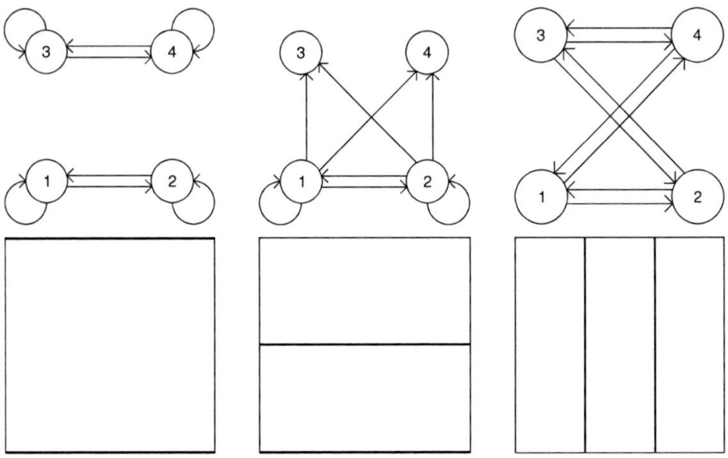

Figure 1.49: Some lines generated by IFS.

The center images of Fig. 1.49 show a different construction: the bottom four arrows of the transition graph generate the line between p_1 and p_2; the arrows $1 \to 3$ and $2 \to 3$ copy this line to the bottom of S_3, the arrows $1 \to 4$ and $2 \to 4$ copy this line to the bottom of S_4. We know that the line between p_1 and p_2 can be generated in isolation, but the line between $(0, 1/2)$ and $(1, 1/2)$ cannot. Do you see how to generate this line from the line between p_3 and p_4?

The right images of Fig. 1.49 are yet another possibility: the line between $(1/3, 0)$ and $(1/3, 1)$ and the line between $(2/3, 0)$ and $(2/3, 1)$ must be generated together. Neither can be produced in isolation.

1.9.3.3 Romes

Consider a transition graph vertex V_i with arrows from each vertex, including V_i, to V_i. Then we say that vertex V_i is a *rome*, in the sense that "every road leads to Rome." This really is the name, from the field called symbolic dynamics. The presence of every arrow $V_j \to V_i$ means every transformation T_j can be followed by T_i and so square S_i contains a scaled copy of the entire attractor. In Fig. 1.50 we see three attractors with romes outlined in boxes.

In the left and middle attractors, copies of the whole attractor are in S_1 and S_4. The corresponding transition graphs show that V_1 and V_4 are romes.

The right attractor is two tall, thin Sierpinski gaskets. Once the gasket is familiar, some people view it as the primary shape and the temptation is to decompose this attractor into a collection of gaskets. But the shapes of interest are copies of the entire attractor, so we decompose this attractor into pairs of gaskets. We see S_1 and S_2 contain copies of the entire attractor. In the transition graph, V_1 and V_2 are romes.

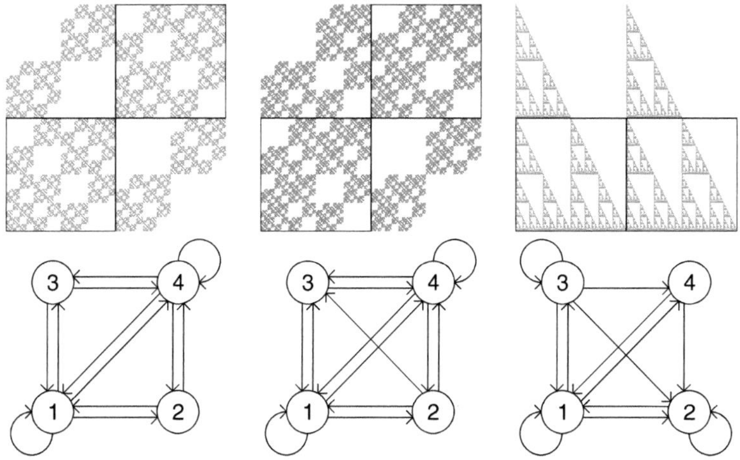

Figure 1.50: Three attractors with romes.

The left attractor is simplest: S_1 and S_4 each contain a copy of the whole attractor scaled by a factor of $1/2$, while S_2 and S_3 each contain two copies of the whole attractor scaled by $1/4$. The arrows $1 \to 2$, $4 \to 2$, $1 \to 3$, and $4 \to 3$ generate these copies. With these we've run out of paths, except those that lead back to the romes, so the attractor consists of two copies scaled by $1/2$ and four copies scaled by $1/4$.

The middle attractor is similar to the left; the main difference is that S_3 of the middle attractor contains two additional copies, each scaled by $1/8$. The arrow $2 \to 3$ accounts for these: whatever is in S_2, two copies of the attractor scaled by $1/4$, is scaled by an additional factor of $1/2$ and translated to S_{32}. This accounts for all the paths in the transition graph, except for those that lead back to the romes, so this attractor consists of two copies scaled by $1/2$, two copies scaled by $1/4$, and two copies scaled by $1/8$.

So far, we've seen that romes correspond to transition graph vertices where the number of incoming arrows equals the number of vertices. Arrows from the romes produce smaller copies of the attractor. The number of these copies is the number of transition graph paths until the path returns to a rome. The scaling factor of each copy is $1/2^{\text{path length }+1}$. The left transition graph has four paths—$1 \to 2$, $1 \to 3$, $4 \to 2$, and $4 \to 3$—before they return to a rome. The middle transition graph has six paths—$1 \to 2$, $1 \to 3$, $4 \to 2$, $4 \to 3$, $1 \to 2 \to 3$ and $4 \to 2 \to 3$—before they return to a rome. Note the path length does indeed give the scaling of these pieces, by the path length formula we mentioned.

The right attractor is more complicated, even though it looks simple. The source of the problem is the loop $3 \to 3$ of the non-rome state 3. This gives rise to an infinite collection of paths through non-romes before the return to a rome:

$$1 \to 3, \quad 1 \to 3 \to 3, \quad 1 \to 3 \to 3 \to 3, \quad \dots$$
$$1 \to 3 \to 4, \quad 1 \to 3 \to 3 \to 4, \quad 1 \to 3 \to 3 \to 3 \to 4, \quad \dots$$

and also $1 \to 4$. The transformations $1 \to 3$ and $1 \to 4$ place copies of the whole fractal scaled by 1/4 in S_{31} and in S_{41}. Then the loop $3 \to 3$ places ever-smaller copies of the contents of S_{31} in $S_{331}, S_{3331}, \dots$. Any number of trips around the $3 \to 3$ loop, followed by $3 \to 4$ places ever smaller copies of the contents of S_{31} in $S_{431}, S_{4331}, S_{43331}, \dots$. This attractor consists of two copies scaled by 1/2, two copies scaled by 1/4, two copies scaled by 1/8, and so on.

The existence of romes shows that part of the attractor is a scaled copy of the whole, paths through non-romes suggest that the rest of the attractor is made of still smaller copies of the whole, and compositions of the T_i determined by the paths define IFS rules that generate those pieces.

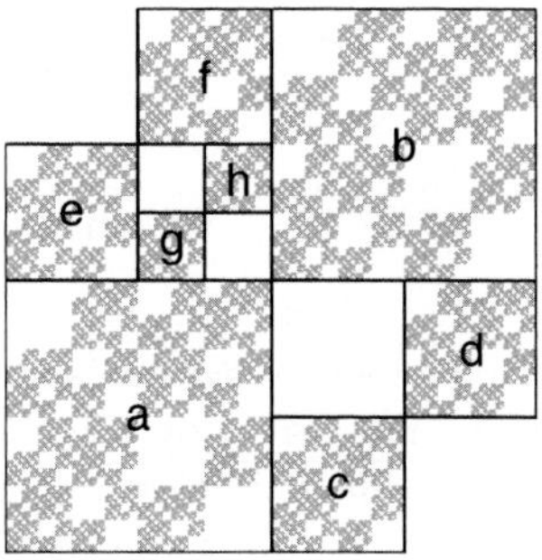

Figure 1.51: A finite IFS decomposition.

This may sound peculiar. After all, we have IFS rules to generate these images, if we include a matrix of allowed combinations. Here we ask a different question: can these images be attractors of IFS without memory? Take the middle attractor of Fig. 1.50, decomposed according to Fig. 1.51, which we'll see is a consequence of the paths in the transition graph. First, though, given this decomposition we can find the IFS rules by inspection.

Here is the IFS table that generates the attractor of Fig. 1.51. We see two pieces, a and b, scaled by 1/2, four pieces, c, d, e, and f, scaled by 1/4, and two pieces, g and h, scaled by 1/8. No piece is reflected or rotated, and the translations can be found by measuring the distance from the the lower left corner of the whole attractor to the lower left corner of each piece. Also, we can find the translations by the composition of the sequence of T_i determined by the corresponding path through the transition graph. For example, note we get to piece h by $b \to d \to h$. That is, the

	r	s	θ	φ	e	f
a	0.5	0.5	0	0	0	0
b	0.5	0.5	0	0	0.5	0.5
c	0.25	0.25	0	0	0.5	0
d	0.25	0.25	0	0	0.75	0.25
e	0.25	0.25	0	0	0	0.5
f	0.25	0.25	0	0	0.25	0.75
g	0.125	0.125	0	0	0.25	0.5
h	0.125	0.125	0	0	0.375	0.625

Table 1.6: IFS rules for Fig. 1.51

composition $T_3(T_2(T_4))$. Let's work through this composition.

$$T_3(T_2(T_4(x,y))) = T_3(T_2(x/2+1/2, y/2+1/2))$$
$$= T_3(x/4+1/4+1/2, y/4+1/4) = T_3(x/4+3/4, y/4+1/4)$$
$$= (x/8+3/8, y/8+1/8+1/2) = (x/8+3/8, y/8+5/8)$$

and we have obtained the translation values for part h.

What happens when we try to apply the same notion to the right attractor of Fig. 1.50? In this example, the loop $3 \to 3$ through non-rome states generated an infinite decomposition of the attractor. Our efforts to find an IFS without memory to generate this attractor will fail, unless you don't mind writing down infinitely many rules.

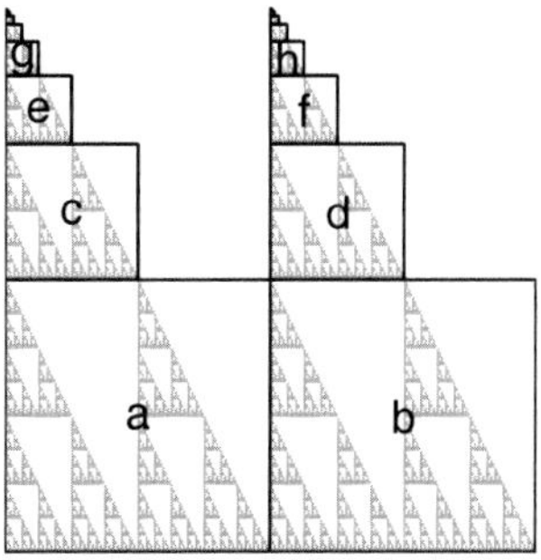

Figure 1.52: An infinite IFS decomposition.

The difference between the transition graphs of the attractors of Figs. 1.51 and 1.52 is that in the first all the lengths of all the paths between non-rome states have an upper bound (two, in this case), while in the second we find arbitrarily long paths between non-rome states.

Here's the result. See [38] and Sect. 2.5 of [17] for the details. Suppose A is the attractor generated by an IFS determined by forbidden pairs. We'll call this a 1-IFS, that is, an IFS with 1-step of memory. The question we ask now is when can A also be generated by an IFS without memory. Usually this will require more, and different, transformations than those for the IFS with memory. In order to do this, the transition graph must

(1) have at least one rome, and

(2) for each non-rome, a path from some rome to that non-rome.

If in addition,

(3) there are only finitely many paths among non-rome states; said another way, there are no loops among only non-rome states,

then A is generated by an IFS without memory and with only finitely many transformations.

This result was Jennifer Lanski's project in my (MF) undergraduate fractal geometry class. Surely she was a math major. Nope, art. But math was in her upbringing: her father is a mathematician. Unknown to us at the time, this result answers a question posed by Layman and Womack [39]. We just thought the question was interesting.

1.9.4 Procedure

Use the idea of addresses (introduced in Lab 1.7) to identify the forbidden pairs in an IFS with memory attractor. To apply the software use the transition matrix $[m_{ij}]$, where m_{ij} is 0 or 1 if $j \to i$ is forbidden or allowed.

Identify romes, draw the transition graph, and determine if the attractor of the IFS with memory can be generated without memory. If it can, and if the required number of transformations is finite, write the IFS table.

1.9.5 Sample 1

(a) Find the transition graph of a 1-step memory IFS with attractor that consists of two lines, one that connects $(0,0)$ to $(1/2,1)$, and one that connects $(1/2,0)$ to $(1,1)$. Draw the transition graph and test that this graph does produce exactly those two lines.
(b) Repeat (a) for a line that connects $(0,0)$ to $(1,1/2)$. and one that connects $(0,1/2)$ to $(1,1)$.
(c) Run the software with the transition graph that consists of all the arrows from the graphs of (a) and (b). Explain what you see.

For (a), first draw the desired lines on a grid of length-2 address squares. This is the left image of Fig. 1.53. The occupied squares have addresses 11, 13, 21, 23, 32, 34, 42, and 44. Consequently, the transition graph has arrows $1 \to 1$, $3 \to 1$, $1 \to 2$, $3 \to 2$, $2 \to 3$, $4 \to 3$, $2 \to 4$, and $4 \to 4$. The transition graph is the right image of Fig. 1.53; the attractor of the 1-step memory IFS with this transition graph is identical to the left image of this figure. So far, so good.

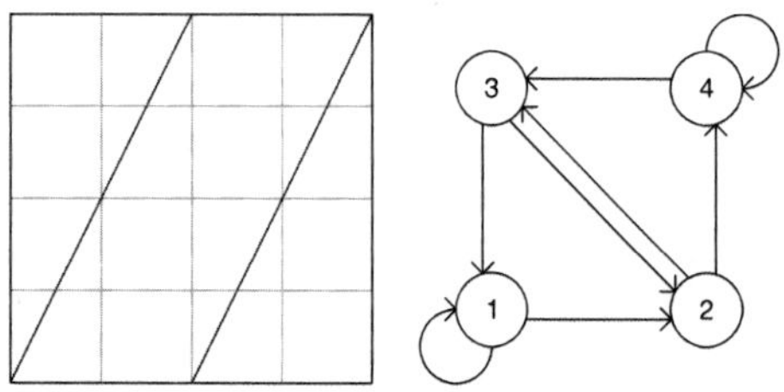

Figure 1.53: Attractor and graph of (a).

For (b) we follow the same approach. The occupied length-2 addresses are 11, 12, 23, 24, 31, 32, 43, and 44. Then the transition graph has arrows $1 \to 1$, $2 \to 1$, $3 \to 2$, $4 \to 2$, $1 \to 3$, $2 \to 3$, $3 \to 4$, and $4 \to 4$. As in (a), with this transition graph the 1-step memory IFS generates these two lines.

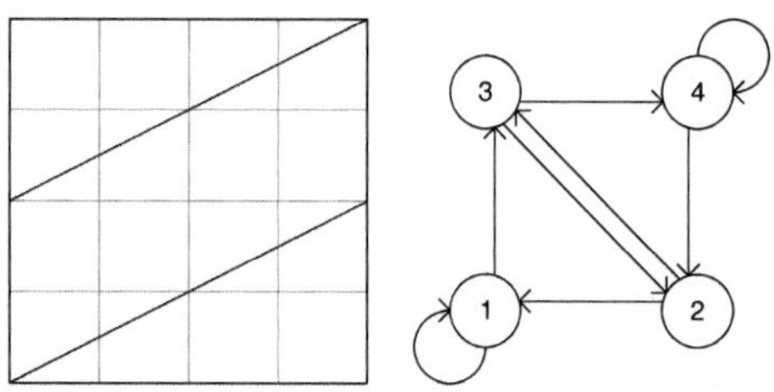

Figure 1.54: Attractor and graph of (b).

In Sect. 1.9.3.2 we saw an example of a pair of lines that could not be generated individually. Do you think the pair of lines of (a) and the pair of lines of (b) behave this way, too? How could you test this?

Building the transition graph that has all the allowed transitions of (a) and of (b) is straightforward. We could say that the graph of (c) is the union of the graphs of (a) and (b). Do you expect that the attractor of (c) is just the union of the attractors of (a) and (b)? Certainly, the attractor of (c) must contain all the lines of (a) and (b), but in Fig. 1.55 we see the attractor is much more than four lines.

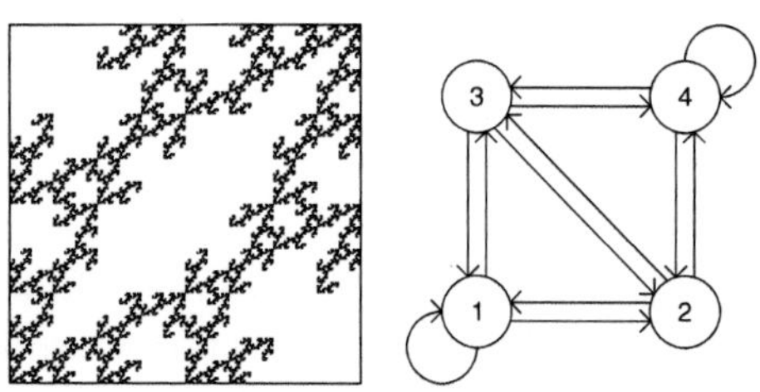

Figure 1.55: Attractor and graph of (c).

Without suggestion by us, some of our summer workshop members viewed these IFS images as Rorschach tests. They called the attractor of Fig. 1.55 "the crab."

1.9.6 Sample 2

For each of these attractors, find the corresponding transition graph. Before you continue, compare the attractor pictured with that of the 1-step memory IFS with that transition graph. For those where the attractors match, determine if they can be generated by an IFS without memory, and if only a finite number of of transformations are needed, find the IFS table.

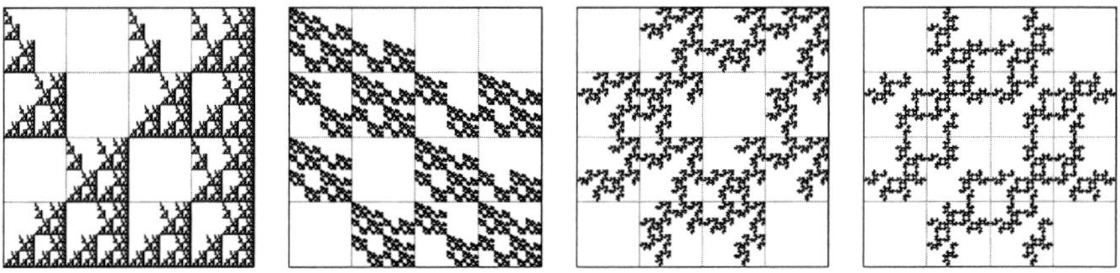

Figure 1.56: Attractors of Sample 2, with length-2 address squares.

First attractor. The occupied addresses are 11, 12, 14, 21, 22 24, 31, 33, 41, 42, 43, and 44. (In cases where so many squares are occupied, it may be simpler to list the empty squares. But we'll stick with occupied squares. If we switch between occupied and empty, then each list must be accompanied by a label of which it is. Simpler is always to use the same convention.) The allowed transitions are $1 \to 1$, $2 \to 1$, $4 \to 1$, $1 \to 2$, $2 \to 2$, $4 \to 2$, $1 \to 3$, $3 \to 3$, $1 \to 4$, $2 \to 4$, $3 \to 4$, and $4 \to 4$. The transition graph is the first of Fig. 1.57. The 1-step memory IFS program with this transition graph does produce this attractor, so the techniques of this lab can be applied.

We see that 4 is a rome because four arrows go to vertex 4. And we see paths $4 \to 1$, $4 \to 2$, and $4 \to 1 \to 3$ from the rome to each non-rome. Consequently, this attractor can be generated by an IFS without memory. But because we also see loops between non-romes—$1 \to 1$, $2 \to 2$, $3 \to 3$, and $1 \to 2 \to 1$—the IFS without memory would require infinitely many transformations.

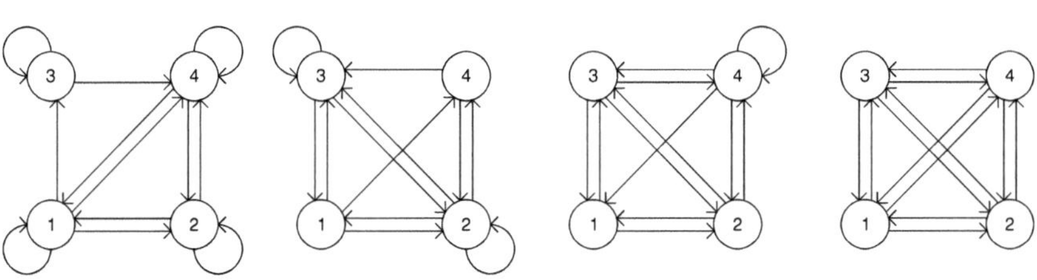

Figure 1.57: Transition graphs from the list of allowed transitions.

Second attractor. The occupied addresses are 12, 13, 21, 22, 23, 24, 31, 32, 33, 34, 41, and 42; the allowed transitions are $2 \to 1$, $3 \to 1$, $1 \to 2$, $2 \to 2$, $3 \to 2$,

$4 \to 2$, $1 \to 3$, $2 \to 3$, $3 \to 3$, $4 \to 3$, $1 \to 4$, and $2 \to 4$. The transition graph is the second of Fig. 1.57. The 1-step memory IFS program with this transition graph does produce the second attractor of Fig. 1.56, so the techniques of this lab can be applied.

From the transition graph, or from the attractor, we see that states 2 and 3 are romes. Also, there are paths $2 \to 1$, $3 \to 1$, and $2 \to 4$ to each non-rome from some rome. (Note there is a path $3 \to 1 \to 4$, so for this IFS we have paths from each rome to each non-rome.) Consequently, this attractor can be produced by an IFS without memory. Because there are no loops among only non-romes (1 and 4 for this attractor), only finitely many transformations are needed. They are T_2, T_3, $T_4(T_2)$, $T_1(T_2)$, $T_1(T_3)$, $T_4(T_1(T_2))$, and $T_4(T_1(T_3))$. Unpacking these, or by inspecting the attractor, we see the IFS table is

	r	s	θ	φ	e	f
T_2	0.5	0.5	0	0	0.5	0
T_3	0.5	0.5	0	0	0	0.5
$T_4(T_2)$	0.25	0.25	0	0	0.75	0.5
$T_1(T_2)$	0.25	0.25	0	0	0.25	0
$T_1(T_3)$	0.25	0.25	0	0	0	0.25
$T_4(T_1(T_2))$	0.125	0.125	0	0	0.625	0.5
$T_4(T_1(T_3))$	0.125	0.125	0	0	0.5	0.625

Third attractor. The occupied addresses are 12, 13, 14, 21, 23, 24, 31, 32, 34, 42, 43, and 44; the allowed transitions are $2 \to 1$, $3 \to 1$, $4 \to 1$, $1 \to 2$, $3 \to 2$, $4 \to 2$, $1 \to 3$, $2 \to 3$, $4 \to 3$, $2 \to 4$, $3 \to 4$ and $4 \to 4$. The transition graph is the third of Fig. 1.57. The 1-step memory IFS program with this transition graph does produce the second attractor of Fig. 1.56, so the techniques of this lab can be applied.

No vertex of the transition graph has four arrows going to that vertex, so this graph has no romes. We can see this from the attractor, too. If each of the squares S_1, S_2, S_3, and S_4 contains part of the attractor, then the presence of all four arrows $1 \to i$, $2 \to i$, $3 \to i$, and $4 \to i$ in the transition graph guarantees that each address $i1$, $i2$, $i3$, and $i4$ contains part of the attractor. (The modification if some of the S_i are empty is straightforward.) Because this attractor has no rome, it can't be generated by an IFS without memory.

Back to the Rorschach tests for a moment. Some people in the workshops saw this image as a "skinny Mandelbrot set." We took some comfort in the fact that one fractal reminded people of another fractal, instead of a crab.

Fourth attractor. The occupied addresses are 12, 13, 14, 21, 23, 24, 31, 32, 34, 41, 42, and 43; the allowed transitions are $2 \to 1$, $3 \to 1$, $4 \to 1$, $1 \to 2$, $3 \to 2$, $4 \to 2$, $1 \to 3$, $2 \to 3$, $4 \to 3$, $1 \to 4$, $2 \to 4$ and $3 \to 4$. The transition graph is the fourth of Fig. 1.57. The 1-step memory IFS program with this transition graph produces the image on the right. While this is close to the fourth attractor of Fig. 1.56, just close is not good enough. The fourth transition graph of Fig. 1.57 does not produce the fourth attractor of Fig. 1.56. If that attractor were produced by an IFS with

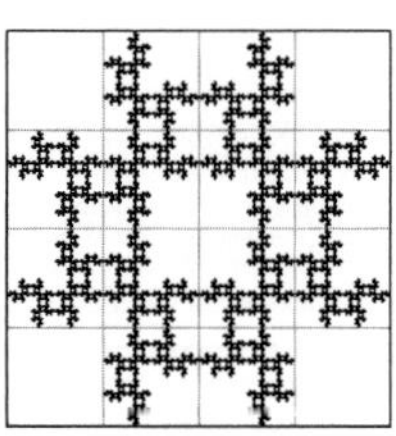

1-step memory, then this transition graph would give that attractor. Consequently, we deduce that more than 1-step of memory is needed to produce the fourth attractor of Fig. 1.56. In Lab 1.10 we'll investigate IFS with longer memory restrictions.

1.9.7 Conclusion

By including memory—specifying which combinations of IFS transformations are allowed—a much wider variety of images can be produced than with standard (memoryless) IFS. Nevertheless, some images generated by IFS with memory can be generated without memory, with the addition of more transformations. The general question of the number of transformations as a trade-off for memory offers many avenues to explore.

1.9.8 Exercises

Prob 1.9.1 Find the transition graph that produces the attractor that consists of the line segment between $(0, 0)$ and $(1, 0)$, and the line segment between $(0, 1/2)$ and $(1/2, 1/2)$.

Prob 1.9.2 Call transition graphs G_1 and G_2 *additive* if A_3, the attractor of the transition graph G_3 whose arrows are those of G_1 and G_2, is the union of the attractors A_1 and A_2 of G_1 and G_2. For example, in the left images of Fig. 1.49 we see that the graphs G_1, all paths between 1 and 2, and G_2, all paths between 3 and 4, are additive.
(a) Suppose G_1 consists of all paths between 1 and 2, and G_2 of all paths between 1 and 3. Are these graphs additive?
(b) Suppose G_1 consists of all paths between 1 and 4, and G_2 of all paths between 2 and 3. Are these graphs additive?
(c) From your observations in (a) and (b) find a condition of G_1 and G_2 that guarantees they are additive.

In Exercises 1.9.3, 1.9.4, and 1.9.5, find the transition graphs. Compare the attractor pictured to that of the 1-step memory IFS with this transition graph. For those where the attractors match, determine if they can be generated by an IFS without memory, and if only a finite number of of transformations are needed, find the IFS table.

Prob 1.9.3

(a) 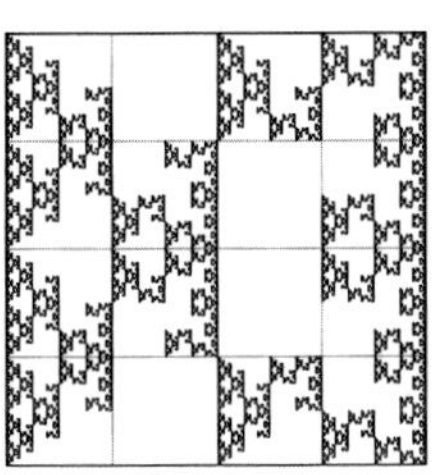 (b) 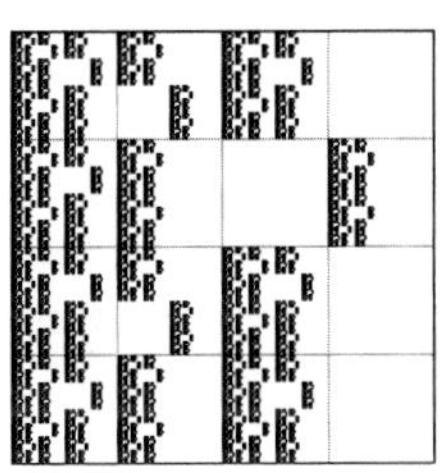 (c) 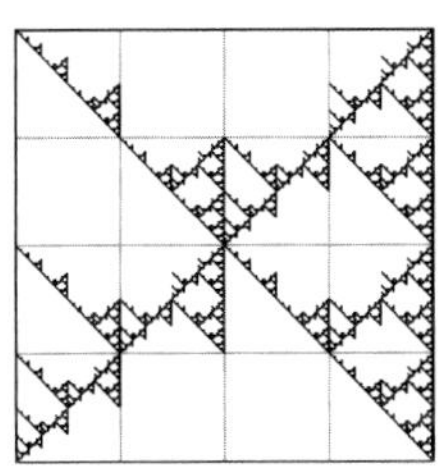

Prob 1.9.4

(a) 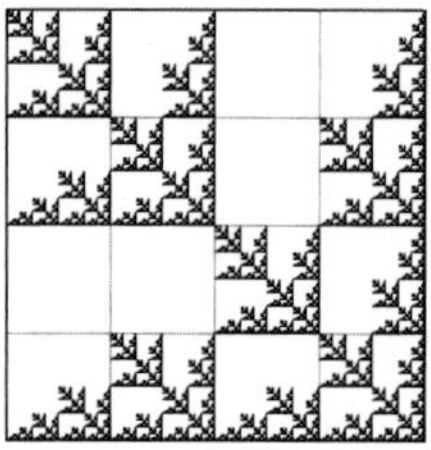 (b) 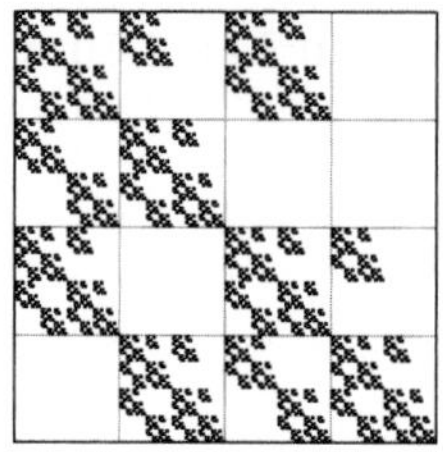 (c) 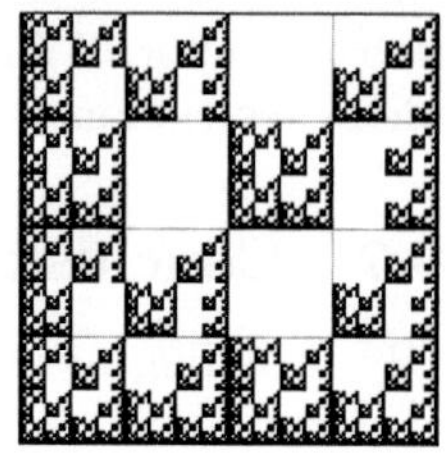

Prob 1.9.5

(a) 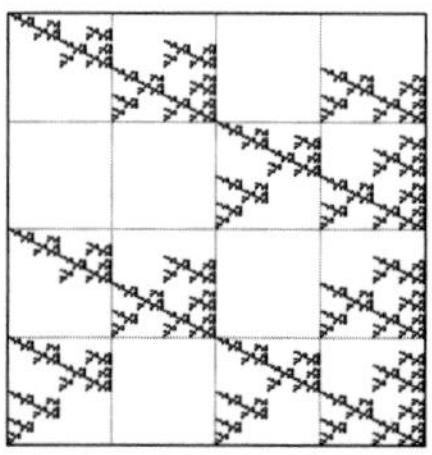 (b) 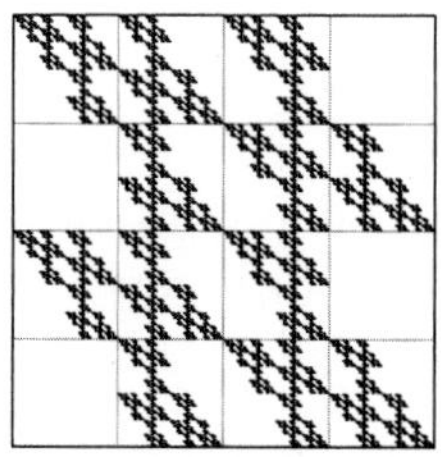 (c) 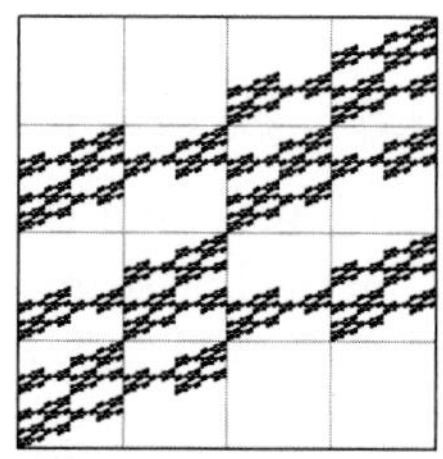

Prob 1.9.6 For each attractor pictured here, find a combination of lines that generate the attractor when the matrix consists of exactly those transitions that produce those lines.

(a) 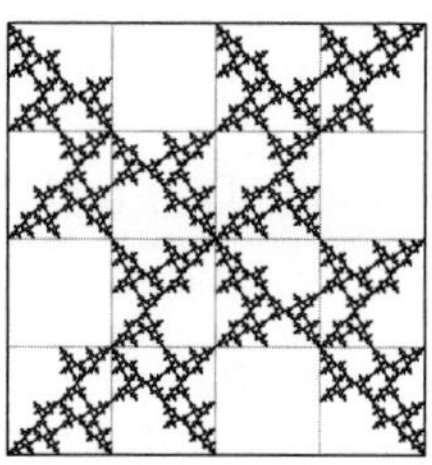 (b) 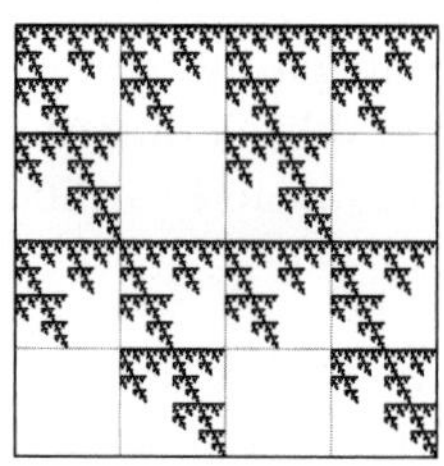 (c) 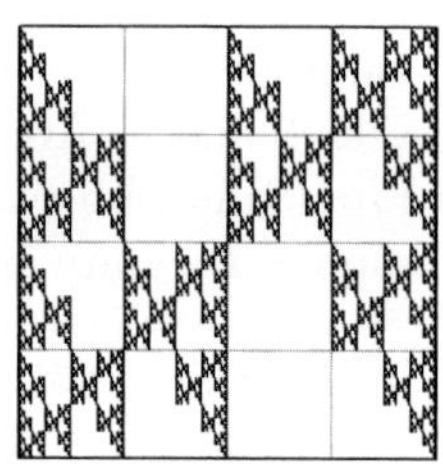

Prob 1.9.7 Pictured here is the attractor of an IFS with 1-step memory. Both address 1 and address 4 are copies of the whole attractor, scaled by 1/2, so in the transition graph 1 and 4 are romes. But also observe that no compositions of the four transformations T_i can turn the whole attractor into the diagonal line segments in addresses 2 and 3. How is this consistent with the statement in Sect. 1.9.3.3?

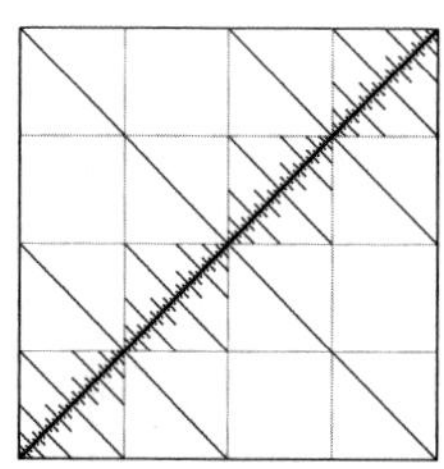

1.10 IFS with more memory

The addition of 2-step memory—that is, a list of allowed compositions of three transformations—we appear to be able to generate a much larger collection of fractals than we can with 1-step memory. But can we? Let's see.

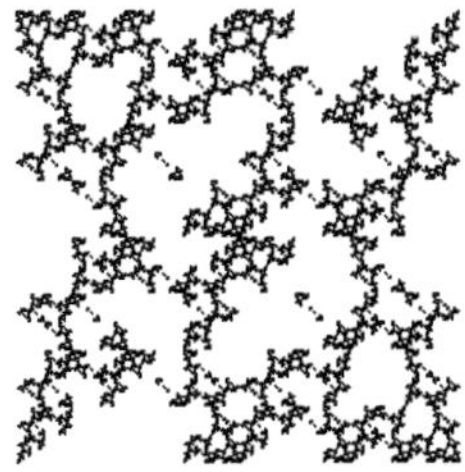

1.10.1 Purpose

We'll learn how images can be generated with 2-step of memory, simple IFS rules on which we prescribe which triples of transformations can be applied. Also, we'll learn how to recognize which 2-step memory attractors can be generated by 1-step memory IFS, and how to determine whether there are n-step memory attractors that are not also the attractors of 1-step memory IFS.

1.10.2 Materials

We'll use IFS with 2-step memory software (1.10 of the Mathematica or Python codes), and IFS with 1-step memory software (1.9 of the Mathematica or Python codes).

1.10.3 Background

For IFS with 1-step memory we represented the allowed compositions by a transition graph. For IFS with 2-step memory, a transition graph would be complicated, so we'll use another representation, the transition matrix, more easily introduced for IFS with 1-step memory. An example is all we need.

We can represent the allowed and forbidden compositions by a transition matrix M. The matrix element m_{ij}, the element in the ith row and the jth column, is 0 if the composition $T_i(T_j)$ is forbidden, and 1 if $T_i(T_j)$ is allowed. For the 1-step memory IFS of Fig. 1.58, the transition matrix is

$$M = \begin{bmatrix} 1 & 1 & 1 & 1 \\ 1 & 1 & 1 & 1 \\ 1 & 1 & 1 & 1 \\ 0 & 1 & 1 & 1 \end{bmatrix}$$

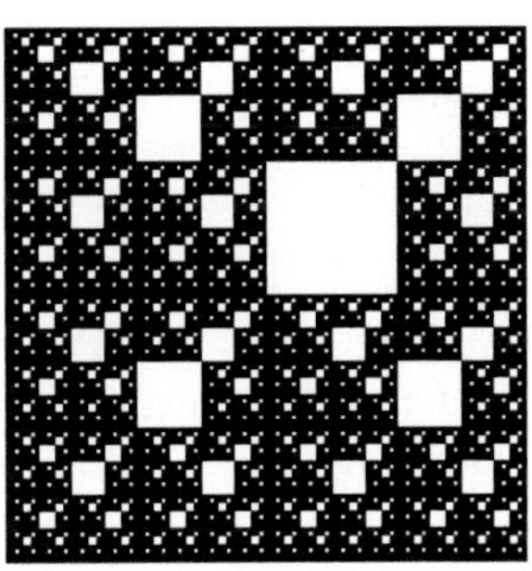

Figure 1.58: IFS of the transition matrix M.

The only forbidden transition is $1 \rightarrow 4$, indicated by the 0 in column 1 ("from") and in row 4 ("to"). This means that address 41 is empty.

For 2-step memory we'll use a an array of four 4×4 matrices. The correspondence between the matrix representation here and the representation in 1.10 of the Mathematica or Python codes offers some opportunity for mistakes, so we'll be especially careful here and in that appendix.

Here we see this array of matrices. Each entry is the address that corresponds to the triple allowed if that matrix entry is 1 and forbidden if that entry is 0.

$$\begin{bmatrix}111 & 112 & 113 & 114\\121 & 122 & 123 & 124\\131 & 132 & 133 & 134\\141 & 142 & 143 & 144\end{bmatrix}\begin{bmatrix}211 & 212 & 213 & 214\\221 & 222 & 223 & 224\\231 & 232 & 233 & 234\\241 & 242 & 243 & 244\end{bmatrix}\begin{bmatrix}311 & 312 & 313 & 314\\321 & 322 & 323 & 324\\331 & 332 & 333 & 334\\341 & 342 & 343 & 344\end{bmatrix}$$
$$\begin{bmatrix}411 & 412 & 413 & 414\\421 & 422 & 423 & 424\\431 & 432 & 433 & 434\\441 & 442 & 443 & 444\end{bmatrix} \tag{1.8}$$

For example, suppose the only 0 in the matrix is entry 312. This means that the only forbidden composition of three transformations is $T_3(T_1(T_2))$. Sometimes we'll represent compositions by arrows, so this composition is $2 \to 1 \to 3$. That is, T_3 follows T_1 follows T_2 is forbidden. The images of Fig. 1.59 show the first, second, and third iterates of the IFS with 2-step memory where the only forbidden composition is $2 \to 1 \to 3$.

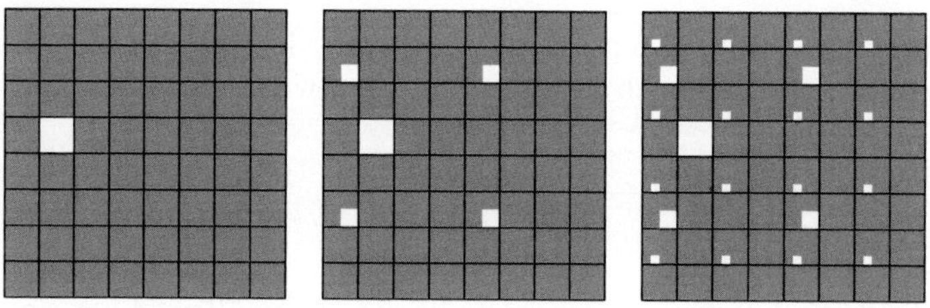

Figure 1.59: An illustration 2-step memory. The triple $2 \to 1 \to 3$ is forbidden, so the address 312 is empty.

For the next example, pictured here is the attractor of the 1-step memory IFS determined by this transition matrix

$$\begin{bmatrix}1 & 1 & 1 & 0\\1 & 1 & 0 & 1\\1 & 0 & 1 & 1\\0 & 1 & 1 & 1\end{bmatrix} \tag{1.9}$$

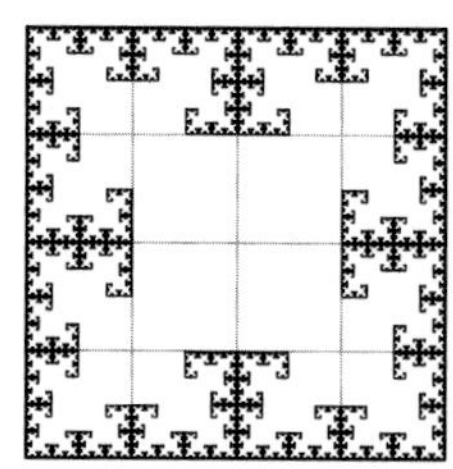

Figure 1.60: The IFS of (1.9).

The forbidden pairs are 14, 23, 32, and 41; the forbidden transitions are $4 \to 1$, $3 \to 2$, $2 \to 3$, and $1 \to 4$. We can generate this image with the software in 1.9 of the Mathematica or Python codes.

There are three obvious ways we can try to insert this matrix in the array (1.8), illustrated in Fig. 1.61.

The first attractor is produced by this array of matrices

$$\begin{bmatrix}1 & 1 & 1 & 1\\1 & 1 & 1 & 1\\1 & 1 & 1 & 1\\0 & 0 & 0 & 0\end{bmatrix}\begin{bmatrix}1 & 1 & 1 & 1\\1 & 1 & 1 & 1\\0 & 0 & 0 & 0\\1 & 1 & 1 & 1\end{bmatrix}\begin{bmatrix}1 & 1 & 1 & 1\\0 & 0 & 0 & 0\\1 & 1 & 1 & 1\\1 & 1 & 1 & 1\end{bmatrix}\begin{bmatrix}0 & 0 & 0 & 0\\1 & 1 & 1 & 1\\1 & 1 & 1 & 1\\1 & 1 & 1 & 1\end{bmatrix}$$

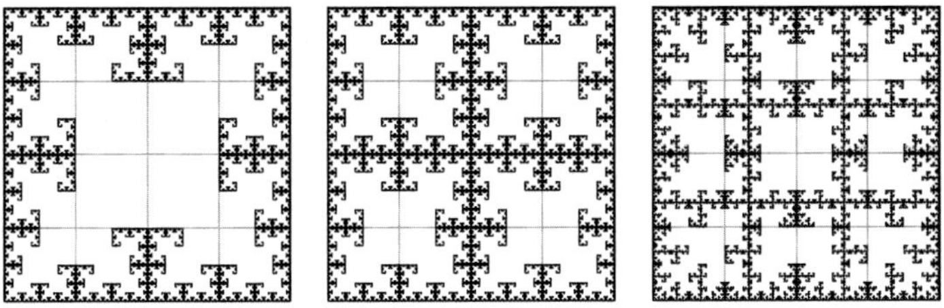

Figure 1.61: Illustrations of 2-step memory

Comparison with (1.8) shows that the forbidden triples are 141, 142, 143, 144, 231, 232, 233, 234, 321, 322, 323, 324, 411, 412, 413, and 414. That is, for each $k = 1, 2, 3, 4$ the compositions $k \to 4 \to 1$, $k \to 3 \to 2$, $k \to 2 \to 3$, and $k \to 1 \to 4$. These are equivalent to the forbidden transitions of Fig. 1.60. To see this, note that because $1 \to 4 \to 1$, $2 \to 4 \to 1$, $3 \to 4 \to 1$, and $4 \to 4 \to 1$, the addresses 141, 142, 143, and 144 are empty. These are the four subsquares of 14, so 14 is empty. That is, $4 \to 1$ is forbidden. There's no surprise that an IFS generated by forbidden pairs also can be generated by forbidden triples. This is how to find the appropriate array of matrices.

The second attractor of Fig. 1.61 is produced by this array

$$\begin{bmatrix} 1 & 1 & 1 & 0 \\ 1 & 1 & 0 & 1 \\ 1 & 0 & 1 & 1 \\ 0 & 1 & 1 & 1 \end{bmatrix} \begin{bmatrix} 1 & 1 & 1 & 0 \\ 1 & 1 & 0 & 1 \\ 1 & 0 & 1 & 1 \\ 0 & 1 & 1 & 1 \end{bmatrix} \begin{bmatrix} 1 & 1 & 1 & 0 \\ 1 & 1 & 0 & 1 \\ 1 & 0 & 1 & 1 \\ 0 & 1 & 1 & 1 \end{bmatrix} \begin{bmatrix} 1 & 1 & 1 & 0 \\ 1 & 1 & 0 & 1 \\ 1 & 0 & 1 & 1 \\ 0 & 1 & 1 & 1 \end{bmatrix}$$

The forbidden triples show that the compositions $4 \to 1 \to k$, $3 \to 2 \to k$, $2 \to 3 \to k$, and $1 \to 4 \to k$ are forbidden for $k = 1, 2, 3$, and 4. That is, in each of the squares 1, 2, 3, and 4 we see the forbidden combinations of Fig. 1.60, and so each of these squares contains a reduced copy of the attractor of that figure.

The third attractor of Fig. 1.61 is produced by this array

$$\begin{bmatrix} 1 & 1 & 1 & 0 \\ 1 & 1 & 1 & 0 \\ 1 & 1 & 1 & 0 \\ 1 & 1 & 1 & 0 \end{bmatrix} \begin{bmatrix} 1 & 1 & 0 & 1 \\ 1 & 1 & 0 & 1 \\ 1 & 1 & 0 & 1 \\ 1 & 1 & 0 & 1 \end{bmatrix} \begin{bmatrix} 1 & 0 & 1 & 1 \\ 1 & 0 & 1 & 1 \\ 1 & 0 & 1 & 1 \\ 1 & 0 & 1 & 1 \end{bmatrix} \begin{bmatrix} 0 & 1 & 1 & 1 \\ 0 & 1 & 1 & 1 \\ 0 & 1 & 1 & 1 \\ 0 & 1 & 1 & 1 \end{bmatrix}$$

Here the forbidden triples are 114, 124, 134, 144, 213, 223, 233, 243, 312, 322, 332, 342, 411, 421, 431, and 441. Then the compositions $1 \to k \to 4$, $2 \to k \to 3$, $3 \to k \to 2$, and $4 \to k \to 1$ are forbidden for $k = 1, 2, 3$, and 4. Because the k is inserted between the first and second entries of the forbidden pairs of Fig. 1.60, we do not expect the third attractor of Fig. 1.61 to consist of copies of the attractor of Fig. 1.60, and indeed it doesn't. It has nine approximate copies of Fig. 1.60, but only approximate, and then the attractor contains vertical lines at $x = 1/3$ and $x = 2/3$, and horizontal lines at $y = 1/3$ and $y = 2/3$.

Another question we can ask is whether the attractor of a given 2-step memory IFS also can be generated by a 1-step memory IFS. From the first attractor of Fig. 1.61 we know sometimes the answer is, "Yes." It turns out that this question has two different answers, because "also can be generated by a 1-step memory IFS" has two different interpretations.

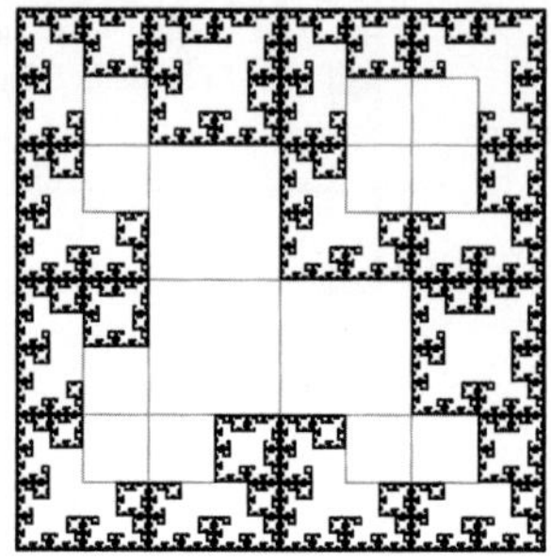

Figure 1.62: How much memory?

If we restrict ourselves to use the four transformations (1.7):

$$
\begin{aligned}
T_1(x, y) &= (x/2, y/2) \\
T_2(x, y) &= (x/2, y/2) + (1/2, 0) \\
T_3(x, y) &= (x/2, y/2) + (0, 1/2) \\
T_4(x, y) &= (x/2, y/2) + (1/2, 1/2)
\end{aligned}
$$

then the answer is straightforward. If every empty length-3 address contains an empty length-2 address, then the answer is "yes;" if not, then the answer is "no."

For instance, in Fig. 1.62 we see that the empty length-2 addresses are 14, 23, and 32; the empty length-3 addresses (ignore those that are contained in an empty length-2 address, for example, 141) are 114, 123, 132, 214, 223, 314, 332, 414, 423, 432, and 441. The last of these, 441, does not contain an empty length-2 address, so this attractor cannot be generated by a 1-step memory IFS with the four transformations (1.7).

Could the attractor of Fig. 1.62 be generated by a 1-step memory IFS with different transformations? In Sect. 1.9.3.3 we showed that the attractors of some IFS with 1-step memory also are the attractors of IFS without memory, but with a different set of transformations. By adapting a construction called a "higher block shift" from symbolic dynamics, we can show that for every $n > 1$, the attractor of *every* IFS with n-step memory also is the attractor of an IFS with 1-step memory. See Appendix B.4 and [40, 41].

1.10.4 Procedure

Three obvious types of problems are

1. Given a 2-step memory array, generate the corresponding IFS.
2. Given the attractor of a 2-step memory IFS, find an array that generates this attractor.
3. Can the attractor of a 2-step memory IFS be generated by a 1-step memory IFS with transformations (1.7)?

Type 1 is simple: enter the array into the 2-step memory IFS software in 1.10 of the Mathematica or Python code and run the program. Unless you have a *lot* of patience, don't take the number of iterations greater than 7, and 4 gives a quick image to check the general form of the attractor.

For type 2, on the unit square superimpose a grid of 64 address length-3 squares, that is, $1/8 \times 1/8$ squares. Record the empty length-3 addresses and translate these into the array of the 2-step memory IFS software.

For type 3, on the unit square superimpose a grid of 16 address length-2 squares and record the addresses of the empty squares. Subdivide the nonempty address length-2 into address length-3 squares and record the addresses of the empty squares. Compare the addresses of the empty length-3 squares with those of the empty length-2 squares.

1.10.5 Sample A

For the attractor (the left image), find the empty length-2 addresses and the empty length-3 addresses. Can this attractor be generated by a 1-step memory IFS with the transformations (1.7)? Find a 2-step memory IFS that generates this attractor.

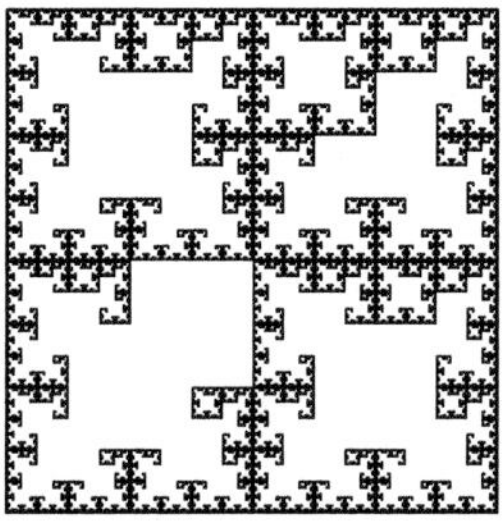

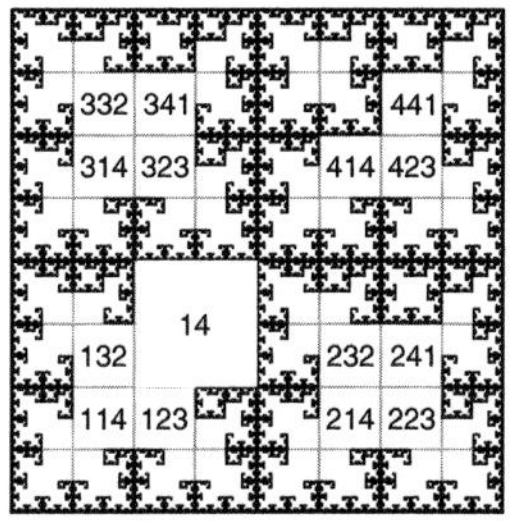

In the right image we see only one empty length-2 address, 14. The empty length-3 addresses not contained in 14 are 114, 123, 132, 214, 223, 232, 241, 314, 323, 332, 341, 414, 423, and 441. Ten of these do not contain 14, so this attractor cannot be generated by a 1-step memory IFS with the transformations (1.7).

To find the array of matrices for the 2-step memory IFS, First note that address 14 has four length-3 components: 141, 142, 143, and 144. Along with the other empty length-3 addresses of the previous paragraph, these give the 0s in the array of matrices for the 2-step memory IFS that generates this attractor. Use (1.8) to map addresses to matrix entries.

$$\begin{bmatrix} 1 & 1 & 1 & 0 \\ 1 & 1 & 0 & 1 \\ 1 & 0 & 1 & 1 \\ 0 & 0 & 0 & 0 \end{bmatrix} \quad \begin{bmatrix} 1 & 1 & 1 & 0 \\ 1 & 1 & 0 & 1 \\ 1 & 0 & 1 & 1 \\ 0 & 1 & 1 & 1 \end{bmatrix} \quad \begin{bmatrix} 1 & 1 & 1 & 0 \\ 1 & 1 & 0 & 1 \\ 1 & 0 & 1 & 1 \\ 0 & 1 & 1 & 1 \end{bmatrix} \quad \begin{bmatrix} 1 & 1 & 1 & 0 \\ 1 & 1 & 0 & 1 \\ 1 & 1 & 1 & 1 \\ 0 & 1 & 1 & 1 \end{bmatrix}$$

Note that the bottom row of 0s in the left-most matrix occupy the length-3 addresses of the empty square 14.

1.10.6 Sample B

In Sect. 1.9.3.2 we saw how to generate lines at $x = 1/3$ and $x = 2/3$ with a 2-step memory IFS. In addition, we observed that these lines must be generated in pairs. One way to understand this is that the line endpoints are periodic: $T_2(T_1(2/3, 0)) = (2/3, 0)$ and $T_1(T_2(1/3, 0)) = (1/3, 0)$. The points $(1/7, 0)$, $(2/7, 0)$, and $(4/7, 0)$ also are periodic, if the transformations are applied in the right order: $T_1(4/7, 0) = (2/7, 0)$, $T_1(2/7, 0) = (1/7, 0)$, and $T_2(1/7, 0) = (4/7, 0)$. Can we generate these three lines with a 3-step memory IFS?

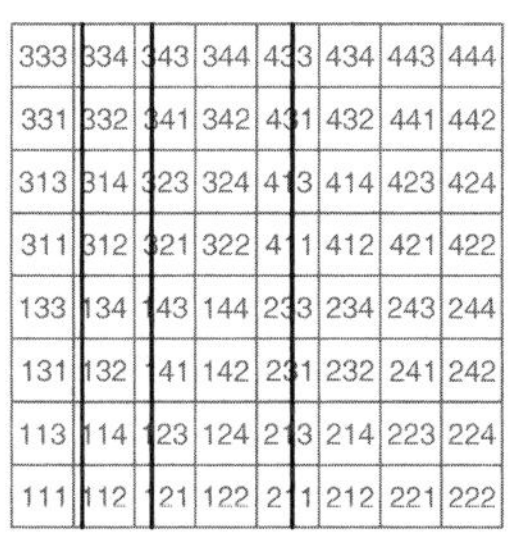

The occupied length-3 addresses can be read from the diagram. With (1.8) this translates to the matrix array

$$\begin{bmatrix} 0&1&0&1\\ 1&0&1&0\\ 0&1&0&1\\ 1&0&1&0 \end{bmatrix} \begin{bmatrix} 1&0&1&0\\ 0&0&0&0\\ 1&0&1&0\\ 0&0&0&0 \end{bmatrix} \begin{bmatrix} 0&1&0&1\\ 1&0&1&0\\ 0&1&0&1\\ 1&0&1&0 \end{bmatrix} \begin{bmatrix} 1&0&1&0\\ 0&0&0&0\\ 1&0&1&0\\ 0&0&0&0 \end{bmatrix}$$

Entering these matrices into the software in 1.10 of the Mathematica or Python code does indeed produce that triple of lines, none of which can be produced independently of the other two.

1.10.7 Conclusion

Adding an additional step of memory to the IFS program enables the generation of more complex images. As we see in Appendix B.4, these can be achieved by 1-step memory IFS, though with more, and perhaps less-transparent, transformations. With a fixed set of transformations (1.7), decoding allowed length-3 compositions is much simpler than finding the transformations and the allowed length-2 compositions. This method increases significantly the collection of attractors for which we can find the IFS.

1.10.8 Exercises

Prob 1.10.1 What is the maximum number of copies of the attractor of Fig. 1.60 that can be produced by a 2-step memory IFS with transformations (1.7)?

Prob 1.10.2 Can we use a 2-step memory IFS to build 16 gaskets placed uniformly throughout the unit square?

Prob 1.10.3 Fig. 1.63 shows attractors of three 1-step memory IFS.
(a) Find 2-step memory IFS that have these attractors.
(b) Find 2-step memory IFS whose attractors are four copies of these attractors.

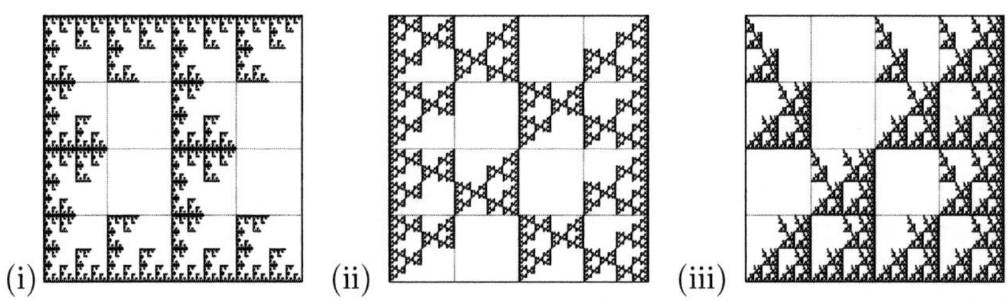

Figure 1.63: Attractors for Exercise 1.10.3.

Prob 1.10.4 For the attractors of Fig. 1.64, find 2-step memory IFS with the smallest number of 0s that generate these attractors. Hint: how many empty length-3 address squares are the consequence of an empty length-2 address square?

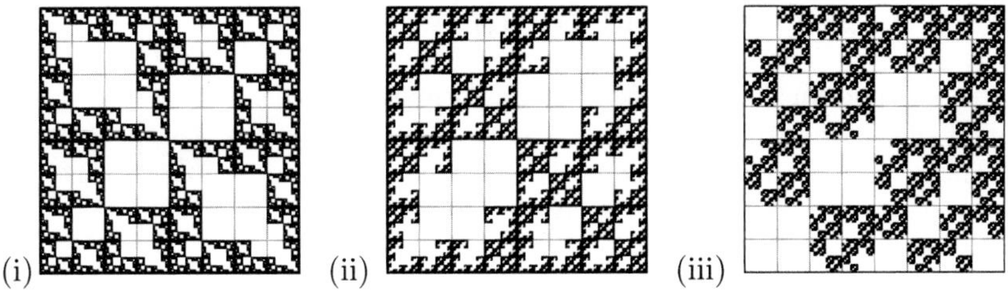

Figure 1.64: Attractors for Exercise 1.10.4.

Prob 1.10.5 For the attractors of Fig. 1.65, find 2-step memory IFS with the smallest number of 0s that generate these attractors. Hint: how many empty length-3 address squares are the consequence of an empty length-2 address square?

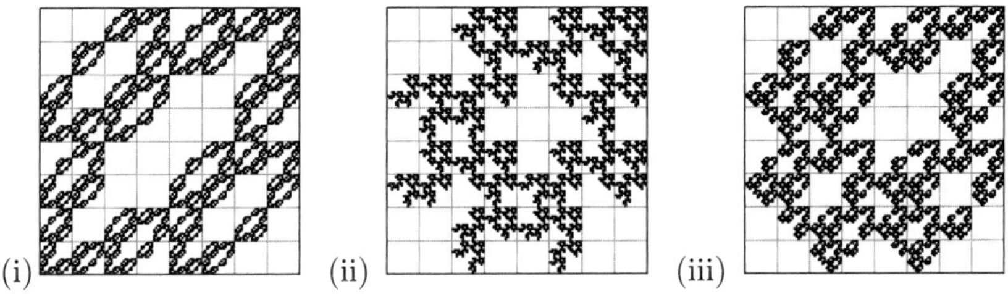

Figure 1.65: Attractors for Exercise 1.10.5.

Prob 1.10.6 Find a 2-step memory IFS that has attractor the vertical lines $x = 3/7$, $x = 5/7$, and $x = 6/7$.

Prob 1.10.7 (a) Can the lines $x = 0$, $x = 1/3$, and $x = 2/3$ be the attractor of a 1-step memory IFS?
(b) Can the lines $x = 0$, $x = 1/3$, and $x = 2/3$ be the attractor of a 2-step memory IFS?

1.11 Data analysis by driven IFS

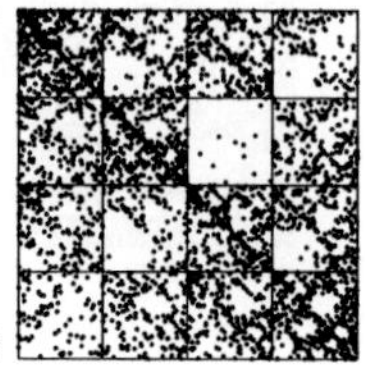

In the random IFS algorithm, transformations are applied one at a time in random order. Randomness guarantees that every (finite-length) address of the attractor eventually will be visited. With a lot more math [43], a more efficient dance of points fills in the attractor more rapidly. In our study of IFS so far, the goal was to generate the fractal. Now we'll reverse this and use the way the fractal fills in to study properties of the sequence of transformations. So far as we know, Ian Stewart [44] first asked what would happen if the random IFS algorithm were driven by a non-random sequence. Stewart used an IFS for a gasket, which already is mostly holes, so he looked for a visual signal in a sparse image. Rather, we'll use the IFS with four transformations

$$\begin{aligned} T_1(x,y) &= (x/2, y/2) & T_2(x,y) &= (x/2, y/2) + (1/2, 0) \\ T_3(x,y) &= (x/2, y/2) + (0, 1/2) & T_4(x,y) &= (x/2, y/2) + (1/2, 1/2) \end{aligned} \tag{1.10}$$

With the deterministic or the random IFS algorithms, these four transformations generate the filled-in unit square. Departures from uniform fill are our diagnostic tool.

Because the IFS transformations are applied in the order determined by, or driven by, a data sequence, we call this *driven IFS*.

1.11.1 Purpose

We'll learn how to find patterns in symbolic sequences (DNA, for example) and numerical sequences (intervals between heartbeats, for example) revealed by driven IFS.

1.11.2 Materials

We'll use driven IFS software (1.11 (a)–(f) of the Mathematica or Python codes).

1.11.3 Background

We'll use a variation of the random IFS algorithm of Lab 1.2 and the notion of addresses from Lab 1.7. Here we'll introduce the driven IFS algorithm and show how to drive an IFS with a symbol sequence. For numerical data we'll discuss binning schemes. We'll learn to interpret some graphical features of driven IFS plots. We'll end with the notion of Markov partitions and driving IFS by iteration of simple functions.

1.11.3.1 The driven IFS algorithm

The IFS with transformations (1.10) generates the filled-in unit square. For the random IFS algorithm, start with a point (x_0, y_0) inside the unit square.

Then the sequence of points $(x_k, y_k) = T_{i_k}(x_{k-1}, y_{k-1})$ uniformly fills in the square if the i_k are independently and uniformly randomly distributed among $\{1, 2, 3, 4\}$. To the extent that the square does not fill in uniformly, we can make deductions about the departures from uniform randomness of the sequence $i_1, i_2, i_3, \dots$. To minimize any bias introduced by the position of the initial point, we always take $(x_0, y_0) = (1/2, 1/2)$.

1.11.3.2 IFS driven by symbol sequences

Any sequence of 1, 2, 3, and 4 can be instructions for the order in which to apply the T_i. No importance is attached to the transformation labels. They may just as well be Homer, Marge, Bart, and Lisa, or maybe C, A, T, and G. These labels underlie H. Joel Jeffrey's work [45, 46] on using IFS to seek patterns in DNA sequences.

For example, the left image of Fig. 1.66 is the IFS driven by the DNA sequence that codes for amylase, a salivary enzyme that converts starch to simple sugars. In the driven IFS, T_1 is applied when C is read, T_2 for A, T_3 for T, and T_4 for G. This sequence has 3957 bases, so the driven IFS has enough points to produce noticeable patterns. Maybe the most apparent is that is that the square with address 41 (See the diagram on the right.) is almost empty.

33	34	43	44
31	32	41	42
13	14	23	24
11	12	21	22

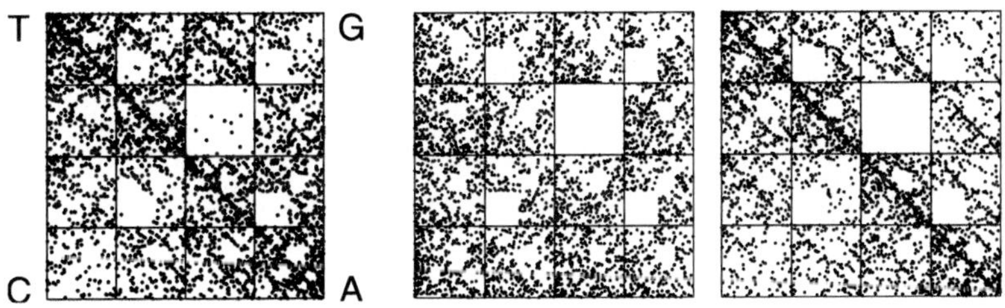

Figure 1.66: Left: IFS driven by the amylase sequence. Middle and right: two surrogates of the amylase-driven IFS.

In the middle image of Fig. 1.66 we plot 3957 points, uniformly randomly except for the restriction that T_4 cannot immediately follow T_1. An obvious difference between the left and middle images is the denser distribution of points along the diagonal of the amylase plot. To approach that issue, we count the number of C, A, T, and G in the amylase plot. We find 589 Cs, 1305 As, 1389 Ts, and 674 Gs. In the right image of Fig. 1.66, we again forbid the composition $T_4(T_1)$, and apply the transformations with probabilities $p_1 = 589/3957 \approx 0.149$, $p_2 \approx 0.330$, $p_3 \approx 0.351$, and $p_4 \approx 0.170$. The right plot is the result. This is a better match than the middle, but still it misses some details.

One way to understand driven IFS is as a tool to find how much of an observed pattern is a consequence of a few simple conditions. Many interesting details swamped by the main aspects of the original plot may be revealed in the residual patterns.

1.11.3.3 Numerical data binning schemes

Not all data come from neat strings of four symbols. More commonly encountered are sequences of numbers, measurements known only to some accuracy. To convert a sequence of decimals into a sequence of symbols, we *bin* or *coarse-grain* the data. We'll describe three bin schemes for a sequence $x_1, \dots, x_N$. Write $m = \min\{x_i\}$, $M = \max\{x_i\}$, and $rng = M - m$. Choose the *bin boundaries* so $b_1 < b_2 < b_3$. Then the bins are

$$B_1 = [m, b_1),\; B_2 = [b_1, b_2),\; B_3 = [b_2, b_3),\; \text{and } B_4 = [b_3, M].$$

The translation from decimals to symbols is straightforward. If x_i lies in bin B_j, then the ith transformation in the sequence is T_j. That is, the order of the transformations is determined by the order of bins visited by the sequence $x_1, \dots, x_N$.

Equal-size bins. The bin boundaries are $b_1 = m + (rng/4)$, $b_2 = m + 2(rng/4)$, and $b_3 = m + 3(rng/4)$. Each bin has length $rng/4$. In the absence of other information, this is a natural choice of bins.

Equal-weight bins. The bin boundaries b_1, b_2, and b_3 are chosen so the four bins have about the same number of points x_i. In Appendix B.5 we'll sketch why equal-weight bins are called a *maximum entropy* partition.

Mean-centered bins. Take the middle bin boundary b_2 to be the mean of the data sequence, $b_2 = (x_1 + \cdots + x_N)/N$. Then for every $\delta > 0$, take $b1 = b_2 - \delta$ and $b_3 = b_2 + \delta$. Of course, for large enough δ all the data points lie in bins B_2 and B_3, so the driven IFS plot will be along the diagonal between $(0, 1)$ and $(1, 0)$. On the other hand, if δ is very close to 0, almost all the data points lie in bins B_1 and B_4, so the driven IFS plot will be along the diagonal between $(0, 0)$ and $(1, 1)$. Unlike the first two binning schemes, here we have a family of bins, parameterized by the number δ. If the distribution of driven IFS points makes an abrupt change, a "geometrical phase transition", the δ of this transition may help us discover interesting features of the underlying dynamics.

1.11.3.4 Interpretation of driven IFS plots

First we'll study some sequences that produce simple driven IFS patterns. In Sect. 1.9.3.2 we've already discussed how lines can be generated by IFS with memory, though we'll see that some of these lines cannot be generated by driven IFS. We'll start with fixed points and cycles, move on to lines, and then to more complex patterns.

Fixed points. Write $T_2(x, y) = (x/2 + 1/2, y/2)$, and verify that

$$T_2^n(x, y) = \left(\frac{x}{2^n} + \frac{2^n - 1}{2^n}, \frac{y}{2^n}\right) \quad \text{so} \quad \lim_{n \to \infty} T_2^n(x, y) = (1, 0)$$

The address of the point $(1, 0)$ is 2^∞. Also note that $(1, 0)$ is the only fixed point of T_2. A sequence of points moving steadily to a corner of the unit square is not so interesting, but it's a first step.

Cycle points. The sequence of points

$T_1(1/2, 1/2)$,
$T_3(T_1(1/2, 1/2))$,
$T_1(T_3(T_1(1/2, 1/2)))$,
$T_3(T_1(T_3(T_1(1/2, 1/2))))$,
...

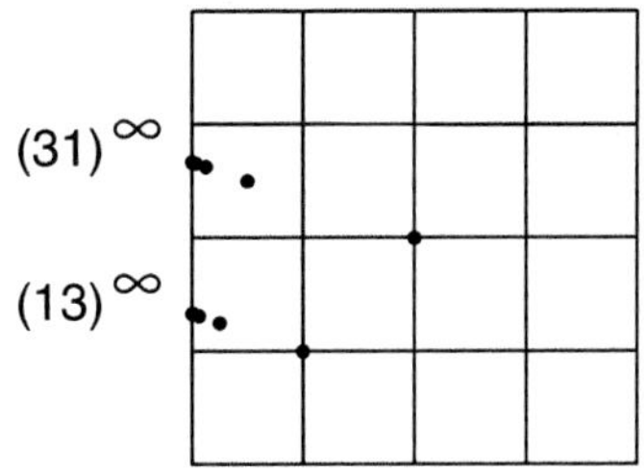

is shown on the right. These points appear to fall into two subsequences, upper points and lower points, and these are produced by these compositions.

$$(T_3(T_1))^n(1/2, 1/2) \qquad \text{upper points}$$
$$(T_1(T_3))^n T_1(1/2, 1/2) = (T_1(T_3))^n(1/4, 1/4) \qquad \text{lower points}$$

Call the limit of the upper sequence (x_*, y_*). The picture suggests that $x_* = 0$, but let's see if we can derive this from the equations. Note

$$(x_*, y_*) = \lim_{n\to\infty} (T_3(T_1))^n(1/2, 1/2)$$

so

$$(T_3(T_1))(x_*, y_*) = (T_3(T_1))\left(\lim_{n\to\infty} (T_3(T_1))^n(1/2, 1/2)\right)$$
$$= \lim_{n\to\infty} (T_3(T_1))^{n+1}(1/2, 1/2) = (x_*, y_*)$$

That is, (x_*, y_*) is the fixed point of $T_3(T_1)$. The solution of $(x_*, y_*) = T_3(T_1(x_*, y_*))$ is $(x_*, y_*) = (0, 2/3)$.

The analogous argument shows that the limit of the lower sequence is $(0, 1/3)$, the fixed point of $T_1(T_3)$.

Every repeated pattern of T_i produces a collection of points that are carried one to another in a cyclic order by application of the T_i in that pattern. Moreover, each of these points is a fixed point of the T_i in this pattern, or of some cyclic shift of the pattern. For example, the 3-cycle points generated by $(T_1(T_2(T_3)))^\infty$ are the fixed points of $T_1(T_2(T_3))$, $T_2(T_3(T_1))$, and $T_3(T_1(T_2))$. And all of these fixed point equations are linear equations, easy to solve.

Points on lines. Some of the lines that we studied in Sect. 1.9.3.2 cannot be constructed with driven IFS. Some can. We'll show an example of each type and invite you to make some general observations.

On the left we see a driven IFS for a random sequence of T_1 and T_2, all combinations of both transformations allowed. This we'll indicate by the transition graph on the right. Unless the line passes through the point $(1/2, 1/2)$, the driven IFS will produce an initial scattering of points that converges to the line.

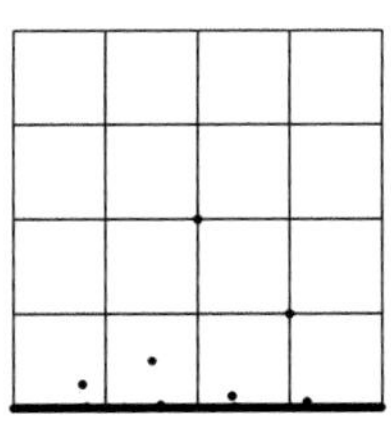

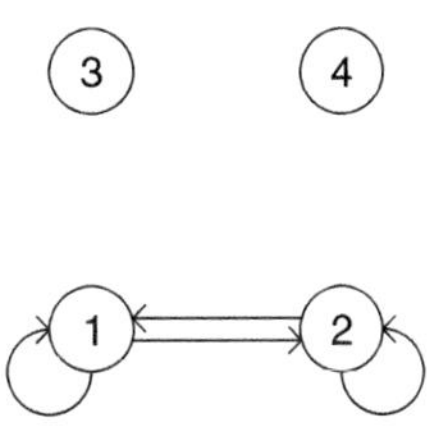

The particular collection of points that converges to the line between (0, 0) and (1, 0), and the dance of driven IFS points along that line, is determined by the particular sequence of bins visited by the data sequence.

We'll mention that the bottom middle pair of lines of Fig. 1.49 of Lab 1.9 cannot be generated by a driven IFS. From the transition graph we see that once a data point is found in bins 3 or 4, no other data points exist, because points in these bins cannot be followed by any point. This may seem to be a silly example, but it shows that driven IFS and IFS with memory can generate different sets of attractors.

Every one of the strings listed here generate points along the lines in the driven IFS plot on the right.

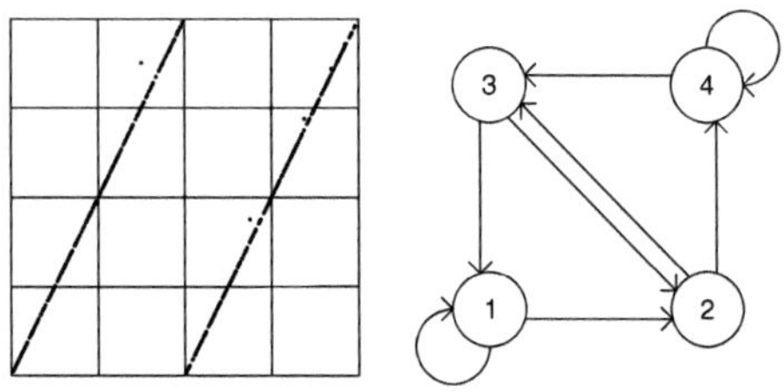

- Arbitrarily long strings of consecutive points in bin 1
- Arbitrarily long strings of consecutive points in bin 4
- Arbitrarily long strings of consecutive points alternating between bins 2 and 3
- Points in bin 1 followed immediately by a point in bin 2
- A point in bin 2 followed immediately by points in bin 4
- Points in bin 4 followed immediately by a point in bin 3
- A point in bin 3 followed immediately by points in bin 1
- Arbitrarily long strings of consecutive points cycling between bins 1, 2 and 3
- Arbitrarily long strings of consecutive points cycling between bins 2, 4 and 3
- Arbitrarily long strings of consecutive points cycling between bins 1, 2, 4, and 3

Certainly, these lines can be generated only by a data sequence of some complexity. Note that not every combination is allowed. For example, two consecutive points cannot both lie in bin 2 or both in bin 3. In the driven IFS this is shown by the emptiness of the squares with address 22 and 33. The other empty length-2 addresses are 12, 14, 24, 31, 41, and 43.

This suggests another direction in our analysis. Rather than note the squares that are occupied, we'll focus on those that are empty. Benoit often said that a fractal can be described just as well by what is left out as by what is included.

Empty squares We've seen an example already in the middle driven IFS plot of Fig. 1.66. Here we'll look at other examples.

On the left side we see a driven IFS. A first step to discover the underlying dynamics is to superimpose the grid of length-2 address squares and note if any are empty. Here we see that 23, 32, and 41 are empty. We know that every length-3 address that contains any of these empty length-2 addresses must be empty. Do these account for all the empty length-3 addresses? On the right

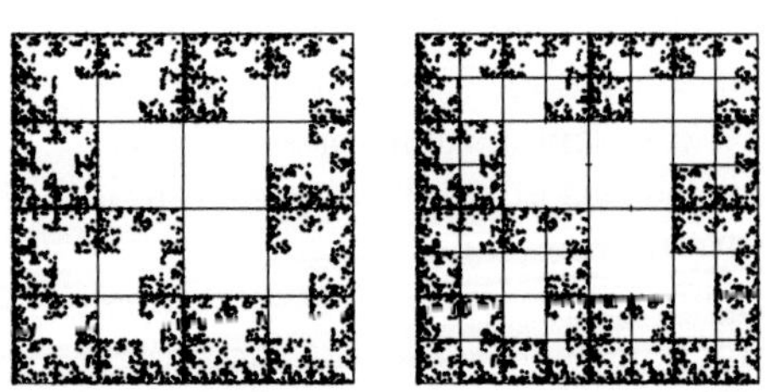

we see the grid of empty length-3 addresses superimposed. The empty squares have addresses 123, 132, 141, 223, 241, 332, 341, 423, 432, and 441. Every length-3 address contains an empty length-2 address. This suggests, though it does not prove, that the system has low complexity: forbidden pairs tell the whole story. In principle, empty addresses of all lengths must be checked, though here we'll consider just length-2 and length-3 addresses.

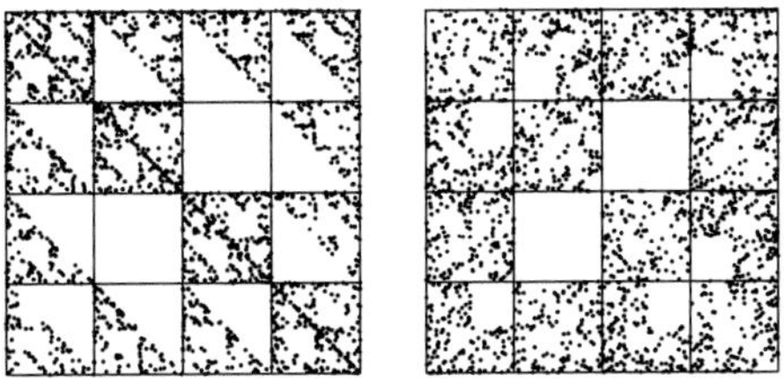

Regime change Sometimes data sequences reveal patterns that do not fit the categories we have established so far. For example, in both driven IFS plots shown here the empty length-2 addresses are 14 and 41. The right image is generated by a sequence that is random, except that 1 never immediately follows 4 and 4 never immediately follows 1. Aside from those long-address squares empty by the finiteness of the data set, the address of every empty square contains 14 or 41.

Something more complicated is involved in the left image. The driven IFS plot appears to be a gasket with corners 1, 2, and 3, and also a gasket with corners 2, 3, and 4. What data sequence could generate such a plot?

Here we see a sequence that generates the "two gaskets" driven IFS plot. To the left of the first vertical line all the data points lie in bins 1, 2, and 3, so the IFS driven with this part of the sequence consists of points on the gasket with corners 1, 2, and 3. Between the vertical lines the data points fall into bins 2 and 3. The IFS driven by this portion of the sequence produces points along the line between (0, 1) and (1, 0), which lies on both gaskets. To the right of the second vertical line, the data points lie in bins 2, 3, and 4, so the driven IFS generates points that lie on the gasket with corners 2, 3, and 4.

If a driven IFS plot appears to be more involved than a straightforward forbidden pairs analysis suggests, decomposing the plot into pieces, each of which can be explained by forbidden pairs, may succeed. So far, this sort of analysis still requires eye and mind. It has not been automated, yet.

1.11.3.5 Markov partitions of data

A partition of the data sequence is called *Markov* (same mathematician as Markov chains, which we'll see in Appendix B.3) if every empty address contains an empty length-2 address. In that sense these are the simplest partitions: the empty length-2 addresses tell the whole story for empty squares.

Both driven IFS pictured here have only one empty length-2 address, 41. For the left image, the empty length-3 addresses are 141, 241, 341, and 441; the right also has address 144 empty. So we might conclude that the left driven IFS was generated by a Markov partition and the right driven IFS was not.

In order to learn the concepts, these are the conclusions we'll draw. But both face some complications.

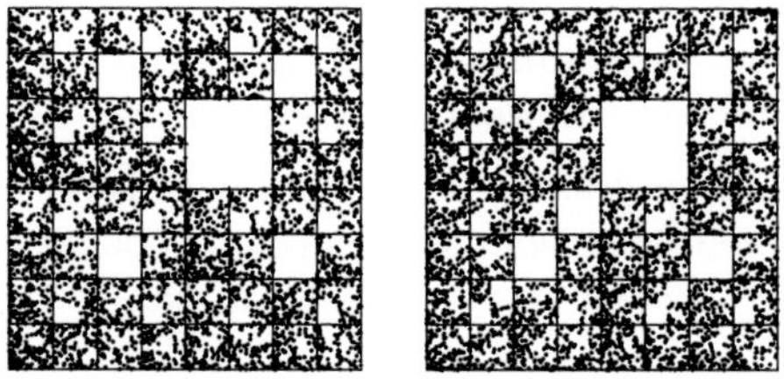

For the left image, to be certain that a Markov partition was used we should check longer and longer addresses. Eventually we'll encounter addresses empty just because of the data set size. In practice, we'll check only through length-3 addresses.

For the right image, to be careful we would need to check that address 144 is empty because of an exclusion in the dynamics, not because the data set is too short. In Appendix A.92 of [17] we describe a way to do this, but the concepts are a bit advanced and the calculations can be tricky. So long as we go only through length-3 addresses, the data sequence consists of at least a few thousand points, and none of the occupied squares is very sparsely occupied compared with the others, we'll assume that an empty length-3 address is empty because of a data exclusion.

1.11.3.6 IFS driven by iteration

Graphical iteration, a simple method to visualize the sequence x, $f(x)$, $f(f(x)) = f^2(x)$, $f(f(f(x))) = f^3(x)$, ..., is the topic of Lab 3.1. In order to relate functional iteration and driven IFS, here we'll give a brief sketch.

The point of graphical iteration is that although all the iterates we want can be computed algebraically, neither long algebraic expressions nor strings of decimals need reveal patterns. Sometimes geometry can help with this.

In Fig. 1.67 we see the graph of a logistic map (also a topic in Lab 3.1), in this case $f(x) = 4x(1-x)$. We've plotted the graph of f only for $0 \leq x \leq 1$, and have included the line $y = x$, used in the graphical iteration method we'll describe next.

Given a point x on the horizontal axis, draw the vertical segment to the graph $y = f(x)$. The point of intersection is $(x, f(x))$. From this, draw the horizontal segment to the diagonal line $y = x$. The point of intersection is $(f(x), f(x))$. Continue. Vertically to the graph, intersecting at $(f(x), f(f(x)))$. Then horizontally to the diagonal, and repeat. This method, vertically to the graph, horizontally to the diagonal, is *graphical iteration*.

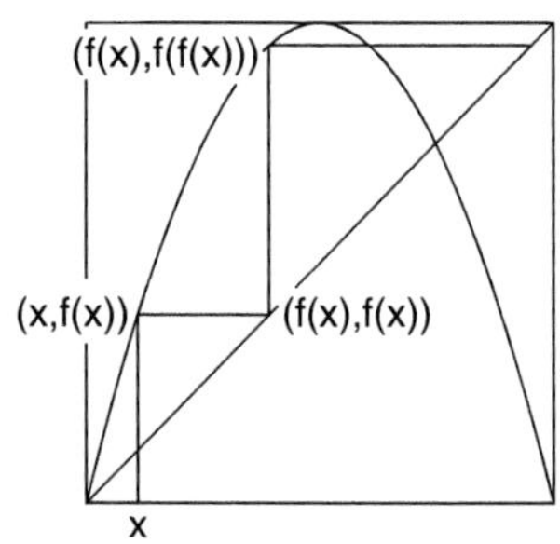

Figure 1.67: A few graphical iteration steps.

Suppose a data sequence $x_1, x_2, x_3, \ldots$ is generated by iterating the function f. That is, $x_{i+1} = f(x_i)$. From the graph of the function and a sketch of the bins, graphical iteration reveals the allowed transitions between bins.

On the left graph of Fig. 1.68, the bins 1, 2, 3, and 4 (equal-size bins in this case) are indicated by labels below the x-axis. From the left side of bin 1 the graphical iteration path stays in bin 1; from the middle of bin 1 the graphical iteration path leads to bin 2; from the right side of bin 1 the graphical iteration path leads to bin 3. That is,

$$1 \to 1,\ 1 \to 2,\ 1 \to 3 \text{ are allowed, while } 1 \to 4 \text{ is forbidden}$$

The right graph of Fig. 1.68 shows a way to streamline these deductions. Here we label the bins along both the x- and y-axes. To find the allowed transitions between bins, inspect the portion of the graph of the function that lies above each bin on the x-axis. Note through which bins along the y-axis this part of the graph passes. For example, above bin 1 the graph passes through bins 1, 2, and 3 along the y-axis. In this way we see that the allowed bin-to-bin transitions are

$$1 \to 1,2,3 \qquad 2 \to 4 \qquad 3 \to 4 \qquad 4 \to 1,2,3 \tag{1.11}$$

On the left side of Fig. 1.69 we see the transition graph that encodes the allowed combinations (1.11). In the middle we see the IFS generated by iterating this logistic map, with the bin boundaries $b_1 = 1/4$, $b_2 = 2/4$, and $b_3 = 3/4$. On the right of Fig. 1.69 we see an IFS plot driven randomly except for the restrictions imposed by the transition graph. That the middle and right driven IFS plots are nearly identical suggests that all the empty addresses of the middle plot are consequences of empty length-2 addresses. In fact, this is true.

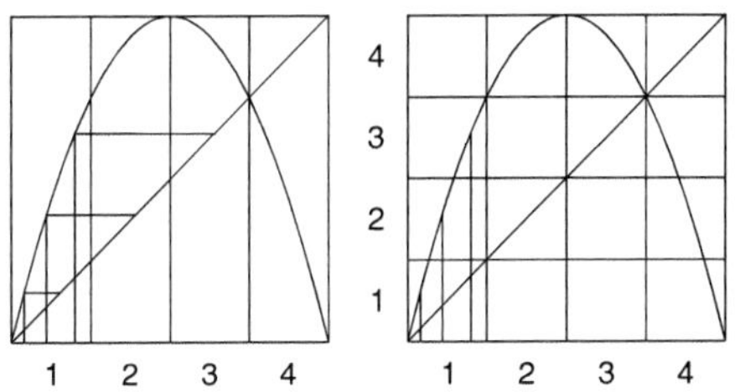

Figure 1.68: Finding the image of a bin.

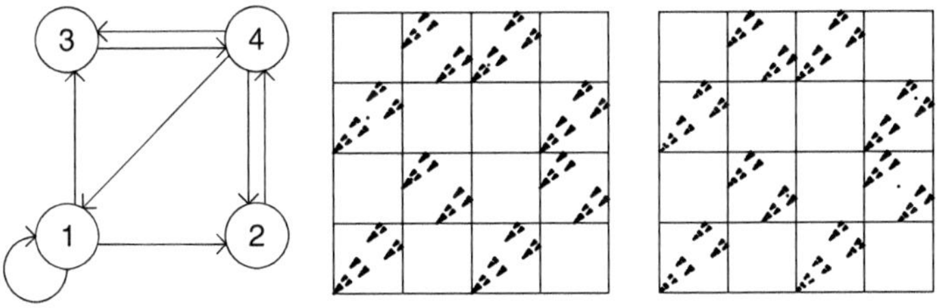

Figure 1.69: Left: the transition graph of (1.11). Middle: an IFS driven by this logistic map with these bins. Right: a driven IFS from a sequence random except for the transitions forbidden by left graph.

For a data sequence generated by iteration of a function $x_{n+1} = f(x_n)$, we say a partition of the domain and range is *Markov* if *above each partition bin along the x-axis, the graph of the function passes through all of each bin it enters along the y-axis.*

To illustrate this observation, the left graph is the logistic map $f(x) = 4x(1-x)$ with $b_1 = 0.25$, $b_2 = 0.5$, and $b_3 = 0.75$. We see that these bins, the equal-size bins, form a Markov partition for this logistic map.

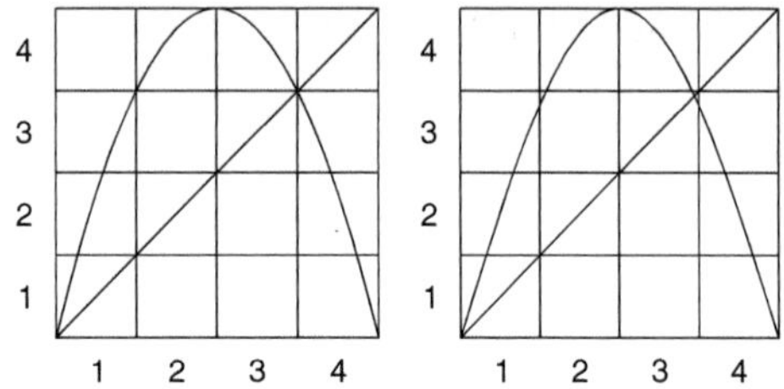

The right graph is the sine map $f(x) = \sin(\pi x)$ with the same bin boundaries. These do not form a Markov partition. For example, the part of the graph above bin 2 covers all of bin 4, but only a tiny part of bin 3.

To illustrate the difference between Markov and non-Markov partitions, on the left we see the IFS driven by the sine map with equal-size bins. The empty length-2 addresses are 12, 13, 22, 23, 41, and 44. The right image is an IFS driven by a sequence that is random, except the transitions $2 \to 1$, $3 \to 1$, $2 \to 2$, $3 \to 2$, $1 \to 4$, and $4 \to 4$ are forbidden. Unlike Fig. 1.69, with this non-Markov partition the driven IFS is not determined by forbidden pairs.

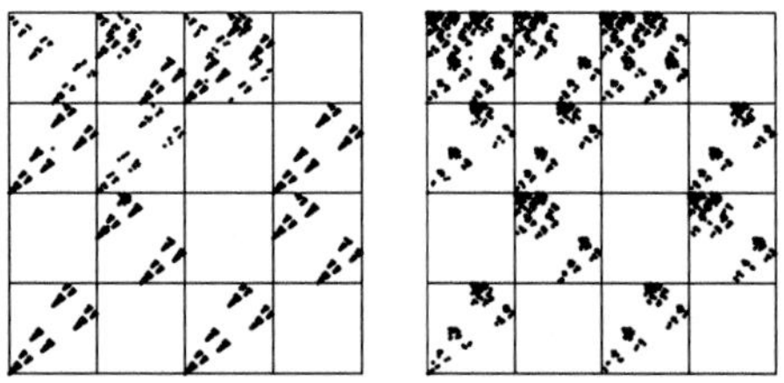

If we move b_3 to the intersection of the sine map and the diagonal, that is, $b_3 \approx 0.736484$, and move b_1 symmetrically to $b_1 \approx 0.263516$, then the partition is (very close to) Markov and the IFS driven by the sine map with this Markov partition is very close to those of Fig. 1.69. On the other hand, if some aspect of the problem requires equal-size bins, then allowed compositions of the sine map will require looking further back in time than the immediate preceding value.

These ideas can be extended in a hierarchical fashion. Given an IFS driven by a data sequence, can we find an N for which all forbidden addresses contain a forbidden address of length $\leq N$? Again, when dealing with (necessarily) finite data sets, we must consider if an empty address is empty because of an exclusion in the process generating the data, or if we haven't enough data. Note there are 4^n squares of address length-n, so an IFS driven by 1,000 data points must of necessity occupy under a quarter of the 4,096 address length-6 squares. None of these need be empty as a result of exclusions in the process generating the data.

These are subtle issues barely introduced here, but by now we have enough information to do the lab exercises.

1.11.4 Procedure

- Obtain a data sequence, comma-separated symbols $S_1, S_2, \ldots, S_M$, or comma-separated decimals $x_1, x_2, \ldots, x_N$. Paste this list between the brackets of the line

$$\text{dlst} = \{\ \};$$

in the appropriate program from 1.11 of Mathematica or Python codes.

Note that to get a sensible result, the data must be sequential in some natural fashion. For data that occur sequentially in time, the order of occurrence is a natural choice. But driven IFS can be applied to non-temporal data, DNA or RNA nucleotide sequences, for example, or molecular weight of proteins in amniotic fluid. Some others—for example, words in a text ordered alphabetically—make no sense at all.

• For numerical data, set the bin boundaries $b_1 < b_2 < b_3$ by entering the values in the line

$$b1 =; \qquad b2 =; \qquad b3 =;$$

• Identify patterns in the driven IFS plot, with special attention to forbidden or unlikely combinations.

• Interpret these patterns in the context of the original data sequence.

1.11.5 Sample

The right image of Fig. 1.70 shows 1,000 points of a time series obtained by iterating the logistic map $x_{n+1} = L_s(x_n) = sx_n(1 - x_n)$ for $s = 3.99$.

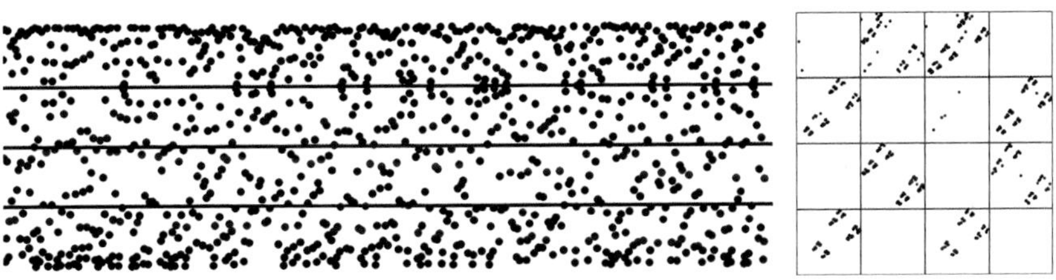

Figure 1.70: Left: An $s = 3.99$ logistic map time series with equal-size bins. Right: The dirven IFS

(a) We begin with equal-size bins. The bin boundaries, b_1, b_2, and b_3 are represented by horizontal lines on the left of Fig. 1.70. On the right we see the corresponding driven IFS. What patterns can we discover?

Quick inspection of the time series reveals few obvious patterns. One is that there appear to be more points in bins 1 and 4 than in bins 2 and 3. With the counters in the program we find bin 1 contains 302 points, bin 2 contains 156, bin 3 contains 192 and bin 4 contains 350.

Also, we see what appear to be clumps of points along the top of bin 4, several smallish clumps along the bin 3-bin 4 boundary. Not much else is easily apparent. What can the driven IFS tell us about these observations? Does the driven IFS show any patterns not evident from the time series?

First, if the clumps of data points along the top of bin 4 really were consecutive, we would see points in the address 44 square of the driven IFS. There are none. Right away, the driven IFS has been useful. Not only are the apparent clumps of data points at the top of bin 4 not consecutive, but *every* data point in bin 4 is followed immediately by a point in bin 1, 2, or 3. We know this because the squares with length-2 addresses 14, 24, and 34 contain points. The mechanics of graphical iteration clarifies the reason.

What about the clumps along the bin 3-bin 4 boundary? To understand this, recall from the cycles example of Sect. 1.11.3.4 that applying T_3 and T_4 infinitely many times in succession produces a sequence converging to two points, one with address $(34)^\infty$ and coordinates $(1/3, 1)$ and the other with $(43)^\infty$ and $(2/3, 1)$. If the alternation between T_3 and T_4 is repeated for a while, but not forever, the driven IFS should contain two sequences of points, one approaching $(1/3, 1)$, the other approaching $(2/3, 1)$. Indeed, we do see this in the diagram.

Now we analyze the distribution of points in the address length-2 and -3 squares. The table shows the populations of points in the length-2 address squares.

2	90	190	0
100	0	4	155
0	180	0	80
122	0	76	0

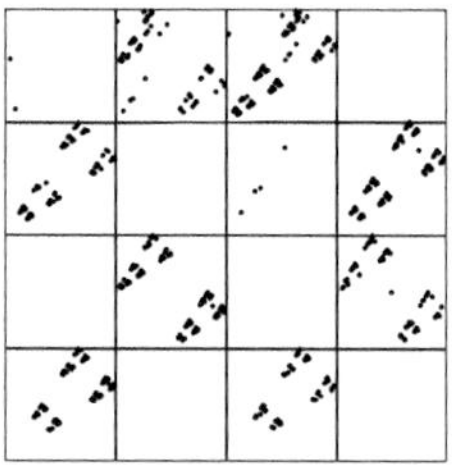

First we notice six length-2 addresses, 12, 13, 22, 23, 32, and 44, are empty, suggesting the transitions $2 \to 1$, $3 \to 1$, $2 \to 2$, $3 \to 2$, $2 \to 3$, and $4 \to 4$ are forbidden, or at least so unlikely that they have not occurred in a run of 1,000 points. In the same way, addresses 33 and 41 are almost empty. Would iterating from a different starting point result in these addresses being empty?

Note the length-2 address populations sum to 999, not 1,000. This is reasonable: the first two transformations must be applied in order to produce a single length-2 address.

Before studying length-3 addresses, we use a comparison of length-1 and -2 addresses to test for correlations. We estimate the probability of an address by the fraction of the total number of points having that address. For example, $p_1 = 0.302$, $p_2 = 0.156$, $p_3 = 0.192$, and $p_4 = 0.350$. If there were no relation between the current bin and the next, we'd have $p_{ij} = p_i \cdot p_j$. We see that this does not hold:

- $p_4 \cdot p_4 = 0.124$ yet we observe no points with address 44.
- $p_4 \cdot p_1 = p_1 \cdot p_4 = 0.106$ yet we observe $p_{14} = 0.180$ and $p_{41} = 0.004$.
- $p_4 \cdot p_3 = p_3 \cdot p_4 = 0.067$ yet we observe $p_{34} = 0.090$ and $p_{43} = 0.190$.

Next we consider the length-3 address squares.

Recall from Sect. 1.11.3.5 that a partition is Markov if every empty address contains an empty length-2 address. The first step of testing if the equal-size bins partition is Markov is to check if every empty length-3 address contains an empty length-2 address.

1	0	53	0	1	90	0	0
1	0	4	33	99	0	0	0
0	45	0	0	0	1	0	80
55	0	0	0	30	0	75	0
0	0	94	0	0	0	43	0
0	0	0	86	0	0	0	36
0	85	0	0	0	49	0	0
37	0	0	0	27	0	0	0

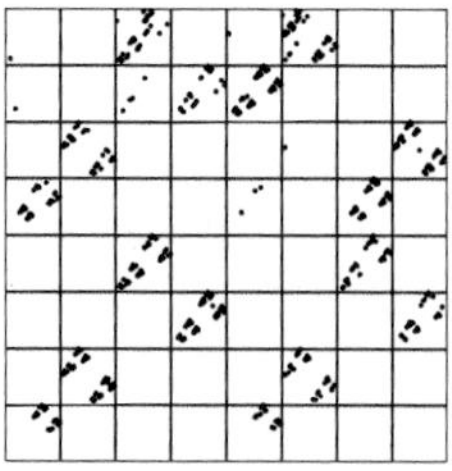

Three length-3 addresses, 141, 241, and 334, do not contain empty length-2 addresses. Consequently, for this logistic map, equal-size bins are not a Markov partition. We must look further than one time step into the past to detect all the excluded combinations.

(b) The data can be binned in other ways; for example, equal-weight bins. Here about one-quarter of the data points lie in each bin. Here is the same logistic map data with equal-weight bins.

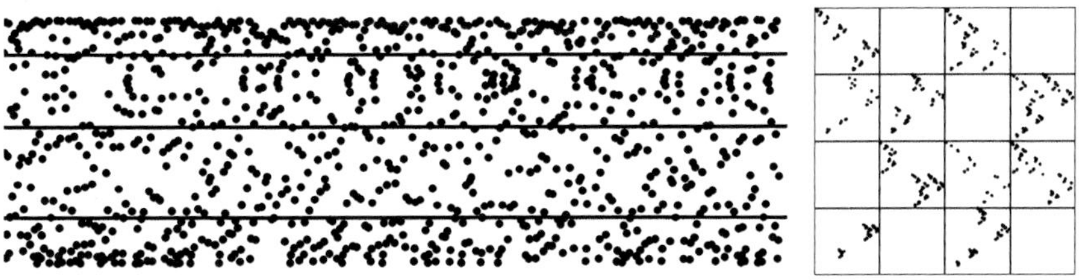

Figure 1.71: Logistic map iterates, and the driven IFS, using equal-weight bins.

Comparing the equal-size and equal-weight bins illustrates just how much the driven IFS depends on how the data are binned.

(c) Sometimes a clearer pattern is observed by differences of data points. From the sequence $x_1, x_2, x_3, \dots x_N$ the one-step differences are

$$x_2 - x_1, x_3 - x_2, x_4 - x_3, \dots, x_N - x_{N-1}$$

In Fig. 1.72 we see the one-step differences from the same logistic map data. The IFS is driven using equal-size bins.

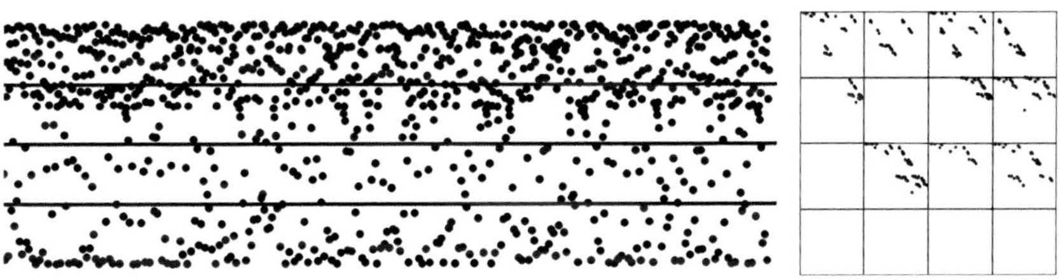

Figure 1.72: Logistic map one-step differences, and the driven IFS, using equal-size bins.

As in (a), studies of bin occupancy counts can test if equal-size bins give a Markov partition of the differences. In Fig. 1.73 we see the same data, using equal-weight bins.

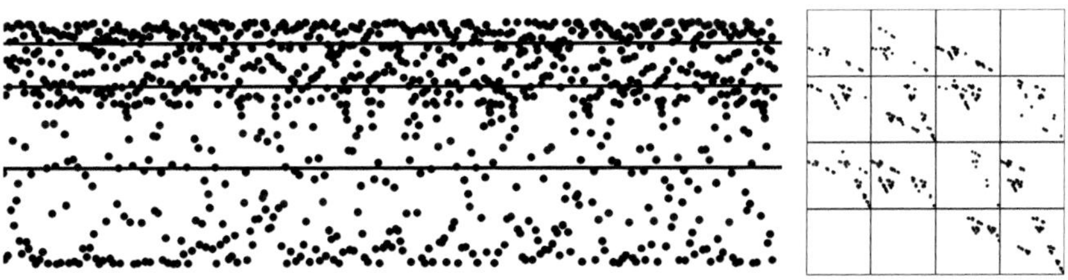

Figure 1.73: Logistic map one-step differences, and the driven IFS, using equal-weight bins.

Points of the difference time series cluster near the top of the plot. Equal-size bins give a clearer picture of the metric relations of the differences, while equal-weight bins give a more detailed picture of movement among the clustered differences. The bin boundaries can be set to any values, and so analyze in detail any part of the dynamics.

1.11.6 Conclusion

Driven IFS can reveal unlikely or forbidden combinations of data values, and also combinations more likely than the average. The kind of information provided by the driven IFS depends delicately on the choice of bin boundaries, so selecting bin boundaries appropriate for the problem requires some care.

1.11.7 Exercises

If we are to use driven IFS as the basis for prediction, we must have a choice of bin boundaries natural for the problem at hand. To build some intuition for how the choice of bin boundaries affects the picture, we begin with some exercises to be done by hand, not using the software.

In Exercises 1.11.1–1.11.5 the horizontal lines locate the bin boundaries b_1, b_2, and b_3.

Prob 1.11.1 Sketch the driven IFS for the data sets of Fig. 1.74, and explain the main features of your sketch.

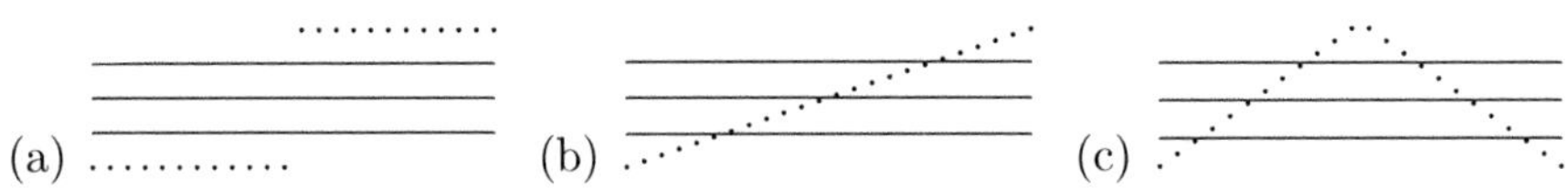

Figure 1.74: Data sequences for Exercise 1.11.1.

Prob 1.11.2 We'll assume that the pattern of data points in Fig. 1.75 continue forever. Locate the points to which the driven IFS converge. Find the coordinates of these points.

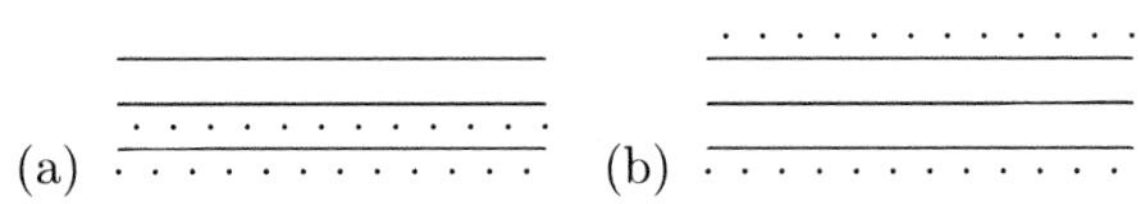

Figure 1.75: Data sequences for Exercise 1.11.2.

Prob 1.11.3 List the empty length-2 addresses in the IFS driven by the data sets of Fig. 1.76.

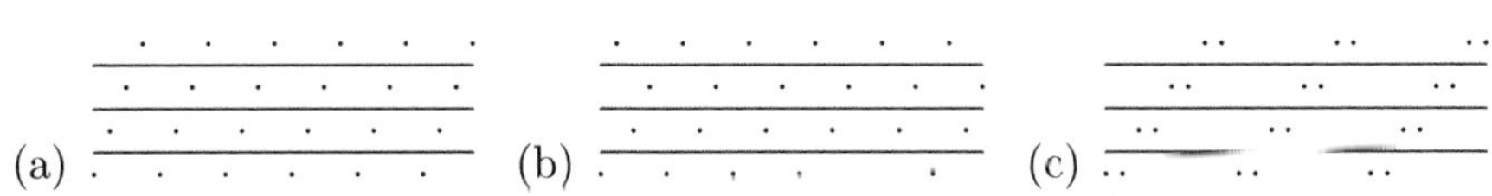

Figure 1.76: Data sequences for Exercise 1.11.3.

Prob 1.11.4 Sketch the driven IFS for the data sets of Fig. 1.77. Explain how you got your results.

(a) 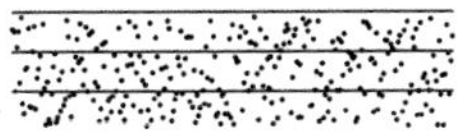 (b) 

Figure 1.77: Data sequences for Exercise 1.11.4.

Prob 1.11.5 Sketch the driven IFS for the data set of Fig. 1.78. Explain how you got your result.

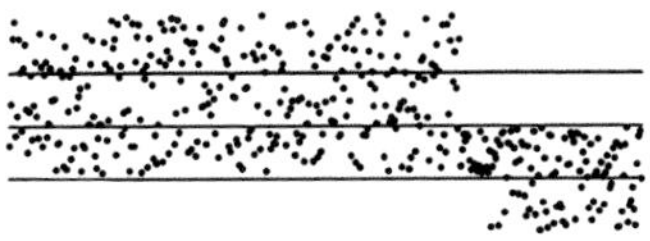

Figure 1.78: Data sequences for Exercise 1.11.5.

Prob 1.11.6 For each of these graphs (plotted in bold, all the functions have domain and range $[0, 1]$) of Fig. 1.79, suppose an IFS is driven by iterating the function graphed, with the bin boundaries as indicated. Determine the empty length-2 addresses in the corresponding driven IFS. Draw the transition graphs. Look carefully at function (a). Describe its driven IFS in more detail.

(a) 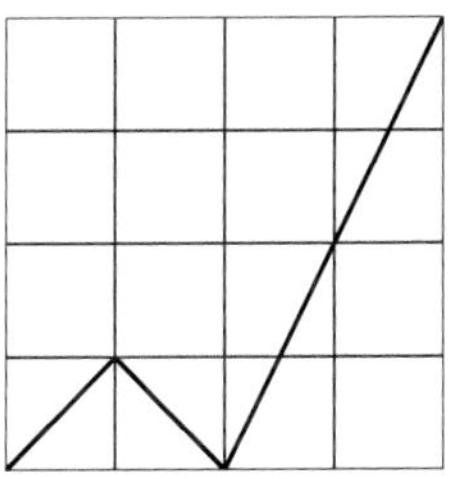 (b) 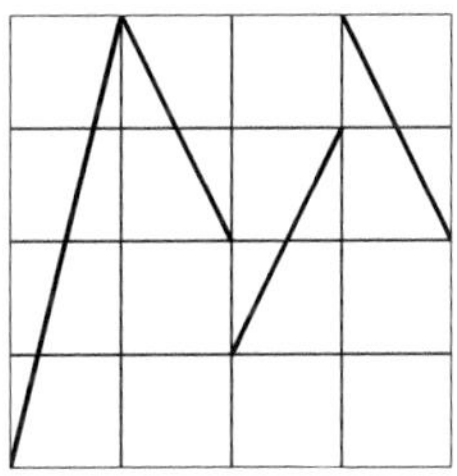 (c) 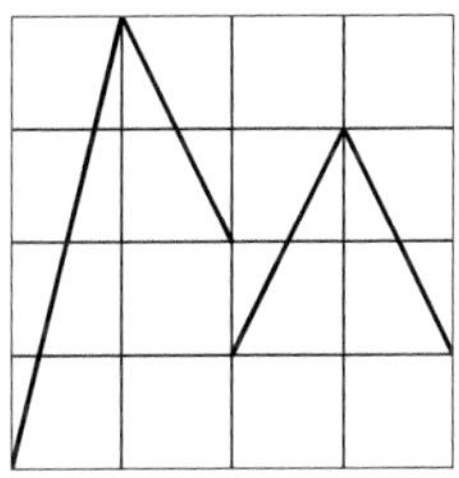

Figure 1.79: Graphs for Exercise 1.11.6.

Prob 1.11.7 (a) In any driven IFS with equal-size bins, why must bin 1 and bin 4 contain at least one point each? Assume the data points do not all have the same value.
(b) For any data set, how would you select bin boundaries so no points land in bin 4? Why must this driven IFS be a subset of a gasket?
(c)For any data set, how would you arrange the bin boundaries so no points land in bin 2 and bin 3? With this choice of bins, how does the driven IFS look?

Prob 1.11.8 Iterating which of the functions of Fig. 1.80 could produce the driven IFS of Fig. 1.81? Give a reason supporting each of your answers.

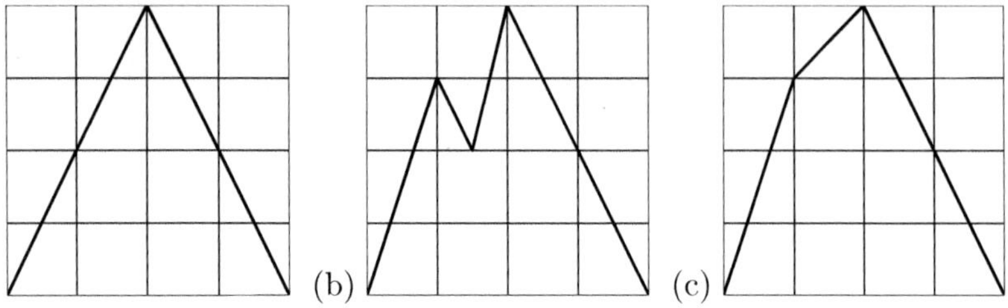

Figure 1.80: Graphs for Exercise 1.11.8.

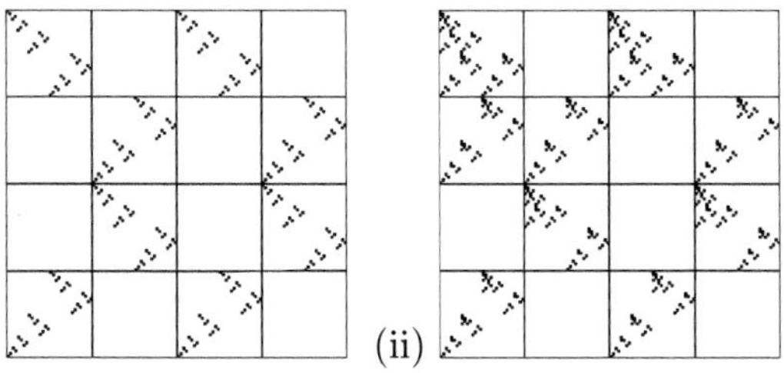

Figure 1.81: Driven IFS for Exercise 1.11.8.

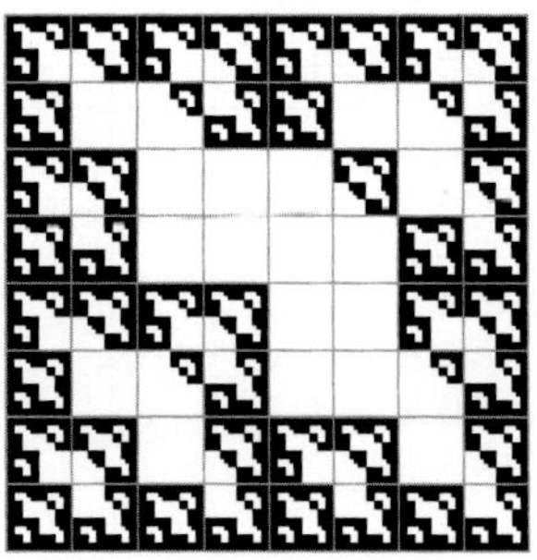

Prob 1.11.9 Is the IFS shown here determined by forbidden pairs, or must forbidden triples be used? So you needn't draw them yourself, a grid of length-3 address squares is superimposed on the fractal. To support your conclusion, list all empty address length-2 squares and all empty address length-3 squares.

Prob 1.11.10 Use driven IFS to analyze the $s = 4.0$ logistic map. Generate 2,000 points and compare the driven IFS plots with equal-size and equal-weight bins. Discuss the similarities and differences. Use the IFS counters to test if $p_{ij} = p_i \cdot p_j$ for the equal-weights statistics.

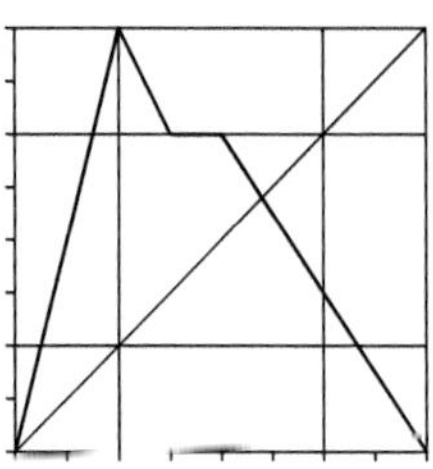

Prob 1.11.11 Pictured here is the graph of a function consisting of straight line segments connecting the points $(0, 0)$, $(1/4, 1)$, $(3/8, 3/4)$, $(1/2, 3/4)$, and $(1, 0)$. The first and third bin boundaries at $b_1 = 1/4$ and $b_3 = 3/4$ are marked on the x- and y-axes.
(a) Locate the middle bin boundary so these bins form a Markov partition. For this problem, define the bins by $[0, b_1)$, $[b_1, b_2)$, $[b_2, b_3)$, and $[b_3, 1]$.
(b) List the allowed transitions and draw the transition graph.

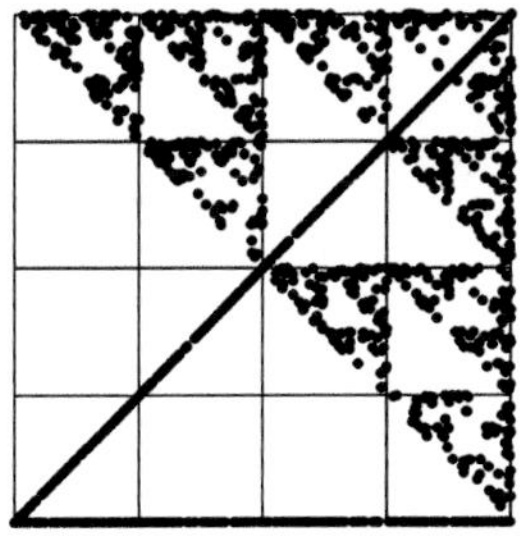

Prob 1.11.12 For the driven IFS pictured here, find a time series (use equal-size bins) that will drive an IFS to produce this image. For reference the bin boundaries are shown along the x-axis and along the y-axis as gray lines.

Can other time series produce similar driven IFS? What criteria must the time series satisfy in order to generate this driven IFS? Remember to include short regions to make clean transitions between the regions that produce the visible features of the driven IFS.

Prob 1.11.13 For this exercise we'll use the function $f(x) = rx(1-x)^2$.
(a) For $r = 27/4$ plot 2,000 points of the driven IFS with equal-size bins.
(b) For $r = 26/4$ plot 2,000 points of the driven IFS with equal-size bins. Record the length-3 addresses that have more than 20 points. What do these tell you about the dynamics of the iterates?

Prob 1.11.14 Find several DNA sequences from GenBank and generate their symbol-driven IFS plots with the software in 1.11 (a) and (b) of the Mathematica or Python codes.

Prob 1.11.15 Find a numerical time series—intervals between successive heartbeats, for example—and generate driven IFS with the software in 1.11 (c) and (d) of the Mathematica or Python codes. If you can, find time series from patients with a disease and from patients without that disease. Do the driven IFS images provide a way to assess the presence of the disease?

Chapter 2

Dimension and Measurement Labs

When fractals began to enter popular culture, a common, though incorrect, description was that fractals are shapes with fractional dimension. This made fractals seem mysterious and therefore interesting, but also made them seem unapproachable. If we understand dimension as the number of independent directions in which we can move, then how can something have dimension 1.585? (It turns out that there is a way to understand movement in that number of directions, though perhaps this isn't obvious. We'll say more about this later.) This apparent unapproachability follows from too narrow a view of the meaning of the word "dimension". A more general interpretation of dimension is the way in which the amount of material in a portion of the object scales with the size of that portion. We'll see that this notion is an extension of the familiar "number of independent directions we can move," but offers many more possibilities to quantify the visual complexity of fractals.

Before we start, we'll clear up a point that can cause some confusion: the difference between the dimension of an object and the dimension of the space in which it lives. For example, even though it is embedded in 3-dimensional space, the surface of a sphere is 2-dimensional. Latitude and longitude are two independent directions in which one can move on a sphere: these two numbers locate every point on a sphere.

Now we'll present a more general notion of dimension. For every $\epsilon > 0$ suppose $N(\epsilon)$ is the minimum number of boxes of side length ϵ needed to cover the shape. The underlying idea is that smaller ϵ boxes detect more detailed structures of the shape. We expect $N(\epsilon)$ to increase as ϵ decreases. If the increase takes a particular form, a power law, $N(\epsilon) \approx K \cdot (1/\epsilon)^d$, then the exponent d is the box-counting dimension of the shape. How to calculate this is the subject of Lab 2.1. Here we estimate the dimensions of natural objects (coastlines), but computations for fractals of known dimension show that convergence can require computations for many levels of substructure.

In Lab 2.2 we explore another variant, mass dimension. We investigate fractal and non-fractal examples and note a visual characteristic of cross-sections that indicates fractality.

In Lab 2.3 we move back to abstract geometry and study a simplification of dimension computation that is based on exact mathematical self-similarity.

In Lab 2.4 we use envelopes to construct a model of a Sierpinski tetraherdon. Inspection of the physical model reveals how very empty it is; calculation gives dimension = 2, helping us to understand that a 2-dimensional object in 3-dimensional space is mostly empty.

In Lab 2.5 we construct a tetrahedral version of the Koch curve. This process yields another shape that is 2-dimensional on the outside, but something different on the inside. That the outside converges to part of a cube already is a surprise; that the inside converges to a shape with a dimension different from that of the outside is another kind of surprise.

One way fractal dimension appears in the physical world is as a measure of roughness. Many, maybe most, surfaces have varying roughness. Some parts are smoother, others are rougher. Multifractal analysis is based on the idea of decomposing a shape into pieces of constant roughness, computing the dimensions of those pieces, and plotting dimension as a function of roughness. For experimental data this has many subtle issues. Rather than fight with this, in Lab 2.7 we present a way to compute multifractal properties for shapes generated by IFS with probabilities assigned to the transformations. While this is not a good model of physical data, it does allow us to develop some intuition about properties of these multifractal plots.

An important step in our understanding of complicated physical processes is the ability to compute something associated with the process. Though some care is needed with this interpretation, dimension as a measure of complexity of fractals is a useful tool.

2.1 Dimension by box-counting

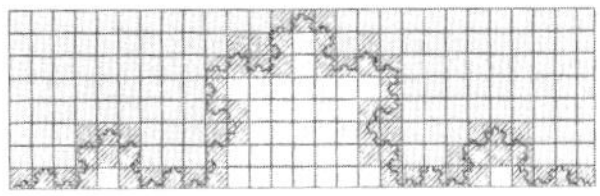

Shapes that appear visually complex—objects we might call rough—may exhibit the same kind of complexity or roughness when viewed under increased magnification. If the scaling range, the range of magnifications over which the same complexity dominates the appearance of the shape, is sufficient (three orders of magnitude for serious studies), then over that range we may measure the complexity with the box-counting dimension.

2.1.1 Purpose

We'll determine the box-counting dimension of natural and mathematical fractals by both the log-log plot and the limit techniques, discover some of the algebraic properties of dimension, understand the importance of a large scaling range in the calculation and validity of the box-counting dimension, and investigate the effect of grid placement on the box-count.

2.1.2 Materials

We'll use pictures of fractals and non-fractals, grid transparencies, a marker pen (experience suggests an erasable, not permanent, marker pen), graph paper, and the linear regression software in Sect. 2.1 of the Mathematica or Python codes.

2.1.3 Background

Natural fractals are not exactly self-similar: they do not contain smaller exact copies of themselves. How can we measure their dimensions without an exact scaling? We know a line segment is 1-dimensional and a filled-in square is 2-dimensional. We'll cover these shapes with cubes to show that we can find the dimension of a shape even if we cover it with boxes of a higher dimension. Fig. 2.1 shows two examples.

Suppose the line segment and square side lengths both are 1. Suppose we denote by $N(r)$ the number of boxes of side length r needed to cover the shape. For the line segment we have

$$N(1/2) = 2 = \left(\frac{1}{1/2}\right)^1, \quad N(1/4) = 4 = \left(\frac{1}{1/4}\right)^1, \quad N(1/8) = 8 = \left(\frac{1}{1/8}\right)^1$$

and for the filled-in square

$$N(1/2) = 4 = \left(\frac{1}{1/2}\right)^2, \quad N(1/4) = 16 = \left(\frac{1}{1/4}\right)^2, \quad N(1/8) = 64 = \left(\frac{1}{1/8}\right)^2$$

Now a line segment is 1-dimensional and the filled-in square is 2-dimensional, so we see that the dimension appears as the exponent. That is, for a d-dimensional

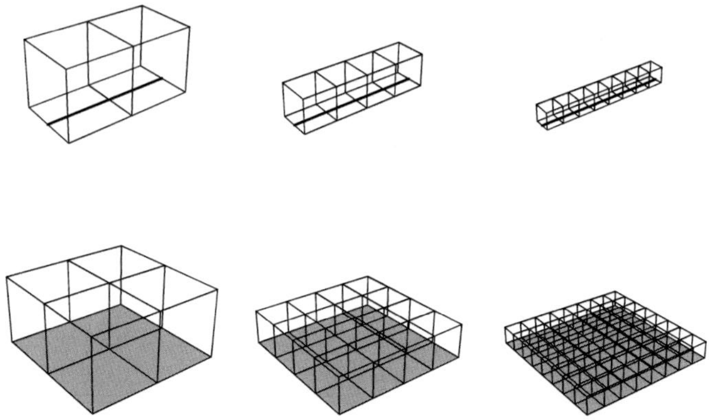

Figure 2.1: Box coverings for a line segment and a filled-in square.

shape, the number of boxes $N(r)$ of side length r scales as

$$N(r) = \left(\frac{1}{r}\right)^d \tag{2.1}$$

Smaller r pick up finer details of the shape. Well, line segments and squares don't have finer details, but rough natural shapes do.

To determine if a shape is a natural fractal, we don't expect a scaling of this simple form. Rather, we'll test if a more general *scaling hypothesis*

$$N(r) \approx k\left(\frac{1}{r}\right)^d \tag{2.2}$$

holds, and if it does, find the range of r values, the *scaling range*, over which Eq. (2.2) holds.

To test the scaling hypothesis and to find d, take the log of both sides of Eq. (2.2).

$$\log(N(r)) \approx d\log(1/r) + \log(k) \tag{2.3}$$

This is just the equation of an approximate straight line with slope d. (Think of $\log(N(r))$ as y, $\log(1/r)$ as x and $\log(k)$ as b.) So for a decreasing sequence of r values, $r_1 > r_2 > \cdots > r_n$, plot the points $(\log(1/r_i), \log(N(r_i)))$. If these points fall close to a straight line, at least for some range of these r_i, then they support the scaling hypothesis, and the slope d is called the *box-counting dimension* of the shape.

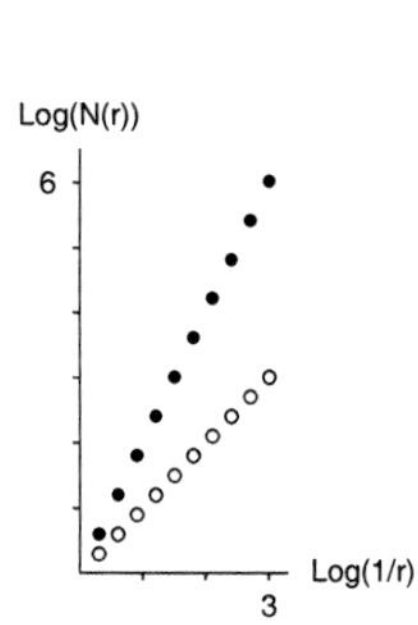

Figure 2.2: Log-log plots for the line segment and square.

In Fig. 2.2 we see the points $(\log(1/r_i), \log(N(r_i)))$ for the boxes to cover the line segment (the circles in the graph) and for the boxes to cover the filled-in square (the dots in the graph), where we've taken $r_i = 1/2^i$ for $i = 1, \ldots, 10$. We find that $d = 1$ for the line segment and $d = 2$ for the filled-in square. These examples suggest that for Euclidean shapes, the box-counting dimension agrees with the Euclidean dimension. This is the *log-log plot approach* to the computation of the box-counting dimension.

These simple examples are particularly clean. In general we can expect some variation (wobble) in the placement of the points: they will not fall exactly on a straight line. For this we'll use a method called linear regression, a way to find the line that best fits a scattering of points. In Appendix B.6 we review this method, derive the formulas for the slope and y-intercept of the line, and the correlation coefficient, which measures how well the line models the data points.

The largest boxes, those that correspond to the left points of the log-log plot, may not pick up the structures revealed by the smaller boxes. Rather than apply linear regression to the entire collection of points, first plot the points and visually identify the scaling range, then apply linear regression to those points.

The log-log plot approach must be applied with some care. In Lab 2.3 we'll see that the exact dimension of the Koch curve is $\log(4)/\log(3) \approx 1.26186$. Lab exercises, admittedly with only a few sizes of grid squares, yielded dimension values ranging from 1.1 to 1.4. A source of this problem is that the number of boxes counted at a specific scale is sensitive to the placement of the grid on the picture. Grid placement has no effect on the limit as $r \to 0$, but it can have a significant impact on the slope of the best-fitting line when only a limited scaling range of r values is used.

For some mathematical fractals, we can find a simple formula for $N(r)$. In this case, we can find the box-counting dimension by the computation of an appropriate limit. Suppose that $N(r)$ exhibits a power law scaling, $N(r) = k(1/r)^d$. Then take the log of both sides and divide by $\log(1/r)$. This gives

$$\frac{\log(N(r))}{\log(1/r)} = d + \frac{\log(k)}{\log(1/r)}$$

As $r \to 0$, $1/r \to \infty$ and $\log(1/r) \to \infty$, though this last divergence is *very* slow. Nevertheless, because k is constant, $\log(k)/\log(1/r) \to 0$ as $r \to \infty$. This gives the *limit approach* to compute the box-counting dimension

$$d = \lim_{r\to 0} \frac{\log(N(r))}{\log(1/r)} \text{ or for a sequence } r_n \to 0,\ d = \lim_{n\to\infty} \frac{\log(N(r_n))}{\log(1/r_n)} \tag{2.4}$$

if the limits exist, of course. The equivalence of these two formulations of the limit is the result of a straightforward calculation. See Appendix A.20 of [17].

For shapes in the plane, we'll use (shaded) squares for boxes. In Fig. 2.3 we see that for the Sierpinski gasket we find $N(1/2) = 3$, $N(1/4) = 9$, $N(1/8) = 27$, and in general $N(1/2^n) = 3^n$, and consequently the box-counting dimension of the gasket is

$$d = \lim_{n\to\infty} \frac{\log(N(1/2^n))}{\log(2^n)} = \lim_{n\to\infty} \frac{\log(3^n)}{\log(2^n)} = \lim_{n\to\infty} \frac{n\log(3))}{n\log(2)} = \frac{\log(3)}{\log(2)} \approx 1.58496$$

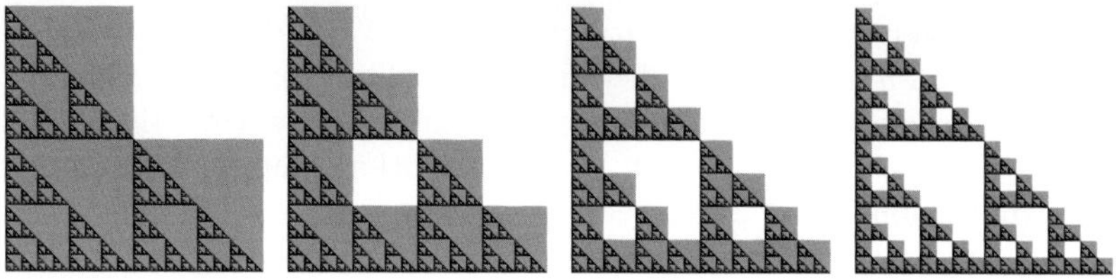

Figure 2.3: Some coverings of the gasket by boxes.

2.1.4 Procedure

These are the basic steps to approximate the box-counting dimension. Copy the five grids of Sect. 2.1.10 onto overhead transparencies and use these to count the boxes.

1. Place the largest grid transparency over the shape so that it is completely covered by the boxes. Secure the transparency to the shape with tape.
2. Count the number of boxes that contain any portion of the curve.
3. Repeat for the next four transparencies. The box sizes are $r_1 = 1, \dots, r_5 = 1/16$.
4. Plot the points $(\log(1/r_i), \log(N(r_i))$.
5. Find the slope of the best fitting line through the points plotted in 4. This can be done in two ways: (a) visually estimate the best-fitting line and (b) use the program in 2.1 of the Mathematica or Python codes to find the slope of the best-fitting line for the points in the scaling range.

2.1.5 Sample A: The dimension of Grenada Lake

We'll use a map of Grenada Lake to illustrate the log-log approach. The range of box sizes is far too small to give a reliable result.

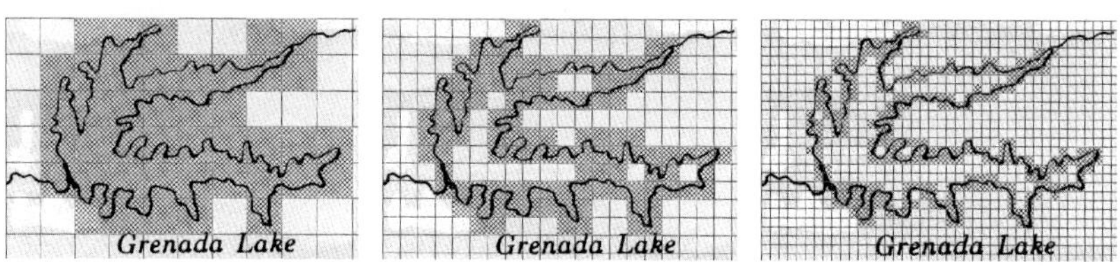

Figure 2.4: The boundary of Grenada lake, covered with boxes of three sizes.

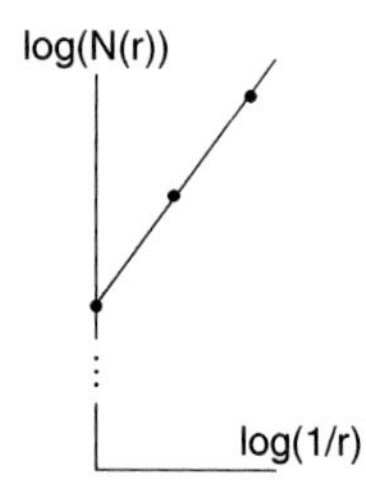

The box sizes are $r_1 = 1$, $r_2 = 0.5$, and $r_3 = 0.25$; the box counts are $N(r_1) = 42$, $N(r_2) = 110$, and $N(r_3) = 264$. The three points $(\log(1/r_i), \log(N(r_i)))$ do appear to lie about along a straight line of slope 1.3. The linear regression program gives $b = 1.62$, $m = 1.33$, and $r^2 = 0.999$, consistent with our visual impression that the three points appear to lie very close to this straight line.

But ... do *not* take this result seriously. While this map is a bit wiggly, in order to make any believable claim of fractality the shape must exhibit structures similar—perhaps approximately or statistically—over at least three orders of magnitude. That is, the smallest boxes have sides 1/1000th as long as the largest boxes.

Google Maps is a powerful tool to explore similar examples. Zoom way out until you can see the entire west coast of Wales, or of Norway, and begin to zoom in. As you do, each little bit of coast will expand to reveal promontories and bays unobserved at the larger scales. Or look at river networks, or edges of glaciers (if there still are glaciers when you read this), or explore dune patterns in the Sahara, or the distribution of islands off the Antarctic Peninsula, or the shore of the Lake of the Woods. The whole world, literally, at your keyboard and touchpad.

2.1.6 Sample B: Measuring in dimensions right and wrong.

Suppose we cover a smooth curve with $N(r)$ boxes of side length r. Then the length of the curve is approximated by $N(r)r$. The approximation is better for smaller r, and we can compute the length by the limit $\lim_{r\to 0} N(r)r$.

Cover a smooth surface with $N(r)$ boxes of side length r. Then the area of the surface is approximated by $N(r)r^2$. The approximation is better for smaller r, and we can compute the area by the limit $\lim_{r\to 0} N(r)r^2$.

Cover a smooth solid region with $N(r)$ boxes of side length r. Then the volume of the region is approximated by $N(r)r^3$. The approximation is better for smaller r, and we can compute the volume by the limit $\lim_{r\to 0} N(r)r^3$.

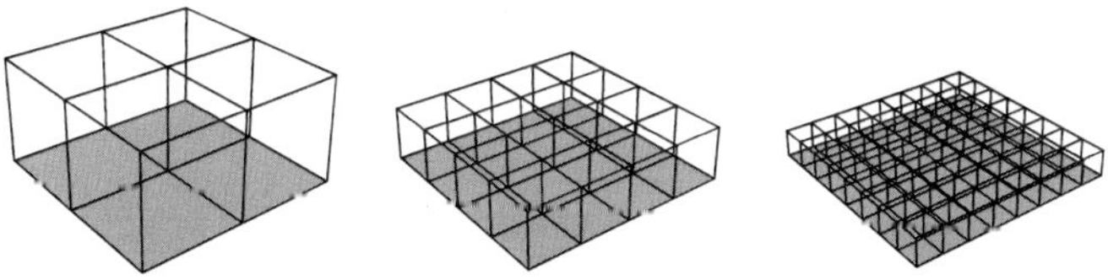

Recall our covering of the filled-in unit square with boxes of side length $r_i = 1/2^i$. We've seen that $N(r_i) = 4^i$.

First, let's try to compute the length of the square. You know this is silly, but be patient and follow the calculation.

$$L = \lim_{r_i\to 0} N(r_i)r_i = \lim_{i\to\infty} 4^i(1/2)^i = \lim_{i\to\infty} (4/2)^i = \infty$$

Second, compute the area of the square.

$$A = \lim_{r_i\to 0} N(r_i)r_i^2 = \lim_{i\to\infty} 4^i((1/2)^i)^2 = \lim_{i\to\infty} (4/4)^i = 1$$

This is the area of the unit square, as expected.

Third, let's try to compute the volume of the square. This looks silly, too, but we are aiming for a particular point.

$$V = \lim_{r_i\to 0} N(r_i)r_i^3 = \lim_{i\to\infty} 4^i((1/2)^i)^3 = \lim_{i\to\infty} (4/8)^i = 0$$

Length is a 1-dimensional measure, area is a 2-dimensional measure, and volume is a 3-dimensional measure. Then these observations are instances of a more general result:

> For a d-dimensional shape, measuring in a dimension $s < d$ gives ∞, while measuring in a dimension $s > d$ gives 0.

By "measuring in a dimension s" we mean summing (diameter)^s for all the collections of sets of diameter $\leq \delta$ that cover the shape, then taking the infimum of that sum for all these covers, and finally taking the limit af these infima as $\delta \to 0$. This remains true for dimensions that are not integers. It is called the s-dimensional Hausdorff measure, and although we sketch a bit of this in Appendix B.9, it's a topic for much more advanced texts. For example, these books by Kenneth Falconer [16, 47, 20] are excellent sources.

Even for simple sets, these calculations are soberingly difficult because they involve not just covers of the set with boxes, but all possible covers of the set.

However, one part of the calculation may be easy: if we can find a sequence of δ-covers $\{U_i\}$ for which $\sum_i \text{diam}(U_i)^s \leq K < \infty$ for all δ, then s is an upper bound for the (Hausdorff) dimension. That is, a single sequence of covers can establish an upper bound for the dimension. To establish a lower bound, we must find some behavior common to *every* possible cover. Falconer's books show just how complicated this can be.

Half of the "measure in the wrong dimension" is perfectly rigorous. If we can find a sequence of coverings by boxes with $\sum \text{diameter}^s \to 0$, then indeed $s \geq$ dimension. That is, "measuring in too high a dimension gives 0" works just fine. On the other hand, application of "measuring in too low a dimension gives ∞" takes much more work than finding a single family of coverings. We'll do this calculation a few times to suggest the general result, but this is just a suggestion, not a proof.

2.1.7 Sample C: The dimension of a product.

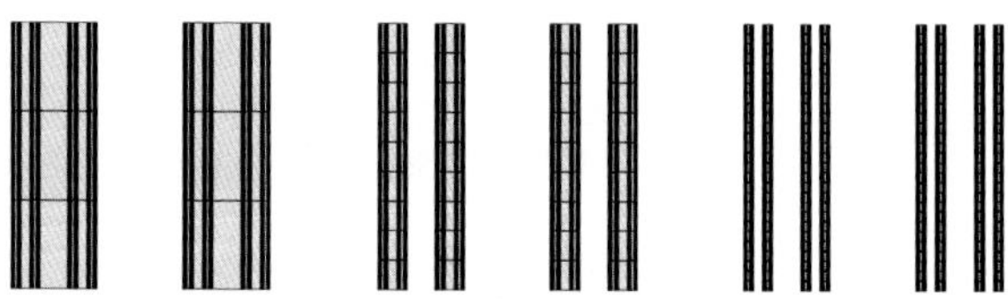

Here we'll compute the box-counting dimension of the product of the a Cantor middle-thirds set, a fractal that consists of 2 copies scaled by 1/3, on the x-axis and the unit interval of the y-axis. Because this Cantor set consists of 2 copies scaled by 1/3, and consequently 4 copies scaled by 1/9, 8 copies scaled by 1/27, and and so on, we see it is covered by 2^n boxes of side length $1/3^n$. From this we see that the dimension of the Cantor middle-thirds set is

$$d = \lim_{r \to 0} \frac{\log(N(r))}{\log(1/r)} = \lim_{n \to \infty} \frac{\log(2^n)}{\log(3^n)} = \frac{\log(2)}{\log(3)}$$

To compute the dimension of the product of this Cantor set and a line segment, we'll cover the product with boxes of side length $1/3$, $1/9 = 1/3^2$, and so on.

From the figure we see

$$N(1/3) = 2 \cdot 3,\ N(1/9) = 4 \cdot 9,\ N(1/27) = 8 \cdot 27,\ \dots, N(1/3^n) = 2^n \cdot 3^n$$

and consequently the box-counting dimension of the product is

$$\begin{aligned} d &= \lim_{r\to 0} \frac{\log(N(r))}{\log(1/r)} = \lim_{n\to\infty} \frac{\log(2^n \cdot 3^n)}{\log(3^n)} = \lim_{n\to\infty} \frac{\log(2^n) + \log(3^n)}{\log(3^n)} \\ &= \lim_{n\to\infty} \frac{n\log(2) + n\log(3)}{n\log(3)} = \frac{\log(2)}{\log(3)} + 1 \end{aligned}$$

where we've used these properties of logarithms: $\log(A \cdot B) = \log(A) + \log(B)$ and $\log(A^B) = B\log(A)$. That is, for this example, the dimension of a product is the sum of the dimensions. That is,

Product rule for dimensions: $d(A \times B) = d(A) + d(B)$.

This result is true for many products. It's an example of what's called the algebra of dimensions. We'll explore some more examples in Exercises 2.1.8, 2.1.9, and 2.1.10, and give a precise formulation of the algebra of dimensions rules in Appendix B.7. More details can be found in Sect. 6.5 of [17].

2.1.8 Conclusion

Box-counting provides a way to estimate the dimension of many natural fractals, but the tedium and possibility of errors in the counting can lead to inaccurate results. In addition, the range of sizes of the boxes is limited by the physicality of the box-counting process. Finally, the result often is sensitive to the placement of the grids. This latter effect is not an issue for mathematical fractals in the box size $\to 0$ limit. For dimensions estimated over a limited range of scales, we must consider the proper counting strategy. Choices include averaging the counts for different grid placements, for each size using the median of the counts for different placements, or using the smallest count of each grid size.

We are sorry to have to mention that many papers published in the 1980s and 1990s reported box-counting dimension calculations based on a range of box sizes that spanned under an order of magnitude. This brings up an important point about when a physical shape should be called fractal. Similar structures, revealed through a power law scaling in the number of boxes as a function of the box size, must dominate the structure for a considerable range of scales. Three orders of magnitude is a decent rule of thumb, and of course a larger range strengthens the claim of fractality. If only a smaller range is observed, fractality is not a significant aspect of the shape. Remember, not every shape is a fractal.

Of course, to gain some insight into the mechanics of box-counting, a much more modest range of box sizes suffices.

2.1.9 Exercises

For Exercises 2.1.1–2.1.4 use the grids of Sect. 2.1.10. Easiest is to copy them onto overhead transparencies.

Prob 2.1.1 Use the log-log plot approach to estimate the box-counting dimension of the portion of the parabola in Fig. 2.6. Align the transparency along the left vertical line.

Prob 2.1.2 Use the log-log plot approach to estimate the box-counting dimension of the portion of the parabola in Fig. 2.6. Align the transparency along the right vertical line. Does your answer differ from that you got in Exercise 2.1.1? If the results differ, this difference is an example of a grid displacement problem.

Prob 2.1.3 Use the log-log plot approach to estimate the box-counting dimension of the Koch curve shown in Fig. 2.7 and described in Exercise 1.8.9.

Prob 2.1.4 Find a coastline map, nice and wiggly, and estimate the box-counting dimension by the log-log plot approach.

Prob 2.1.5 Use the method of Sample 2.1.6 to suggest that the dimension d of the Sierpinski carpet of Fig. 2.5 satisfies $1 < d < 2$. That is, find the 2-dimensional measure and the 1-dimensional measure of the carpet and use these results to show the dimension of the carpet is between 1 and 2.

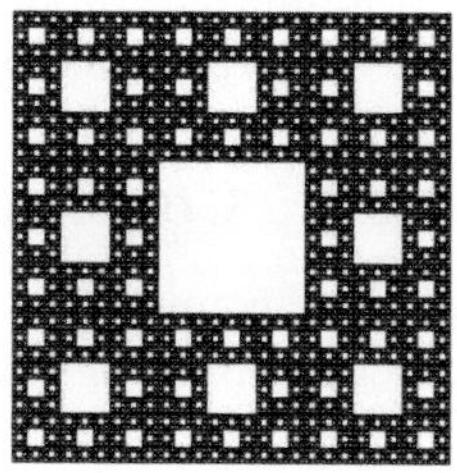

Figure 2.5: Sierpinski carpet

Prob 2.1.6 Use the method of Sample 2.1.6 to suggest that the dimension d of the Sierpinski gasket satisfies $1 < d < 2$.

Prob 2.1.7 Compute the dimension of a "fat Cantor set". For any q, $0 < q < 1$, from the unit interval remove the middle interval of length $q/3$, from the two remaining intervals remove the middle $q/3^2$, and so on. Compute the total length of the removed intervals, deduce that the fat Cantor set has positive 1-dimensional measure, and apply the approach of Sample 2.1.6.

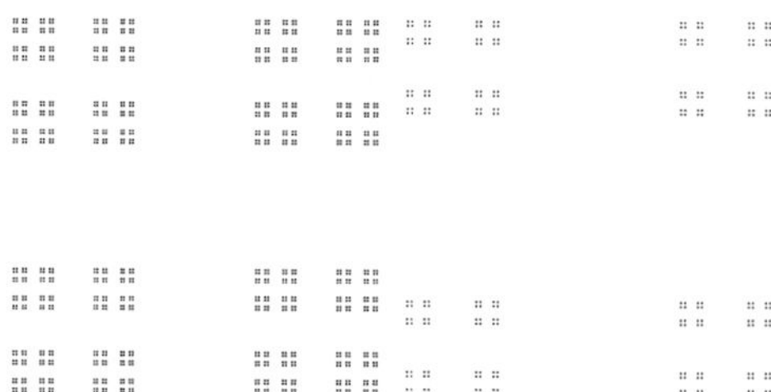

Prob 2.1.8 Compute the box-counting dimension of
(a) the product (left) of two Cantor middle-thirds sets (remove the middle third of the unit line segment, the middle thirds of the remaining segments, and so on), and
(b) the product (right) of two Cantor middle-halves sets (remove the middle half of the unit line segment, the middle halves of the remaining segments, and so on).

Prob 2.1.9 Compute the box-counting dimension of the union of

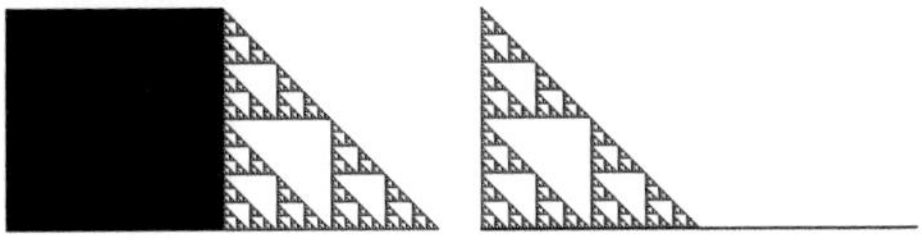

(a) a Sierpinski gasket and a filled-in square, and
(b) a Sierpinski gasket and a line segment.

Prob 2.1.10 Compute the box-counting dimension of the union of

(a) a Cantor middle-thirds set and a line segment on the x-axis, and
(b) a Cantor middle-thirds set on the x-axis and a line segment on the y-axis.

Figures 2.6 and 2.7 are for Exercises 2.1.1, 2.1.2, and 2.1.3.

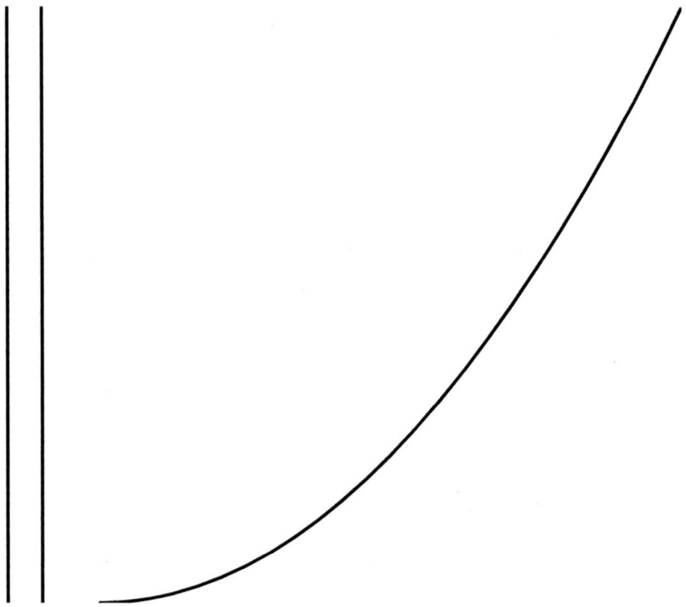

Figure 2.6: A portion of a parabola.

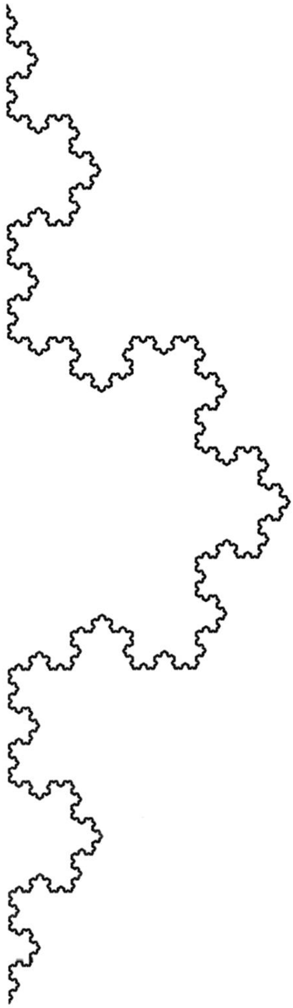

Figure 2.7: The Koch curve.

2.1.10 Grids

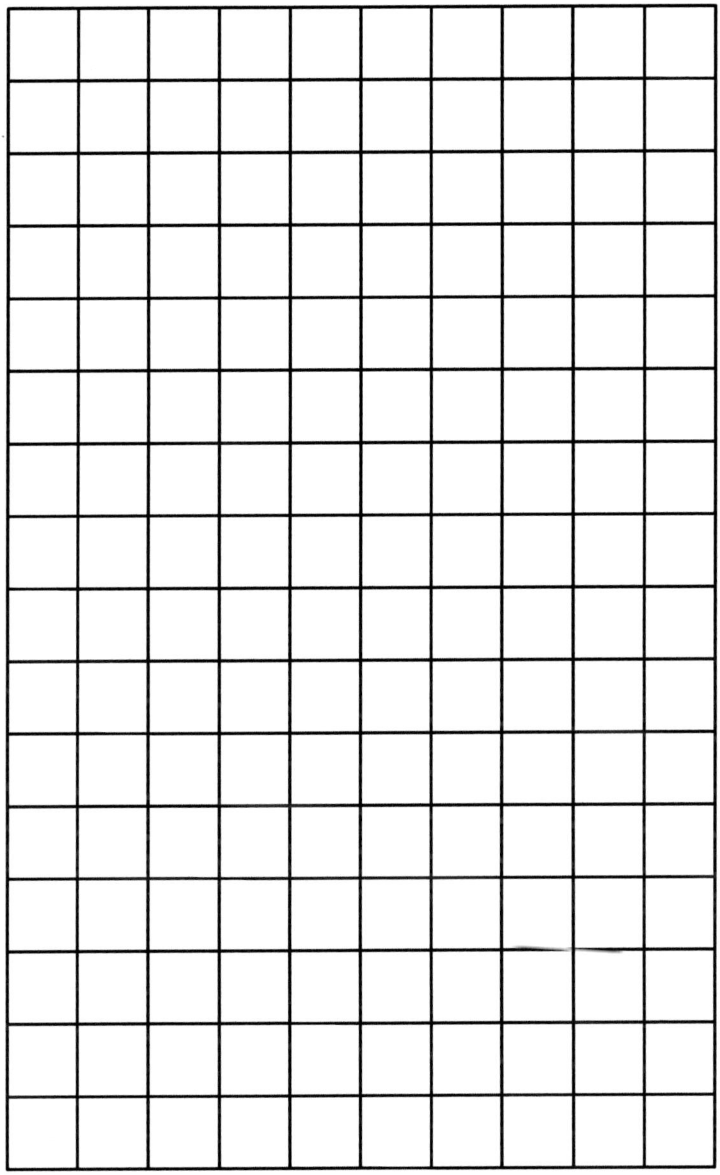

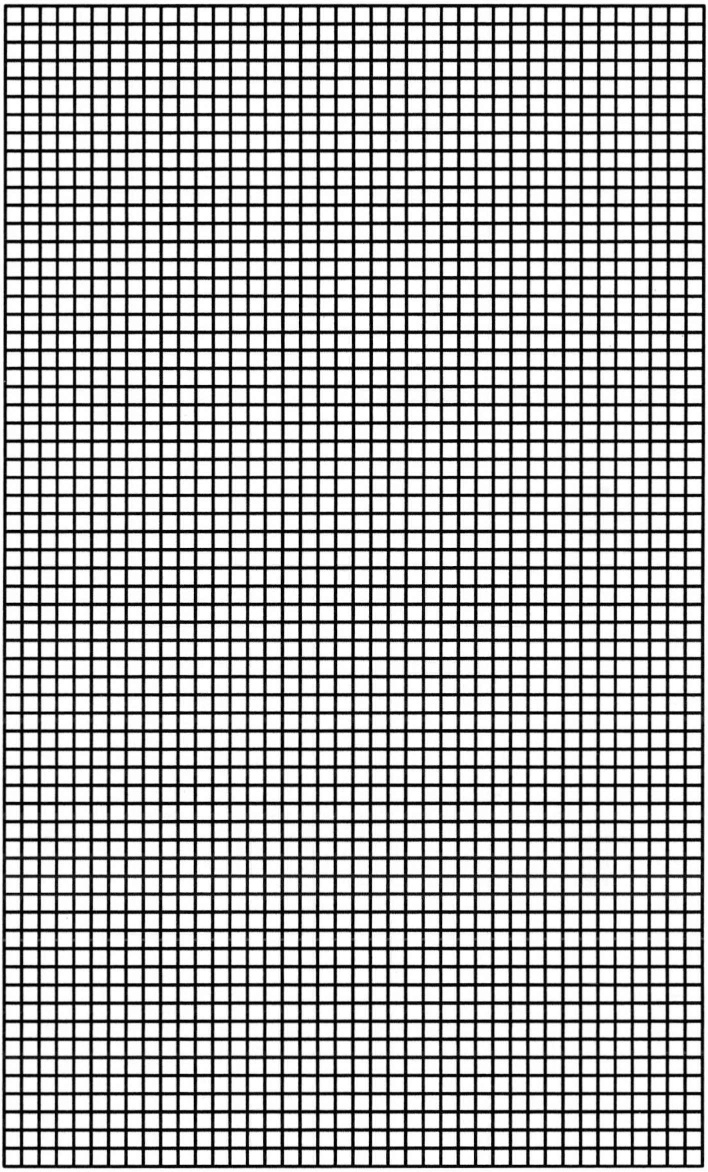

2.2 Paper ball and bean bag dimensions

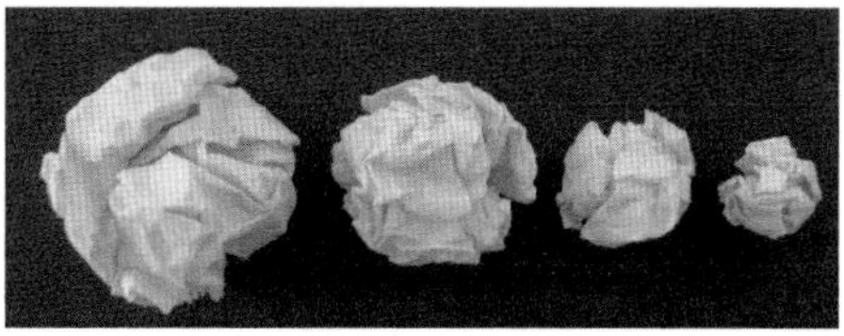

Here we'll investigate the notion that most of us have been making fractals since we were old enough to draw pictures we didn't like. Early on this led some jokers to comment that "Fractals can be found in wastebaskets," though as recounted in Chapter 14 (page 151) of Benoit Mandelbrot's memoir [3], this statement is historically accurate. The paper that inspired Benoit's study of word frequency power laws, and marked the start of his path to fractal geometry, began with a paper retrieved from his uncle Szolem's wastebasket.

2.2.1 Purpose

We'll show that crumpling paper introduces cavities of a range of sizes and this distribution of holes reveals itself trough a power law relation between mass and diameter. We explore another aspect of this in Lab 5.6. As a control we'll study the power law relation between between mass and diameter of roughly spherical bags of beans. The beans pack together tightly, so bean clusters do not exhibit holes over a range of sizes. Here the log-log plot reveals a 3-dimensional object.

2.2.2 Materials

Two sheets each of $8.5'' \times 11''$ paper of three different compositions and weights, a ruler, 64 small dry beans (such as garbanzo or navy beans), clear plastic wrap, a method to approximate the slope of the best-fitting line in the log-log plot (graph paper and ruler, or a graphing calculator, or the program in Sect. 2.1 of the Mathematica or Python codes), a compass, (optional) an accurate weighing device such as a scale or balance capable of weighing small amounts. A sharp serrated knife For Exercise 2.1.6.

2.2.3 Background

First we'll define and show how to calculate the mass dimension. Then we'll add another case to our rules for the algebra of dimensions, how dimension behiaves under intersection.

2.2.3.1 Mass dimension

We'll adopt the working assumption that the mass M of an object has a power law relation to the size (radius, for example) r of the object. Our interpretation of radius can be a bit loose. Half the largest diameter is a good choice; for squares and cubes we'll take a side length because the geometry is a bit clearer. The difference can be absorbed in the proportionality constant. Here's why: suppose s is the side length of a square and r its radius (half its diameter), so $r = s\sqrt{2}/2$. If $M(r) = ar^d$, then $M(s) = a(s\sqrt{2}/2)^d = (a(\sqrt{2}/2)^d))s^d$. So $M(r)$ and $M(s)$ have the same exponent. In the relation

$$M(r) = ar^d \tag{2.5}$$

the exponent d is called the *mass dimension* of the shape. The examples of Fig. 2.8 illustrate the relation between mass dimension and other dimensions. A piece of wire is effectively 1-dimensional if the wire is very very thin. A metal square is effectively 2-dimensional if it is very very thin. A metal cube is 3-dimensional. (Why don't we need a quantifier for the cube?)

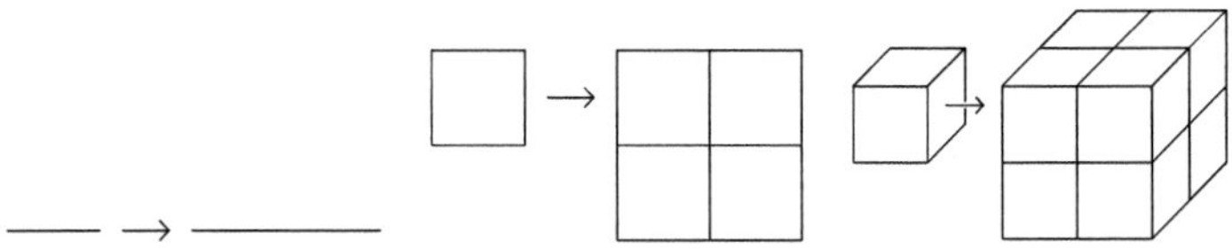

Figure 2.8: The effect of doubling three shapes.

If we keep the density constant, how does the mass of each object change if we double the size of the object?

• For the wire, if we double the size (length s) of the wire, the mass doubles. That is, $M(s) = k_1 s^1$ because then

$$M(2s) = k_1(2s)^1 = 2k_1 s^1 = 2M(s).$$

For a 1-dimensional shape, the power law exponent is 1.

• For the square, if we double the side length s of the square, the mass is multiplied by $4 = 2^2$. That is, $M(s) = k_2 s^2$ because then

$$M(2s) = k_2(2s)^2 = 2^2 k_2 s^2 = 2^2 M(s)$$

For a 2-dimensional shape, the power law exponent is 2.

• For the cube, if we double the side length s of the cube, the mass is multiplied by $8 = 2^3$. That is, $M(s) = k_3 s^3$ because then

$$M(2s) = k_3(2s)^3 = 2^3 k_3 s^2 = 2^3 M(s)$$

For a 3-dimensional shape, the power law exponent is 3.

If the power law (2.5), with r the distance from a point near the center of the shape, holds over a significant range of r values, then the exponent d is the mass dimension of the shape. As with the box-counting dimension, this range of r values where (2.5) holds is called the scaling range. An important difference between the box-counting dimension d_b and the mass dimension d_m is that while d_b is revealed through the detection of similar fine structure patterns for *ever smaller* r values, for d_m we see similar patterns with *ever larger* pieces of the whole shape.

Note that because all physical objects are of finite extent, for large enough r the entire object is contained in a sphere of radius r and $M(r)$ remains constant with further increase of r.

For physical objects we can estimate d_m, and verify the power law hypothesis (2.5), by a log-log plot. Take the log of both sides of (2.5) to obtain

$$\log(M(r)) = d\log(r) + \log(a) \tag{2.6}$$

When we measure $M(r)$ for a set of r values, if the points $(\log(r), \log(M(r)))$ are about on a straight line for some range of r values, that range is the scaling range and there the power law relation holds.

To emphasize again the difference between the *ever smaller* r values for d_b and the *ever larger* values for d_m, note that in Eq. (2.3) the horizontal axis variable is $\log(1/r)$, while in Eq. (2.6) the horizontal axis variable is $\log(r)$.

Example 2.2.1 Suppose for $r = 1, \dots, 20$ we measure these values of $M(r)$: 1.018, 1.366, 1.541, 1.810, 1.978, 2.045, 2.190, 2.202, 2.290, 2.464, 2.643, 2.652, 2.770, 2.950, 3.046, 3.059, 3.150, 3.191, 3.268, and 3.318. Plotting $(\log(r), \log(M(r)))$ for these values gives this graph. We see all the points fall reasonably along the best-fit line. Linear regression gives a slope of 0.393 with correlation coefficient $r^2 = 0.994$ so we can say the data support the claim this object has mass dimension $d_m \approx 0.4$. □

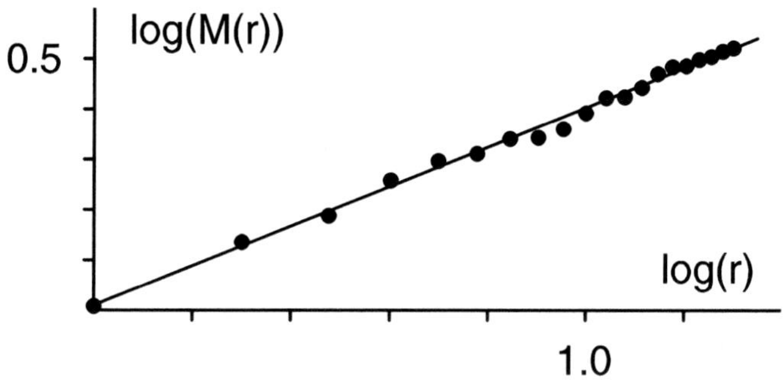

Example 2.2.2 Suppose for $r = 1, \dots, 20$ we measure these values of $M(r)$: 1.018, 1.020, 1.035, 1.051, 1.978, 2.045, 2.055, 2.070, 2.095, 2.905, 3.103, 3.510, 4.321, 5.919, 5.993, 6.435, 6.501, 6.609, 6.621, and 6.687. Plotting $(\log(r), \log(M(r)))$ for these values gives this graph. Do these points look well-modeled by a straight line? No, not really. Yet when we use this data in the linear regression program, it gives $r^2 = 0.854$, which is not particularly low, and draws the line with m and b determined by the data. This is yet another example of the importance of looking at graphs. □

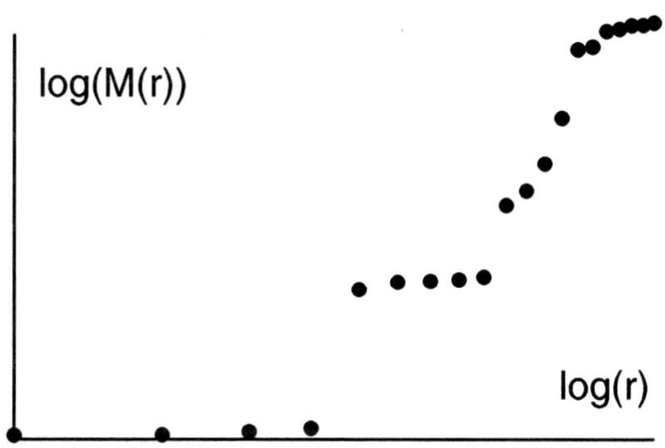

Finally, though this shouldn't need to be mentioned, we must recall that no deduction at all can be drawn from a graph having only two points. For example, suppose the underlying data are the circles in the graph on the right, but only the two filled-in dots have been measured. Given these two points, certainly we can draw a line passing through them. Does it follow that all subsequently measured data points will lie near this line? Not at all.

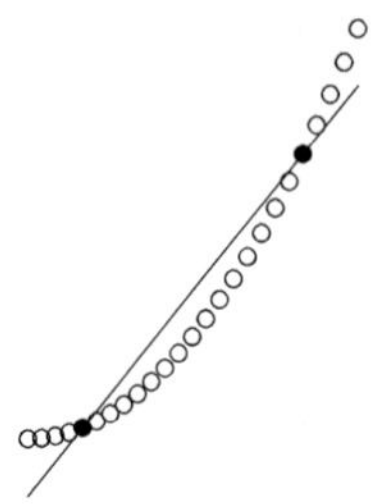

In general, the larger the range of r-values over which a power law scaling is observed, the better the claim that the underling process is fractal. And the more data, the better.

2.2.3.2 Dimensions and intersections

In Sample 2.1.7 and in Exercises 2.1.8–2.1.10 we illustrated two rules of the algebra of dimensions, the *product rule* $d(A \times B) = d(A) + d(B)$ and the *union rule* $d(A \cup B) = \max\{d(A), d(B)\}$. A consequence of the union rule is the *monotonicity rule* $A \subseteq B$ implies $d(A) \leq d(B)$.

To get a view of the distribution of the sizes of cavities in crumpled paper, we'll bisect a crumpled paper ball and look at the surface along the cut. Geometrically this surface is the intersection of the crumpled paper ball with a plane. So we'd like to know how dimension behaves under intersections. This, it turns out, is a more subtle rule.

We'll begin with intersections in Euclidean geometry. Typically two lines A and B in the plane intersect in a point. Two other intersections are possible: parallel lines that do not coincide do not intersect at all; parallel lines that coincide intersect in a line. These are the possibilities: no intersection, intersect in a single point, or intersect in a line. For no intersection, the lines must be parallel, so of all the possible slopes of B, only one will work, the slope of A. If A and B are placed randomly, parallel is very unlikely. Coincident is more unlikely still: the lines must have the same slope, and also the same y-intercept. Almost every placement of the lines result in a single point intersection.

Now note $d(\mathbb{R}^2) = n = 2$, $d(A) = d(B) = 1$, and $d(\text{point}) = 0$, so for typical placements of two lines in the plane, we have

$$d(A \cap B) = 0 = 1 + 1 - 2 = d(A) + d(B) - n$$

Two planes in space (dimension $n = 3$) typically intersect in a line, and again we see

$$d(A \cap B) = 1 = 2 + 2 - 3 = d(A) + d(B) - n$$

What about two lines in space? Typically they don't intersect. The formula that we've seen, $d(A \cap B) = d(A) + d(B) - n$, gives

$$d(A \cap B) = d(A) + d(B) - n = 1 + 1 - 3 = -1$$

The usual interpretation of a negative dimension for the intersection is that the spaces don't intersect, though more nuanced interpretations have been proposed in [49], and the references mentioned there.

While the proof is quite complicated (Chapter 8 of [16], for example), the intersection formula holds for fractals as well as Euclidean objects. That is, *Intersection rule* Suppose A and B have dimensions $d(A)$ and $d(B)$ and are contained in n-dimensional space. Then for almost all placements of A and B in this space,

$$d(A \cap B) = d(A) + d(B) - n$$

with the understanding that a negative dimension signals no intersection of A and B.

Here's an example. Suppose A and B both are Cantor sets of dimension $d < 1/2$. In Lab 2.3 we'll see that any Cantor set formed by a process that begins with the removal of a middle interval of length $> 1/2$ has dimension $d < 1/2$. Then for almost all placements of A and B in the line,

$$d(A \cap B) = d(A) + d(B) - 1 < 0$$

That is, if you randomly place these Cantor sets (near one another, not miles apart) in a line, for almost all placements the Cantor set A falls entirely in the gaps of B.

A list of rules for the algebra of dimensions, along with precise statements of which relations hold always and which hold in most cases, is given in Appendix B.7.

2.2.4 Procedure

Crumpled paper balls. Here we'll approximate the mass dimension of crumpled paper balls. In the construction of the paper balls, try to crumple the different sizes of pieces of paper in about the same way. In particular, *do not* fold the larger pieces of paper before you crumple them.

- Select two sheets of paper of the same size ($8.5'' \times 11''$ inches) and same type (thickness and stiffness).
- Crumple one sheet of paper into an approximately spherical ball. Take the mass of this ball to be 8 because the smallest sheet of paper we'll crumple is 1/8 of a whole sheet.

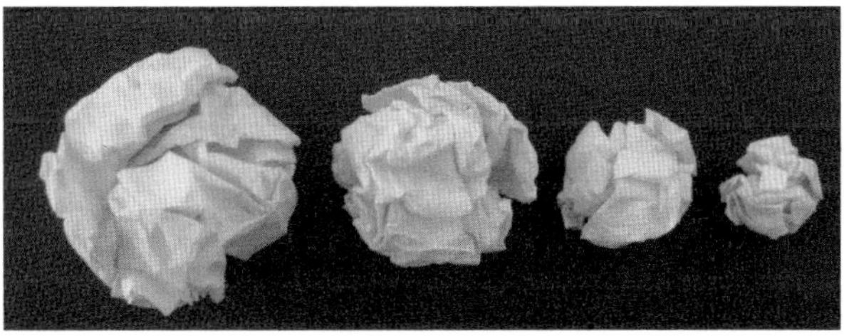

Figure 2.9: Four balls of crumpled stiff paper.

- Cut the other sheet of paper in half. Always cut the paper perpendicular to its longer side. Crumple one piece into a ball. Because we take the largest piece of paper to have mass 8, this piece will have mass 4.
- Cut the remaining piece in half. Crumple one piece into a ball. This piece will have mass 2.
- Cut the remaining piece in half. Crumple one piece into a ball. This piece will have mass 1.

• Measure the diameters of these paper balls, using a compass as a caliper. For balls that are not very close to spherical, the average of several diameters can be used. Record the masses and diameters.

• Plot the points (log(radius), log(mass)) for these paper balls. If the points appear to lie on a line, the slope of the best-fitting line is an estimate of the mass dimension.

Bean bags. A similar measurement can be done using dry beans (such as garbanzos), or any other hard approximately spherical objects, marbles or ball bearings, for example. Best is to weigh each cluster on a scale, but even without a laboratory scale a scale, we can estimate the mass by counting the number of beans or marbles or ball bearings. Denote by N the number of beans in the smallest cluster; take the mass of that cluster to be 1. A cluster of M beans is taken to have mass M/N. Here are specific steps with $N = 8$.

• Count out 64 of the beans into a piece of the clear plastic wrap. Tighten the wrap around the beans to approximate a sphere. Because the smallest bean cluster will consist of 8 beans, we take the mass of this cluster to be 8.

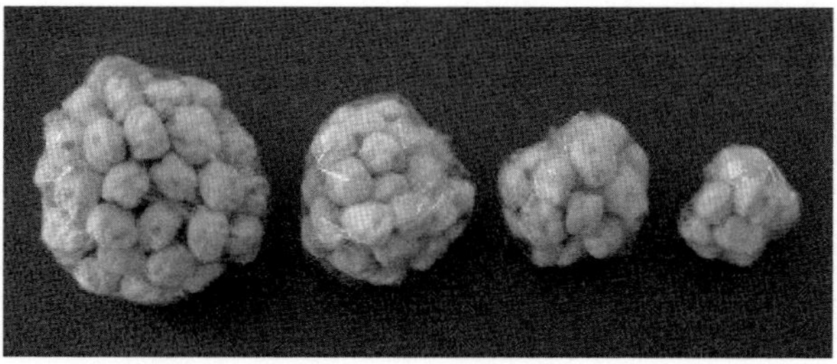

Figure 2.10: Four bags of beans.

• Count out 32 of the beans into a piece of the clear plastic wrap. Tighten the wrap around the beans to approximate a sphere. Take the mass to be 4.

• Count out 16 of the beans into a piece of the clear plastic wrap. Tighten the wrap around the beans to approximate a sphere. Take the mass to be 2.

• Count out 8 of the beans into a piece of the clear plastic wrap. Tighten the wrap around the beans to approximate a sphere. Take the mass to be 1.

• Measure the diameters of these paper balls, using a compass as a caliper. For balls that are not very close to spherical, the average of several diameters can be used. Record the masses and diameters.

• Plot the points (log(radius), log(mass)) for these bean bags. If the points appear to lie on a line, the slope of the best-fitting line estimates the mass dimension.

Alternately, we could take the mass of each cluster to be the number of beans in that cluster. We used this "fraction of the largest cluster" approach because it is similar to the procedure for crumpled paper balls.

2.2.5 Sample A. Crumpled paper balls.

We'll do two experiments. First, make four paper balls from two sheets of stiff copy paper. Second, make four paper balls from two sheets of (less stiff) composition paper. In these graphs, and the graph of Sample 2.2.6, r stands for the diameter of the shape.

Stiff copy paper

Here the measured values (diameters are in mm) are $(r, M(r)) = (22, 1)$, $(29, 2)$, $(40, 4)$, and $(53, 8)$. Note that the diameter is the first coordinate and the mass is the second. In our workshops several groups reversed the coordinates and obtained dimensions less than 0.5, clearly at odds with the fact that something made of a crumpled 2-dimensional sheet of paper must have dimension at least 2. In general, if you get a nonsensical result, look for simple mistakes.

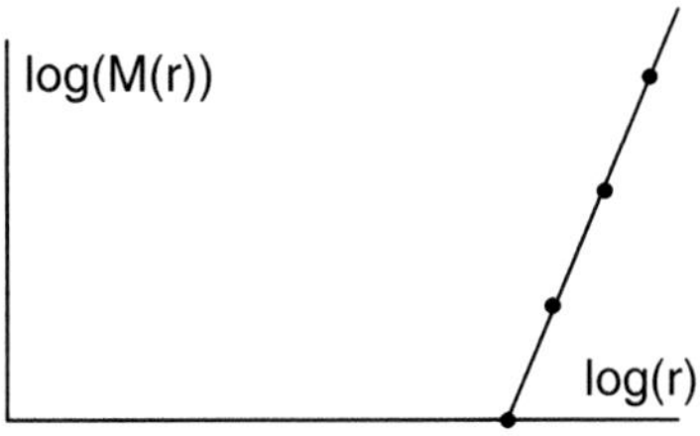

With these measurements, the points do appear to fall close to a straight line. Linear regression gives $m = 2.340$ and $b = -3.136$. (Do you see why the y-intercept is negative?) Moreover $r^2 = 0.999$, suggesting that the data points fit the line very well. With the usual warning that the range of measurements is too small to take any deduction very seriously, the data do support the claim that crumpled paper balls have dimension about 2.3.

Composition paper

With crumpled composition paper we measure $(r, M(r)) = (18, 1)$, $(22, 2)$, $(30, 4)$, and $(37, 8)$. From the graph we see that the points fall close to a straight line. Linear regression gives $m = 2.784$ and $b = -3.476$. The correlation is $r^2 = 0.993$ so again we observe that data points are well-modeled by this straight line.

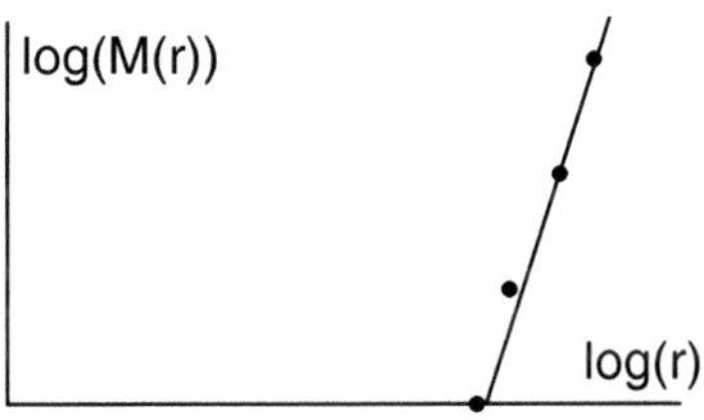

We see that the composition paper crumples into a ball with higher dimension than does the stiffer copy paper. Why should this be? One hypothesis is that less stiff paper crumples more tightly, more compactly, so crumpling forms smaller gaps, smaller holes. You can pursue this line of thought a bit further in Exercises 2.2.1–2.2.4.

2.2.6 Sample B

For these clusters of beans we measure $(r, M(r)) = (25, 1)$, $(33, 2)$, $(40, 4)$, and $(49, 8)$. From the graph we see that the points fall close to a straight line. Linear regression gives $m = 3.111$ and $b = -4.377$. The correlation is $r^2 = 0.992$, reinforcing the visual closeness of the fit.

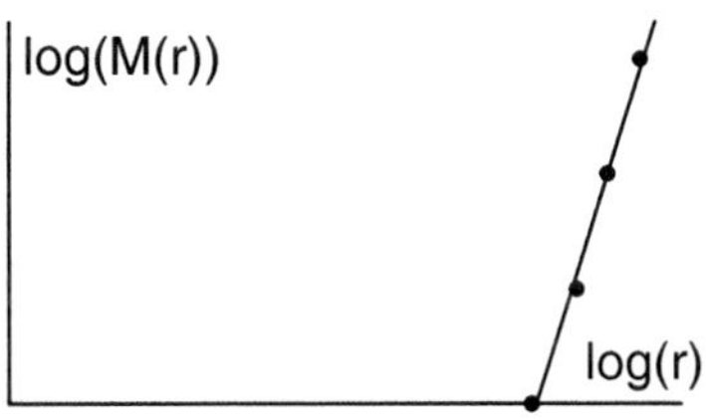

However, by the monotonicity rule for dimensions, or for that matter just by common sense, something (a cluster of beans for instance) that fits into 3-dimensional space cannot have a dimension larger than 3. The estimated value of 3.1 for the dimension of bean clusters makes crystal clear that the range of measurements is too short for any serious deductions. These experiments are meant to explore methodology, and to suggest a few general directions, not to tell us anything about real clusters of beans or crumpled paper.

2.2.7 Conclusion

The logs of the data values from each experiment should fall approximately on a straight line whose slope is the mass dimension of the object. That the data points fall near a straight line in the log-log plot vindicates the power law hypothesis $M(r) = k \cdot r^d$.

For the paper balls, the slope of the line, and therefore the dimension, should be between 2 and 3. This value depends on the stiffness of the paper and whether it was reprocessed (crumpled, flattened, and recrumpled), among other factors.

For bean bags, the slope should be close to 3. Why don't the spaces between the beans lower the dimension of the bean bags? Because the spaces are all about the same size. Unlike crumpled paper, bean bags don't have a hierarchy of gap sizes, a necessary component of fractal constructions. The bean bag spaces lower the measured average density (mass/volume), but not the dimension.

2.2.8 Exercises

Prob 2.2.1 Do the paper ball experiment with a light weight paper. Repeat the experiment three times. Do you find a significant change mass dimension estimate?

Prob 2.2.2 Do the paper ball experiment with a medium weight paper. Repeat the experiment three times. Do you find a significant change mass dimension estimate?

Prob 2.2.3 Do the paper ball experiment with a heavy weight paper. Repeat the experiment three times. Do you find a significant change mass dimension estimate?

Prob 2.2.4 Do the paper ball experiment using a medium weight paper, but crumple and recrumple the paper balls several times before recording the diameters. Do multiple recrumplings change the estimate?

Prob 2.2.5 Compare the dimensions obtained in Exercises 2.2.1–2.2.4. Explain the differences. Is it possible that a crumpled paper ball has dimension greater than 3 or less than 2? Explain.

Prob 2.2.6 With a very sharp serrated knife bisect the largest paper ball. Take care to minimize the compression or distortion of the paper ball while slicing it. What do the revealed edges illustrate? Estimate the dimension of the revealed edges. For example, put ink or paint on a flat surface, rest the sliced surface of the paper ball on this fluid for a moment, then rest it on a blank piece of paper for a moment and remove the sliced paper ball. When the paint has tried, estimate the dimension with the box-counting grids. How is the dimension of the revealed edges related to that of the paper ball?

Prob 2.2.7 Do the bean bag experiment by counting beans. Do the experiment three times. Do you obtain significantly different values for the mass dimension? Save these bean bags for Exercise 2.2.8.

Prob 2.2.8 Do the bean bag experiment by weighing the bean bags. Use the bean bags of Exercise 2.2.7. Compare the dimension obtained by bean weighing with that obtained by bean counting.

Prob 2.2.9 Do the paper ball experiment with plastic wrap. Do you expect the mass dimension to be higher or lower than those of crumpled paper?

Prob 2.2.10 Do the bean bag experiment with objects that are far from spherical. Elbow macaroni for example. Try objects of several different shapes. Does the shape have an effect on the dimension? Does the shape have an effect on the range of cluster sizes needed to get a reasonable estimate of the dimension?

2.3 Calculating similarity dimension

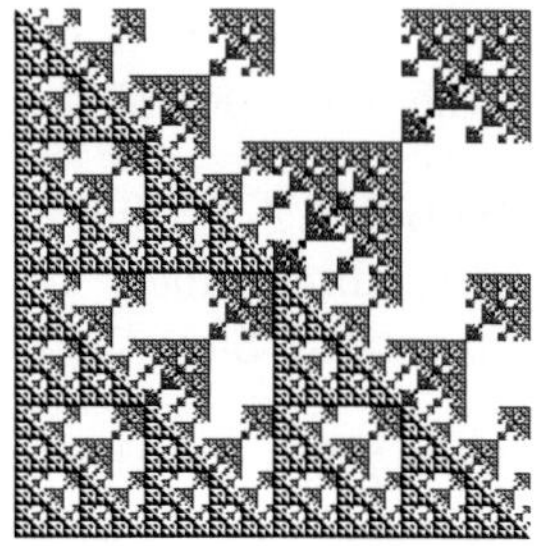

When we compute the box-counting dimension by the limit approach, we might notice a pattern. Recall the Cantor middle-thirds set consists of 2 copies scaled by 1/3, 4 copies scaled by 1/9, ..., 2^n copies scaled by $1/3^n$, and so for this Cantor set

$$d_b = \lim_{n\to\infty} \frac{\log(2^n)}{\log(3^n)} = \frac{\log(2)}{\log(3)}$$

The pattern revealed in the first step, 2 pieces scaled by 1/3, is all we need to compute the dimension. Self-similarity implies this pattern is repeated across a cascade of ever smaller scales; the first step in the construction of the Cantor set encodes all the subsequent substructure. Is this always enough to compute the dimension? In this lab we'll explore when this works, find a simple formula, the Moran equation, to compute the dimension of self-similar fractals. Because this dimension is based on self-similarity, we call it the similarity dimension, d_s. We'll see how to extend the Moran equation to fractals generated by IFS with memory, and to some random fractals.

The Moran equation extends to far more general settings, but these require much more sophisticated math.

2.3.1 Purpose

Here we'll learn a quick way to calculate the dimension of self-similar sets and extend this to fractals generated by IFS with memory and to some random fractals.

2.3.2 Materials

A mechanism for finding the roots of polynomials and the eigenvalues of 4×4 matrices. Some graphing calculators can find roots of polynomials; 2.3 (a) and (b) of the Mathematica or Python codes find roots of polynomials and eigenvalues of matrices.

2.3.3 Background

For a self-similar shape that consists of N copies, each scaled by a factor of r, the *similarity dimension* is defined by

$$d_s = \frac{\log(N)}{\log(1/r)} \qquad (2.7)$$

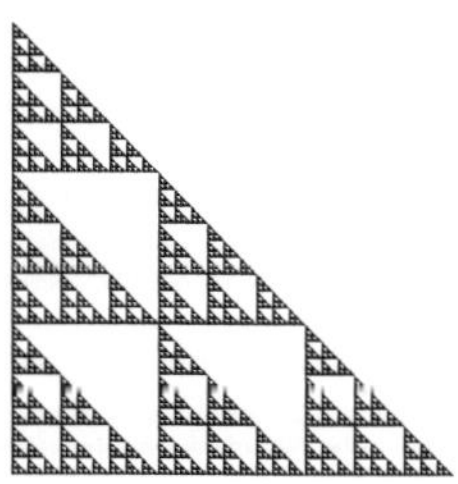

The first fractal on the right is composed of $N = 6$ pieces each scaled by a factor of $r = 1/3$, and so has dimension $d_s = \log(6)/\log(3) \approx 1.63093$.

The second fractal is composed of $N = 3$ pieces each scaled by a factor of $r = 1/2$, and so has dimension $d_s = \log(3)/\log(2) \approx 1.58496$. The familiar Sierpinski gasket also has $N = 3$ and $r = 1/2$, so also has $d_s = \log(3)/\log(2)$. The second fractal here does not look much like the gasket, so we see that dimension does not characterize fractals. That a single number does not uniquely determine a fractal is hardly a surprise.

For fractals that consist of pieces all scaled by the same factor, Eq. (2.7) gives the dimension.

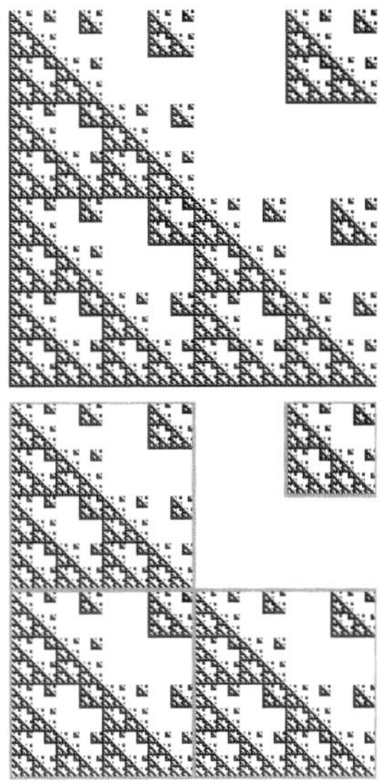

This fractal is more complicated. The squares outlined in the lower figure show it consists to 3 copies scaled by $r = 1/2$ and 1 copy scaled by $r = 1/4$. How can we apply Eq. (2.7) with two different values of r? As it is written, we can't. So we need to work with the equation a bit. Multiply both sides by $\log(1/r)$ and follow the algebra:

$$\begin{aligned} d\log(1/r) &= \log(N) \\ \log((1/r)^d) &= \log(N) \\ (1/r)^d &= N \\ 1 &= Nr^d \\ 1 &= r^d + \cdots + r^d \end{aligned}$$

Now this looks promising, because the fractal has N pieces and this relation has N separate copies of r^d. If the pieces of the fractal are scaled by $r_1, \dots, r_N$, then the dimension d is given by the *Moran equation*

$$r_1^d + \cdots + r_N^d = 1 \tag{2.8}$$

For this fractal the Moran equation is

$$\left(\frac{1}{2}\right)^d + \left(\frac{1}{2}\right)^d + \left(\frac{1}{2}\right)^d + \left(\frac{1}{4}\right)^d = 1$$

This doesn't look too inviting, but we can rewrite the last term in a way that will prove useful

$$\left(\frac{1}{4}\right)^d = \left(\frac{1}{2^2}\right)^d = \left(\left(\frac{1}{2}\right)^2\right)^d = \left(\left(\frac{1}{2}\right)^d\right)^2$$

Now take $x = (1/2)^d$. Then the Moran equation for this fractal is

$$x + x + x + x^2 = 1$$

This is a quadratic equation with solutions $x = (-3 \pm \sqrt{13})/2$. Because $x = (1/2)^d$, x must be positive. Then solve $(-3 + \sqrt{13})/2 = (1/2)^d$ for d by taking the log of both sides and dividing both sides by $\log(1/2)$:

$$d = \frac{\log((-3 + \sqrt{13})/2)}{\log(1/2)} \approx 1.72368$$

Will this always work? Why should we expect that Eq. (2.8) has any solution at all, or if it has a solution, why should it have only one? The math that underlies many of the properties of dimensions is fairly sophisticated and relies on measure theory, a part of mathematical analysis often encountered in the first year of graduate school. But in this case, the proof, in Appendix B.8.2, requires just a tiny bit of first-semester calculus.

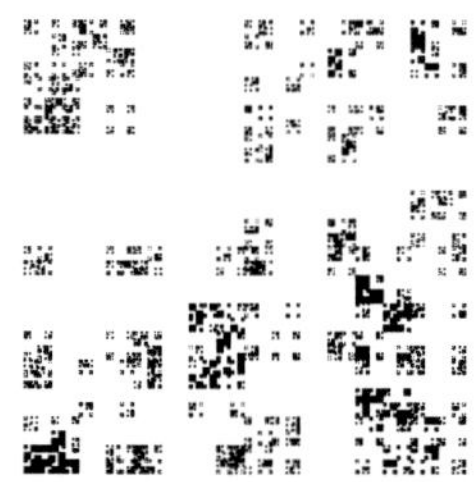

We can extend the Moran equation to find the dimensions of certain types of random fractals. What are random fractals? One way to construct random fractals is this: rather than a fixed scaling factor for each transformation, suppose that with each iteration the scaling factors of each transformation can take on one of several values, selected with prescribed probabilities. For example, here we see a randomized Cantor set product, similar to the product of Exercise 2.1.8, but here the scaling factor of each piece is 1/2 or 1/4, selected randomly and independently. In this setting, the Moran equation becomes the *random Moran equation*

$$\mathbb{E}(r_1^d) + \cdots + \mathbb{E}(r_N^d) = 1 \tag{2.9}$$

where $\mathbb{E}(x)$ is the expected value of x. For instance, if $x = 1/2$ with probability 1/3 and $x = 1/4$ with probability 2/3, then

$$\mathbb{E}(x) = (1/2)(1/3) + (1/4)(2/3) = 1/3$$

In general, if x can take on values $x_1, \dots, x_m$ and $P(x = x_i)$ is the probability that $x = x_i$, then the expected value is

$$\mathbb{E}(x) = x_1 P(x = x_1) + \cdots + x_m P(x = x_m)$$

The expected value is another name for the average.

Because these constructions involve randomness, very unlikely combinations can occur. For example, if at every iterate, all $r_i = 1/4$, then $N = 4$ and $r = 1/4$ so $d_s = 1$ by Eq. (2.7). On the other hand, if all $r_i = 1/2$, then $d_s = 2$. Both these are very unlikely so all we can say is that almost surely (technically, with probability 1) the dimension is the solution of Eq. (2.9). A proof is in Sect. 15.1 of [16].

Back to the example. Suppose for each i, $P(r_i = 1/2) = 1/3$ and $P(r_i = 1/4) = 2/3$. Then $P(r_i^d = (1/2)^d) = 1/3$ and $P(r_i^d = (1/4)^d) = 2/3$, and so

$$\begin{aligned}\mathbb{E}(r_i^d) &= (1/2)^d P(r_i^d = (1/2)^d) + (1/4)^d P(r_i^d = (1/4)^d) \\ &= (1/2)^d (1/3) + (1/4)^d (2/3)\end{aligned}$$

Consequently, the random Moran equation becomes

$$4(1/2)^d(1/3) + 4(1/4)^d(2/3) = 1$$

We've seen a trick to solve this equation. Note $(1/4)^d = ((1/2)^2)^d = ((1/2)^d)^2$, so if we take $x = (1/2)^d$ we recognize that the random Moran equation is the quadratic equation

$$(4/3)x + (8/3)x^2 = 1$$

The solutions are $x = (-1 \pm \sqrt{7})/4$. Because $x = (1/2)^d$ is positive for all real d, we must take $x = (-1 + \sqrt{7})/4$. Then we find

$$d = \frac{\log((-1 + \sqrt{7})/4)}{\log(1/2)} \approx 1.28125$$

We'll use another extension of the Moran equation to find the dimension of the shape pictured on the right, the attractor of Fig. 1.55 from Lab 1.9. The transition matrix of the IFS with memory with this attractor is

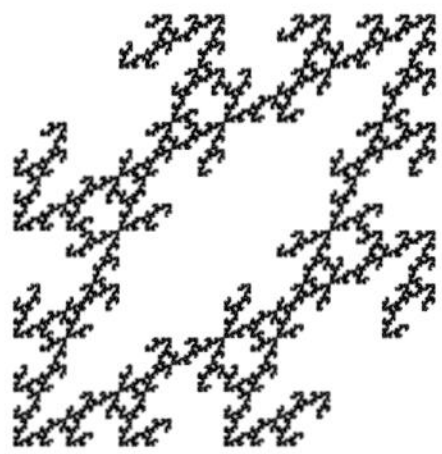

$$M = \begin{bmatrix} 1 & 1 & 1 & 0 \\ 1 & 0 & 1 & 1 \\ 1 & 1 & 0 & 1 \\ 0 & 1 & 1 & 1 \end{bmatrix}$$

In a moment we'll state the general formula for the dimension of the attractors of IFS with memory. In the examples of Lab 1.9, where each scaling factor r is the the same, $r = 1/2$, the dimension d is the solution of the *Moran equation with memory*

$$r^d \rho[M] = 1 \tag{2.10}$$

Recall that $\rho(M)$ is the spectral radius of M, the eigenvalue of M with the largest absolute value. In Appendix B.8 we'll give a bit of background for computing eigenvalues. We'll compute eigenvalues numerically with the commands in 2.3 (b) of the Mathematica or Python code.

For this transition matrix M the eigenvalues are 3, 1, -1, and -1. Then $\rho(M) = 3$ and the dimension d is the solution of $(1/2)^d 3 = 1$, that is, $d = \log(3)/\log(2)$, the dimension of the Sierpinski gasket.

The general formula, where r_i is the scaling factor of T_i, is

$$\rho \begin{bmatrix} m_{11}r_1^d & m_{12}r_1^d & m_{13}r_1^d & m_{14}r_1^d \\ m_{21}r_2^d & m_{22}r_2^d & m_{23}r_2^d & m_{24}r_2^d \\ m_{31}r_3^d & m_{32}r_3^d & m_{33}r_3^d & m_{34}r_3^d \\ m_{41}r_4^d & m_{42}r_4^d & m_{43}r_4^d & m_{44}r_4^d \end{bmatrix} = 1 \tag{2.11}$$

The proof [50] is complicated; some more detail is in Sect. 6.4.2 and Appendix A.85 of [17].

The Moran equation extends to other cases as well, through much more complex math. For some nonlinear fractals (the scaling factor varies with location in the fractal), this involves concepts called the thermodynamic formalism and topological pressure, but we won't pursue these here. Chapter 5 of [20] is a good source, if you know a lot of math.

Finally, we'll mention another class of fractals, self-affine fractals, for which scaling factors are different in different directions, say 1/2 in the x-direction and 1/3 in the y-direction. For these fractals we have nothing like a Moran equation. Some special cases have been solved [51, 52, 53, 54, 55, 56] for example, but the general problem of finding the dimensions of self-affine fractals is soberingly difficult.

2.3.4 Procedure

Determine if a fractal is exactly self-similar with each piece scaled by the same factor r, with different scaling factors for different pieces, if the scaling factors vary randomly, or if only certain combinations of transformations are allowed. Identify and apply the relevant version of the Moran equation. (Eq. (2.7) is equivalent to the Moran equation (2.8) with all $r_i = r$. Do you see why?) If several scaling factors are involved and all are powers of a single factor, the Moran equation reduces to a polynomial equation. If the scaling factors do not have this relationship, numerical solutions can be applied. For fractals generated by IFS with memory, eigenvalues can be found with the software in 2.3 (b) of the Mathematica of Python code.

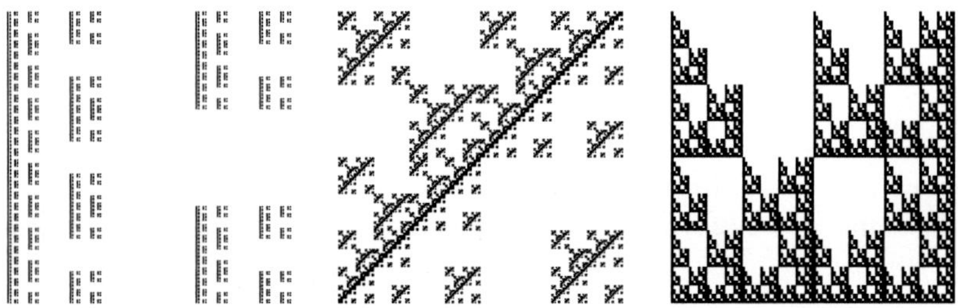

Figure 2.11: Left to right: figures for (a) and (b) of Sample 2.3.5, and for (a) of Sample 2.3.6.

2.3.5 Sample A

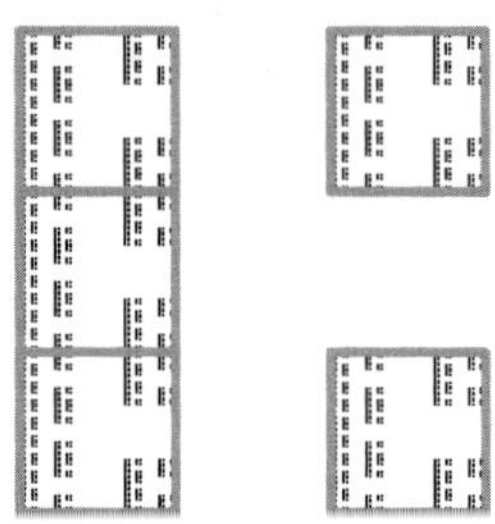

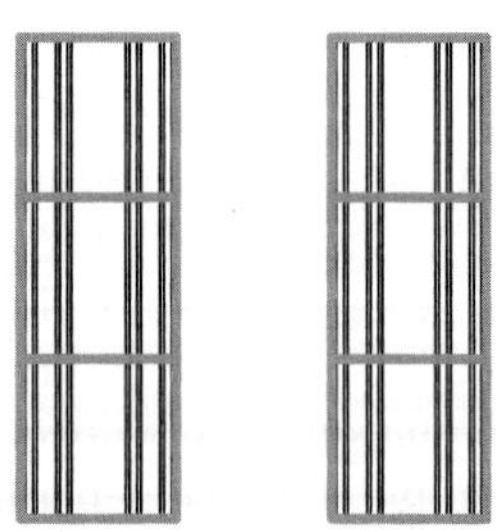

(a) Find the similarity dimension of the left image of Fig. 2.11. Remember we look to decompose the fractal into small copies of the entire shape. The two pieces on the right of the fractal are clear. If the left side is a bit harder to decompose, in the figure on the right we have outlined $N = 5$ copies, each scaled by $r = 1/3$. Then by the similarity dimension formula (2.7) the similarity dimension of this fractal is $d_s = \log(5)/\log(3) \approx 1.46497$. On the left edge of the fractal we see a vertical line, echoed in infinitely many, smaller and smaller lines.

If we add another copy scaled by 1/3 and with its lower left corner at $(2/3, 1/3)$, not only do we fill in the middle box of the right column of the shape, we fill in the middle right boxes of all parts of the shape. The single vertical length 1 line on the left of the original fractal becomes an infinite collection of length 1 vertical lines, one above each point of a Cantor middle-thirds set on the x-axis. The dimension is $\log(6)/\log(3) = (\log(3) + \log(2))/\log(3) = 1 + \log(2)/\log(3)$, the dimension of the line segment plus that of the Cantor middle-thirds set. Do you see why this result is sensible? Think of the product rule $d(A \times B) = d(A) + d(B)$.

(b) Find the similarity dimension of the middle image of Fig. 2.11. This shape is more complicated than the fractal of (a), so we'll suggest some steps to find a decomposition of the fractal. The lower right and upper left corners are copies of the whole shape, scaled by 1/4. This suggests that we try a 4×4 grid. We do find seven copies of the shape scaled by 1/4, but the squares with lower left corners $(1/4, 0)$, $(0, 1/4)$, $(3/4, 1/2)$, and $(1/2, 3/4)$ do not contain copies of the whole shape scaled by 1/4. The lower left and upper right $1/2 \times 1/2$ squares are copies scaled by 1/2.

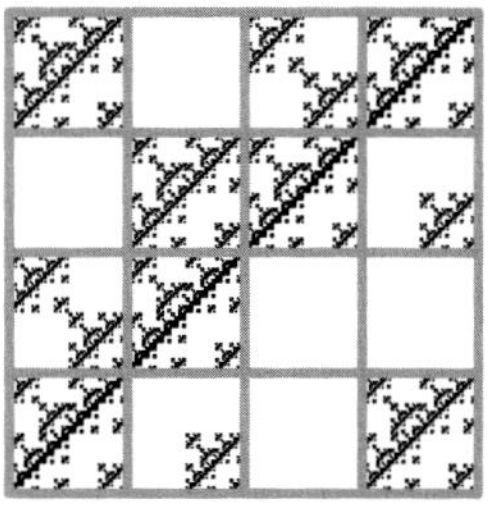

This suggests another decomposition, 2 copies scaled by 1/2 and 3 copies scaled by 1/4. Because we have two scaling factors, we need the Moran equation (2.8) to find the dimension. For this fractal the Moran equation becomes

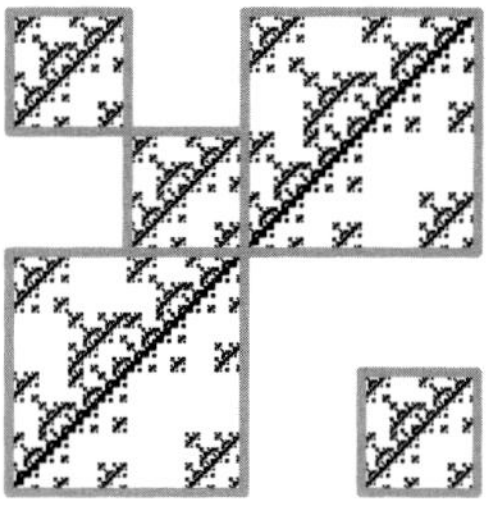

$$2(1/2)^d + 3(1/4)^d = 1$$

Because $(1/4)^d = ((1/2)^2)^d = ((1/2)^d)^2$, take $x = (1/2)^d$ and the Moran equation can be written

$$2x + 3x^2 = 1$$

The solutions are $x = -1$ and $x = 1/3$. Recall x must be positive, so we have $1/3 = (1/2)^d$. Take the log of both sides and solve for d to obtain $d = \log(3)/\log(2)$.

Do you see why this fractal has the same dimension as the gasket? The gasket consists of 3 copies scaled by 1/2. Self-similarity guarantees that each of these three copies consists of 3 copies scaled by 1/4. So we can decompose the gasket as 2 copies scaled by 1/2 and 3 copies scaled by 1/4. So long as the pieces overlap only along edges, the number and size of the pieces, not their positions, determine the similarity dimension of the fractal.

(c) Suppose a fractal consists of 1 copy scaled by 1/3, 2 copies scaled by 1/9, 2 copies scaled by 1/27, and so on. That is, 1 copy scaled by 1/3 and 2 copies scaled by $1/3^n$ for $n = 2, 3, \ldots$. Assume the Moran equation can be generalized to infinite series (It can. See Appendix A.77 of [17].) and write the Moran infinite series equation for this fractal. Find the similarity dimension.

For this fractal the Moran infinite series equation is

$$(1/3)^d + 2(1/9)^d + 2(1/27)^d + 2(1/81)^d + \cdots = 1$$

Because each scaling factor is a power of 1/3, take $x = (1/3)^d$ and write the Moran infinite series equation as

$$x + 2x^2 + 2x^3 + 2x^4 + \cdots = 1$$
$$x + 2x^2(1 + x + x^2 + \cdots) = 1$$
$$x + 2x^2\left(\frac{1}{1-x}\right) = 1 \qquad (2.12)$$

where in (2.12) we used the sum of the geometric series $1 + x + x^2 + \cdots = 1/(1-x)$. To sum this series we must have $|x| < 1$. Here $x = (1/3)^d$, so $|x| < 1$ so long as $d > 0$.

Eq. (2.12) simplifies to $x^2+2x-1 = 0$. The positive solution is $x = -1+\sqrt{2}$ and so the dimension of this fractal is $d = \log(-1+\sqrt{2})/\log(1/3) \approx 0.802261$.

2.3.6 Sample B

(a) Find the similarity dimension of the right fractal of Fig. 2.11.

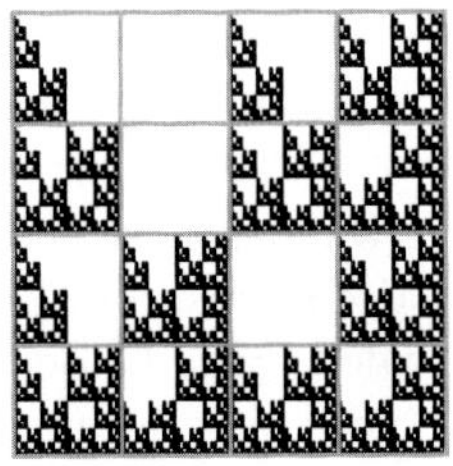

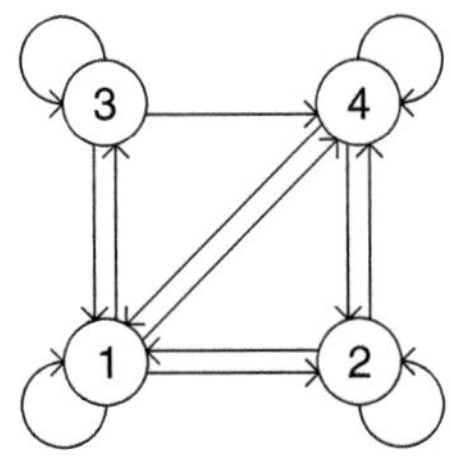

Here we see the fractal with a 4×4 grid superimposed. We see that the empty length-2 addresses are 23, 32, and 34. The transition graph is shown beside the fractal. From the graph, or from the fractal with grid, we see that the transition matrix is

$$M = \begin{bmatrix} 1 & 1 & 1 & 1 \\ 1 & 1 & 0 & 1 \\ 1 & 0 & 1 & 0 \\ 1 & 1 & 1 & 1 \end{bmatrix}$$

The eigenvalues of M are $\frac{3 \pm \sqrt{13}}{2}$, 1, and 0. Then the spectral radius is $(3+\sqrt{13})/2$ and by Eq. (2.10) $d_s = \dfrac{\log((3+\sqrt{13})/2)}{\log(2)} \approx 1.72368$.

(b) Suppose a fractal consists of 1 copy scaled by 1/2, 6 copies scaled by 1/4, and 4 copies scaled by 1/8. Find the dimension of this fractal.

Different scaling factors mean we use the Moran equation,

$$(1/2)^d + 6(1/4)^d + 4(1/8)^d = 1$$

Because the scaling factors are powers of 1/2, take $x = (1/2)^d$. Then the Moran equation becomes

$$4x^3 + 6x^2 + x - 1 = 0$$

There is a cubic formula, and for that matter, a quartic formula, analogous to the quadratic formula, but far more complicated. (There are no corresponding formulas for higher-order polynomials. It's not that we don't know the formulas; we know that such formulas are impossible. Google Évariste Galois to learn about the person who crafted the proof. The proof uses techniques from abstract algebra, far beyond the level of math in this book.) A simpler approach is to try to find a linear factor. Because the constant term is -1, the first choice is to divide $4x^3+6x^2+x-1$ by $x-1$ or $x+1$. Dividing by $x-1$ gives $4x^2+10x+11$ with a remainder of 10, so $x-1$ isn't a factor. Dividing by $x+1$ gives $4x^2+2x-1$ without remainder, so

$$4x^3 + 6x^2 + x - 1 = (x+1)(4x^2+2x-1)$$

Consequently, the solutions of the Moran equation are $x = -1$ and $x = (-1 \pm \sqrt{5})/4$. The only positive solution is $x = (-1 + \sqrt{5})/4$, so the dimension is $d = \log((-1 + \sqrt{5})/4)/\log(1/2) \approx 1.69424$.

(c) Suppose a randomized gasket has these three scaling factors

$$
\begin{aligned}
P(r_1 = 1/3) &= 1/2 & \qquad P(r_1 = 1/9) &= 1/2 \\
P(r_2 = 1/3) &= 1/4 & \qquad P(r_2 = 1/9) &= 3/4 \\
P(r_3 = 1/3) &= 1/8 & \qquad P(r_3 = 1/9) &= 7/8
\end{aligned}
$$

Find the exact maximum and exact minimum values of the dimension of this gasket. Find the exact expected value of the dimension of this randomized gasket.

The maximum value of the dimension occurs when all the scaling factors take on their maximum value, $r_i = 1/3$. Then we can apply the similarity dimension formula (2.7) to find $d_s = \log(N)/\log(1/r) = \log(3)/\log(3) = 1$. Similarly, the minimum dimension occurs when all the scaling factors take their minimum value $r_i = 1/9$. This gives $d_s = \log(3)/\log(9) = 1/2$.

To find the expected value of the dimension, use the randomized Moran equation (2.9), $\mathbb{E}(r_1^d) + \mathbb{E}(r_2^d) + \mathbb{E}(r_3^d) = 1$. These expected values are

$$
\begin{aligned}
\mathbb{E}(r_1^d) &= (1/2)(1/3)^d + (1/2)(1/9)^d \\
\mathbb{E}(r_2^d) &= (1/4)(1/3)^d + (3/4)(1/9)^d \\
\mathbb{E}(r_3^d) &= (1/8)(1/3)^d + (7/8)(1/9)^d
\end{aligned}
$$

Combining like terms the random Moran equation (2.9) is

$$(7/8)(1/3)^d + (17/8)(1/9)^d = 1$$

Substitute $x = (1/3)^d$, so $(1/9)^d = x^2$, to write the random Moran equation as $(17/8)x^2 + (7/8)x - 1 - 0$. The solutions are $x = (-7 + \sqrt{593})/34$, so the expected value of the dimension is

$$d = \frac{\log((-7 + \sqrt{593})/34)}{\log(1/3)} \approx 0.612296$$

2.3.7 Conclusion

The Moran equation is the principal tool to compute similarity dimensions. For mathematical fractals generated by IFS, IFS with memory, and IFS with randomly varying scaling factors, the similarity dimension can be computed by variations of the Moran equation (2.8), $r_1^d + \cdots + r_N^d = 1$. The basic similarity dimension formula (2.7), $d = \log(N)/\log(1/r)$, is a special case of the Moran equation with all $r_i = r$. When the scaling factors vary randomly, we use the random Moran equation (2.9), $\mathbb{E}(r_1^d) + \cdots + \mathbb{E}(r_N^d) = 1$. For fractals generated by IFS with memory, the Moran equation becomes the memory Moran equation (2.10) $r^d \rho[M] = 1$ if all the scaling factors $r_i = r$. For fractals generated with different scaling factors, the dimension is the solution of the general memory

Moran equation (2.11). If each scaling factor $r_i = r^k$, the Moran equation can be recast as a polynomial equation. If not, the Moran equation must be solved numerically. We see that the Moran equation is a fundamental construct for the computation of dimensions of self-similar shapes.

We'll generalize the Moran equation in a different way to compute multifractal curves in Lab 2.7.

2.3.8 Exercises

Prob 2.3.1 The Sierpinski gasket consists of $N = 3$ pieces, each scaled by $r = 1/2$. But self-similarity implies that each of these three pieces consists of 3 still smaller pieces, each of which consists of 3 even smaller pieces, and so on. That is the gasket can be described by $N = 3, r = 1/2$, $N = 9, r = 1/4$, $N = 27, r = 1/8$, and so on. What happens if we apply the similarity dimension formula (2.7) to these decompositions? Does the dimension depend on the level of the decomposition? What about decompositions into pieces of different sizes?

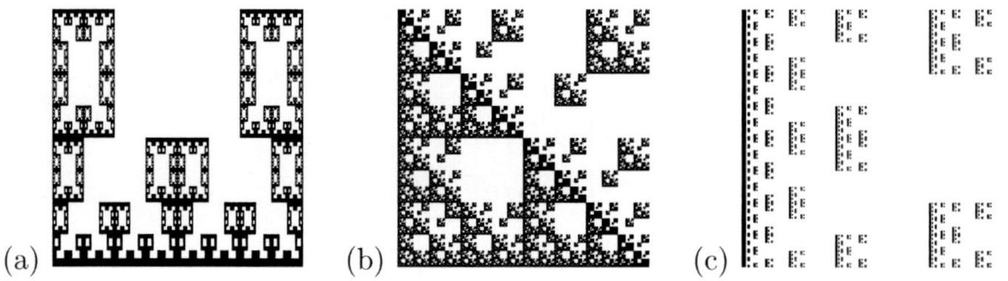

Figure 2.12: Fractals for Exercise 2.3.2

Prob 2.3.2 Compute the similarity dimensions of the fractals of Fig. 2.12.

Prob 2.3.3 Compute the similarity dimension of a fractal that consists of
(a) 1 copy scaled by 1/4, and 3 copies scaled by 1/16, and
(b) 3 copies scaled by 1/2, 3 copies scaled by 1/8, and 1 copy scaled by 1/16.

Prob 2.3.4 Compute the similarity dimension of the fractal that consists of these pieces. For (b) find a numerical solution to 3 places to the right of the decimal. For (c) remember that to sum a geometric series, the ratio x of consecutive terms must satisfy $|x| < 1$.
(a) 1 copy scaled by $1/3^n$ for $n = 1, 2, 3, \ldots$,
(b) 1 copy scaled by 1/2 and 1 copy scaled by $1/2^n$ for $n = 3, 4, 5, \ldots$, and
(c) 2 copies scaled by 1/2 and 1 copy scaled by $1/2^n$ for $n = 2, 3, 4, \ldots$.

Prob 2.3.5 Compute the similarity dimension of the fractal that consists of
(a) 2 copies scaled by $1/5^n$ for $n = 1, 2, 3, \ldots$,
(b) 4 copies scaled by 1/3 and 2 copies scaled by $1/3^n$ for $n = 2, 3, 4, \ldots$, and
(c) 4 copies scaled by 1/2, 6 copies scaled by 1/4, and 1 copy scaled by 1/8. Can this fractal be a subset of the plane?

(a) (b) (c)

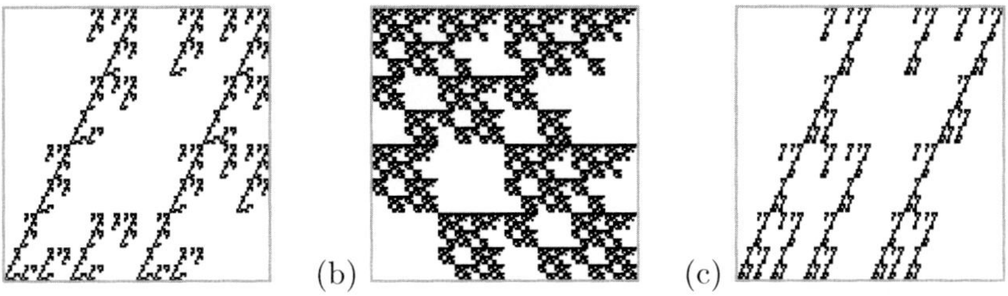

Figure 2.13: Fractals for Exercise 2.3.6

Prob 2.3.6 Compute the similarity dimensions of the fractals generated by 1-step memory IFS of Fig. 2.13. Hint: for each, find the empty length-2 addresses.

Prob 2.3.7 Suppose we construct a randomized gasket by $r_1 = r_2 = 1/2$ and $P(r_3 = 1/2) = 1/5$, $P(r_3 = 1/4) = 4/5$.
(a) Find the maximum and minimum possible dimensions for this randomized gasket.
(b) Find the expected value for the dimension of this randomized gasket.

Prob 2.3.8 Construct a randomized Cantor set with two transformations, with scaling factors r_1 and r_2 that have $P(r = 1/2) = 1/4$ and $P(r = 1/4) = 3/4$. Find the expected value of the dimension of this randomized Cantor set.

Prob 2.3.9 Suppose a random Cantor set is constructed with two scaling factors r_1 and r_2, where

$$P(r_1 = 1/2) = P(r_2 = 1/2) = p \quad \text{and } P(r_1 = 1/4) = P(r_2 = 1/4) = 1 - p$$

Find an expression for the expected value of the dimension d of the random Cantor set as a function of p. Don't expect a simple expression. Plot $d(p)$ for $0 < p < 1$.

Prob 2.3.10 Suppose $z[M]$ denotes the number of 0s in the matrix M, and $\dim[M]$ denotes the dimension of the fractal generated by the IFS with memory with transition matrix M and all scaling factors equal. If $z[M_1] < z[M_2]$, must $\dim[M_1] > \dim[M_2]$? Note that by the memory Moran equation (2.10) it suffices to show $z[M_1] < z[M_2]$ implies $\rho[M_1] > \rho[M_2]$. If you think this is true, give an argument to support this; if you think it is false, give a counter example. Exercise 2.3.6 is a step in this problem.

2.4 Sierpinski tetrahedron

Except for the crumpled paper balls, all the fractals we've constructed so far are subsets of the plane. Now we'll show how to build a version of the Sierpinski gasket but based on a tetrahedron rather than on a triangle. You won't be surprised that it's called the Sierpinski tetrahedron.

Pictured here is the result of one of these labs, with Benoit peering through the central gap in the Sierpinski tetrahedron.

2.4.1 Purpose

We'll build a model of a fractal tetrahedron in 3-dimensional space to illustrate scaling at different levels, to calculate its dimension, and to develop some understansing of the relation of the dimension of an object to that of the space in which it is embedded. In particular, construction of a physical model reveals how very empty the Sierpinski tetrahedron is. This provides another example of the consequences of measurement in the wrong dimension, a notion we introduced in Sample 2.1.6.

2.4.2 Materials

We'll use 256 envelopes of size $3\frac{5}{8}'' \times 6\frac{1}{2}''$, rulers, scissors, pencils, and tape.

2.4.3 Background

In the early years of my (MF) fractal geometry courses, all mathematical fractals were subsets of the plane, more-or-less a consequence of comparatively primitive 3-dimensional graphics software available in student computer labs in the mid 1980s, while all natural fractals—cauliflower, broccoli, Queen Anne's Lace, crumpled paper balls—inhabited 3-dimensional space. In one final examination question students were asked to characterize the difference between mathematical and natural fractals. The answer I expected was "Mathematical fractals are exactly self-similar on all scales, while natural fractals are approximately self-similar over a limited range of scales." While some students gave answers like that, many more replied, "Mathematical fractals live on pieces of paper or computer screens, natural fractals live in the 3-dimensional world." Based on the examples they'd seen in class, this answer was correct. So I began to include the Sierpinski tetrahedron as a physical representation of a mathematical fractal that lives in 3-dimensional space. Still, at first they were computer models, though I could rotate them on the screen. Eventually, it became apparent that a physical model was needed to make clear that mathematical fractals can live in any dimension.

2.4.3.1 The tetrahedron

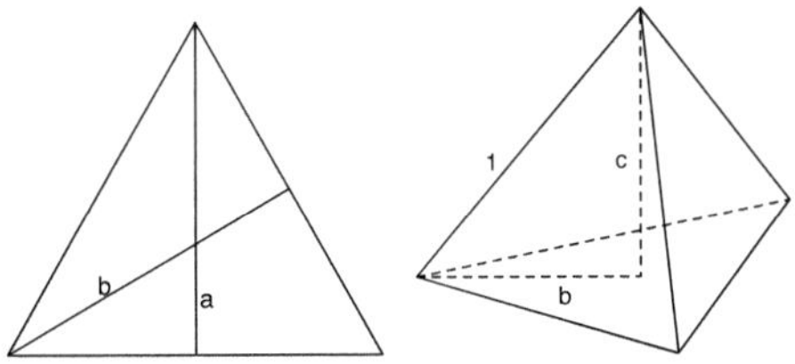

The tetrahedron consists of four faces, all congruent equilateral triangles. Taking the triangles to have side length 1, the area of each face is $\sqrt{3}/4$. Symmetry considerations show the apex of the tetrahedron lies above the intersection of the angle bisectors of the base. With some geometry guided by the left image, we see $a = \sqrt{3}/6$ and $b = \sqrt{3}/3$. Then from the right image, we see the altitude of the tetrahedron is $c = \sqrt{6}/3$.

The tetrahedron is a cone with base an equilateral triangle, so has volume

$$\frac{1}{3} \cdot (\text{area of base}) \cdot \text{altitude} = \frac{1}{3} \cdot \frac{\sqrt{3}}{4} \cdot \frac{\sqrt{6}}{3} = \frac{\sqrt{2}}{12}$$

2.4.3.2 The Sierpinski tetrahedron

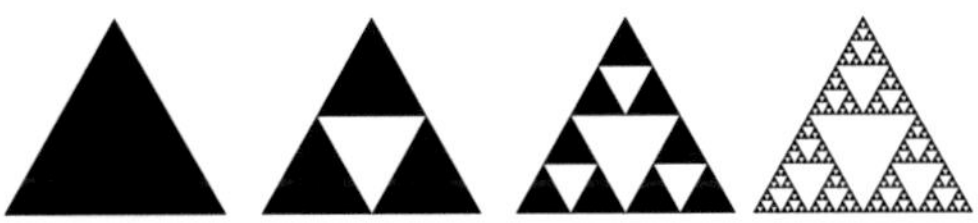

One way to generate the equilateral Sierpinski gasket is to begin with a filled-in equilateral triangle G_0. Call its vertices V_1, V_2, and V_3. Now shrink G_0 by 1/2 toward each vertex V_1, V_2, and V_3 to obtain G_1, the second figure above. Now shrink G_1 by 1/2 toward each vertex V_1, V_2, and V_3 to obtain G_2, the third figure. Three more iterations gives G_5, the fourth figure. Infinitely many more iterations gives the equilateral Sierpinski gasket.

The Sierpinski tetrahedron is grown in an analogous fashion. Begin with T_0, a filled-in (equilateral) tetrahedron. Then T_{n+1} is obtained by shrinking T_n by 1/2 toward each of the four vertices of T_0. In Fig. 2.14 we see T_0, T_1, T_2, and T_5. The Sierpinski tetrahedron is $\lim_{n\to\infty} T_n$.

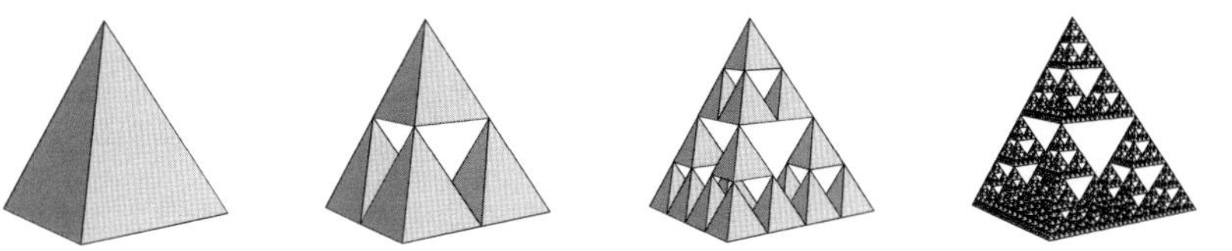

Figure 2.14: Steps T_0, T_1, T_2, and T_5 in growing the Sierpinski tetrahedron.

We see the Sierpinski tetrahedron is self-similar and consists of $N = 4$ pieces scaled by a factor of $r = 1/2$. Then by the basic similarity dimension formula (Eq. 2.7) we see that $d_s = \log(4)/\log(2) = 2$. Recall the argument we introduced in Sample 2.1.6: measured in a dimension higher than that of the shape, the measure is 0. Because the Sierpinski tetrahedron has dimension 2, its volume is 0. To follow the calculation is one thing; to look at a physical model, stick your head into the big gap, see how insubstantial it is, is something more visceral. It's just a wisp, hardly there at all. This we cannot learn from a calculation alone. Physical models tell the right stories, sing the right songs.

The construction process described here involves scaling successive generations by a factor of $r = 1/2$, and assembling the copies in the shape of the original. This is impractical for the construction of a physical model. Instead, we build small tetrahedra and assemble them into increasingly large models of the mathematical fractal. Growth by accretion rather than by erosion. Think of trees or corals rather than coastlines.

If we imagine each successive physical stage reduced by a factor of 1/2, then we retrieve the stages of the mathematical construction.

2.4.3.3 The Sierpinski tetrahedron complement

Recall T_0 is a tetrahedron and T_1 consists of four tetrahedra, each half the size of T_0 (Fig. 2.15, left). The complement, denoted $T_1 - T_0$ is an octahedron (Fig. 2.15, center). The complement $T_2 - T_1$ consists of four octahedra (Fig. 2.15, right), each half the size of the octahedron $T_1 - T_0$.

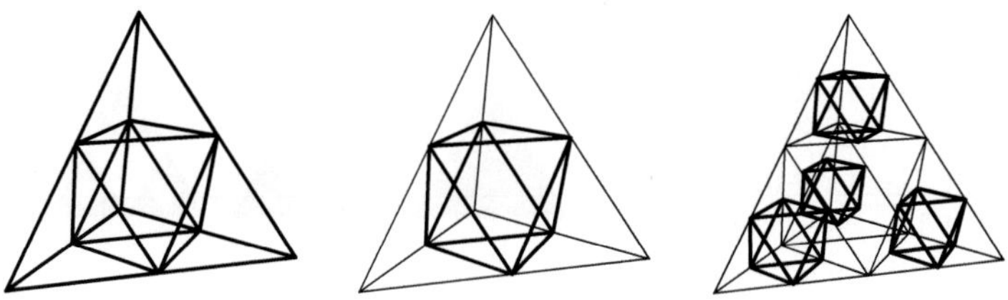

Figure 2.15: Left: the four tetrahedra of T_1. Center: the edges of the octahedron $T_0 - T_1$. Right: the four octahedra of $T_1 - T_2$.

The complement of the Sierpinski tetrahedron is the union of an infinite collection of octahedra, 1 of side length 1/2, 4 of side length 1/4, 16 of side length 1/8, and so on. See Exercise 2.4.2 (b).

2.4.4 Procedure

The most complicated step is to make a tetrahedron from an envelope. Once we have a bunch of tetrahedra, the construction of an approximation of the Sierpinski tetrahedron is a simple recursive process.

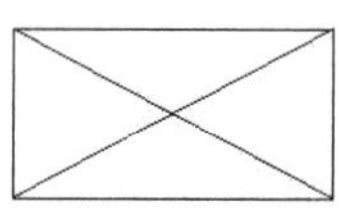

1. Draw diagonal lines on one side of the envelope. Be careful to align the ruler at opposite corners of the envelope. Sloppiness here can produce lopsided tetrahedra.

2. Fold the envelope both ways along both diagonals. Fold against a ruler to make crisp folds. Do this several times so the next steps can be done more easily.

3. Position the envelope so that the longer sides are at the top and bottom and the flap is along the bottom. Cut along the diagonals from the bottom corners of the envelope to where the diagonals meet. This separates the envelope into two pieces. Discard and recycle the smaller piece.

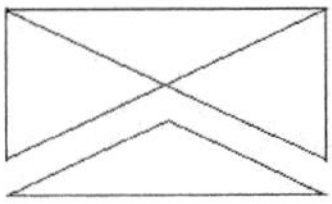

4. Keep the longer, uncut side of the remaining piece on top. Make a vertical fold by folding the left side to the right side. Open the piece and flatten it. Fold the right side to the left side. Repeat this several times. Be careful not to tear the piece along this fold.

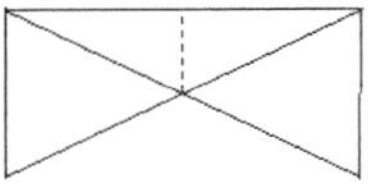

5. Open up what remains of the envelope. Fold along the fold made in step 4. Place one side into the other. For example, here the left half is folded into the right half. The face of the left half that does not go into the right half is the fourth face of the tetrahedron. The more care used with this step, the less crinkling and crumpling of the envelope, the sharper and cleaner the tetrahedron will be.

6. Tape the edges along the bottom to secure the tetrahedron shape. Sharply crease the edges to produce a sturdy tetrahedron.

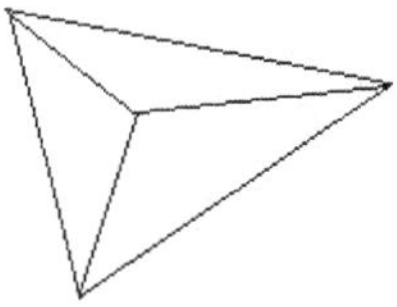

Once we've made individual tetrahedra, the next step is to assemble these in groups of four to make the first stage of the Sierpinski tetrahedron. Then assemble four of these to produce the second stage. Continue until you run out of tetrahedra. So: to avoid leftovers the number of tetrahedra should be a power of 4. With 256 tetrahedra we can build the fourth stage of the Sierpinski tetrahedron.

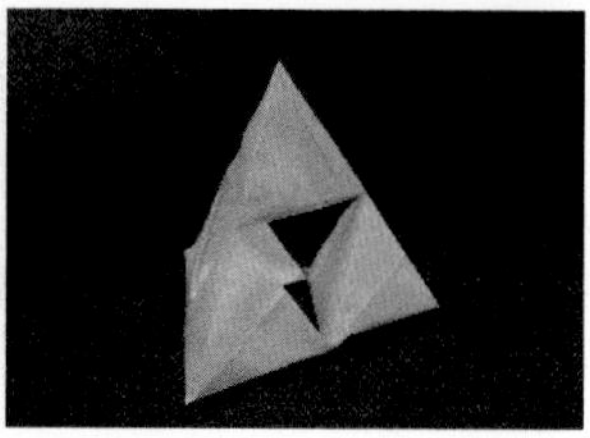

To be sure the assembly process is clear, here we see the steps to build the first stage. First, tape three tetrahedra together to form the base of the first stage. Individual tetrahedra are taped at their corners, one corner of one tetrahedron taped to one corner of another tetrahedron. Then place the fourth tetrahedron on top of these three. Tape each corner of base of the top tetrahedron to the top of one of the base tetrahedra.

Tape four of these in the same pattern. Tape four of those on the same pattern. Continue.

2.4.5 Sample

Here are some pictures of this construction from 256 tetrahedra. Left: construct the second stage. Right: construct the third stage.

And here are some steps in the construction of the fourth stage. Note that the entire weight of the top piece is supported at only three points. With tetrahedra made from standard envelopes, the next stage of the construction might collapse.

2.4.6 Conclusion

The stage four of the Sierpinski tetrahedron is impressive and relatively sturdy, but subject to the familiar effects of humidity on paper. A permanent record can be obtained by photographing the object before gravity and time take their inevitable toll.

Certainly, this shape exhibits some element of fractality, though to be sure only over a limited range of scales. Building this model is a good bridge between mathematical fractals, exhibiting similar structure over infinitely small scales, and natural fractals, exhibiting similar structure over only a limited range of scales. The Sierpinski tetrahedron is a straightforward way to get some insight into the emptiness of a fractal embedded in a higher-dimensional space.

2.4.7 Exercises

Here by the Sierpinski tetrahedron we mean the limiting shape, that is, the mathematical fractal modeled by the construction of this lab.

Prob 2.4.1 Find IFS rules (with functions of 3 variables) to generate the equilateral Sierpinski tetrahedron.

Prob 2.4.2 Calculate the volume of the Sierpinski tetrahedron. This can be done in (at least) two ways.
(a) With the determinisitc IFS algorithm, the Sierpinski tetrahedon is the limit of a sequence $T_0, T_1, T_2, \dots$, where T_0 is the original tetrahedron, T_1 consists of 4 copies of this tetrahedron, each scaled by 1/2, and so on. Compute $\lim_{n\to\infty} \text{vol}(T_n)$.
(b) In Sect. 2.4.3.3 we observed that the complement of T_1 in T_0 is an octahedron, the compliment of T_2 in T_1 is four copies of this octahedron, each scaled by 1/2, and so on. Sum the volumes of these complementary octahedra and compare that sum to the volume of T_0.

Prob 2.4.3 Interpret the answer of Exercise 2.4.2 in terms of the similarity dimension of the Sierpinski tetrahedron.

Prob 2.4.4 (a) In the plane, how many pieces scaled by $r = 1/2$ are needed to produce a shape with dimension $d = 2$? Assume the pieces overlap at most along edges.
(b) In 3-dimensional space, how many pieces scaled by $r = 1/2$ are needed to produce a shape with dimension $d = 3$? Assume the pieces overlap at most along faces.
(c) In the calculations of (a) and (b), does the placement of the pieces affect the number of pieces? This is a tricky point. Explore your answer carefully.

Prob 2.4.5 If we don't impose the overlap conditions of Exercise 2.4.4, can $N > 4$ pieces with $r = 1/2$ still give shapes of dimension 2?

2.5 Koch tetrahedron

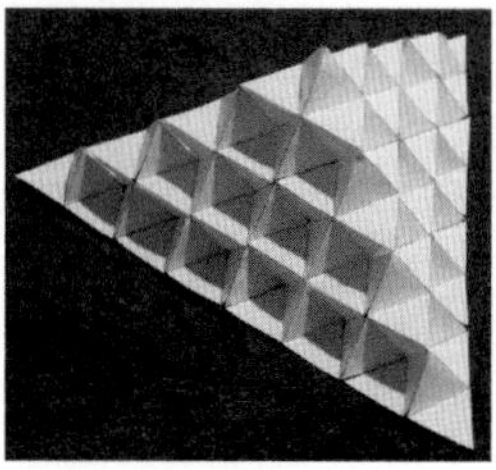

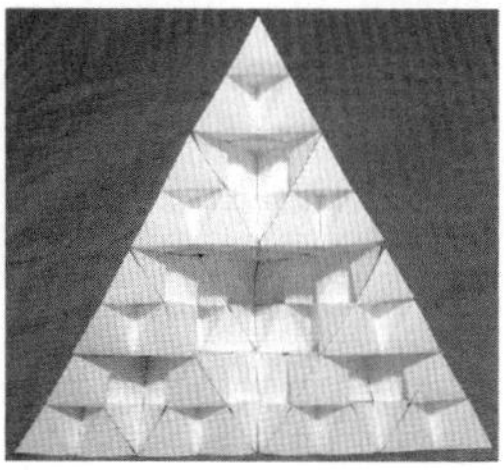

Here we'll adapt the approach of Lab 2.4 to investigate the Koch tetrahedron, a shape which is to the Koch curve as the Sierpinski tetrahedron is to the Sierpinski gasket. This is not just another example of a recursive physical construction. It offers a beguiling puzzle: even though the shape is made of little triangles of paper, the dimension of the inside is higher than the dimension of the outside. Several workshop participants suggested we call this the "Tardis Lab," evidently a reference to the Dr. Who mythology.

2.5.1 Purpose

We'll construct a model of one face of the third stage of a Koch tetrahedron and find the dimension of the inside and the outside of the Koch tetrahedron fractal.

2.5.2 Materials

Eighteen copies of the template of Sect. 2.5.8, scissors, a straightedge for folding, and clear tape. The construction is more sturdy if the template is copied to heavy paper, 24 lb. or more.

2.5.3 Background

One of the oldest mathematical fractals is the Koch curve, introduced in 1904 by Helge von Koch[57] as a geometrical construction of a continuous curve without tangents. A curve has no tangent at a corner; each stage of the Koch curve construction adds more corners. In the limit, the curve has corners arbitrarily close to every point, so no tangent anywhere. A translation of von Koch's paper is Chapter 3 of Gerald Edgar's very useful book [58]. The construction is recursive. The image on the left of Fig. 2.16, four segments of length 1/3, is the first step. Replace each segment with a scaled copy of the left image, obtaining the curve in the middle of 2.16. Iterate. The limit of this process of the Koch curve; a representation is shown on the right of 2.16.

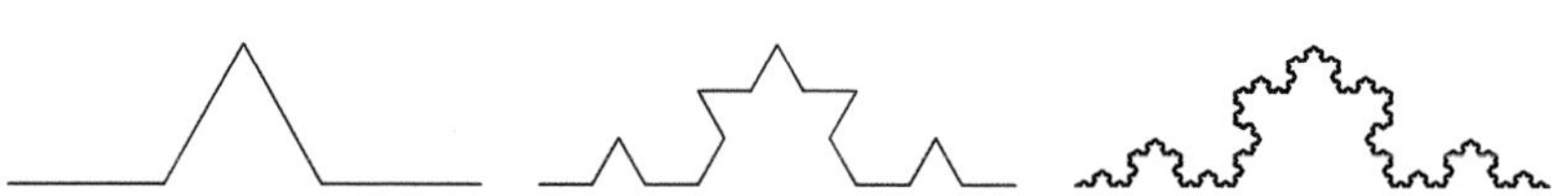

Figure 2.16: The first, second, and fifth stages in the Koch curve construction.

An extension of the Koch curve is the *Koch snowflake*. As illustrated in Fig. 2.17, apply the Koch curve construction to each edge of an equilateral triangle. Now with the initial definition of self-similarity, the Koch snowflake is not self-similar. It is not made up of smaller copies of the entire snowflake. So we'll relax our definition a bit and say that a shape is self-similar if each piece of the shape consists of arbitrarily small copies of a large piece of the shape. We see that the Koch snowflake is made up of arbitrarily small copies of the Koch curve.

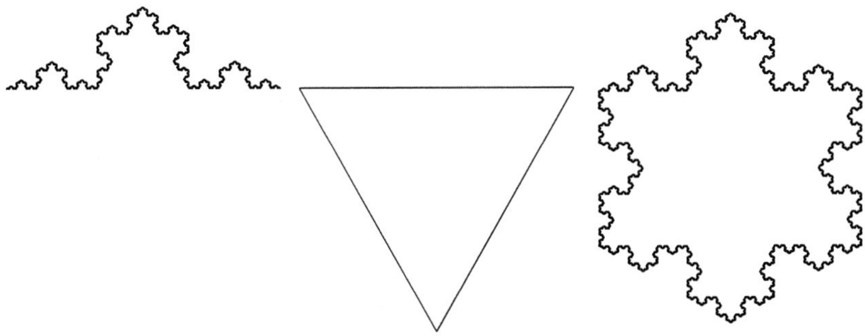

Figure 2.17: Construction of the Koch snowflake.

From this definition of self-similarity, we expect that the similarity dimension of the Koch snowflake equals that of the Koch curve, $d_s = \log(4)/\log(3)$ because the Koch curve consists of $N = 4$ pieces each scaled by $r = 1/3$. But let's be careful and apply some of the dimension rules, specifically the union rule $d(A \cup B) = \max\{d(A), d(B)\}$ and the invariance rule $d(f(A)) = d(A)$ for any similarity transformation f.

Denote the Koch snowflake by S and the three Koch curves that make up S by K_1, K_2, and K_3. The K_i are rotations and translations of the standard Koch curve, so by the invariance rule $d(K_1) = d(K_2) = d(K_3) = \log(4)/\log(3)$. Then by the union rule $d(S) = \max\{d(K_1), d(K_2), d(K_3)\} = d(K_i) = \log(4)/\log(3)$. That is, the Koch snowflake has the same dimension as the Koch curve. This illustrates a reason behind our extension of the definition of self-similarity.

2.5.4 Procedure

Make 18 copies of the template sheet and cut out both patterns on the eighteen sheets. For heavy paper, use the straightedge to aid in making clean folds. Step 4 works best with the template placed on a table or other flat surface.

Be careful with both the cuts and the folds. The more precise these are, the easier will be the later stages in the assembly of the Koch tetrahedron.

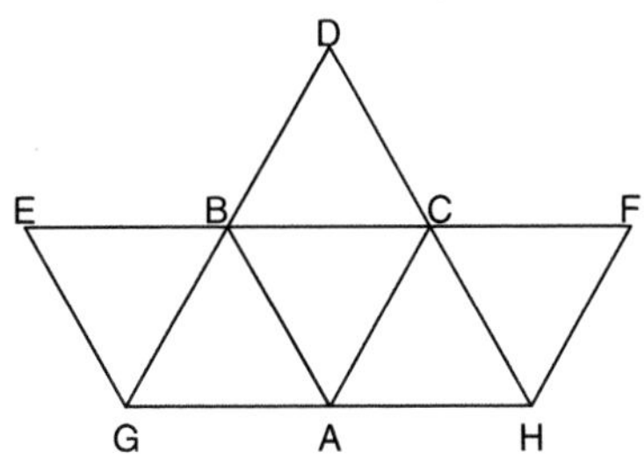

Figure 2.18: A Koch tetrahedron template.

The construction has nine steps, the first five of which must be done to the 36 copies of the template. These steps of the lab would be a real bother if you have make all the folds alone. Nine people can do this fairly easily and fairly quickly.

Step 1. Fold back along the segment AB so the back of vertex E coincides with the back of vertex D, as illustrated in the left drawing. Then fold back along segment AC so the back of vertex F coincides with the back of vertex D, as illustrated in the right image.

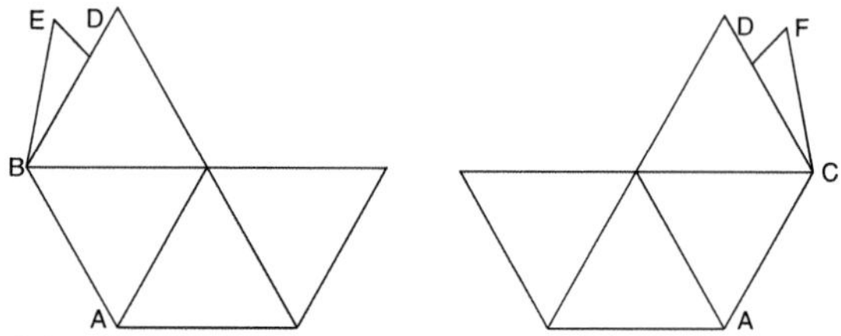

Step 2. Fold forward along segment GB so vertex E meets vertex A, as illustrated in the left drawing. Then fold forward along segment HC so vertex F meets vertex A, as illustrated in the right image.

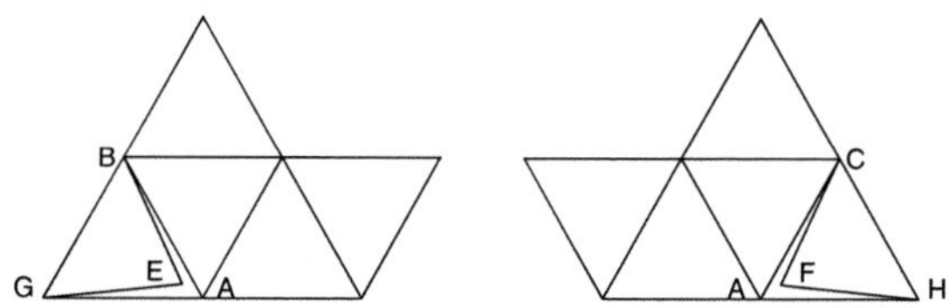

Step 3. Fold forward along segment BC so vertex D meets vertex A.

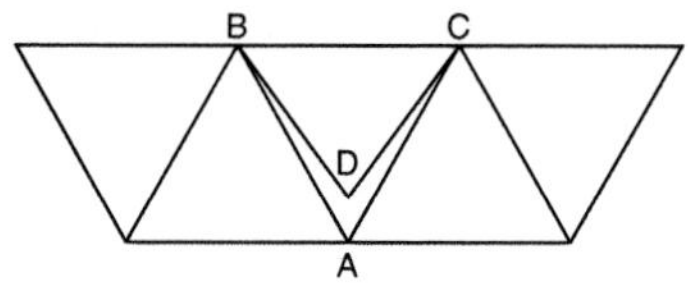

All of the folds we've done should be repeated several times, foreward and backward, to ease how the shape folds along these lines. The next step involves the movement of parts of the shape that induces some folds. We must be sure the paper folds easily.

Step 4. Now the shape is on the configuration of Fig. 2.18. The point of Steps 1, 2, and 3 is to make the folds at GB, BA, AC, CH, and BC flexible enough that the move of this step can raise the vertex A above the table. Now push E and F away from D, keeping the points D, B, E, G, H, F, and C on the table. When this is done, A rises above the plane of the table, so that G meets H and edge AG coincides with edge AH.

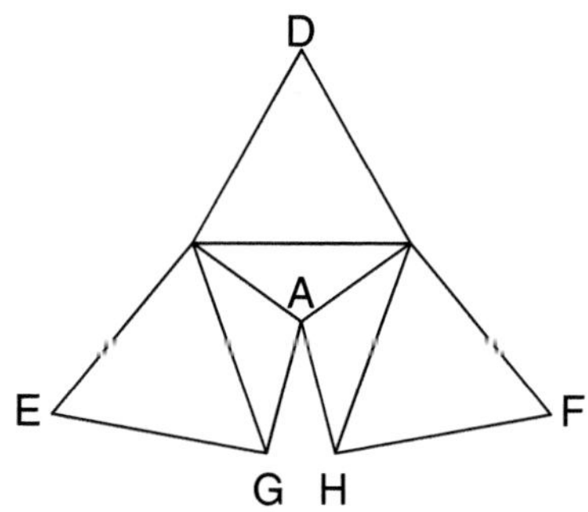

Step 5. Tape the edges AG and AH together. This forms a *stage 1 face* of a Koch tetrahedron, the left image of Fig. 2.19.

Repeat steps 1–5 for all 36 copies of the Koch tetrahedron template. This is where you will benefit from having many friends.

Step 6. Assemble six stage 1 faces. Tape them together so the stage 1 faces are oriented as the triangles of Fig. 2.18.

Step 7. Apply Step 4 to this assembly of six stage 1 faces. Tape together the edges corresponding to AG and AH. This produces a *stage 2 face*, the middle image of Fig. 2.19.

Step 8. Assemble six stage 2 faces. Tape them together so the stage 2 faces are oriented as the triangles of Fig. 2.18.

Step 9. Apply Step 4 to this assembly of six stage 2 faces. Tape together the edges corresponding to AG and AH. This produces a *stage 3 face*, the right image of Fig. 2.19.

Figure 2.19: The first, second, and third stages in the Koch tetrahedron construction.

2.5.5 Sample

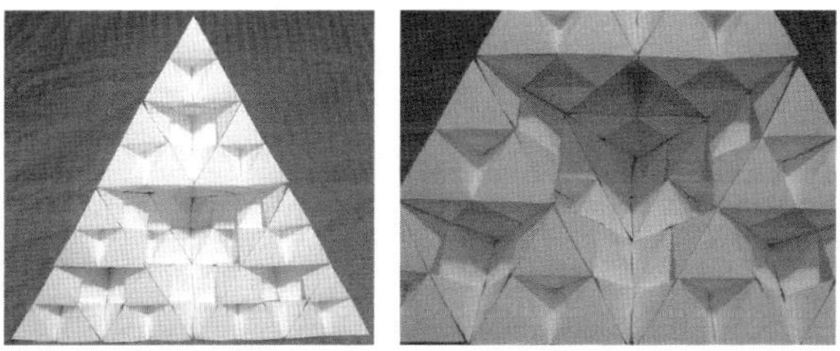

The inside of the Koch tetrahedron is interesting and complicated: three smaller chambers open off each larger chamber. "Big chambers have small chambers, a hint of crystallinity. And small chambers have smaller chambers, and so on to infinity." Sorry. All apologies to Jonathan Swift and his cascade of fleas upon fleas. The network of chambers contains no loops, but it does have an unlimited supply of spiral paths, smaller spirals branching off of larger spirals at each new chamber.

The outside is a different story. To see this more clearly, we arranged four copies of the stage 1 face along the faces of a tetrahedron. In principle, we'd take four copies of the stage 2 face along the faces of a tetrahedron, then four copies of the stage 3 face, and so on. These collapsed, or at least sagged, under their own weight. So in Fig. 2.20 we turn to computer images to see that the mathematical limit of four copies of the Koch tetrahedron is a ... cube.

In the early days of 3-dimensional computer graphics, the images of Fig. 2.20 were thought to be the surprise of this construction, that the spiky images on the left could turn into a cube. Images of this sort are part of the cover art of Stan Wagon's book[59]. Stan called this the "Koch planet." Certainly that

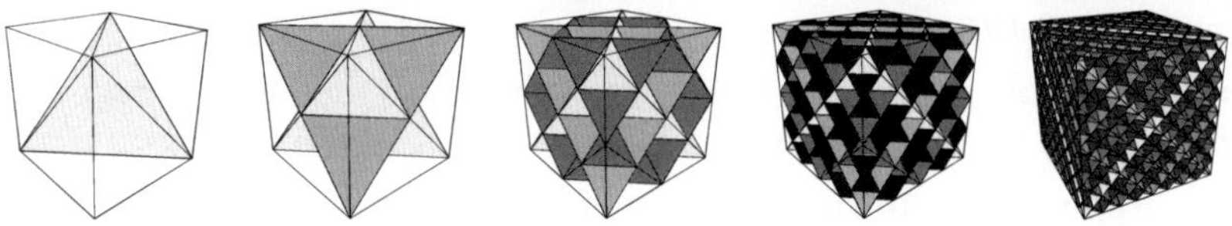

Figure 2.20: The first through fifth stages in the outside of the Koch tetrahedron construction.

a cube appears in the limit is a bit of a surprise, but we found the inside to be more surprising, and the difference between the outside and the inside to be disorienting.

2.5.6 Conclusion

In the Exercises we'll show that the dimension of the outside of the Koch tetrahedron is 2, while the dimension of the inside is $\log(6)/\log(2)$. How can one side of a surface have a higher dimension than the other?

To understand this, recall $\log(6)/\log(2)$ is the dimension of the whole object, not just of the outside. The inside of the construction is made of chambers within chambers within chambers. The interior walls twist and turn in a convoluted maze of nooks and crannies. Here is where the fractal lies.

The outside of the Koch tetrahedron can have dimension lower than the inside, and of the whole shape, by the monotonicity rule: the dimension of any piece is $\leq$ the dimension of the whole.

The fractal shown in Fig. 2.21 gives simpler example of this effect. Stage 1 is pictured on the left. To produce stage 2, pictured in the center, replace each line segment in stage 1 with a scaled version of stage 1, oriented appropriately. Stage 6 is shown on the right. Note the outside edges of the fractal have become two diagonal line segments, so the outside has dimension 1. Yet the whole fractal has dimension $\log(5)/\log(3)$.

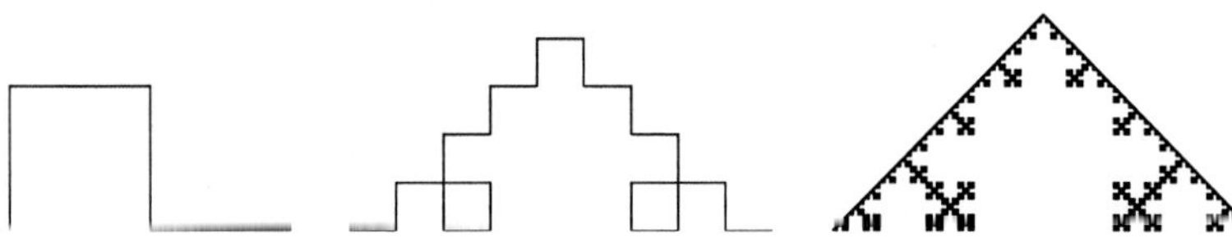

Figure 2.21: Construction of another shape whose outside and inside differ considerably.

Interestingly, the outside of the Koch tetrahedron can be painted with a finite amount of paint, but the inside cannot be painted with any finite amount of paint. Investigate the model for insight into why this is true.

2.5.7 Exercises

In the model just assembled, note each stage became larger than the previous one. In contrast to this, the mathematical construction of the Koch tetrahedron proceeds by filling in ever finer detail. Specifically, the process consists of these steps.

(i) Draw lines connecting the midpoints of the edges of each triangular face.

(ii) Remove the middle triangle formed this way.

(iii) Attach the top three faces of a tetrahedron whose base would have been the triangle just removed.

The difficulty of performing these operations physically is clear. The model is meant to serve as a readily constructed surrogate for this mathematical process. In these exercises, think of the mathematical fractal suggested by the physical model.

To clarify this point, suppose a stage 1 face consists of six equilateral triangles of side length 1. Then in the mathematical process a stage 2 face consists of 36 equilateral triangles of side length 1/2.

Prob 2.5.1 (a) Find the similarity dimension of a face of the Koch tetrahedron fractal.
(b) Find the similarity dimension of the Koch tetrahedron fractal.

Prob 2.5.2 In a stage 1 face, consider the tip of the three triangles on the tetrahedron (the other three triangles of this face are coplanar). Where is this tip located in the stage 2 face that results from applying the construction to the stage 1 face? Where on the stage 3 face? On the stage n face for all n?

Prob 2.5.3 The outer surface of the stage 3 face has many indentations. What happens to the size of these indentations as the stage number goes to infinity? What is the outer surface of this limiting shape formed by the sequence of stage n faces for all n?

Consider a stage 1 face. Here we see line segments drawn from the tip of the three faces of the tetrahedron to each of the vertices of the base.

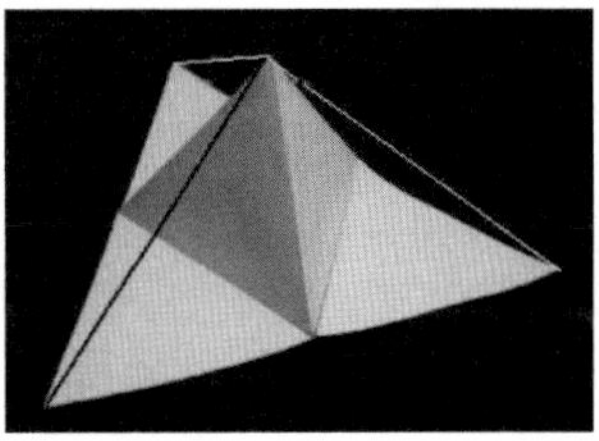

Prob 2.5.4 (a) Assume the six triangles making up the stage 1 face have side length 1. Find the lengths of the segments drawn here.
(b) Find the angle between two of these segments.

Prob 2.5.5 Using the information from Exercises 2.5.2, 2.5.3, and 2.5.4, what familiar shape is the outside of the Koch tetrahedron?

Prob 2.5.6 (a) Using the solution to Exercise 2.5.5, find the outside surface area of the limiting shape.
(b) From the center of the Koch tetrahedron, we see walls made of a collection of triangles. Find the area of these interior walls.
(c) From the solutions to (a) and (b) along with what we know about measuring in the wrong dimension (introduced in Sample 2.1.6), comment on the relation between the dimensions of the exterior walls and the interior walls of the Koch tetrahedron.

2.5.8 Templates

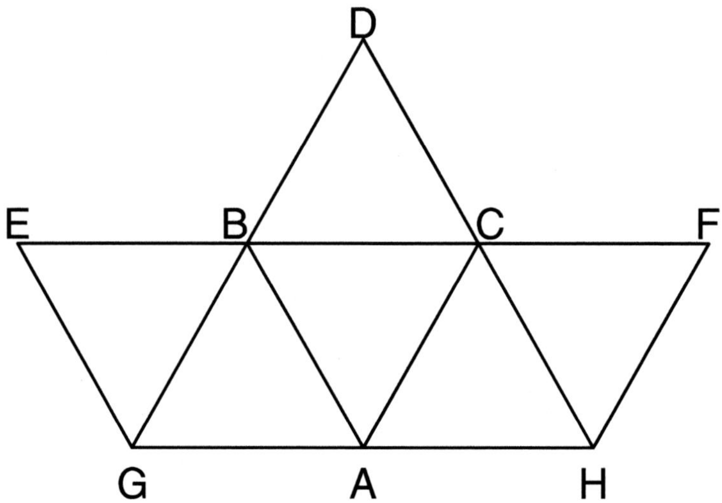

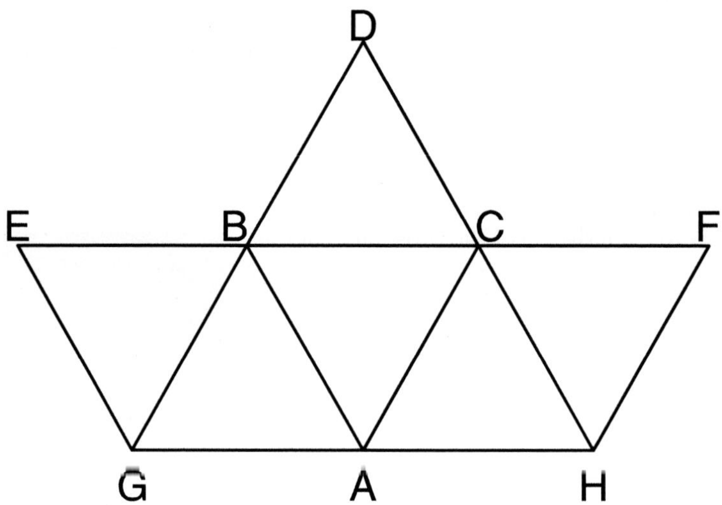

2.6 Sierpinski hypertetrahedron

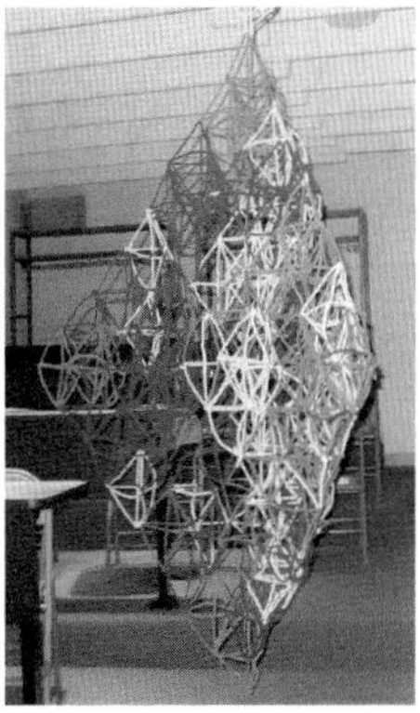

The Sierpinski tetrahedron is a natrual extension of the Sierpinski gasket. In Lab 2.4 we constructed a physical model of the Sierpinski tetrahedron. From this we learned (at least) three things.

• The self-similarity of the Sierpinski tetrahedron is expressed in the scale-invariant nature of the building instructions. Build four tetrahedra, attach them at their corners. Make four of these collections, attach them at the corresponding corners. Continue.

• Being a 2-dimensional subset of 3-dimensional space, the Sierpinski tetrahedron is quite empty.

• The complement is the union of a collection of octahedra, whose volumes sum to the volume of the tetrahedron. Because the dimension of the Sierpinski tetrahedron is 2, this is another instance of the result that measuring an object in too high dimension (volume is a 3-dimensional measure) gives 0.

The simplest 4-dimensional analog of a regular polyhedron is the 4-dimensional tetrahedron, or hypertetrahedron. After studying several ways to represent the hypertetrahedron, we'll find instructions for how to assemble hypertetrahedra into stages of the Sierpinski hypertetrahedron. With a pipecleaner model we explore some properties of this projection in 3 dimensions of a fractal in 4 dimensions.

2.6.1 Purpose

By analogy with how 3-dimensional objects project to 2 dimensions, we'll learn several ways that a 4-dimensional tetrahedron projects into 3 dimensions, and then project that projection into 2 dimensions. Guided by this, we'll bypass the last step and build a model of a Sierpinski hypertetrahedron in 3 dimensions.

2.6.2 Materials

A whole lot of pipe cleaners, and scissors to divide some of them. To build the third stage of the Sierpinski hypertetrahedron, we'll use 750 pipe cleaners. To allow for mistakes, around 800 is a safe number.

2.6.3 Background

Some people think 4-dimensional geometry is too difficult to understand. An encyclopedia article one of us (MF) read in childhood stated that 4-dimensional geometry "requires extreme feats of mental gymnastics." While the development of intuition for 4-dimensions analogous to our ability to manipulate 3-dimensional shapes in our minds does require considerable discipline, a basic comprehension of 4-dimensional geometry can be obtained easily, with the right approach. Thomas Banchoff's delightful book [60] provides much more information than we do here. We'll present what we think is the minimum amount of geometry needed to understand this lab.

Of all the 4-dimensional shapes, the hypercube may be the easiest to understand. We'll build up to the hypercube dimension by dimension.

We'll start with the unit line segment

$$L = \{x : 0 \leq x \leq 1\}.$$

The boundary of L is two points, $x = 0$ and $x = 1$. That is, the boundary of L is where the one variable, x takes one of its extreme values.

Now add a dimension. A unit line segment in both the x and y directions is the filled-in unit square

$$A = \{(x, y) : 0 \leq x, y \leq 1\}$$

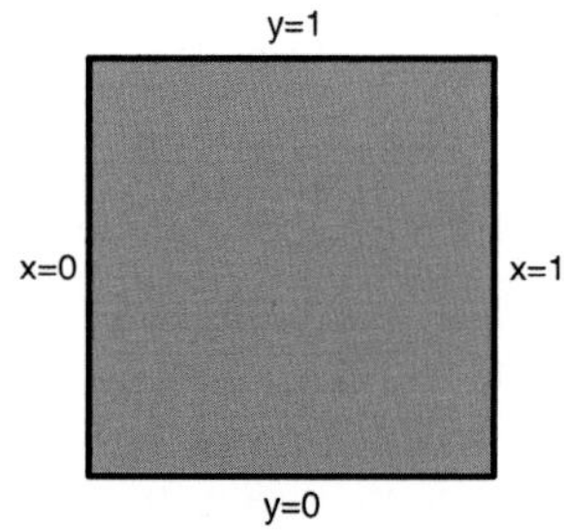

That is, all points (x, y) with $0 \leq x \leq 1$ and $0 \leq y \leq 1$. The boundary of the square is four line segments:

- *left*: $x = 0$ and $0 \leq y \leq 1$,
- *right*: $x = 1$ and $0 \leq y \leq 1$,
- *bottom*: $y = 0$ and $0 \leq x \leq 1$, and
- *top*: $y = 1$ and $0 \leq x \leq 1$.

That is, the boundary of S is where one variable takes one of its extreme values and the other variable takes its whole range. Two variables, each with two extreme values, so four line segments.

Now add another dimension to obtain the filled-in unit cube

$$C = \{(x, y, z) : 0 \leq x, y, z \leq 1\}$$

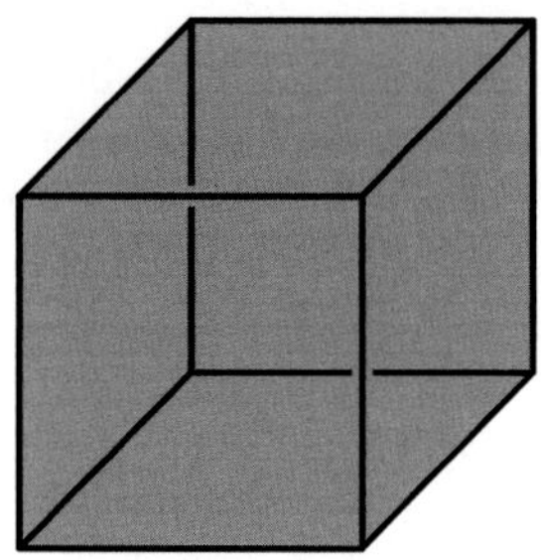

The boundary of the cube is six squares:

- *left*: $x = 0$ and $0 \leq y, z \leq 1$
- *right*: $x = 1$ and $0 \leq y, z \leq 1$
- *bottom*: $y = 0$ and $0 \leq x, z \leq 1$
- *top*: $y = 1$ and $0 \leq x, z \leq 1$
- *back*: $z = 0$ and $0 \leq x, y \leq 1$
- *front*: $z = 1$ and $0 \leq x, y \leq 1$

That is, the boundary of C is where one variable takes one of its extreme values and the other two variables take their whole range. Early art classes prepare us to sort out the squares that make up the cube boundary. They are in Fig. 2.22

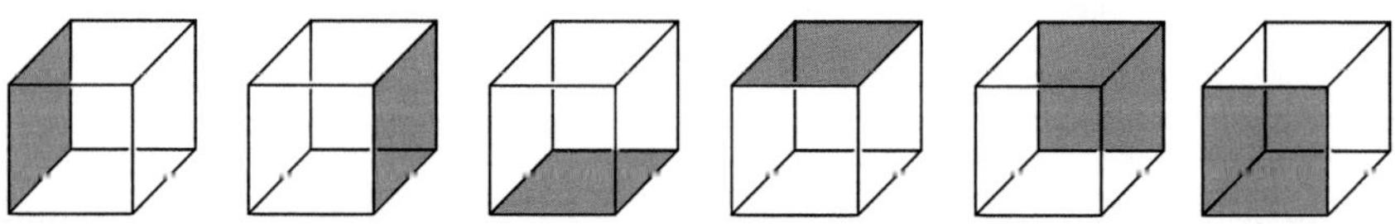

Figure 2.22: Cube faces: left, right, bottom, top, back, and front.

Now we're ready to find an algebraic expression for a hypercube. In our opinion, geometry gives true comprehension, but occasionally we must be led for a few moment by algebra. This is one of those moments. Look at our definitions of the unit line segment L, the unit square S, and the unit cube C. For the next step in this progression, the unit hypercube, we'll need a space with four coordinates, which we'll call w, x, y, and z. Then algebraically the unit hypercube is

$$H = \{(w, x, y, z) : 0 \leq w, x, y, z \leq 1\}$$

Pretty simple, but how can we draw this? Here's one way.

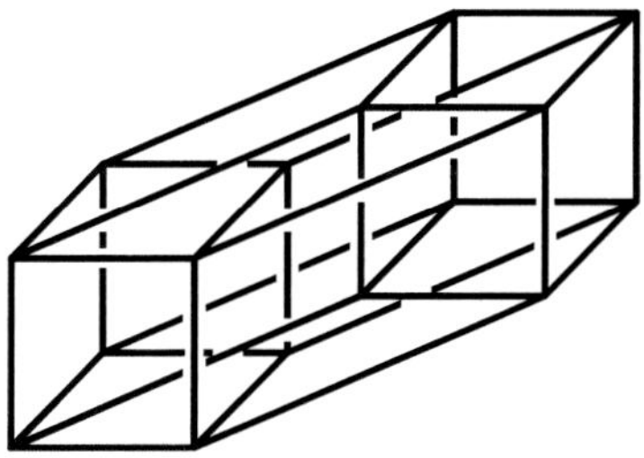

This might not have been your first guess, though it might have been if you really thought about the picture of the cube. We see the front and back squares, then the square formed by connecting the top of the front square to the top of the back square, the square formed by connecting the left side of the front square to the left side of the back square, and so on.

What is the analogous construction for the hypercube? Start with two cubes, which we'll call forward and backward. Then the left cube is formed by connecting the left square of the forward cube to the left square of the backward square. Similarly connect the other five squares that bound the forward cube to the corresponding squares that bound the backward cube.

Algebraically, the boundary of the hypercube consists of eight cubes:

- *left*: $x = 0$ and $0 \leq w, y, z \leq 1$
- *right*: $x = 1$ and $0 \leq w, y, z \leq 1$
- *bottom*: $y = 0$ and $0 \leq w, x, z \leq 1$
- *top*: $y = 1$ and $0 \leq w, x, z \leq 1$
- *back*: $z = 0$ and $0 \leq w, x, y \leq 1$
- *front*: $z = 1$ and $0 \leq w, x, y \leq 1$
- *backward*: $w = 0$ and $0 \leq x, y, z \leq 1$
- *forward*: $w = 1$ and $0 \leq x, y, z \leq 1$

The boundary of H is where one variable takes one of its extreme values and the other three variables take their whole range. That is, they define a (filled-in) cube. The eight boundary cubes of the hypercube are shown in Fig. 2.23

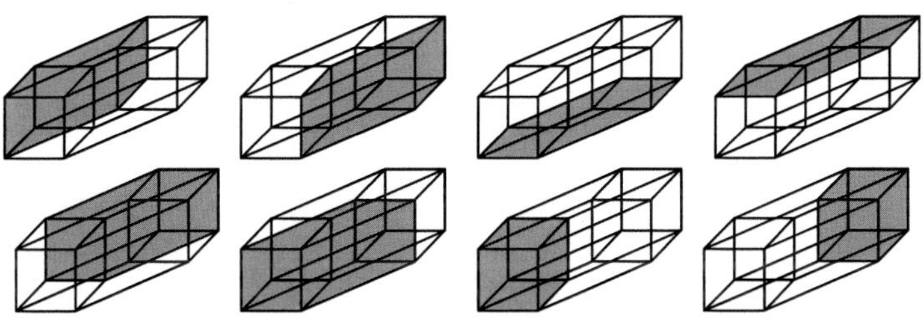

Figure 2.23: The cubes that bound the hypercube.

Now maybe you think, "Hold on. Only the last two of these look like cubes. The others are some sort of solid with parallel faces." That's certainly how the picture presents itself in the plane of the page, but you didn't complain about Fig. 2.22, and there only two of the cube faces are squares. Projected into the plane, four of the faces are parallelograms. Experience with our 3-dimensional surroundings presented through 2-dimensional images has taught us to parse these parallelograms as squares presented by perspective. The same holds for the cubes bounding the hypercube, though clear visualization of this does take some practice.

What fractal could we make that corresponds to a hypercube? Some version of the Sierpinski carpet pushed from 2 dimensions to 4? This would have a lot of pieces, too many for a model. Think of the carpet pictured here. Eight squares scaled by 1/3 doesn't hint at fractality. To see even the barest hint of self-similarity we'll need at least one more level, 64 copies scaled by 1/9. And that's just in 2 dimensions. In Exercise 2.6.2 you'll do the corresponding calculation in 3 and 4 dimensions. These numbers are far too large unless you have enough helpers to fill the Meteropolitan Opera House, twice, or Battell Chapel at Yale eight times.

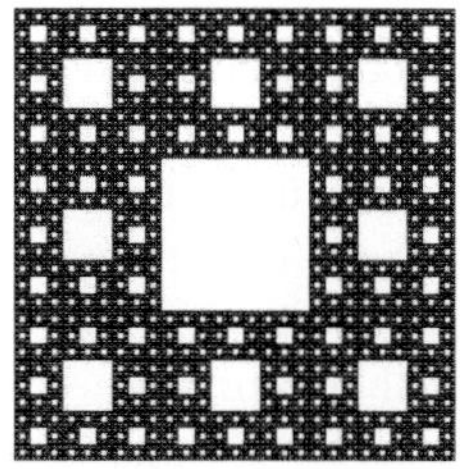

So we'll go to a simpler example, Sierpinski gasket to Sierpinski tetrahedron to Sierpinski hypertetrahedron. If the construction is clear enough, extension to even higher dimensions will be straightforward, if more than a little tedious. These models are just of the edges; no surfaces or solids. The higher the dimension, the more thoughtful we must be to interpret the model.

Our first step is to find a convenient projection of the hypertetrahedron. Algebraically simplest is to base the hypertetrahedron on a right isosceles triangle $\{(x,y) : 0 \le x, 0 \le y, x+y \le 1\}$. The right isosceles tetrahedron is $\{(x,y,z) : 0 \le x, 0 \le y, 0 \le z, x+y+z \le 1\}$. Then the algebraic description of the corresponding hypertetrahedron is straightforward:

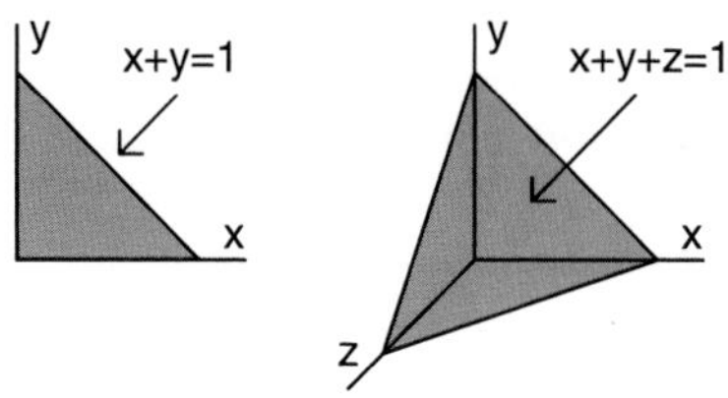

$$\{(w,x,y,z); 0 \le w, 0 \le x, 0 \le y, 0 \le z, w+x+y+z \le 1\}$$

Before we move on to geometry, let's count boundary pieces. We'll find the formula for the number of tetrahedra that make the boundary of a hypertetrahedron that is analogous to the formula for the number of cubes that make the boundary of a hypercube.

For the filled-in right isosceles triangle we can find the boundary by setting each condition of the definition to its extreme value: $x = 0$ is the extreme value of $x \le 0$ and gives the bottom edge of the triangle, $y = 0$ gives the left edge, and $x + y = 1$ is the extreme value of $x + y \le 1$ and gives the diagonal edge of the triangle. The boundary of the triangle is three line segments. This is obvious, in fact it's the definition of a triangle. We start this way to build a way to count the number of tetrahedra that make the boundary of a hypertetrahedron.

The extreme values of the tetrahedron definition are $x = 0$, $y = 0$, $z = 0$, and $x + y + z = 1$. These determine the left triangle, the bottom triangle, the back triangle, and the diagonal triangle that make up the boundary. The tetrahedron boundary is four triangles. Again, we knew this already. The point is the observation that each condition in the definition of a tetrahedron gives a piece of the tetrahedron boundary. The hypertetrahedron definition consists of five inequalities, so its boundary consists of five tetrahedra. For example, $w = 0$ gives $0 \leq x$, $0 \leq y$, $0 \leq z$, and $x + y + z \leq 1$, which defines a tetrahedron the xyz-space.

The same count of boundary components works for hypertetrahedra based on triangles of any shape, not just right isosceles.

Back to geometry. To ease the complexity of model construction, we'll use a more symmetric representation. We'll define a process that can be applied again and again with each additional dimension. The same rule, applied over and over, produces tetrahedra in ever-higher dimensions.

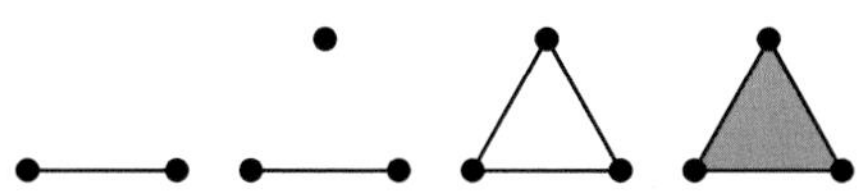

Begin with two points a unit distance apart on the x-axis. Add another point in the xy-plane, a unit distance from each of the initial points. Connect the points with line segments and fill in the triangle. This is a pretty complicated way to build an equilateral triangle. The point is that we can extend this construction to higher dimensions.

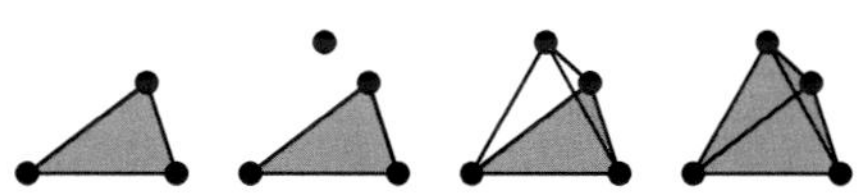

Take this equilateral triangle in the xy-plane. In xyz-space find a point that has distance 1 from each of the three vertices of the equilateral triangle. There are two such points, one with $z > 0$, the other with $z < 0$. Pick one. Connect the new point to the three vertices of the triangle. Fill in the triangular faces, and fill in the tetrahedron.

How do we build a 4-dimensional tetrahedron? Find a point in $wxyz$-space that is distance 1 from the four vertices of the tetrahedron. Connect this new point with the four vertices. Fill in the triangles, tetrahedra, and the hypertetrahedron we've just formed. This is called a symmetric hypertetrahedorn; the formation process is illustrated in Fig. 2.24.

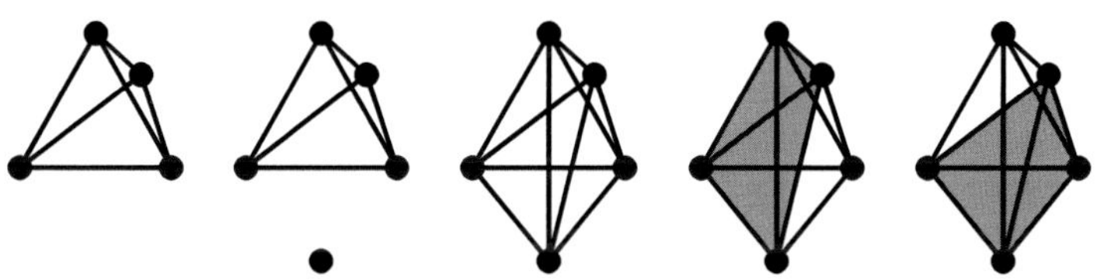

Figure 2.24: Construction of a symmetric hypertetrahedron. Two of the component tetrahedra are shaded.

Can we describe how to extend this process to higher dimensions? Certainly, but for this lab we'll stop at 4 dimensions.

Now maybe one step bothered you: "In xyz-space find a point that has distance 1 from each of the three vertices of the equilateral triangle." Without access to 4-dimensional space we must do this by a projection into 3-dimensional space. In four dimensions the segment from the top vertex to the bottom vertex has the same length as the other segments. In the symmetric projection this top to bottom segment is longer than the others.

Here we'll explore the first few steps in the construction of a Sierpinski hypertetrahedron based on the symmetric projection of the hypertetrahedron. In Exercise 2.6.4 you'll derive the construction of the Sierpinski hypertetrahedron based on the pyramid projection and the central projection of the hypertetrahedron.

2.6.4 Procedure

The procedure is based on iteration of the pattern used to form the symmetric projection of the hypertetrahedron. Recall the steps: first, place a point at each vertex of an equilateral triangle in the xy-plane. Second, above and below the centroid of the triangle place two points symmetrically. Connect these points to the triangle vertices with segments of length 1. Then connect these two points with a segment, of length 1 in 4-dimensional space but of necessity longer in this projection.

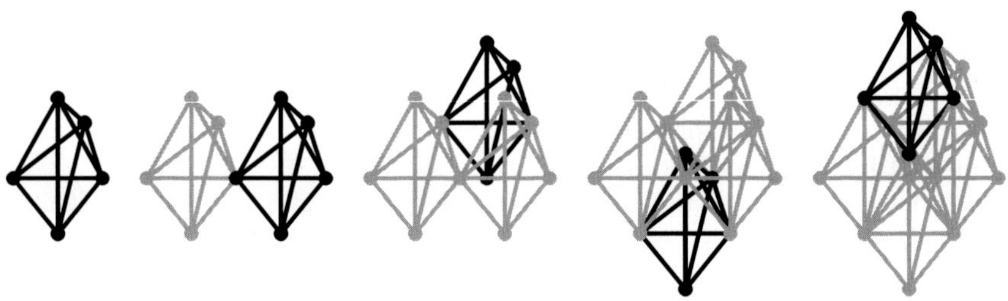

Figure 2.25: Construction of a symmetric Sierpinski hypertetrahedron.

Make five copies of the symmetric hypertetrahedron. Fig. 2.25 illustrates the steps of the construction. In each of the five images, the shape added in the step an image illustrates is darker than the others. Left to right: place one copy of the symmetric hypertetrahedron, add another copy to the right and connected at a vertex to the previous copy. Add a third copy behind these two, again connected at vertices. Add a copy below these three, the three middle vertices of this copy attached to the bottom vertices of the first three copies. Add a copy above these three, the three middle vertices of this copy attached to the top vertices of the first three copies.

How to continue is clear. Assemble five copies of this and connect them in the same pattern.

How can we do this in physical space?

Begin with 6 pipe cleaners. As shown in the left image, bend 5 at their midpoints, at about 60° angles. To attach one pipe cleaner to another, twist about a quarter inch of their ends together. Connect two of the bent pipe cleaners as shown on the right.

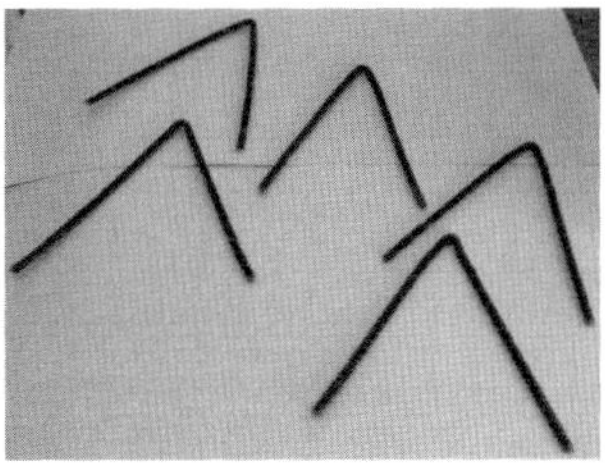
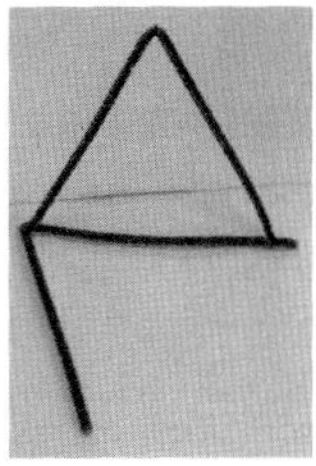

Next, attach a third bent pipe cleaner, forming a tetrahedron, shown in the left image. Note the leftover quarter inch of pipe cleaner extending to the right. This will be used to attach other pipe cleaners.

Now attach another bent pipe cleaner along a side of the base of the tetrahedron, as shown on the right.

Attach one edge of the remaining bent pipecleaner to the tetrahedron base corner not used in the previous step. Attach the bend to the bend of the pipecleaner attaqched in the previous step and direct the other leg of the pipe cleaner toward the apex of the tetrahedron, shown in the left image. This leg of the pipe cleaner will not reach that apex, so use a portion of the unbent pipecleaner to complete the edge to that apex, shown on the right.

This is a model of the edges of the symmetric projection of the hypertetrahedron. To build the first stage of the Sierpinski hypertetrahedron, attach five of these in the positions suggested by the vertices, with edges corresponding to attachments of the hypertetrahedra. Portions of the remainder of the unbent pipe cleaners can be used to make these attachments.

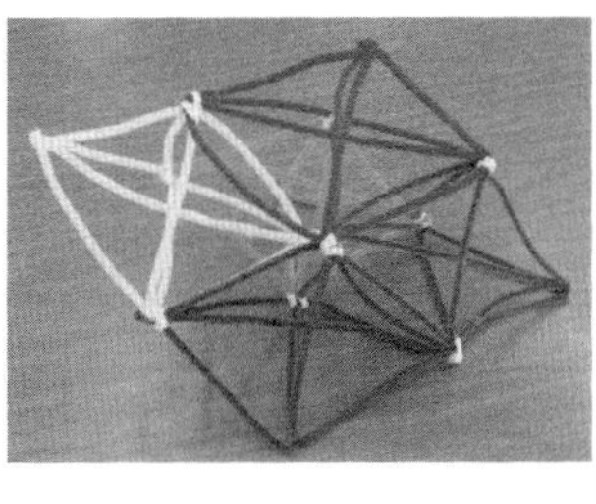

For the second stage, assemble five of these in the corresponding pattern. For the third stage, assemble five copies of the second stage in the corresponding pattern. And surely that's enough.

2.6.5 Sample A

Here is a picture of a model of the edges of the third stage in the construction of the symmetric projection of the Sierpinski hypertetrahedron. The first stage uses five copies of the hypertetrahedron, the second stage uses five copies of the first, and so 25 copies of the hypertetradedron. The third stage uses five copies of the second stage, hence 125 copies of the hypertetrahedron. So the first step is to make 125 hypertetrahedron skeleta, the collection of hypertetrahedron edges. This is the reason this lab, along with Labs 2.4 and 2.5, are better as group projects. While every group follows the same procedure, for the last step all the models are assembled together.

The first stage of the Sierpinski hypertetrahedron is adequately rigid to support itself on a desktop. When five of these are assembled into the second stage, gravity causes an inevitable sag of the lower bits under the weight of the upper bits. The suggestion that we assemble the third stage in the microgravity environment of the ISS was not taken seriously. Of course, the suggestion wasn't made seriously. Instead, we hung one copy of the second stage from the ceiling, attached the top vertex of three copies to the side vertices of this top copy, and then attached the side vertices of another copy (the bottom copy) to the bottom vetex of each of the three middle copies. The final connection is the top vertex of the bottom copy to the bottom vertex of the top copy.

Then we stood around and looked at the model. Of course everyone photographed the model with their cellphones. Everyone except one of your Luddite authors (MF) who doesn't have a cellphone. Then looks of mild puzzlement gave way to understanding as the projection unpacked itself and revealed a hint of how the Sierpinski hypertetrahedron inhabits four dimensions. Not everyone succeeded in seeing this, but many did, and for those the most common expression was surprise.

2.6.6 Sample B

Here we'll shade the five tetrahedra that make up the boundary of the symmetric projection of the hypertetrahedron. This is straightforward. A hypertetrahedron has five vertices, a tetrahedron has four vertices. To find the tetrahedra that make up the boundary of a hypertetrahedron, take all collections of four of the five hypertetrahedron vertices. (There are five of these collections. Do you see why?) Then shade the tetrahedron determined by each collection of four vertices. For the symmetric projection, these are shown in Fig. 2.26.

If you can see algebra more clearly than geometry, the hypertetrahedron developed from right isosceles triangles may be clearer. Remember this formu-

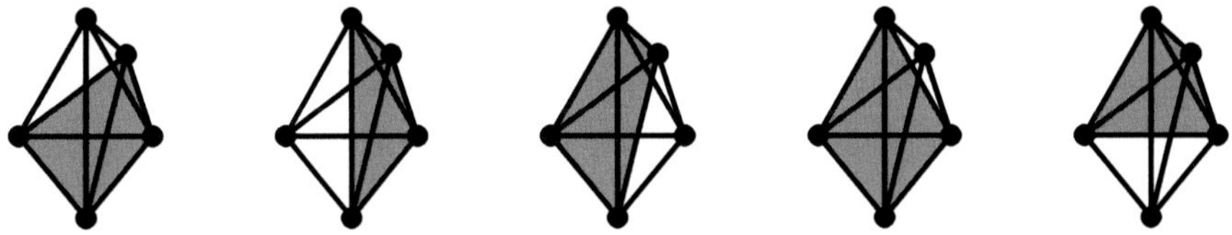

Figure 2.26: The five tetrahedra that make up the symmetric projection of the hypertetrahedron.

lation is

$$\{(w,x,y,z) : 0 \leq w, 0 \leq x, 0 \leq y, 0 \leq z, w+x+y+z \leq 1\}$$

Then the five tetrahedra that make up the boundary of the hypertetrahedron are determined by restricting coordinates to their extreme values. These are
$w = 0$, $0 \leq x, y, z$, $x + y + z \leq 1$,
$x = 0$, $0 \leq w, y, z$, $w + y + z \leq 1$,
$y = 0$, $0 \leq w, x, z$, $w + x + z \leq 1$,
$z = 0$, $0 \leq w, x, y$, $w + x + y \leq 1$, and
$0 \leq w, x, y, z$, $w + x + y + z = 1$.

The first four are filled-in tetrahedra in xyz-space, wyz-space, wxz-space, and wxy-space. To see that the fifth is a filled-in tetrahedron, observe that is the region bounded by four triangles: $w = 0$ and $x + y + z = 1$,
$x = 0$ and $w + y + z = 1$,
$y = 0$ and $w + x + z = 1$, and
$z = 0$ and $w + x + y = 1$.
These four triangles bound the fifth tetrahedron.

2.6.7 Conclusion

Along with Labs 2.4 and 2.5, this lab is an example of a recursive construction: the same steps applied to the result of the previous application of those steps. The product is fractal because the process is fractal, the same steps repeated across different scales.

But also, this lab is an effective way to visualize some aspects of 4-dimensional geometry. Is the physicality of 3-dimensional space better understood through a perfectly smooth sphere or through a realistic globe rough with mountains and valleys? Because scaling occurs so often in our surroundings, our minds and eyes are trained to recognize scaling and exploit it to parse our surroundings. The self-similarity of a 4-dimensional fractal can help us see the new direction.

2.6.8 Exercises

Prob 2.6.1 (a) How many hypercubes make up the boundary of a 5-dimensional cube? We'll call a 5-dimensional cube a 5-cube.
(b) How many 5-cubes make up the boundary of a 6-cube?

(c) For all integers $n \geq 1$, how many $(n-1)$-cubes make up the boundary of an n-cube?

Prob 2.6.2 (a) For the analog of the Sierpinski carpet in 3 dimensions—remove the middle $1/3 \times 1/3 \times 1/3$ cube, then iterate—how many $1/9 \times 1/9 \times 1/9$ cubes are in the second iterate?
(b) For the analogue of the Sierpinski carpet 4 dimensions, how many $1/9 \times 1/9 \times 1/9 \times 1/9$ hypercubes are in the second iterate?

The next two exercises use two other projections of the hypertetrahedron, the *central projection* (Fig. 2.27) and the *pyramid projection* (Fig. 2.28). In both, edges intersect only at vertices, the dots in these images. All other apparent crossings of edges are illusions that result from projection. Also, in 4-dimensional space, the proper home of these hypertetrahedra, all edges have length 1. When projected into 3 dimensions—really, into the 2 dimensions of this page—the lengths of some edges appear to depart from 1.

To build a central projection of a hypertetrahedron, begin with a regular tetrahedron in 3-dimensional space (left image), add a fifth vertex near the centroid of the tetrahedron (middle image), and add edges that connect the fifth vertex to the four vertices of the original tetrahedron. In 4 dimensions all the edges have length 1.

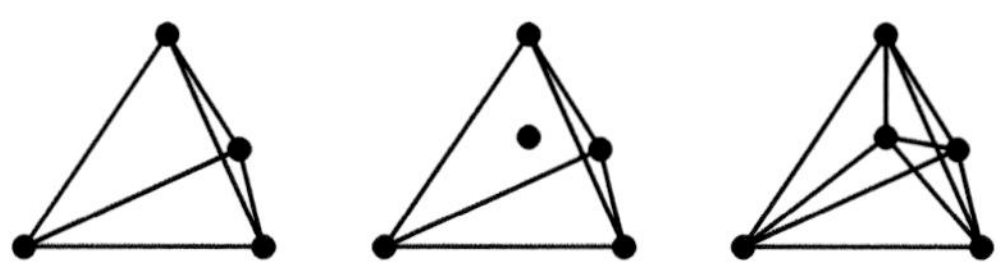

Figure 2.27: Central projection

To build a pyramid projection, begin with a regular tetrahedron projected onto a plane so its vertices project to the vertices of a square. This is the left image. Remember that the two edges that appear to cross inside the square don't intersect. This apparent crossing is a result of the projection. Now add a vertex above the center of the square (middle image), and connect that vertex to each vertex of the original tetrahedron.

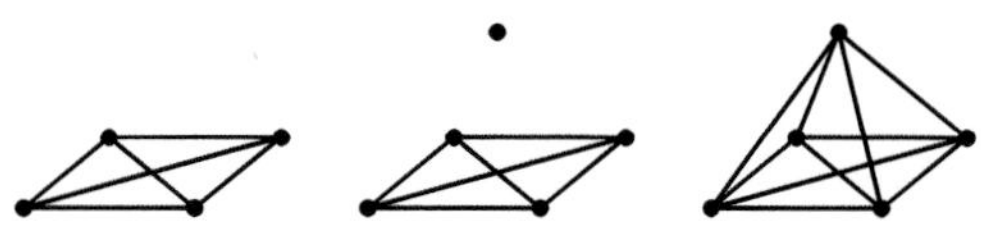

Figure 2.28: Pyramid projection

Prob 2.6.3 Shade the five tetrahedra that make up the boundary of the hypertetrahedron for
(a) the central projection and
(b) the pyramid projection

Prob 2.6.4 Describe the process for the construction of the second stage of the Sierpinski hypertetrahedron with
(a) the pyramid projection and
(b) the central projection.

Prob 2.6.5 (a) Compute the similarity dimension of the Sierpinksi hypertetrahedron.

(b) Find the general formula for the dimension $d(n)$ of the Sierpinski gasket analog $SGA(n)$ in n dimensions. What do we mean by n here? The gasket has $n = 2$, the Sierpinski tetrahedron has $n = 3$, the Sierpinski hypertetrahedron has $n = 4$, and so on.
(c) For $n = 2, \dots 100$, plot the points $(n, d(n)/n)$. Interpret the graph as an indicator of the wispiness, the emptiness, of $SGA(n)$ as the ambient dimension n increases.

2.7 Basic multifractals: $f(\alpha)$ curves

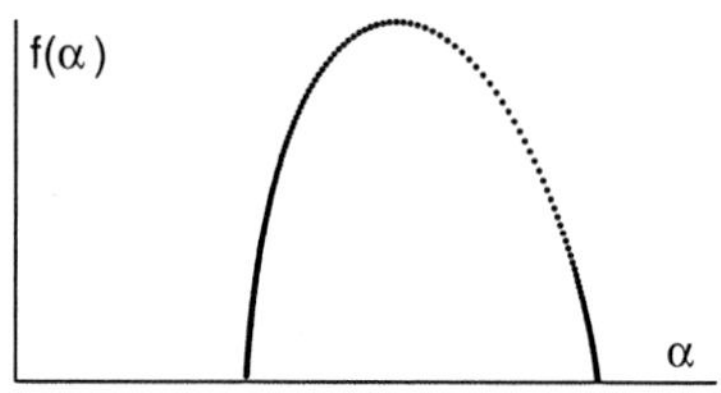

As Benoit developed interpretations of fractal dimension in the sciences, eventually he saw that dimension is a measure of roughness [61]. So far, dimension appears to be the only repeatable measure of roughness. This filled a gap in the relation between our senses and sensations we can quantify. The sensation of warmth is measured by degrees of temperature, loudness by decibels, sourness by pH, heaviness by pounds, and so on. But until scientists developed methods to estimate fractal dimension of physical objects, the sensation of roughness could not be quantified.

This, it turns out, is just the start of the story. Run your fingers lightly across the bark of a tree or the surface of a stone. Often the sensation of roughness is not constant. And often a more careful analysis shows that the roughness is not distributed in a simple fashion, nothing like the left side is rough and the right side is smooth. Regions of one roughness are mixed into regions of another in complicated ways. One way to understand this mixing is to measure the dimension of areas of constant roughness. If these "iso-roughness" regions are fractals, then the surface is made of a multitude of fractals. It is a *multifractal*.

That this technique is useful can be seen by the number of publications that use multifractals. On March 26, 2021 a Google Scholar search for the word "multifractal" returned about 811,000 hits. That this topic is worthy of our efforts is undeniable. But multifractals are complicated. We'll focus on a simple model, based on a generalization of the Moran equation (2.8).

2.7.1 Purpose

Here we'll learn some of the basic concepts of multifractal analysis. Specifically, we'll introduce the Hölder exponents α, the $f(\alpha)$ curve (the multifractal spectrum) through the Moran equation generalized to include probabilities. Other approaches include the method of moments and the method of histograms to compute multifractal spectra, and ways to use multifractals to study time series. But these involve many subtle issues, and some math considerably more advanced than we'll pursue. Our main emphasis will be what we can learn from the generalized Moran equation.

2.7.2 Materials

We'll use multifractal software in 2.7 (a), (b), and (c) of the Mathematica or Python codes.

2.7.3 Background

We'll use the IFS of Lab 1.1 and the IFS with probability of Lab 1.2, the notion of addresses from Lab 1.7, and the Moran equation from Lab 2.3.

In addition, we'll need some concepts specific to this lab. We'll need the Hölder exponent α, how to compute the multifractal spectrum, and some familiarity with what information we can read from the $f(\alpha)$ curve. Here we'll develop concepts and state results. The longer derivations are presented in Appendix B.9.

2.7.3.1 The generalized Moran equation

For an IFS with similarity transformations $T_1, \dots, T_n$ and with attractor A, if A is rendered by the random IFS algorithm of Lab 1.2, then the T_i are applied with probability p_i. Denote the contraction factor of T_i by r_i. If the pieces $T_i(A)$ don't overlap much, then the similarity dimension d of this fractal is the (unique) solution of the Moran equation $r_1^d + \cdots + r_n^d = 1$.

How can we modify this equation to account for the probabilities? We can't just multiply each r_i^d by p_i, because unless all the p_i are equal, what would go on the right-hand side of the equation? A single equation to capture the variety of several interacting probabilities seems unlikely.

If one equation won't work, let's try a lot of equations. In fact, let's try infinitely many. For each real number q, multiply r_i^d by p_i^q. The reason for this approach is that positive q magnify the effect of the larger p_i, the larger the q, the larger the effect; while negative q magnify the effect of the smaller p_i. But there's a problem, because $p_1^q r_1^d + \cdots + p_n^d r_n^d$ needn't equal 1, unless $q = 1$. To get the left-hand side equal to 1, we must take the exponent of r_i to be $\beta(q)$, a function of q. Then the generalized Moran equations are

$$p_1^q r_1^{\beta(q)} + \cdots + p_n^q r_n^{\beta(q)} = 1 \tag{2.13}$$

In Appendix B.9 we'll see why this equation has a unique solution for each q.

2.7.3.2 The Hölder exponent

For each term of the left-hand side of Eq. (2.13),

$$\frac{d}{dq}\left(p_i^q r_i^{\beta(q)}\right) = p_i^q \ln(p_i) r_i^{\beta(q)} + p_i^q r_i^{\beta(q)} \ln(r_i) \frac{d\beta}{dq}$$

So when we differentiate Eq. (2.13) with respect to q and solve for $d\beta/dq$ we obtain

$$\frac{d\beta}{dq} = -\frac{p_1^q r_1^{\beta(q)} \ln(p_1) + \cdots + p_n^q r_n^{\beta(q)} \ln(p_n)}{p_1^q r_1^{\beta(q)} \ln(r_1) + \cdots + p_n^q r_n^{\beta(q)} \ln(r_n)} \tag{2.14}$$

Now we define α, the Hölder exponent, by

$$\alpha = -\frac{d\beta}{dq} \tag{2.15}$$

If all the IFS scaling factors take on a common value $r_i = r$, we can solve Eq. (2.13) for $\beta(q)$:

$$\beta(q) = -\frac{\ln\left(p_1^q + \cdots + p_n^d\right)}{\ln(r)} \tag{2.16}$$

and we can simplify Eq. (2.14),

$$\alpha = -\frac{d\beta}{dq} = \frac{p_1^q \ln(p_1) + \cdots + p_n^q \ln(p_n)}{\ln(r)\left(p_1^q + \cdots + p_n^q\right)} \tag{2.17}$$

Because $0 < r < 1$ and $0 < p_i < 1$, we see that $\ln(r) < 0$ and $\ln(p_i) < 0$, so $\alpha > 0$. In the situation that all $r_i = r$, the range of α values is easy to find. Here's how.

Start with Eq. (B.20) and adapt it to the situation all $r_i = r$.

$$\alpha(i_1 \dots i_k) = \frac{\log(p_{i_1} \cdots p_{i_k})}{\log(r_{i_1} \cdots r_{i_k})} = \frac{\log(p_{i_1} \cdots p_{i_k})}{\log(r^k)}$$

A relation between this formulation of α and that of Eq. (2.15) is in Sect. 17.3 of [16]. Now

$$(\min(p_i))^k \leq p_{i_1} \cdots p_{i_k} \leq (\max(p_i))^k$$

Then because log is an increasing function,

$$\log((\min(p_i))^k) \leq \log(p_{i_1} \cdots p_{i_k}) \leq \log((\max(p_i))^k)$$

Finally, because $\log(r^k)$ is negative,

$$\frac{\log((\min(p_i))^k)}{\log(r^k)} \geq \frac{\log(p_{i_1} \cdots p_{i_k})}{\log(r^k)} \geq \frac{\log((\max(p_i))^k)}{\log(r^k)}$$

Consequently, for all k and for all $i_1, \dots, i_k$,

$$\max(\alpha) = \frac{\log(\min(p_i))}{\log(r)} \geq \alpha(i_1 \dots i_k) \geq \frac{\log(\max(p_i))}{\log(r)} = \min(\alpha) \tag{2.18}$$

Note that the maximum value of α corresponds to the minimum probability.

The more general case, where the r_i are not all identical, is this

$$\max\left(\frac{\log(p_i)}{\log(r_i)}\right) \geq \alpha(i_1 \dots i_k) \geq \min\left(\frac{\log(p_i)}{\log(r_i)}\right) \tag{2.19}$$

2.7.3.3 The $f(\alpha)$ curve

For each value of α, $f(\alpha)$ is the dimension of the subset of those points of A for which the Hölder exponent is α. This is not at all obvious, but $f(\alpha)$ is given by this formula:

$$f(\alpha) = q\alpha + \beta(q) \tag{2.20}$$

To see this requires a considerable amount of work. We'll see a rough sketch in Appendix B.9; more details are in Appendices A.97–A.99 of [17], and complete analyses are in Chapter 17 of [16] and Chapters 10 and 11 of [20].

In Appendix B.9 we'll sketch proofs of these two properties of $f(\alpha)$ curves.

1. The $f(\alpha)$ curve is concave down if the $\log(p_i)/\log(r_i)$ are not all equal.
2. The maximum value of the $f(\alpha)$ curve is the dimension of the attractor.

The random IFS algorithm can be extended to IFS with memory (Lab 1.9) by a modification of the transition matrix to $[p_{ij}]$ where p_{ij} is the probability that transformation T_i follows T_j. Then the generalized Moran equation becomes

$$\rho([p_{ij}^q r_i^{\beta(q)}]) = 1 \tag{2.21}$$

where $\rho[M]$ is the spectral radius of the matrix M. For fractals generated by random IFS with memory, once we have an expression for $\beta(q)$ from Eq. (2.21), α can be computed by Eq. (2.17) and $f(\alpha)$ by Eq. (2.20).

Before we can do any calculations, we must address one more issue. How can we plot the $f(\alpha)$ curve if we can't write f as an explicit function of α? In fact, we can find a formula for f as a function of α only occasionally. So what can we do?

If all the scaling factors take on a common value, $r_i = r$, then Eq. (2.17) gives an expression for α as a function of q. Together with Eq. (2.16), from Eq. (2.20) we can write f as a function of q. Then we plot the points $(\alpha(q), f(q))$ for a range of q values, say $q = -20$ to $q = 20$, though some experimentation may be needed to find a range that covers most of the $f(\alpha)$ curve. Connect the dots to get an approximation of the curve.

2.7.4 Procedure

The input is an IFS with probabilities. Sometimes we can determine properties of the $f(\alpha)$ curve from first principles. Other times we plot points on these curves with the software in 2.7 of the Mathematica or Python codes.

We'll assume that the translations are arranged so the pieces of the attractor overlap at most along edges. Then the computation of both the dimension and the multifractal spectrum depend on only the scaling factors and probabilities. So this is all we'll specify in exercises.

2.7.5 Sample A

Without the software, for these IFS find the minimum and maximum values, $\alpha_{\min}$ and $\alpha_{\max}$ of α, and find $f(\alpha_{\min})$ and $f(\alpha_{\max})$, and find the maximum height of the $f(\alpha)$ curve.
(a) $r_1 = r_2 = r_3 = r_4 = 0.5$; $p_1 = 0.1$, $p_2 = 0.2$, $p_3 = p_4 = 0.35$.

Because all the r_i are equal, we can use (2.18) to compute the bounds on α. We find

$$\alpha_{\min} = \frac{\log(\max(p_i))}{\log(0.5)} = \frac{\log(0.35)}{\log(0.5)} \approx 1.51457$$

$$\alpha_{\max} = \frac{\log(\min(p_i))}{\log(0.5)} = \frac{\log(0.1)}{\log(0.5)} \approx 3.32193$$

The minimum value of α occurs on the attractor of the IFS $\{T_3, T_4\}$. For these transformations $r = 1/2$, so $f(\alpha_{\min}) = \dim\{T_3, T_4\} = \log(N)/\log(1/r) = \log(2)/\log(2) = 1$. Here $\dim\{IFS\}$ is the dimension of the IFS attractor.

The maximum value of α occurs on the attractor of the IFS $\{T_1\}$. For this transformation $r = 1/2$, so $f(\alpha_{\min}) = \dim\{T_1\} = \log(1)/\log(2) = 0$.

The maximum value of the $f(\alpha)$ curve is the dimension of the attractor. Then $N = 4$ and $r = 1/2$, so $\max\{f(\alpha)\} = \log(4)/\log(2) = 2$.

(b) $r_1 = r_2 = r_3 = 0.5$, $r_4 = r_5 = r_6 = r_7 = 0.25$; $p_1 = p_2 = p_3 = 0.3$, $p_4 = p_5 = p_6 = p_7 = 0.025$.

Because the r_i have different values, we must use (2.19) to compute the bounds on α. This IFS has only two values of r and only two values of p, larger p associated with larger r, smaller p with smaller r. We find

$$\alpha_{\min} = \frac{\log(0.3)}{\log(0.5)} \approx 0.173697 \qquad \alpha_{\max} = \frac{\log(0.025)}{\log(0.25)} \approx 2.66096$$

The minimum value of α occurs on the attractor of the IFS $\{T_1, T_2, T_3\}$. That is, the set with $\alpha = \alpha_{\min}$ consists of $N = 3$ pieces scaled by $r = 1/2$, so $f(\alpha_{\min}) = \dim\{T_1, T_2, T_3\} = \log(3)/\log(2) \approx 1.58496$.

The maximum value of α occurs on the attractor of the IFS $\{T_4, T_5, T_6, T_7\}$. For these, $r = 1/4$and so $f(\alpha_{\min}) = \dim\{T_4, T_5, T_6, T_7\} = \log(4)/\log(4) = 1$.

The maximum value of the $f(\alpha)$ curve is the dimension of the attractor. This has $N = 3$ pieces scaled by $r = 1/2$ and $N = 4$ pieces scaled by $r = 1/4$, so we'll use the Moran equation, although after that calculation we'll see a simpler way to find the dimension of the attractor. The Moran equation becomes

$$3(1/2)^d + 4(1/4)^d = 1 \quad \text{or } 3x + 4x^2 = 1 \text{ if we take } x = (1/2)^d$$

The solutions are $x = -1$, impossible because $x = (1/2)^d$ is positive, and $x = 1/4$. Then $1/4 = (1/2)^d$ gives $d = 2$.

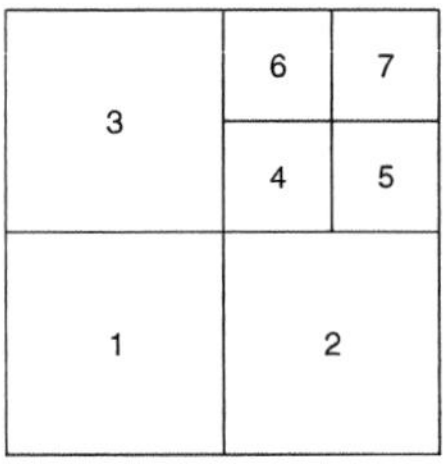

Another way to see that the dimension of the attractor is 2 is to recall a corollary of all our dimension formulas: so long as the pieces don't overlap too much (overlapping at most along edges suffices), then the dimension is determined by the number of pieces and the size of each piece. The position of the pieces doesn't change the dimension. So as shown here, we can decompose a filled-in square into three pieces scaled by 1/2 and four pieces scaled by 1/4. The filled-in square has dimension 2, so any shape that can be decomposed this way has $d = 2$.

2.7.6 Sample B

Use the software of 2.7 in the Mathematica or Python codes to plot the multifractal spectra for Sample 2.7.5 (a) and (b). What can you read from the comparison of these curves?

The software gives parametric plots, that is, points that lie on the curves. By plotting enough points, and points close enough together, we can get a good picture of the actual $f(\alpha)$ curve. The software plots points $(\alpha(q), f(\alpha(q)))$ for q evenly spaced between a negative bound q_{neg} and a positive bound q_{pos}. As a first step, $q_{\text{neg}} = -10$ and $q_{\text{pos}} = 10$ are good choices, but bounds of larger magnitude will fill more points at the left and right ends of the $f(\alpha)$ curve.

Equal spacing in q doen't give equal spacing in α. Typically, the points clump together at the ends of the graph and spread out near the middle.

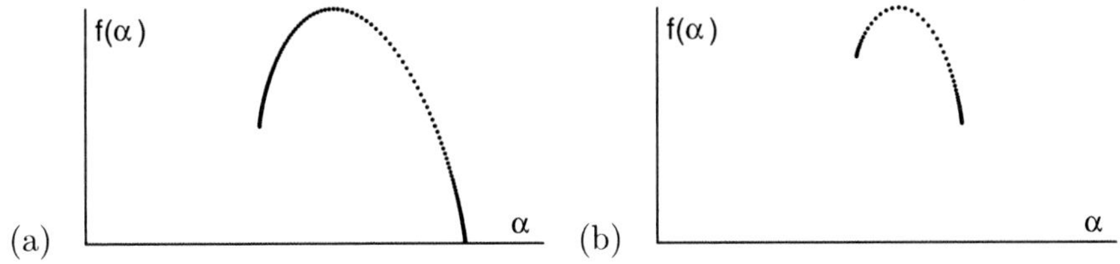

Figure 2.29: Points on $f(\alpha)$ curves for (a) and (b) of Sample 2.7.5.

What do we see? The (b) curve looks like a copy of the (a) curve, reduced in size and translated up. Of course, that is just a rough impression. Careful measurement would reveal differences. If $f(\alpha_{\min}) > 0$ or $f(\alpha_{\max}) > 0$, then more than one IFS transformation generates the part of the attractor with $\alpha = \alpha_{\min}$ or $\alpha = \alpha_{\max}$. The maximum of both curves occurs at the α for which $f(\alpha) = d(A)$, the dimension of the attractor of the IFS. This α is far away from $\alpha_{\min}$ and $\alpha_{\max}$. Which combinations of transformations give points in the region with this dimension? Both graphs look approximately symmetrical about a vertical line through their maximum, if we imagine extending the curves to the α-axis. If the distribution of probabilities were more skewed—some very small values, some very large values—would the $f(\alpha)$ curve be less symmetrical? Questions abound. The exercises will provide examples that may suggest some answers.

2.7.7 Sample C

We'll use the four transformations (1.7) that generate the filled-in unit square, together with the probability transition matrix $P = [p_{ij}]$ and the corresponding topological transition matrix $M = [m_{ij}]$

$$P = \begin{bmatrix} 0 & 0.50 & 0.33 & 0.33 \\ 0.50 & 0.25 & 0 & 0.33 \\ 0.25 & 0 & 0.33 & 0.34 \\ 0.25 & 0.25 & 0.34 & 0 \end{bmatrix} \qquad M = \begin{bmatrix} 0 & 1 & 1 & 1 \\ 1 & 1 & 1 & 0 \\ 1 & 0 & 1 & 1 \\ 1 & 1 & 1 & 0 \end{bmatrix}$$

Because all the scaling factors are $r_i = r = 1/2$, and because the eigenvalues of $[kp_{ij}]$ are k times the eigenvalues of $[p_{ij}]$ (an example or two will show why this is true), the generalized Moran equation (2.21) simplifies to

$$r^{\beta(q)}\rho([p_{ij}^q]) = 1 \qquad (2.22)$$

which we can solve for $\beta(q)$

$$\beta(q) = \frac{\rho([p_{ij}^q])}{\log(2)} \qquad (2.23)$$

Only very rarely can we find explicit formulas for the eigenvalues of $[p_{ij}^q]$, so we'll rely on numerical solutions with the software of 2.7 in the Mathematica or Python codes. Because we don't have an explicit formula for β as a function of q, we can't find an expression for $\alpha = -d\beta/dq$. This, too, we'll approximate numerically.

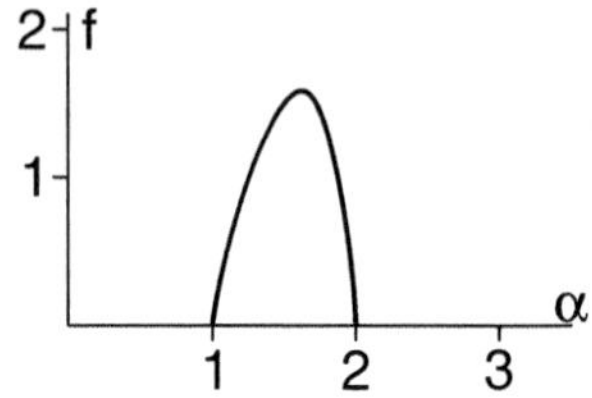

We can check the maximum height of the $f(\alpha)$ curve by an application of the Moran equation with memory (2.11) to the topological transition matrix M. Here that equation becomes $(1/2)^d \rho(M) = 1$, which gives $d = \log(\rho(M))/\log(2)$. The eigenvalues of M are $3, 1, -1, -1$ so $\rho[M] = 3$ and $\max\{f(\alpha)\} = d = \log(3)/\log(2) \approx 1.58496$.

2.7.8 Sample D

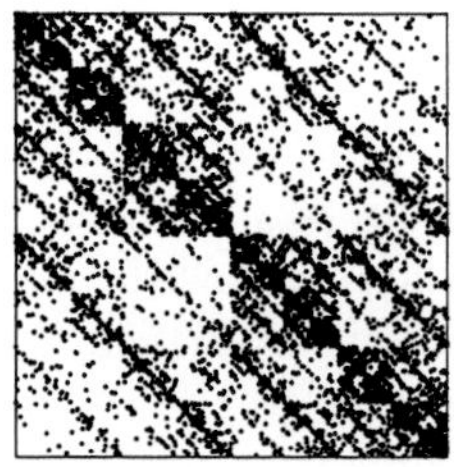

We'll use the random IFS algorithm with the four transformations (1.7) that generate the filled-in unit square. The probabilities p_1, p_2, p_3, and p_4 of applying T_1, T_2, T_3, and T_4 determine the rate of fill of the subsquares with address 1, 2, 3, and 4. Visual inspection of the IFS image suggests the relative likelihood of landing in each subsquare. With an estimate for the p_i, we can generate the $f(\alpha)$ curve with Eqs. (2.16), (2.17), and (2.20). What does this random IFS image suggest about the relative likelihood of a point falling in each of the four length-1 address subsquares?

Subsquares 2 and 3 are much more fully filled than are subsquares 1 and 4. If we wanted to be careful about this, we'd count the number of points in each square. How would we do that? Visual inspection is out of the question, because so many points are piled up along the diagonal. If the IFS were driven by a time series as in Lab 1.11, we'd add four counters to the program, one counter for each T_i. Then each time T_i is applied, add 1 to the count for address i.

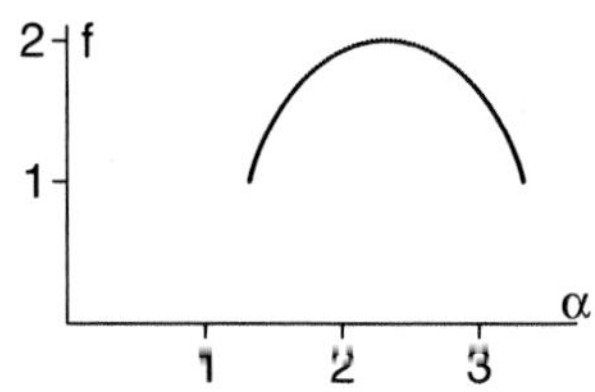

Here our goal is less mechanical. What do you see? Subsquares 2 and 3 appear to be filled to about the same density, so we'll take $p_2 = p_3$. Also, subsquares 1 and 4 appear to be filled to about the same density, a lower density than that of subsquares 2 and 3. So we'll take $p_1 = p_4 < p_2 = p_3$.

What's the relation between, say, p_1 and p_2? We'll assume a simple relation, and $p_1 = p_2/4$ looks about right. This gives $p_1 = p_4 = 0.1$ and $p_2 = p_3 = 0.4$, and we get this $f(\alpha)$ curve.

2.7.9 Conclusion

Multifractals are a powerful generalization of fractals. A fractal made of fractals, stratified by like levels of roughness or something similar. Multifractals have found application in many fields. A very short list includes cardiac dynamics [62], climate [63], dynamical systems [64], earthquakes [65], financial markets [189, 66], geometric group theory [67] the internet [68], literature [69], medical imaging [70], number theory [71], solar flares [72], topology [73], and turbulence [74].

The computation of multifractal spectra from experimental data is more subtle than the IFS-based approach of this lab. Approaches are based on histograms, on moments, and on wavelets. Here our goal is more modest: a simple way to generate $f(\alpha)$ curves. Examples and experiments can give some intuition about the implications of the shape of the curve.

2.7.10 Exercises

Prob 2.7.1 For the multifractals generated by these IFS, find $\alpha_{\min}$, $\alpha_{\max}$, $f(\alpha_{\min})$, $f(\alpha_{\max})$, and the maximum height of the $f(\alpha)$ curve.
(a) $r_1 = r_2 = r_3 = r_4 = 0.5$; $p_1 = 0.1$, $p_2 = 0.4$, $p_3 = 0.4$, $p_4 = 0.1$.
(b) $r_1 = r_2 = r_3 = r_4 = 0.5$; $p_1 = 0.1$, $p_2 = 0.2$, $p_3 = 0.3$, $p_4 = 0.4$.
(c) $r_1 = r_2 = r_3 = r_4 = 0.5$; $p_1 = 0.1$, $p_2 = 0.1$, $p_3 = 0.1$, $p_4 = 0.7$.
(d) $r_1 = r_2 = r_3 = r_4 = 0.25$; $p_1 = 0.1$, $p_2 = 0.1$, $p_3 = 0.1$, $p_4 = 0.7$.

Prob 2.7.2 Sketch the $f(\alpha)$ curves for the IFS of Exercise 2.7.1. First make a sketch by hand, using the data from the solutions of that exercise. Then generate a parametric plot with programs of 2.7 in the Mathematica or Python codes.

Prob 2.7.3 For the multifractals generated by these IFS, find $\alpha_{\min}$, $\alpha_{\max}$, $f(\alpha_{\min})$, $f(\alpha_{\max})$, and the maximum height of the $f(\alpha)$ curve. All these IFS have $r_1 = r_2 = r_3 = 0.5$, $r_4 = r_5 = 0.25$.
(a) $p_1 = p_2 = p_3 = 0.1$, $p_4 = p_5 = 0.35$.
(b) $p_1 = p_2 = 0.1$, $p_3 = 0.4$, $p_4 = p_5 = 0.2$.
(c) $p_1 = 0.52$, $p_2 = p_3 = 0.2$, $p_4 = p_5 = 0.04$.
(d) $p_1 - p_2 - 0.36$, $p_3 = 0.2$, $p_4 = p_5 = 0.04$.

Prob 2.7.4 Sketch the $f(\alpha)$ curves for the IFS of Exercise 2.7.3. First make a sketch by hand, using the data from the solutions of that exercise. Then generate a parametric plot with programs in 2.7 of the Mathematica or Python codes.

Prob 2.7.5 (a) For a 10 function IFS with $r_1 = \cdots = r_{10} = .25$ and with $p_1 = p_2 = p_3 = 0.2$, $p_4 = p_5 = p_6 = 0.1$, and $p_7 = p_8 = p_9 = p_{10} = 0.025$, find $\alpha_{\min}$, $\alpha_{\max}$, $f(\alpha_{\min})$, $f(\alpha_{\max})$, and the maximum height of the $f(\alpha)$ curve. Plot the $f(\alpha)$ curve.
(b) For an IFS with 9 transformations, suppose the scaling factors are $r_1 = 1/2$, $r_2 = \cdots = r_6 = 1/4$, and $r_7 = r_8 = r_9 = 1/8$, and the probabilities are $p_1 = 0.1$, $p_2 = \cdots = p_6 = 0.15$, and $p_7 = p_8 = p_9 = 0.05$. Find $\alpha_{\min}$, $\alpha_{\max}$, $f(\alpha_{\min})$, $f(\alpha_{\max})$, and the maximum height of the $f(\alpha)$ curve.

Prob 2.7.6 For an IFS with 6 transformations, suppose $r_1 = \cdots = r_6 = 0.25$. find probabilities $p_1, \ldots, p_6$ so $f(\alpha_{\min}) = 1$ and $f(\alpha_{\max}) = 1/2$.

$$P_1 = \begin{bmatrix} 0.25 & 0.33 & 0.33 & 0.33 \\ 0.25 & 0.34 & 0 & 0.33 \\ 0.25 & 0 & 0.34 & 0 \\ 0.25 & 0.33 & 0.33 & 0.34 \end{bmatrix} \qquad P_2 = \begin{bmatrix} 0.70 & 0.15 & 0.15 & 0.15 \\ 0.10 & 0.70 & 0 & 0.15 \\ 0.10 & 0 & 0.70 & 0 \\ 0.10 & 0.15 & 0.15 & 0.15 \end{bmatrix} \tag{2.24}$$

Prob 2.7.7 With the four transformations (1.7), sketch the $f(\alpha)$ curve for the IFS with memory that has transition probabilities P_1 of Eq. (2.24).

Prob 2.7.8 (a) With the four transformations (1.7), sketch the $f(\alpha)$ curve for the IFS with memory that has transition probabilities P_2 of Eq. (2.24).
(b) What is the most obvious difference between the $f(\alpha)$ curves of Exercises 2.7.7 and 2.7.8? What difference between the matrices P_1 and P_2 accounts for this difference? Give a reason to support your answer.

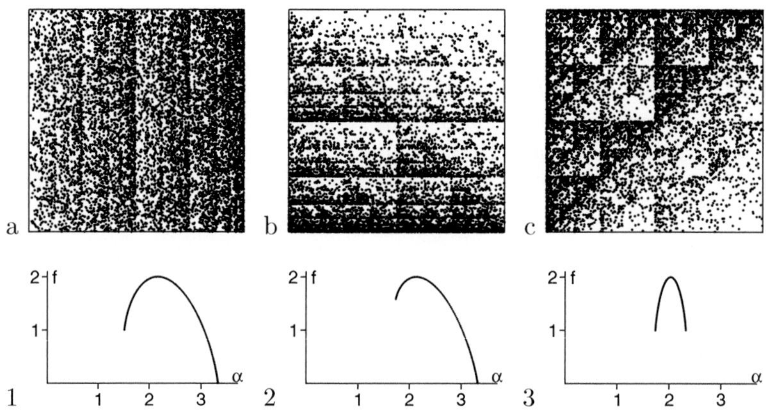

Figure 2.30: Images for Exercise 2.7.9.

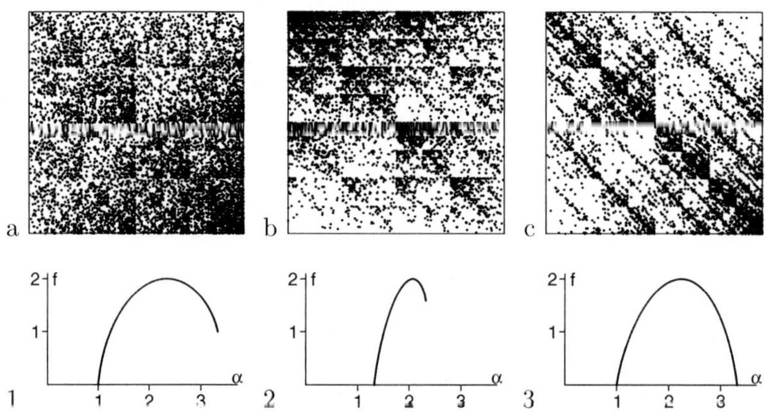

Figure 2.31: Images for Exercise 2.7.10.

Prob 2.7.9 The top three images of Fig. 2.30 are generated by random IFS. The bottom three images are $f(\alpha)$ curves. Which $f(\alpha)$ curve corresponds to which random IFS? Explain your choice for each. Do not give two reasons and deduce the third by the "process of excretion," as one student suggested for these problems. Of course, he may have been describing how he felt about these problems.

Prob 2.7.10 The top three images of Fig. 2.31 are generated by random IFS. The bottom three images are $f(\alpha)$ curves. Which $f(\alpha)$ curve corresponds to which random IFS? Explain your choices.

Chapter 3

Iteration Labs

Fractals are built by iteration, from geometrical transformations to paper-folding. In these labs we'll study iteration in some other settings.

In Lab 3.1 we'll continue our study of graphical iteration introduced in Lab 1.11, and we'll explore other visualization tools: time series, occupancy histograms, return maps, and Kelly plots, a technique developed by David Peak and first presented in [14]. These techniques are applied in Lab 3.2 to chaotic dynamical systems, to develop some familiarity with visual signatures of synchronized chaos.

Graphical iteration is based on the repeated application of the same function, but the underlying geometry can be adapted to study the composition of functions. In Lab 3.3 we apply this approach and obtain a simple visual way to find the domain and the range of the composition of functions.

Another iteration technique can be derived from the familiar Pascal's triangle. With increasingly general rules, in Lab 3.4 we explore properties of groups through visual patterns of these relatives of Pascal's triangle. Many of these patterns will be fractals.

Probably the best-known, or at least the most often-viewed, images generated by iteration are the Mandelbrot set and Julia sets. Much of the structure of the Mandelbrot set is thoroughly understood, although one major question remains unanswered. In Lab 3.6 we'll explore the Mandelbrot set and Julia sets for functions different from the familiar $z^2 + c$. We'll find some similarities with, and also some differences from, the standard quadratic Mandelbrot set.

In Lab 3.7 we study circle inversions and develop a nonlinear version of IFS. We'll look at the attractors of these IFS and find a new class of fractals.

Finally, in Lab 3.8 we apply iteration in space, rather than in time, to generate fractal tilings of the plane.

In this book we've omitted some iteration topics including L-systems, fractal music, cellular automata, and fractals in finance. We ran labs on these topics, at least once, and some several times. Our implementations of labs for these topics were not so successful. The ratio of classroom-transferrable skills to required background was too low. So far as we know, nothing prohibits these topics from being the foundations of effective labs. Our efforts didn't work; you may find more useful approaches.

3.1 Visualizing iteration patterns

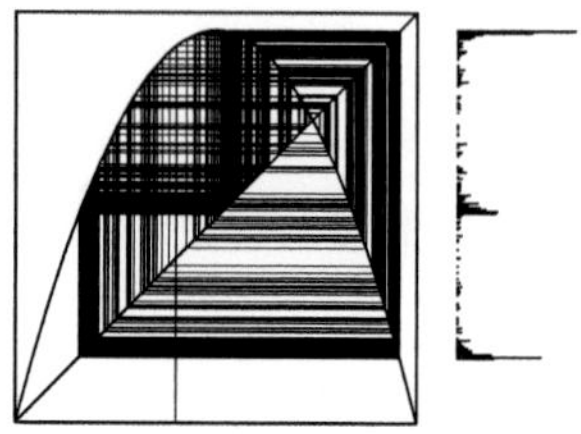

Here we'll expand our study of graphical iteration, which was popularized in [75] and which we encountered first in Lab 1.11. This is a way to visualize the iteration process; it is not so useful for the analysis of a given data sequence. For this we'll introduce four graphical techniques: time series plots, occupancy histograms, return maps, and Kelly plots. Our goal in this lab is to develop some familiarity with these methods. In Lab 3.2 we'll apply some of these methods to study synchronization of chaotic dynamical systems. One tool, graphical iteration, is the main ingredient of the visual computation of compositions of functions, used in Lab 3.3 to find domains and ranges of compositions

3.1.1 Purpose

We'll find that some properties of a sequence of function iterates can be deduced by graphical iteration. The other methods of this Lab can can reveal different attributes of a data sequence. Taken together, they can provide a nuanced sketch of a sequence. We note that driven IFS also can help us understand dynamics. This is a subtle technique that we studied in Lab 1.11.

3.1.2 Materials

We'll use 3.1 (a)–(e) of the Mathematica or Python codes. Some of this work, as long as the number of iterates is not too high, can be done by hand. For this, a ruler and pencil usually suffice.

3.1.3 Background

3.1.3.1 Graphical iteration

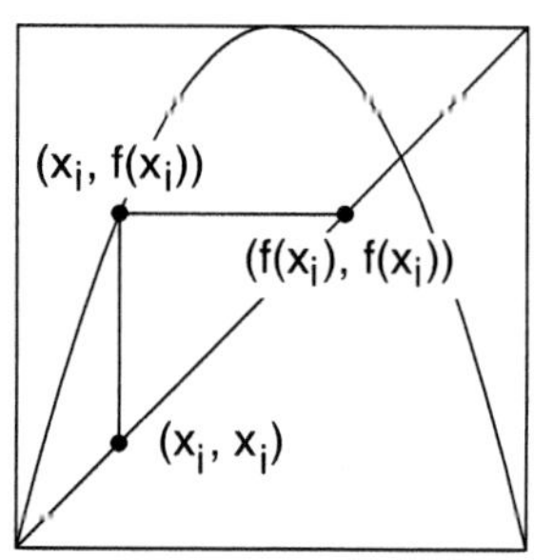

For a function f and initial value x_0, we are interested in patterns exhibited by the iterates x_0, $x_1 = f(x_0)$, $x_2 = f(x_1)$, ... We could do this algebraically, but that approach would not reveal complex patterns. Instead, we'll rely on geometry.

From Sect. 1.11.3.6 recall the basic mechanics of graphical iteration. Suppose we have a point (x_i, x_i) on the diagonal line. Then graphical iteration of a function f consists of two steps:
1. Draw a vertical line from the point (x_i, x_i) to the point $(x_i, f(x_i)) = (x_i, x_{i+1})$, and
2. Draw a horizontal line from the point (x_i, x_{i+1}) to the point $(f(x_i), f(x_i)) = (x_{i+1}, x_{i+1})$.
Repeat this until you've generated as many points as you wish.

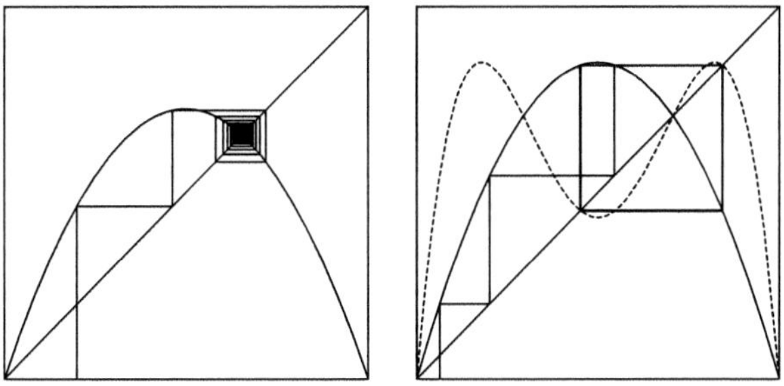

Figure 3.1: Fixed point and 2-cycle.

Graphical iteration handily reveals some dynamical behaviors. For example, a point x_a is a fixed point of f if $f(x_a) = x_a$. Consequently, the point (x_a, x_a) on the diagonal also is a point $(x_a, f(x_a))$ on the graph of f. That is, the fixed points of f are the intersections of the graph of f with the diagonal. See the left image of Fig. 3.1.

To find cycles, we could try to locate complicated geometric patterns with vertices alternately on the diagonal and the graph. While this is true, it is more difficult than identifying fixed points as the intersection of the graph and the diagonal. This intersection approach can locate cycles, because cycle points are fixed points of a different function. Suppose x_a is a fixed point of f, and that x_b and x_c are a 2-cycle for f. That is, $f(x_b) = x_c$ and $f(x_c) = x_b$. Then

$$f^2(x_b) = f(f(x_b)) = f(x_c) = x_b \quad \text{and } f^2(x_c) = f(f(x_c)) = f(x_b) = x_c$$

That is, the 2-cycle points of f are the fixed points of the composition $f^2 = f(f)$. See the right image of Fig. 3.1.

One final item to mention is that fixed points of f also are fixed points of f^2: $f^2(x_a) = f(f(x_a)) = f(x_a) = x_a$. So to locate the 2-cycle points of f, locate the fixed points of f^2 and from that list remove the fixed points of f.

With adequate care, this approach can be generalized to locate cycles of any order. What do we mean by "with adequate care"? This example tells the whole story: to locate the 6-cycles of f, find the fixed points of the composition f^6 and from that list remove the fixed points, the 2-cycle points of f, and the 3-cycle points of f.

In fact, graphical iteration can reveal a lot more about the dynamics of f, but for now this is enough.

3.1.3.2 Time series plots

Graphical iteration is a simple way to generate a series of values from a function. But more often we are presented with a sequence of values and asked to predict some aspects of the long-term behavior of the sequence. Time series plots, occupancy histograms (Sect. 3.1.3.3), return maps (Sect. 3.1.3.4), and Kelly plots (Sect. 3.1.3.5) can help in the prediction of different kinds of behaviors.

Time series can reveal low-order cycles or approximate cycles ("noisy" cycles), and can suggest the presence of chaos. Occupancy histograms tell which regions are visited more often. They approximate the "invariant measure", the limiting distribution of points across the attractor. Return maps show

how close the series is to being generated by iteration of a function, and with a bit more work, can estimate how many iterates in the past we need to predict the next value. Kelly plots can detect noisy cycles and also "intermittency", a temporary clumping of values that suggests a parameter change can produce a cycle. Happily, all these methods are based on simple geometry.

Consider a sequence $x_0, x_1, x_2, \ldots$ of measurements ordered in some natural way, for example, by time. If the x_i are successive measurements made at equal time intervals, then the *time series* is a plot of the points $(0, x_0)$, $(1, x_1)$, $(2, x_2)$, and so on. The top graph of Fig. 3.2 shows points of a time series. These points are connected by line segments in the bottom graph.

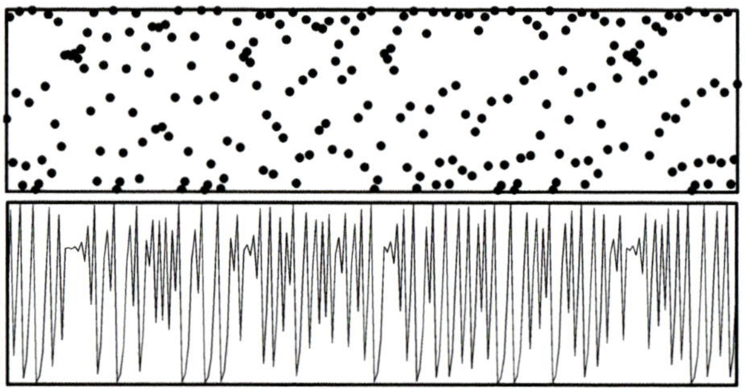

Figure 3.2: Time series (top), dots connected (bottom).

For plots with many points, the time order can be difficult to discern from the points alone. Connecting the dots is useful in this situation. On the other hand, if a few points fall between a low value and a high value, connecting the dots may not reveal the presence of these points. Consequently, usually we'll take the "belt *and* suspenders" approach: plot the points *and* connect the dots.

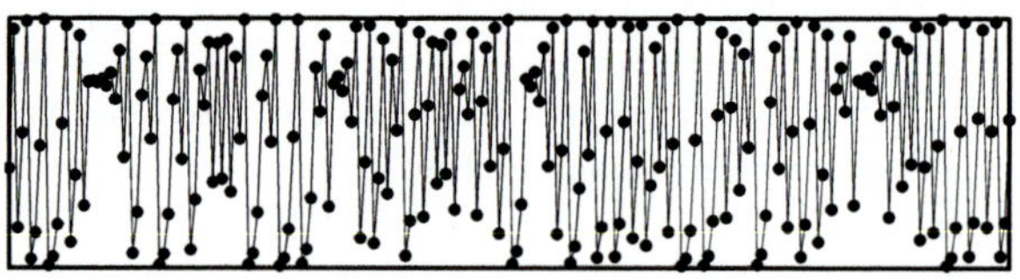

3.1.3.3 Occupancy histograms

All measurement involves some uncertainty, so points in a time series of measurements really are little bars, not little dots. To allow for this uncertainty we should *coarse-grain* or *bin* the data. Divide the range of values into some number of boxes and count how many data points lie in each box. From these counts we construct a *histogram* of the time series.

Histograms lose all information about the ordering of the data, but then visual inspection of a time series plot is not an effective way to determine how many points lie in each bin.

For comparison with the graphical iteration plot generating this data, histograms are drawn horizontally. That is, the range on the y-axis is binned and the extent in the x-direction of the histogram bars indicates the number of points in each bin. We'll use the same orientation with all time series data, a natural choice because the height of each histogram bar corresponds to the measured value, plotted in the y-direction, of the data.

Unless a natural binning is suggested by the source or an application of the data, the bin size can be varied, sometimes revealing different information. Care must be exercised: small bins may give only one or two points per bin, large bins may hide substructure. In Fig. 3.3 the data of Fig. 3.2 is presented, in histograms of 25 (left) and 50 (right) bins.

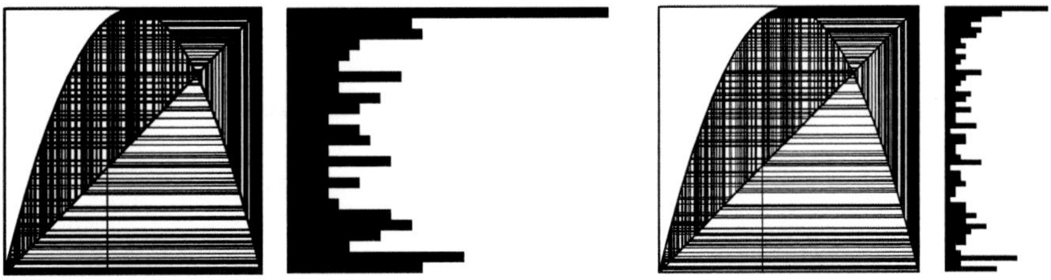

Figure 3.3: Occupancy histograms: 25 bins (left), 50 bins (right)

To compare these histograms, note they are made of little squares. Each additional point in a bin adds another square to the histogram bar that corresponds to that bin. The right histogram squares have half the height, and consequently half the width, of those on the left. So to compare the number of points in the top bin of these histograms, twice the length of the top bar of the right histogram should be compared with the length of the top bar of the left histogram.

3.1.3.4 Return maps

If the sequence $x_0, x_1, x_2, \ldots$ is generated by iteration, $x_{i+1} = f(x_i)$, then the points (x_{i-1}, x_i) lie on the graph of f. The *return map* is a method for testing if a time series is generated by iteration. Simply plot the points (x_0, x_1), (x_1, x_2), $(x_2, x_3), \ldots$. The left image of Fig. 3.4 is the time series, the right image is the return map. This looks like the graph of a function, so we conclude that the point on the graph above x_i on the x-axis is our prediction of x_{i+1}. The current value predicts the next value. This is pretty straightforward.

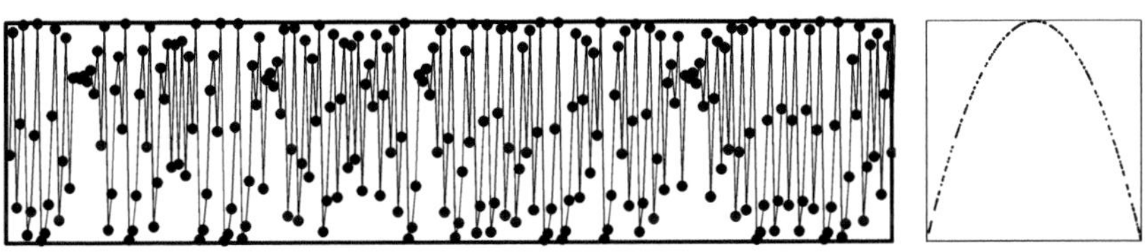

Figure 3.4: Time series (left) and return map (right).

If the return map appears to be a structureless blob, there's little hope of prediction. But what about a return map that is something between a simple curve and a blob? What if it has some structure but is not single-valued above each point of the x-axis?

The left image of Fig. 3.5 is 250 points of the time series, in appearance quite different from the time series of Fig. 3.4. It almost looks like a low-frequency oscillation with a high-frequency signal superimposed. In Lab 3.2 we'll see the mechanism that generated this time series.

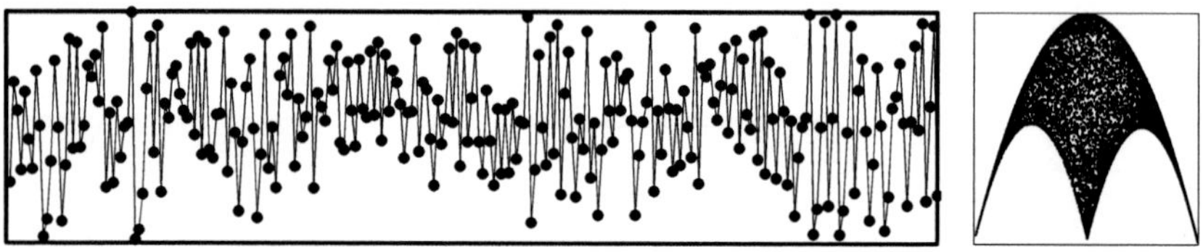

Figure 3.5: Time series (left) and return map (right).

The return map (10,000 points are plotted on the right of Fig. 3.5) is interesting. It is not the graph of a function because above each point on the x-axis we see many points. But neither is it a shapeless blob. We see clear structures: a large parabola above, two smaller parabolas below.

Here we see the return map in 3 dimensions, plots of the points (x_i, x_{i+1}, x_{i+2}). If we view this from several perspectives (a 3-dimensional plot in Mathematica can be rotated freely with mouse or touchpad), the views suggest that the return map lies on a surface. Put another way, above a point $(x_i, x_{i+1}, 0)$ in the $x_i x_{i+1}$-plane, we find a unique point on the return map. If we know the previous *two* values, we can predict the next. A bit more detail is presented in [76]. Certainly, this isn't an easy thing to see from the time series.

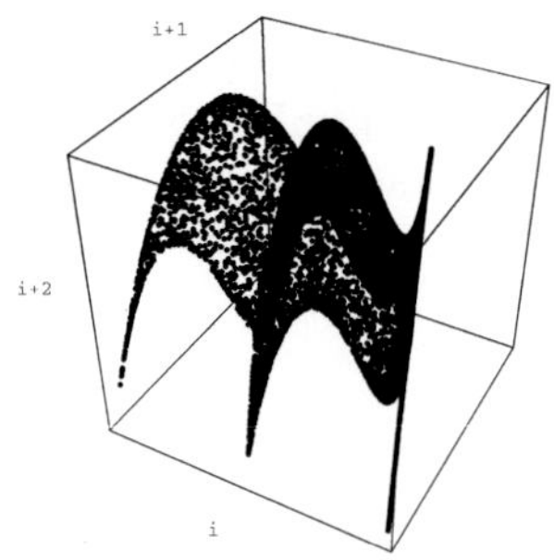

3.1.3.5 Kelly plots

Inspired by Ellsworth Kelly's paintings *Spectrum of Colors arrenged by Chance*, (search for "spectrum of colors arranged by chance" on Google images) Kelly plots are a way to visualize strings of repeated patterns, or rare or forbidden combinations. As with occupancy histograms, the range of data values is binned. Unlike histograms, Kelly plots retain the sequence data.

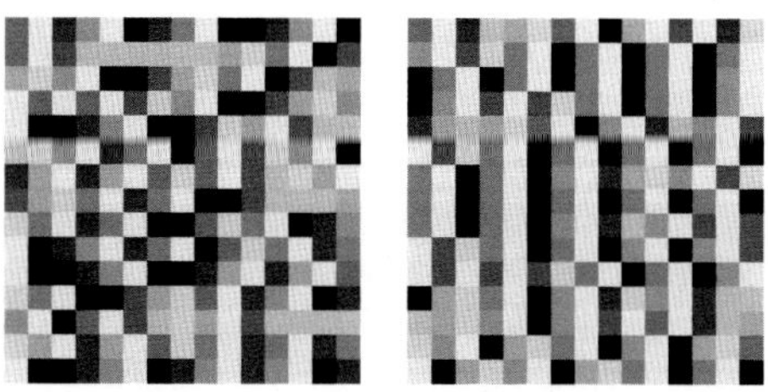

Figure 3.6: Kelly plots.

Assign a color to each data bin. For a sequence of N data points, make a grid of about $\sqrt{N}$ by $\sqrt{N}$ squares. Assign the color of the bin in which each time series point lies to a grid square, proceeding left to right, top to bottom. The left side of Fig. 3.6 shows the Kelly plot (ten bins, using grayscales instead of colors) of the time series of Sect. 3.1.3.2.

The left side of Fig. 3.6 reveals few obvious patterns other than some long horizontal bands of constant color, though even these may have some instructive interpretations. The right side is more interesting. We see a pattern of length three—dark grey, light grey, black—that is repeated often, but interspersed with excursions of other shades. The time series stays near a noisy 3-cycle for a while, then does something else for a while, and returns again and again to the 3-cycle. The distribution of repeated pattern lengths, and the distribution of the lengths of these excursions, is a rich source of information about the dynamics generating the time series.

3.1.4 Procedure

Paste the data to be analyzed in the time series, occupation histogram, return map, and Kelly plot programs, 3.1, (b)–(e) of the Mathematica or Python code.

3.1.5 Sample A

With the Kelly plot, time series, and occupancy histogram study the long-term dynamics of the logistic map $x_{n+1} = 3.7381x(1-x)$. Because these time series are generated by iterating the logistic map, the return map would generate points on the parabola. We'll skip these plots.

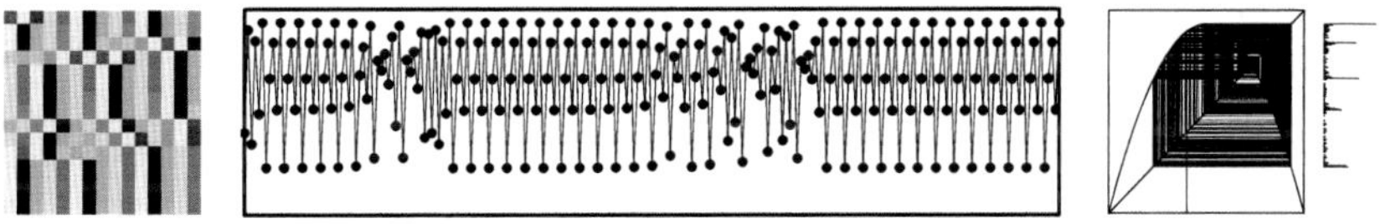

Figure 3.7: Plots form Sample 3.1.5.

Kelly plot. In three separate time regimes we see a repeating pattern of five grey shades. The pattern is read most easily along the bottom row of the Kelly plot. Despite the prevalence of this pattern, the occasional departures from it suggests this logistic map does not exhibit a 5-cycle.

Time series. The Kelly plot suggests a 5-cycle for a while and occasionally. The binning of the Kelly plot could give the same result for a noisy 5-cycle. Here's a magnification of a small portion of the time series. The horizontal dashed lines suggest this truly is a 5-cycle, at least for a while.

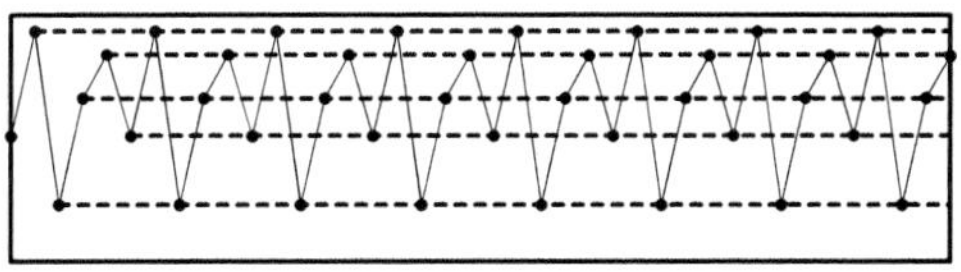

Occupancy histogram. The points seem to cluster about five bars, but with many more points scattered between. Maybe more iterates would reveal eventual convergence to the 5-cycle? No, regardless of how long the program runs, how many points are generated, various lengths of 5-cycle copies will be interrupted by excursions away from the 5-cycle, only to return again to the cycle.

This is called *intermittency.* It signals that for some nearby parameter values, the logistic map iterates converge to a 5-cycle and stay there.

3.1.6 Sample B

Here we generate a time series $z_0, z_1, z_2, \ldots$ from two sequences:

$$z_n = \frac{x_n + y_n}{2} \quad \text{where } x_{n+1} = 4x_n(1 - x_n) \text{ and } y_{n+1} = 3.5y_n(1 - y_n)$$

The x_n sequence is chaotic, the y_n sequence exhibits a 4-cycle. What's the average of chaos and a 4-cycle?

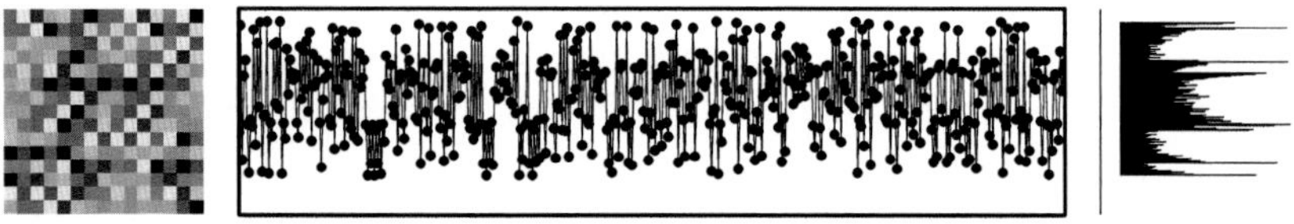

Figure 3.8: Plots for Sample 3.1.6.

Kelly plot. Unlike our other examples of Kelly plots, here we don't see significant sections with repeating patterns of gray scales. We do see some lighter regions and some darker regions, and a few short horizontal stripes of alternately light and dark squares, but nothing points to a recognizable behavior. No surprise really: this Kelly plot is too short to reveal structure in truly complicated dynamics.

Time series. This graph shows some regions of rapid alternation between low values and high values, the sections with steep lines that connect points well below the middle with points well above the middle. In other regions the oscillation is of a more modest scale. This gives the appearance of a narrow band undulating through the graph, punctuated by larger excursions. Still nothing that suggests a hybrid of chaos and a 4-cycle.

Occupancy histogram. In order to see enough of the histogram bars to make patterns noticeable, here we've run 10,000 iterates. No point in displaying graphical iteration, so we show just the histogram. Also, we use 200 bins. The vertical line extends through the range $0 \leq z_n \leq 1$, and we see that a bit of the range near 1 is unoccupied, as is a larger bit of the range near 0. Not surprisingly, these minimum and maximum values of z_n match with the lowest and highest points of the time series.

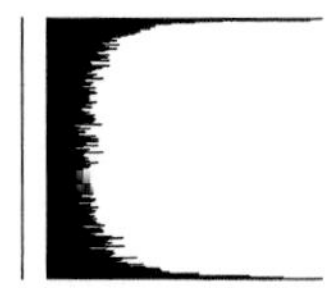

Figure 3.9: Logistic histogram

The most noticeable features of the histogram are the eight long horizontal spikes. These signal eight small regions that iteration visits more often. Also, the middle of the graph exhibits a "plateau" of histogram bars. These correspond to the narrow band that undulates through the time series plot.

The histogram does give a hint of how chaos and a 4-cycle can be averaged. In Fig. 3.9 we see the 200 bin histogram for a single chaotic logistic map, $x_{n+1} = 4x_n(1 - x_n)$. Note the large horizontal spikes at the top and at the bottom of the graph. So, the eight spikes in the histogram of Fig. 3.8 could be the four top spikes and four bottom spikes of four chaotic logistic parabolas. To test this, we'll look at the return map.

Return map. Here we see the return map in 2 and 3 dimensions. As expected, the plot in 2 dimensions shows points that fall on four logistic parabolas. The plot in 3 dimensions suggests that the apparent crossings of these parabolas are the result of projection into a too low-dimensional, in this case, 2-dimensional, space. Unfolded into 3 dimensions, the crossings disappear.

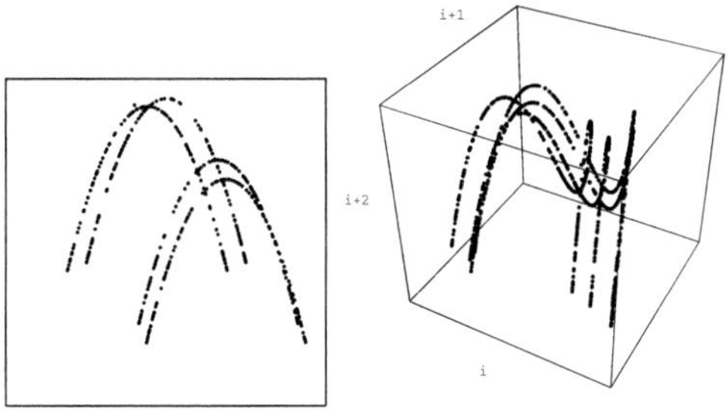

If we combine the 2-dimensional return map with the time series, we learn a bit more. Iterates don't fill in one parabola and then hop to another. Rather, it appears that a point on one parabola is followed by a point on another. For example, we don't see two consecutive time series points near 0, as we do in the time series for a single chaotic logistic map, shown here. We see that the average of a chaotic logistic map and logistic 4-cycle is a dance between four logistic parabolas.

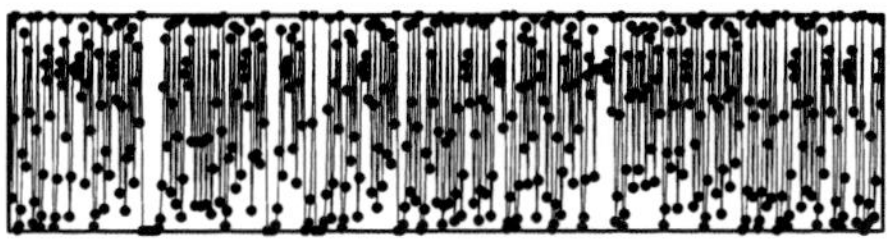

3.1.7 Conclusion

Graphical methods help generate numerical sequences, and also aid in the discovery of patterns. Different tools reveal different kinds of features in these patterns. Time series and Kelly plots preserve the sequence data. Both uncover recurring patterns, though in different ways, while forbidden combinations are more easily detected with Kelly plots (or driven IFS). Occupancy histograms erase sequencing information, but make transparent how often each bin has been visited. Return maps show how close the data points are to the graph of a function. Sometimes return maps can show how far into the past we must look in order to predict the next iterate. Used in combination these tools can give a detailed portrait of the process that generates some time series.

3.1.8 Exercises

Prob 3.1.1 In this problem we use the tent map

$$T_s(x) = \begin{cases} sx & \text{for } 0 \le x \le 1/2 \\ -sx + s & \text{for } 1/2 \le x \le 1 \end{cases}$$

Plot the time series, occupancy histograms, and Kelly plots for the tent map with
(a) $s = 1.1$ and
(b) $s = 1.75$.

Prob 3.1.2 Suppose some portion of a time series alternates between two values. Must the corresponding portion of the Kelly plot alternate between two shades? Why or why not?

Prob 3.1.3 Which of the Kelly plots of Fig. 3.10 corresponds to this time series? Why?

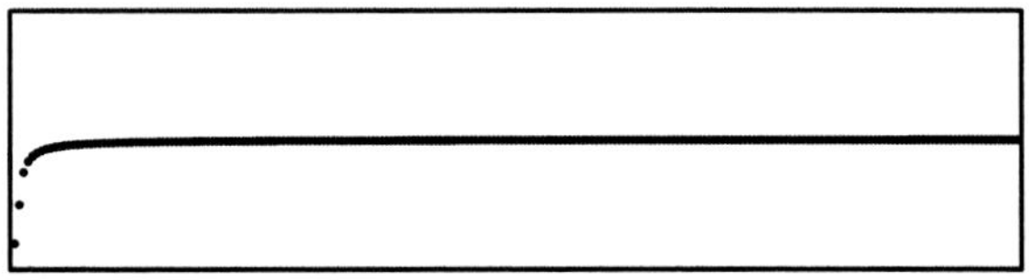

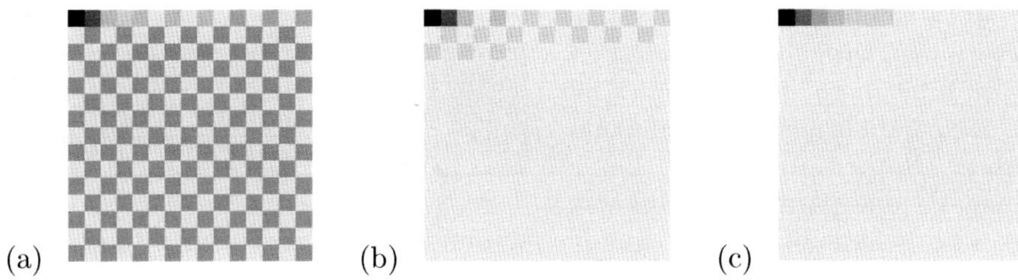

Figure 3.10: Kelly plots for Exercise 3.1.3.

Prob 3.1.4 Which time series of Fig. 3.11 can produce the return map pictured on the right? Why? Hint: the curvature of regions of the graph may give some subtle information about the distribution of the iterates, but here we look for simpler deductions. Specifically, what can you tell from the fact that the return map is split into two connected pieces?

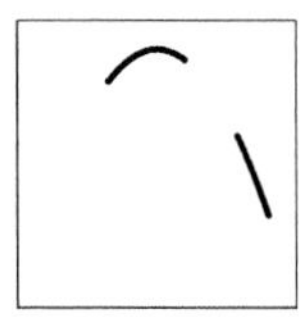

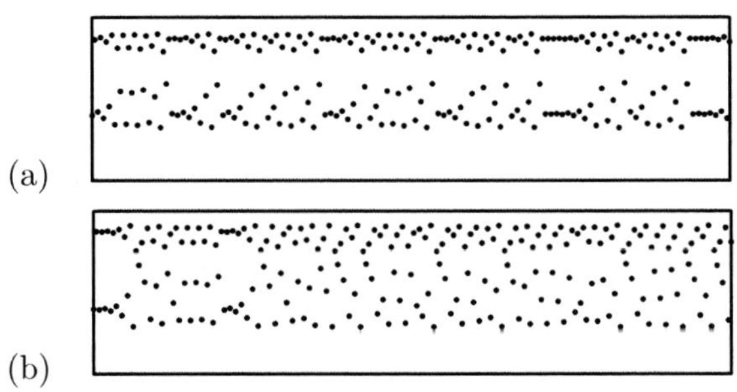

Figure 3.11: Time series graphs for Exercise 3.1.4.

The remaining exercises deal with coupled maps. Suppose f and g are functions, $f : [0,1] \to [0,1]$ and $g : [0,1] \to [0,1]$. The time series will consist of $z_n = (x_n + y_n)/2$ where

$$\begin{aligned} x_{n+1} &= (1-c)f(x_n) + cg(y_n) \\ y_{n+1} &= (1-c)g(y_n) + cf(x_n) \end{aligned}$$

The constant c is called the *coupling constant.* The factors $1-c$ and c are arranged in this fashion so for $c = 0$ the system uncouples, that is, $x_{n+1} = f(x_n)$ and $y_{n+1} = g(y_n)$. For these functions we'll use the logistic map $L_s(x)$, the tent map $T_s(x)$, and the sine map $Si_s(x) = s\sin(\pi x)$. For each exercise, plot the time series and the 2-dimensional return map for (a) $c = 0.1$ and (b) $c = 0.3$. Plot 1,000 iterates. To ease the interpretation of the time series plot, also plot iterates 800 through 1,000. What do you see?

Prob 3.1.5 Take $f(x) = L_4(x)$ and $g(x) = L_{3.7}(x)$.

Prob 3.1.6 Take $f(x) = L_4(x)$ and $g(x) = T_{1.7}(x)$.

Prob 3.1.7 Take $f(x) = T_2(x)$ and $g(x) = T_{1.7}(x)$.

Prob 3.1.8 Take $f(x) = L_4(x)$ and $g(x) = Si_{0.9}(x)$.

Prob 3.1.9 Take $f(x) = T_2(x)$ and $g(x) = Si_{0.9}(x)$.

Prob 3.1.10 Take $f(x) = Si_1(x)$ and $g(x) = Si_{0.9}(x)$.

3.2 Synchronized chaos

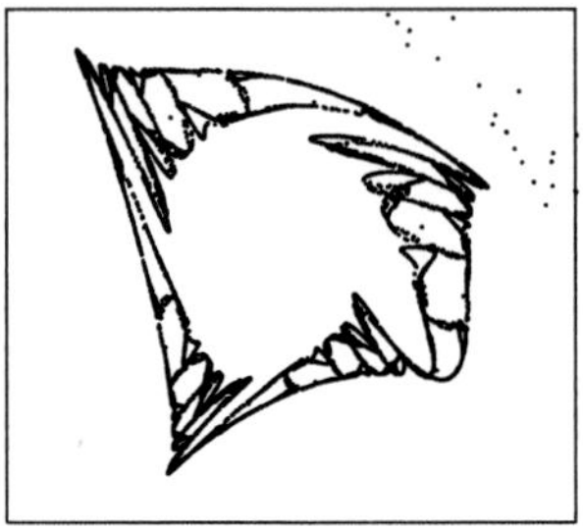

Synchronization is thoroughly folded into our world. Think of marching bands, cardiac pacemaker cells, fireflies in southeast Asia (and in one area of the Great Smokey Mountains National Park), some traffic flow patterns, and on and on. There is little surprise that periodic behaviors can synchronize—think of the musicians in a marching band—but in the late 1980s scientists were surprised to learn that chaotic processes can synchronize. This story, and much more, is told in Steven Strogatz's delightful book *Sync* [77, 78]. With this in mind, can we combine mathematical chaotic oscillators in such a way that their time series graphs synchronize? These combined or coupled chaotic maps form a network of chaotic oscillators. In this lab we'll study some simple networks of chaotic oscillators.

3.2.1 Purpose

We'll study some simple networks, as well as a more complicated network with fractal topology. In these networks we'll use logistic maps for the chaotic oscillators, and use return maps to detect synchronization. Some particularly interesting return maps we'll investigate with driven IFS.

3.2.2 Materials

We'll use 3.2 (a), (b), and (c) of the Mathematica or Python codes.

3.2.3 Background

Recall that a single logistic map is $x_{t+1} = L_s(x_t) = sx_t(1 - x_t)$. We use subscripts t and $t + 1$ to emphasize these are the values of x at successive time steps. To guarantee that all iterates x_t stay between 0 and 1, we take $0 \leq s \leq 4$. If x_t represents the fraction of the maximum population present in generation t, then certainly we must have $0 \leq x_t \leq 1$. For interpretations other than populations, perhaps we can relax the requirement $0 \leq x_t \leq 1$. Take any $x_t < 0$ or any $x_t > 1$ and try a few iterates by hand. You'll see why we don't want to relax the condition $0 \leq x_t \leq 1$.

We'll sketch networks of two and of three maps, and a network inspired by the Sierpinski gasket topology. For two map networks, a simple analytic argument gives a relation between the logistic map parameter s and the coupling constant c that guarantees the maps will synchronize. For larger networks, we rely on simulations.

3.2.3.1 Two maps

Call the two map variables x_t and y_t. The 2-map network is

$$\begin{aligned} x_{t+1} &= (1-c)sx_t(1-x_t) + csy_t(1-y_t) \\ y_{t+1} &= (1-c)sy_t(1-y_t) + csx_t(1-x_t) \end{aligned} \tag{3.1}$$

Here c, the *coupling constant*, determines how strongly each variable is influenced by the other. We arrange c and $1-c$ this way so $c=0$ is the *uncoupled* case. Also note that we take $0 \le s \le 4$ and $0 \le c \le 1$, so $0 \le x_t \le 1$ and $0 \le y_t \le 1$ implies that $0 \le x_{t+1} \le 1$ and $0 \le y_{t+1} \le 1$. That is, we can iterate as many times as we wish.

Before we look at some graphs, let's find conditions that guarantee these two processes synchronize, even when both are chaotic. We'll find conditions that guarantee

$$|x_{t+1} - y_{t+1}| < \delta |x_t - y_t|$$

for some $\delta < 1$. Then

$$|x_{t+k} - y_{t+k}| < \delta^k |x_t - y_t| \to 0 \text{ as } k \to \infty$$

First, apply Eq. (3.1) and simplify a bit to obtain

$$\begin{aligned} |x_{t+1} - y_{t+1}| &= \left|(1-2c)s\big(x_t(1-x_t) - y_t(1-y_t)\big)\right| \\ &= |1-2c|s\left|(x_t - y_t) - (x_t^2 - y_t^2)\right| \\ &= |1-2c|s|x_t - y_t||1-(x_t+y_t)| \\ &\le |1-2c|s|x_t - y_t| \end{aligned}$$

where the last line follows because $0 \le x_t + y_t \le 2$ implies $|1-(x_t+y_t)| \le 1$.

Now pick any $\delta < 1$. Then synchronization is guaranteed if $|1-2c|s \le \delta$. For example, if $s=4$, certainly a parameter where the logistic map exhibits chaos, then these two logistic maps will converge to the same chaotic sequence if $3/8 < c < 5/8$.

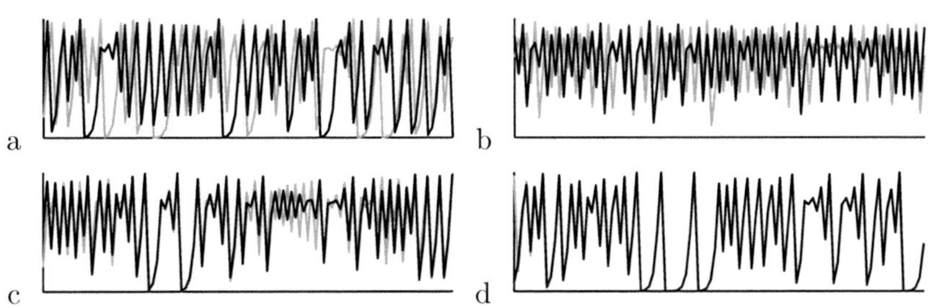

Figure 3.12: Time series for x_t and y_t. The coupling constant is (a) $c=0$, (b) $c=0.1$, (c) $c=0.2$, and (d) $c=0.3$.

In Fig. 3.12 we plot time series of x_t and y_t for several values of the coupling constant c. In (a) we have $c=0$. The logistic maps are uncoupled and their time series appear to have no relation to one another. In (b) we have $c=0.1$.

The logistic map dynamics still differ, but not as much as in (a). In (c) we have $c = 0.2$. For a while the maps behave similarly, then not so much, then again are similar. They appear to drift in and out of synchronization. In (d) we have $c = 0.3$ and here the maps synchronize almost immediately. The complex appearance of the time series certainly suggests that they exhibit chaotic dynamics, and yet they have synchronized. Think of what this implies about how we interpret sensitivity to initial conditions (the butterfly effect), the characteristic of chaos where initially nearby points iterate away from each other. Technical examples are given in Sect. 1.8 of [75] and Sect. 7.2 of [17]; popular discussions in James Gleick's popular book *Chaos: Making a New Science* [79] and the "Time and Punishment" segment of the "Treehouse of Horror V" episode of *The Simpsons*, where Grandpa Simpson gives a clear formulation of the butterfly effect, "If you ever go back in time, don't step on anything, because even the tiniest change can alter the future in ways you can't imagine."

Here's a specific example. For the logistic map $x_{n+1} = 4x_n(1 - x_n)$, with $x_0 = 0.2195$ we find $x_{20} = 0.394406$, and with $x_0 = 0.2196$ we find $x_{20} = 0.0958195$. That is, a change of 0.0001 in the initial value after 20 iterates grows to a change of about 0.3. Yet this growth does not continue forever because all x_i lie in the interval $[0, 1]$. The differences between successive iterates increase, then decrease, then increase, without apparent pattern. Little surprise this behavior is called "chaos."

Note also that $c = 0.3$ is a bit lower than the $c = 3/8$ bound we just derived. But remember, what we showed is that in the range $3/8 < c < 5/8$ two $c = 4$ logistic maps will synchronize. Our calculation does not show synchronization occurs only on this range. In fact, plot (d) of Fig. 3.12 shows the synchronization range is larger than that given by our calculation.

For more than two logistic maps, we'd have to plot a lot of time series to deduce synchronization. And there's another issue: in many physical systems what we measure is an average behavior, not that actions of individual agents. Many measurements reflect this: temperature is the average energy of the molecules that come into contact with the measuring device. So in a network of N logistic variables $x_t(1), \dots, x_t(N)$, we'll suppose we don't have access to individual variables, but rather to the averages $z_t = (x_t(1) + \cdots + x_t(N))/N$.

Here we see the time series plots for $z_t = (x_t + y_t)/2$ for $c = 0.2$ (top) and $c = 0.3$ (bottom). These are not so informative. The return maps are more useful. The $c = 0.2$ return map shows some complicated behavior; the $c = 0.3$ return map has collapsed to that of a single logistic map.

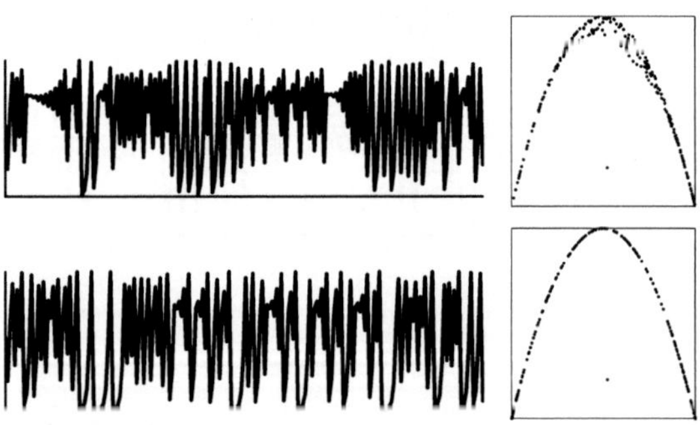

This is our visual test for synchronization, at least when all the logistic maps have the same s value. With different s values the situation can be more complicated.

3.2.3.2 Three maps

The 3-map network we'll use is the obvious extension of the 2-map network (3.1),

$$\begin{aligned} x_{t+1} &= (1-c)L_s(x_t) + (c/2)\big(L_s(y_t) + L_s(z_t)\big) \\ y_{t+1} &= (1-c)L_s(y_t) + (c/2)\big(L_s(x_t) + L_s(z_t)\big) \\ z_{t+1} &= (1-c)L_s(z_t) + (c/2)\big(L_s(x_t) + L_s(y_t)\big) \end{aligned} \tag{3.2}$$

The coupling factor $c/2$ is chosen because the bracketed expression it multiplies can take values between 0 and 2.

Before examples, we'll mention the simplifying assumptions of this model. Why do we take the same logistic parameter s, and why the same coupling constant c, for each individual map? The c value specifies how strongly the network elements interact, so the same c everywhere says that the same physical laws govern the interactions of all the elements. While this need not be true, in many circumstances it is. One of our goals is to explore the wealth of behaviors we can find in simple networks, that is, what dynamics are due to the network topology alone. The same s value for each map is a bit harder to justify. If we think of the s-value as a representation of the energy of that network element, then a more natural choice would be s-values distributed over some range. We'll investigate this a bit in the Exercises, but again we'll start with what we can find with the simplest model.

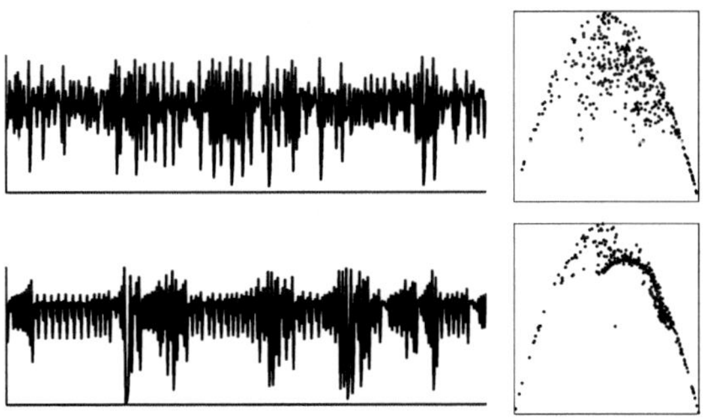

The top row shows 400 points of the time series and return map for the average of three $s = 4$ logistic maps with coupling constant $c = 0.1$; the bottom row for $c = 0.2$. The $c = 0.1$ return map appears to be related in some way to the return map of Fig. 3.5, maybe with three lower parabolas rather than two. The $c = 0.2$ return map shows points concentrated in a smaller parabola, but the time series shows that the larger parabola isn't transients. The network continues to make sporadic large excursions.

3.2.3.3 A fractal network

For this network we'll couple logistic maps in an arrangement suggested by the Sierpinski gasket. Our network extends over only three levels of the gasket construction, so calling this a fractal network takes some liberty with our usual understanding of what constitutes a fractal. Still, this network may suggest behavior that could be seen in networks carried to more levels.

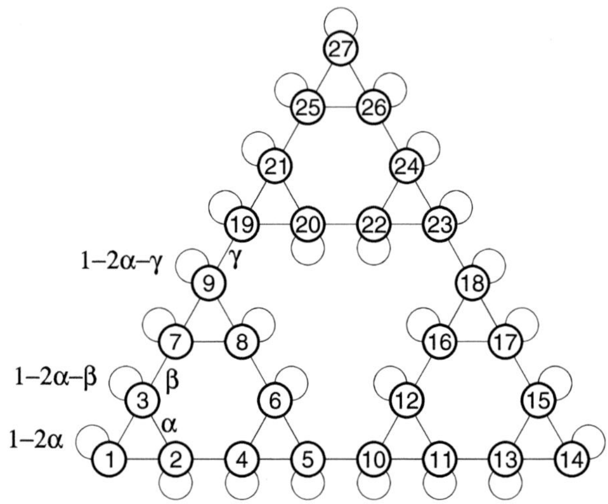

Figure 3.13: A fractal coupling graph.

Each of the 27 vertices of this graph contains a logistic map. To keep from overcrowding the diagram, we show only representative coupling strengths. For example, in each group of three logistic maps, $\{1, 2, 3\}$, $\ldots$, $\{25, 26, 27\}$, the coupling strength between any two distinct maps in a triple is α. The coupling strength between groups of three maps is β and between groups of nine maps is γ. The couplings of a map with itself is adjusted so the coupling from each map sums to 1.

To illustrate the coupling between groups, the updating rule for map 3 is

$$x_{t+1}(3) = (1 - 2\alpha - \beta)L_{s_3}(x_t(3)) + \alpha\big(L_{s_1}(x_t(1)) + L_{s_2}(x_t(2))\big) + \beta L_{s_7}(x_t(7))$$

This network models systems in which subsystems interact only through specific channels. For example, maps 1 through 9 interact with maps 19 through 27 only by the coupling between maps 9 and 19. Other arrangements are possible. For example, the interaction between groups could be through the average value of the maps in a group.

The time series can be the average value of the variables of maps within a group, or the average over the whole network.

3.2.4 Procedure

Select the 2-map, 3-map, or fractal network software, set the logistic parameter and coupling constants, and interpret what you see.

3.2.5 Sample A

For the 3-map network with all $s = 4.0$, plot 1000 points of the return map for the average logistic values, for $c = 0.0, 0.1, \ldots, 1.0$. For which c values does the return map suggest synchronization? For the highest c-value for which the return map suggests synchronization, plot the differences $x_t - y_t$, $x_t - z_t$, and $y_t - z_t$. Do the difference plots support synchronization?

The return maps are shown in Fig. 3.14.

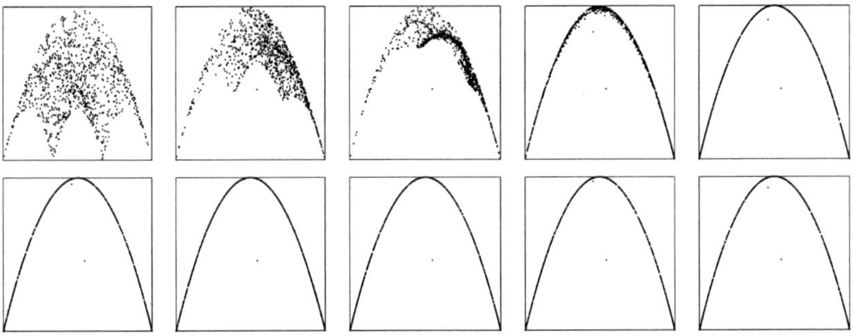

Figure 3.14: Return maps for Sample 3.2.5. Top row, left to right, $c = 0.0$, $c = 0.1$, $c = 0.2$, $c = 0.3$, and $c = 0.4$. Bottom row, left to right, $c = 0.5$, $c = 0.6$, $c = 0.7$, $c = 0.8$, and $c = 0.9$.

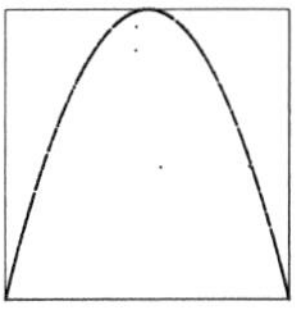

Finally, here is the $c = 1.0$ return map. The return map of the average of the three logistic values appears to collapse to a parabola, which suggests that the logistic maps have synchronized, first at $c = 0.4$. This collapse to a parabola seems to continue through $c = 1.0$. Can we believe this? For the coupling constant $c = 1.0$, x_{t+1} depends with equal weights on y_t and z_t, but not at all on itself. To test if the logistic maps truly have synchronized, in Fig. 3.15 we'll plot the differences $x_t - y_t$, $x_t - z_t$, and $y_t - z_t$. All three differences go to 0, so indeed these three logistic maps synchronize, even when $c = 1.0$.

Figure 3.15: Difference maps for the $c = 1.0$ system of Sample 3.2.5. Left to right these are $x_t - y_t$, $x_t - z_t$, and $y_t - z_t$.

So it appears that this network does synchronize, even with $c = 1$.

3.2.6 Sample B

In the fractal network of Sect. 3.2.3.3 set all logistic parameters s to 4 and the coupling constants $\alpha = 0$, $\beta = 0.15$, and $\gamma = 0.5$. Plot the return map of the averages $\langle x_t(1), x_t(2), x_t(3)\rangle$, $\langle x_t(4), x_t(5), x_t(6)\rangle$, $\langle x_t(1), \dots, x_t(9)\rangle$, and the average $\langle x_t(1), \dots, x_t(27)\rangle$ of the entire network. Interpret these graphs.

With these parameters, in all of our simulations the return maps of the nine triples $\langle x_t(1), x_t(2), x_t(3)\rangle, \dots, \langle x_t(25), x_t(26), x_t(27)\rangle$, return either one or two clean parabolas, after some initial transient noise—steps in the convergence to the attractor—is dropped. The number of single or paired parabolas that occur among these nine return maps depends on the initial values $x_0(i)$. For the paired parabolas, shorter duration plots always show points on both parabolas. Evidently the return map hops between these two parabolas.

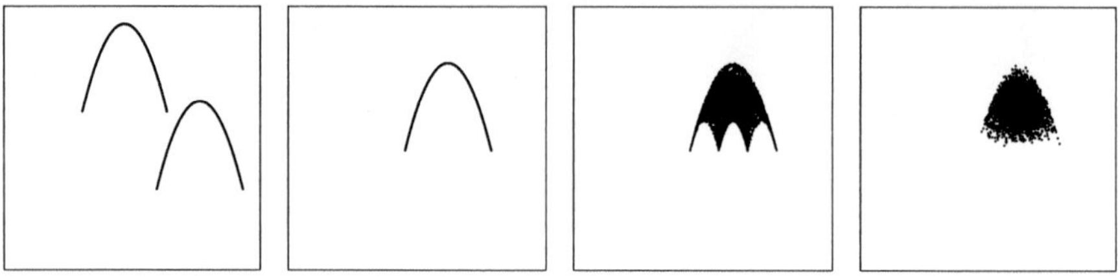

Figure 3.16: Return maps for Sample 3.2.6. Left to right: the average of the first three maps, of the second three, of the first nine, of the entire network.

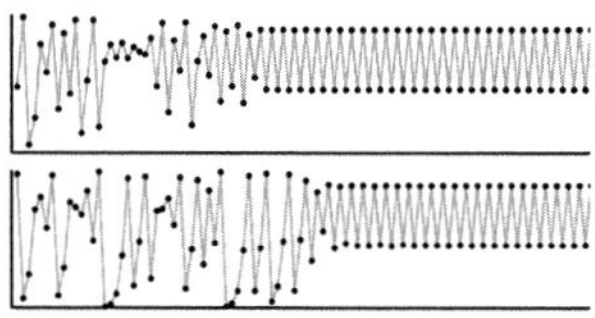

We've interpreted clean parabolas in return maps as indicators of synchronization, but there's a problem with these parameters: because $\alpha = 0$, $x_t(1)$ is uncoupled from all the other variables, so how can the return map of the average $\langle x_t(1), x_t(2), x_t(3)\rangle$ give two clean parabolas? Here we plot the time series for $x_t(2)$ and $x_t(3)$. Because of their non-zero coupling, they have collapsed to 2-cycles, so the two parabolas in the return map of $\langle x_t(1), x_t(2), x_t(3)\rangle$ are generated by the parabola of the $x_t(1)$ return map, twinned by the 2-cycle.

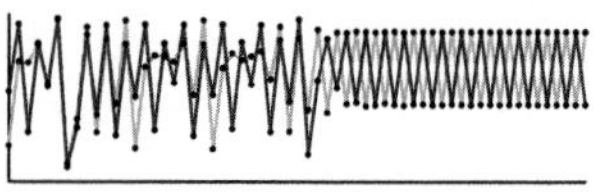

But how do $x_t(2)$ and $x_t(3)$ collapse to 2-cycles? Because $\alpha = 0$, the network breaks into three isolated elements—$x_t(1)$, $x_t(14)$, and $x_t(27)$—and 12 coupled pairs. For example, $x_t(2)$ is coupled to $x_t(4)$, and $x_t(3)$ to $x_t(7)$, both with coupling strength 0.15. As we see here, in a 2-map network with coupling constant 0.15 both maps collapse to a 2-cycle. The length of the initial noise before this collapse depends on the starting values. Also, sometimes $x_t(2)$ and $x_t(3)$ are synchronized, sometimes as picture here, they are anti-synchronized. The average of the logistic parabola of the uncoupled $x_t(1)$ and a synchronized 2-cycle for $x_t(2)$ and $x_t(3)$ gives two parabolas in the return map. If the 2-cycle is anti-synchronized, the return map of the average is a single parabola.

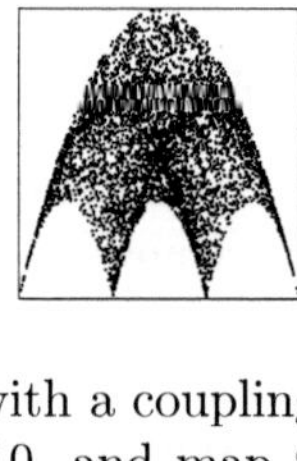

The third image of Fig. 3.16 is similar to a scaled and shifted return map for the average of three uncoupled logistic maps, shown here. How would the average of these nine logistic maps give a return map of three uncoupled logistic maps? Look again at the fractal network graph of Fig. 3.13 and focus on logistic maps 1 through 9. Map 1 is uncoupled and contributes a logistic parabola to the average of these nine maps. Map 2 is coupled to 4, map 3 to 7, and map 6 to 8, all with a coupling constant that produces a 2-cycle. Map 5 is coupled to map 10, and map 9 to 19, both with a coupling constant ($\gamma = 0.5$) that gives synchronized chaos and so a logistic parabola for their return maps. Maps 5 and 9 are uncoupled from map 1, and from one another. In the fractal network of Fig. 3.13 with $\alpha = 0$ there is no path along non-zero couplings that connects maps 5 and 9. So in $\langle x_t(1), \ldots, x_t(9)\rangle$ we average three uncoupled chaotic logistic maps and three 2-cycles, evidently adequately anti-synchronized that these parabolas do not twin. The result, then, is the average of three uncoupled chaotic logistic maps. If all three 2-cycles were synchronized, would the average of the first nine maps give a return map with six lower parabolas rather than three?

The fourth image of Fig. 3.16 is a blob, but the top of the blob is roughly parabolic, and the bottom of the blob might have some smaller parabolic arcs, though we haven't generated nearly enough data points to test this. How many arcs do we expect? The uncoupled logistic parabolas come form maps 1, 14, and 27 and from the synchronized pairs 5 and 10, 9 and 19, and 18 and 23. So we expect the return map of the whole network average should resemble the return map of the average of six uncoupled logistic maps. Here we see the return map for the average of six uncoupled chaotic logistic maps. The six parabolas at the bottom of the blob are more apparent than along the bottom of the fourth image of Fig. 3.16, so maybe in the whole network some of the 2-cycles have synchronized.

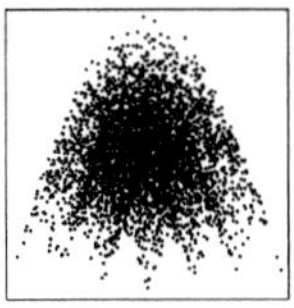

Certainly some features of these plots can be understood, but others offer more opportunities for exploration.

3.2.7 Conclusion

Coupled logistic maps can synchronize. They exhibit chaotic motion but in lockstep, even though they start at different values and sensitivity to initial conditions suggests future iterates are uncorrelated. With the proper coupling, chaotic logistic maps can dance together perfectly. With more complex networks we can see a range be behaviors not seen before. An interesting project, one we won't pursue here, is to map coupled network behavior to observed dynamics. Close matches could reveal ways to group the system agents to account for some features of the data.

3.2.8 Exercises

Prob 3.2.1 In the 2-map system with $s_1 = s_2 = 4$, plot the return map of the average for $c = 0, c = 0.1, \dots, c = 1$. Find the collection of these c values for which this system synchronizes.

Prob 3.2.2 In a 2-map system, take $s_1 = 3.44$ and $s_2 = 3.88$.
(a) In the uncoupled case $c = 0$, what are the dynamics of the individual logistic maps? Drop the first 200 iterates and plot the return maps for the next 500.
For parts (b) and (c) use the plot range $0.58 \leq x \leq 0.73$ and $0.58 \leq y \leq 0.73$.
(b) For $c = 0.89$ drop the first 200 points and plot the return map for the next 2,000.
(c) To understand how the averages trace out the figure of (b), in another return map drop the first 200 points and plot the next 200.

Prob 3.2.3 With the plot range of Exercise 3.2.3, set $c = 0.9$, drop 200 points, and plot the next 10,000. Use PointSize[0.003] for your plot. Look at the return map for a while. What does this suggest?

Is this return map pattern a consequence of specific choices of initial values?

Repeat this a few times with different ransom number seeds. Do you get (approximately) the same return map?

Prob 3.2.4 For a 3-map network, set $s_x = s_y = 4$ and $c = 0.5$. Drop 200 points and plot the next 1,000 for $s_z = 4, 3$, and 2. Speculate on the cause of the $s_z = 3$ and $s_z = 2$ return maps. Hint: if the return map appears to be a parabola that doesn't reach the top of the unit square, recall that the maximum height of $sx(1-x)$ is $s/4$. Can you find this "effective s-value" from a simple combination of the values of s_x, s_y, and s_z?

Prob 3.2.5 For a 3-map network, set $s_x = 4$ and $s_y = s_z = 2$. Drop the first 200 points and plot the next 1,000 of the return map for $c = 0$, 0.10, 0.15, 0.16, 0.17, 0.18, 0.25, and 0.50. Describe how the dynamics change as the coupling value c increases.

Prob 3.2.6 Modify the 3-map network by changing the z_{t+1} equation to

$$z_{t+1} = (1 - (c/2))L_{s_z}(z_t) + (c/2)L_{s_x}(x_t)$$

Set $s_x = s_y = s_z = 4$. Drop the first 200 points and plot the next 5,000 for $c = 0.260, 0.255, 0.251, 0.250, 0.248, 0.247$. Plot the return maps in the region $[0.5, 0.8] \times [0.5, 0.8]$. Describe what you see.

For the fractal network of Sect. 3.2.3.3 recall that all the coupling constants must lie between 0 and 1. That is, $0 \leq \alpha, \beta, \gamma \leq 1$, $0 \leq 1 - 2\alpha \leq 1$, $0 \leq 1 - 2\alpha - \beta \leq 1$, and $0 \leq 1 - 2\alpha - \gamma \leq 1$.

Prob 3.2.7 In the fractal network take $s = 3.8$ for all logistic maps. For the couplings, take $\alpha = .25$, $\beta = 0$, and $\gamma = 0.5$. Drop 200 points and generate the next 1,000 points for this network.
(a) Plot the return map for $\langle x_t(1), x_t(2), x_t(3)\rangle$ and for $\langle x_t(4), x_t(5), x_t(6)\rangle$.
(b) Plot the return map for $\langle x_t(1), \dots, x_t(9)\rangle$.
(c) Explain these plots in light of the condition $\beta = 0$.

Prob 3.2.8 In the fractal network take $s = 3.8$ for all logistic maps, and for the couplings take $\alpha = \beta = \gamma = 0.2$. Drop the first 200 iterates and generate the next 2,000. Plot the return map for the average of the whole network. If the return map points are restricted to a small portion of the unit square, magnify the plot. Do this several times, changing the random number seed with each new try. Do you always get the same, or very similar, return maps?

Prob 3.2.9 In the fractal network take all $s_i = 3.6$, and $\alpha = 0.3$, $\beta = 0.1$, $\gamma = 0.1$.
(a) Drop the first 200 iterates and plot the return map of the network average for the next 2,000 iterates.
(b) Set $s_5 = 4$. How does the return map for the network average change?
(c) Set $s_4 = s_5 = s_6 = 4$. How does the return map for the network average change?

Prob 3.2.10 Go exploring. Find an interesting return map for a 2-map network, for a 3-map network, or for a fractal network.

3.3 Domains of compositions

The method of graphical iteration was introduced in Labs 1.11 and 3.1. It is a simple visual method to generate repeated applications of a function, the "iteration" of "graphical iteration."

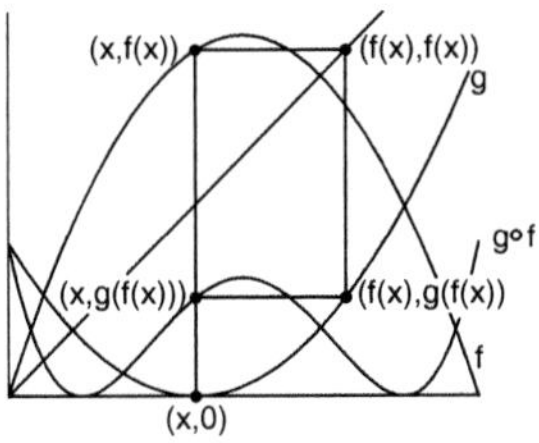

The basic step, "vertically to the graph, horizontally to the diagonal," need not always be applied to the same function. In fact, if we go vertically to the graph of f and horizontally to the diagonal, the vertically to the graph of g and horizontally to the diagonal, we have a way to visualize the composition $g(f)$.

3.3.1 Purpose

We'll apply the method of graphical iteration to investigate the composition of functions. With this, we'll develop a simple way to determine the range and the domain of a composition.

3.3.2 Materials

No computers here, just graph paper, pencil, and a ruler.

3.3.3 Background

First we'll review graphical iteration and show how it can be generalized slightly to allow easy visualization of the composition of functions.

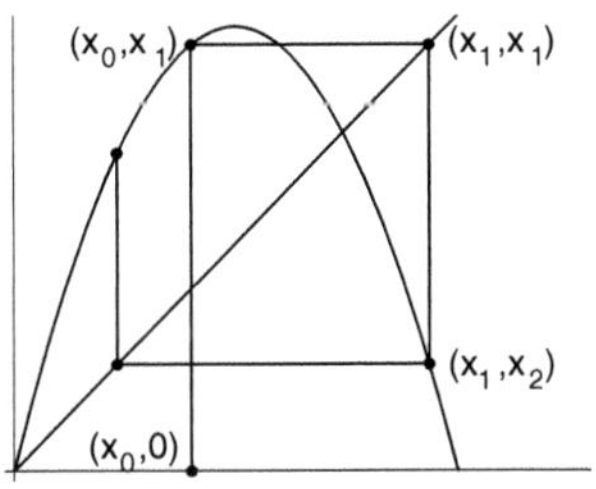

Given an initial value x_0 and a function f, we can generate the sequence x_0, $x_1 = f(x_0)$, $x_2 = f(x_1)$, $x_3 = f(x_2), \dots$ algebraically, but a geometric approach is quicker and can make more apparent some important behaviors. From the initial point $(x_0, 0)$ we draw a vertical line to the graph of f. This line crosses the graph at $(x_0, f(x_0)) = (x_0, x_1)$. From this point draw a horizontal line to the diagonal, intersecting at the point (x_1, x_1). Then vertically to (x_1, x_2) on the graph, and horizontally to (x_2, x_2) on the diagonal.

To modify graphical iteration to visualize the composition $g \circ f$ is straightforward: simply replace the outer (left) f in $f \circ f$ by g. Only one step (4 in the list below) needs to be added to the graphical iteration recipe. For any x in the domain of $g \circ f$, *graphical composition* involves four steps.

1. Draw the vertical segment from $(x, 0)$ to intersect the graph $y = f(x)$ at the point $(x, f(x))$.
2. Draw the horizontal segment from $(x, f(x))$ to intersect the diagoanal $y = x$ at the point $(f(x), f(x))$.
3. Draw the vertical segment from $(f(x), f(x))$ to intersect the graph $y = g(x)$ at the point $(f(x), g(f(x)))$.

4. Draw the horizontal line from $(f(x), g(f(x)))$ to intersect the vertical line through $(x, 0)$ at the point $(x, g(f(x)))$.

In Fig. 3.17 we illustrate the graphical composition process with f a concave down parabola and g a concave up parabola. The composition $g \circ f$ is quartic. This is not a surprise: in the composition of a quadratic with a quadratic, the highest power term is x^4.

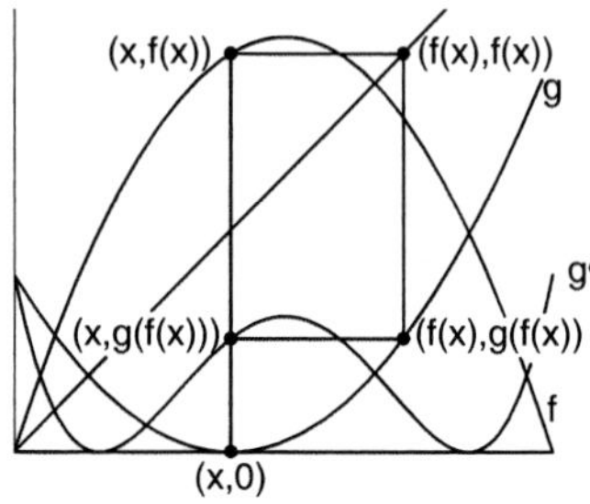

Figure 3.17: Graphical composition.

For practice, use the same graphs of f and g to sketch the composition $f \circ g$. For many examples, composition of functions is not commutative, that is, for some x,

$$(f \circ g)(x) = f(g(x)) \neq g(f(x)) = (g \circ f)(x).$$

Plenty of operations in the physical world aren't commutative. "Put on your socks" and "put on your shoes" are a clear example.

This application of graphical iteration to visualize compositions and find their domains is described in [80].

3.3.4 Procedure

Finding the domain and range of a composition $g(f)$ are more challenging aspects of studying compositions, at least when the algebraic approach is used. These become straightforward with graphical composition. Project the range of f and the domain of g to the diagonal line. Form the intersection of these sets. Reverse graphical iteration of f gives the domain of $g(f)$; graphical iteration of g gives the range. This is most clearly seen through the examples in the sample.

3.3.5 Sample

Find the domain of $g(f)$ for the functions f and g pictured in Fig. 3.18. The function f is on the left and is a portion of a concave up parabola. The function g is on the right and is a portion of a concave down parabola.

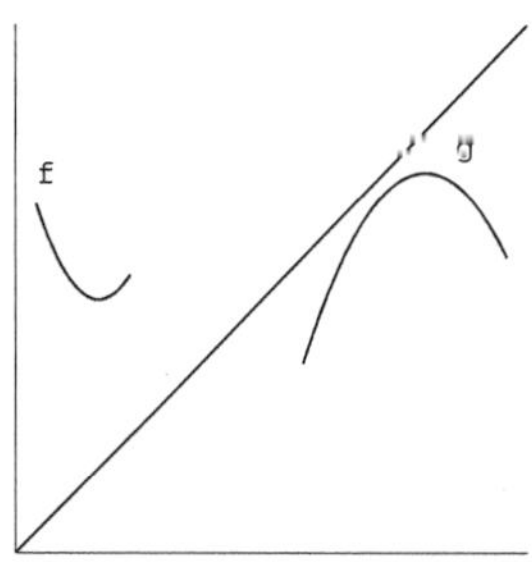

Figure 3.18: Left, f; right g.

Here are the steps to find the domain of $g(f)$.

1. Project the range of f horizontally to the diagonal line $y = x$. This is illustrated in the first graph of Fig. 3.19.
2. Project the domain of g vertically to the diagonal line $y = x$. This is illustrated in the second graph of Fig. 3.19.
3. On the line $y = x$ find the intersection of the range of f and the domain of g. This is illustrated in the third graph of Fig. 3.19.

4. Reverse the process of graphical iteration of f (horizontally to the graph of f, vertically to the diagonal): project the intersection to the x-axis to obtain the domain of $g(f)$. In the fourth graph of Fig. 3.19 we extend the vertical projection to the x-axis, the more familiar location for representations of domains.

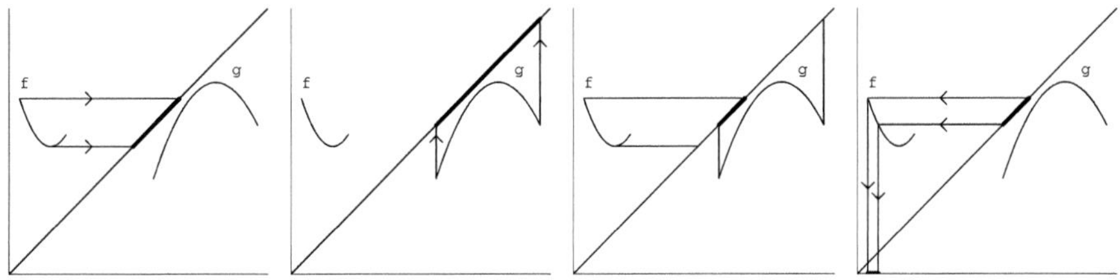

Figure 3.19: Steps to find the domain of the composition.

Here are the steps to find the range of the composition $g(f)$.

From the intersection found in step 3 of the domain calculation, apply graphical iteration of g to find the range of $g(f)$. In the left graph of Fig. 3.20 we extend the projected image to the y-axis, the more familiar location for representations of ranges.

In the right graph of Fig. 3.20 we assemble the four steps from the domain of $g(f)$ to the range of $g(f)$.

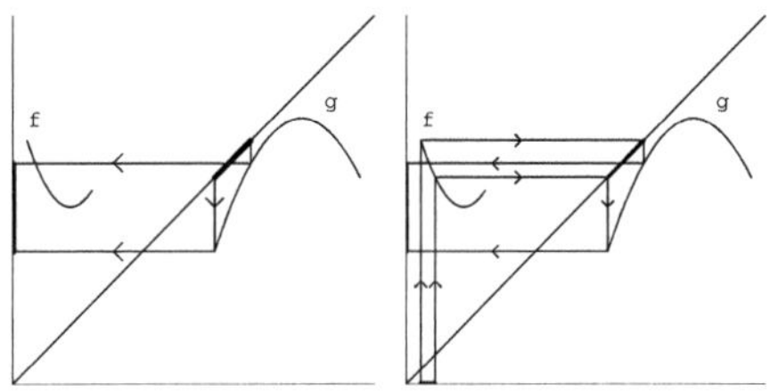

Figure 3.20: Range of the composition.

3.3.6 Conclusion

Graphical iteration can be applied in a straightforward fashion to visualize the composition of functions. With this method, graphical composition, the domain and range of a composition can be found.

3.3.7 Exercises

Find the domain and range of the composition $g(f)$ for Exercises 3.3.1 and 3.3.2; find the domain and range of the composition $f(g)$ for Exercises 3.3.3 and 3.3.4. Use the equations defining f and g to find numerical values of the endpoints of these intervals.

Prob 3.3.1 $f(x) = (x-2)^2 + 8$ for $1 \le x \le 5$; $g(x) = (1/2)(x-10)^2 + 3$ for $6 \le x \le 12$. See the left graph of Fig. 3.21.

Prob 3.3.2 $f(x) = x^2 + 1$ for all real x; $g(x) = \sqrt{16 - x^2}$ for $-4 \le x \le 4$. See the right graph of Fig. 3.21.

Prob 3.3.3 $f(x) = |-x^2 + 4|$ for $-3 \le x \le 3$; $g(x) = x^2 + 2$ for all real x. See the left graph of Fig. 3.22.

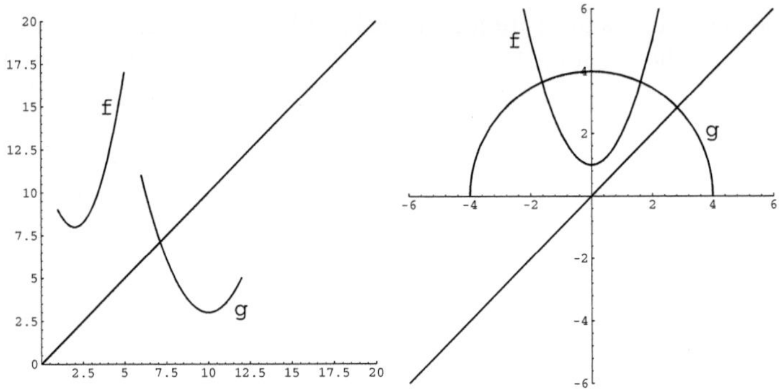

Figure 3.21: Graphs for Exercise 3.3.1 (left) and 3.3.2 (right).

Prob 3.3.4 $f(x) = \sqrt{x+1}$, for $-1 \leq x \leq 3$; $g(x) = 1/x$ for all real x except 0. See the right graph of Fig. 3.22.

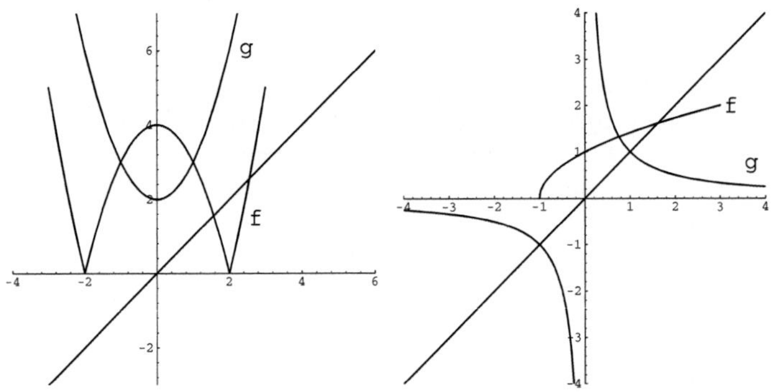

Figure 3.22: Graphs for Exercise 3.3.3 (left) and 3.3.4 (right).

3.4 Fractals and Pascal's triangles

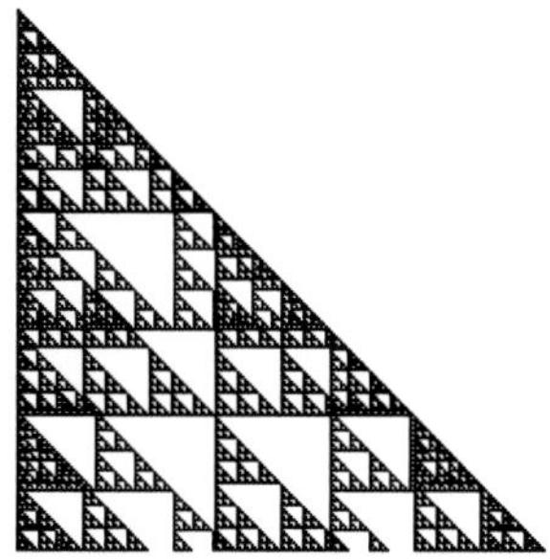

Pascal's triangle is a familiar way to generate coefficients of terms in binomial expansions. With an appropriate filter—for example, shade boxes with odd numbers and leave blank boxes with even numbers—Pascal's triangle can generate fractals. A consequence of the triangular grid of cells in Pascal's triangle is that most of the fractals produced are relatives of the Sierpinski gasket, though not relatives in the sense that Tara Taylor used the term. Still, different filters give different fractals, and some give no fractals at all. This observation will motivate a brief introduction to group theory.

In Lab 3.5 we'll use variations on Pascal's triangle to explore some more topics in group theory.

3.4.1 Purpose

We'll see how Pascal's triangle can generate fractals, and how to use this construction to illuminate some basic concepts from group theory.

3.4.2 Materials

Copies of the Pascal's triangle template, a pencil and an eraser, and Pascal's triangle software, 3.4 (a) and (b) of the Mathematica or Python codes.

3.4.3 Background

First we'll recall the familiar relation between Pascal's triangle and binomial coefficients, then show how Pascal's triangle can give rise to an approximation of the Sierpinski gasket. We'll find what Pascal's triangle produces with other number patterns, and end with a tiny bit of group theory.

3.4.3.1 Pascal's triangle and binomial coefficients

Pascal's triangle is familiar from the study of algebra and combinatorics. Though sometimes presented as an equilateral triangle, for our purposes a right isosceles format is better. Make a triangular array of boxes and fill each box of the left and right sides of the triangle with 1s. Starting from the top, fill all the other boxes with the sum of the entries in the box immediately above, and immediately above and to the left. If $p_{n,i}$ is the number in row n and column i, $i < n$, the rule can be written $p_{n,i} = p_{n-1,i-1} + p_{n-1,i}$. Here are the first few rows.

1								
1	1							
1	2	1						
1	3	3	1					
1	4	6	4	1				
1	5	10	10	5	1			
1	6	15	20	15	6	1		
1	7	21	35	35	21	7	1	
1	8	28	56	70	56	28	8	1

Figure 3.23: The first 8 rows.

Why should we care about this arrangement of numbers? The common explanation is based on the expansion of the binomial $(x+1)^n$:

$$
\begin{aligned}
(x+1)^0 &= 1\\
(x+1)^1 &= x+1\\
(x+1)^2 &= x^2+2x+1\\
(x+1)^3 &= x^3+3x^2+3x+1\\
(x+1)^4 &= x^4+4x^3+6x^2+4x+1
\end{aligned}
$$

and so on. The coefficients are are the entries of the corresponding rows of Pascal's triangle. To see the source of the rule $p_{n,i} = p_{n-1,i-1} + p_{n-1,i}$, note $(x+1)^n = (x+1)^{n-1}(x+1)$ and observe the effect of multiplication by $(x+1)$ on the coefficients of $(x+1)^{n-1}$. The $n = 3$ case should suffice to make the point.

3.4.3.2 Pascal's triangle and the Sierpinski gasket

One of the simplest distinctions between numbers is *even* or *odd*. Suppose we paint black any Pascal's triangle square containing an odd number, and paint white any square containing an even number. From the pattern in Fig. 3.23 we obtain the left image of Fig. 3.24. The right image is the first 64 rows with the same painting scheme, rescaled to the same size as the left image.

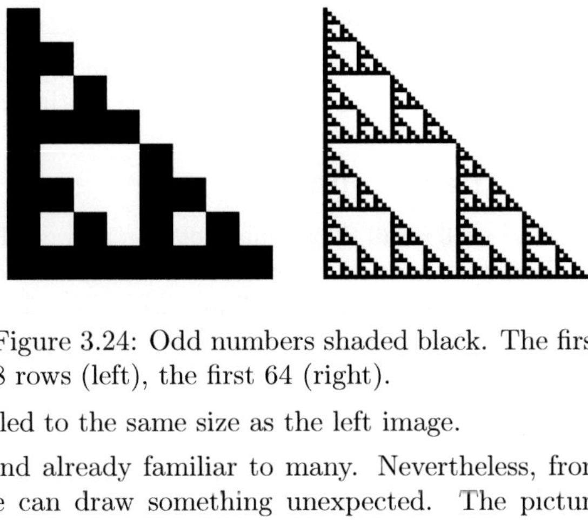

Figure 3.24: Odd numbers shaded black. The first 8 rows (left), the first 64 (right).

This is not a surprise, and already familiar to many. Nevertheless, from even this simple picture we can draw something unexpected. The picture suggests a natural rescaling as we look at the coefficients of $(x+1)^n$ for ever larger n. Take the square size to be $1/2^n$. Then all the pictures have the same height and width. The limit of these rescaled pictures corresponds to binomials of all orders, and we see the odd coefficients form a Sierpinski gasket. (The picture may be all the evidence you need, but if you want more detail, note that the part of the pattern between row 2^n+1 and 2^{n+1} is two copies of the part of the pattern between row 1 and 2^n.) Even and odd together form a right isosceles triangle, so the odd coefficients form a set of dimension $\log(3)/\log(2)$ in a set of dimension 2. Consequently, almost all binomial coefficients are even [81]. This is not so apparent from inspection of the first few rows of Pascal's triangle.

3.4.3.3 Pascal's triangle and other number patterns

The even or odd distinction is an interpretation of the result of dividing a number by 2: odd numbers give the remainder 1, even numbers the remainder 0. Numbers giving the same remainders when divided by 2 are called *congruent* (mod 2), so every even number is congruent to 0 (mod 2) and every odd number is congruent to 1 (mod 2). We write this as $6 \equiv 0 \pmod 2$ and $7 \equiv 1 \pmod 2$, for example.

The *congruence class* of $a \pmod 2$ is all the numbers congruent to $a \pmod 2$. We'll denote this congruence class as $[a]_2$, so $[1]_2 = [3]_2 = [17]_2$ is all odd numbers, and $[0]_2 = [2]_2 = [514]_2$ is all even numbers.

For any integer $p > 1$, in Pascal's triangle we can group integers congruent $\pmod p$. In Fig. 3.25 we paint black the numbers congruent to 0 (mod 3), or to 1 (mod 3), or to 2 (mod 3), or to 1 or 2 (mod 3). Because now we group Pascal's triangle coefficients by their remainder when divided by 3, we expect a cleaner pattern if the number of rows is a power of 3, so in each picture of Fig. 3.25 we see 81 rows.

Figure 3.25: The first 81 rows of Pascal's triangle. Left to right the painted points correspond to numbers congruent to 0, 1, 2 and 1 or 2 (mod 3).

In the rescaled limit as the number of rows goes to infinity, the right image of Fig. 3.25, coefficients $\equiv 1$ or $2 \pmod 3$, approaches a fractal of similarity dimension $\log(6)/\log(3) < 2$. Then by the union rule for dimensions, the coefficients $\equiv 0 \pmod 3$ form a set of dimension 2 and consequently, almost all binomial coefficients are multiples of 3. What else can we see?

If we think of, say, numbers (mod 3), we see a simple pattern in addition:

$$\begin{array}{lll} 0+0 \equiv 0 \pmod 3 & 1+0 \equiv 1 \pmod 3 & 2+0 \equiv 2 \pmod 3 \\ 0+1 \equiv 1 \pmod 3 & 1+1 \equiv 2 \pmod 3 & 2+1 \equiv 0 \pmod 3 \\ 0+2 \equiv 2 \pmod 3 & 1+2 \equiv 0 \pmod 3 & 2+2 \equiv 1 \pmod 3 \end{array}$$

That is, the numbers (mod 3) *cycle* under addition. For any integer $n > 1$, this is true of the numbers $\pmod n$. We say these numbers form the *cyclic group of order n*, denoted $\mathbb{Z}_n$. To understand which combinations give fractals, we need some basic concepts from group theory.

3.4.3.4 Some group theory

A *group* is a set of elements $G = \{a, b, c, \dots\}$, together with an operation, usually denoted $+$ or $\cdot$, that takes pairs of elements of G and returns elements of G. We say G is *closed* under the operation. The operation is *associative*:

$a + (b + c) = (a + b) + c$ for all elements a, b, and c of G. The group G has an *identity element* e having the property that $e + a = a + e = a$ for all $a \in G$. Finally, each $a \in G$ has an *inverse*, $b \in G$ having the property that $a + b = b + a = e$.

If G consists of a finite collection of elements, it is called a finite group and the number of elements is the *order* of G, denoted $|G|$. The order of an element $g \in G$ is the smallest number n for which $g + \cdots + g = e$, where n is the number of copies of g in the sum.

Consider $\mathbb{Z}_6$, the integers $(\text{mod } 6)$, as an illustration. Closure and associativity are easy to verify. The identity element is 0: for example, $0+1 = 1+0 = 1$. The inverse of each element a is $6 - a$: for example, $1 + 5 \equiv 0 \pmod 6$. So $\mathbb{Z}_6$ is a group under the operation of addition (mod 6). Its order is 6. In $\mathbb{Z}_6$ the element 2 has order three: $2 + 2 + 2 \equiv 0 \pmod 6$.

A subset H of a group G is a *subgroup* if it is itself a group, under the operation of G. Consequently, H must contain the identity element, the inverse of each of its elements, and must be closed. For example, it is easy to see that $H_1 = \{0, 3\}$ and $H_2 = \{0, 2, 4\}$ are subgroups of $\mathbb{Z}_6$, while $\{0, 1, 5\}$ is not.

Does $\mathbb{Z}_6$ have other subgroups? Of course, $\{0\}$ and $\mathbb{Z}_6$ itself are subgroups. The subgroup $\{0\}$ is called the trivial subgroup. In fact every group G has the corresponding subgroups, the identity element and all of G. All subgroups other than G itself are called *proper* subgroups, so the question is does $\mathbb{Z}_6$ have any proper subgroups in addition to $\{0\}$, H_1 and H_2?

The answer, no, is fairly easy to see, especially using *Lagrange's Theorem*: if H is a subgroup of a finite group G, then the order of H must be a divisor of the order of G. The divisors of 6 are 1, 2, 3, and 6, so proper nontrivial subgroups must have order 2 or 3. A subgroup of order 2 contains only one element in addition to the identity element. In $\mathbb{Z}_6$ the only element that is its own inverse is 3, so the only subgroup of order 2 is $\{0, 3\}$. Because 2 and 4 are inverses of one another, a subgroup H of order 3 that is different from H_2 must not contain 2 or 4. Suppose H contains 1. Closure implies H contains $1 + 1 \equiv 2 \pmod 6$, impossible if $H \neq H_2$. Suppose H contains 5. Closure implies H contains $5 + 5 \equiv 4 \pmod 6$, impossible if $H \neq H_2$. This leaves the only possible elements of H as 0 and 3, but we have seen these form the group H_1, of order 2.

Another easy consequence of Lagrange's theorem is that groups of prime order have no proper nontrivial subgroups.

We have seen two groups of order 2: $\mathbb{Z}_2$ and the subgroup $H_1 = \{0, 3\}$ of $\mathbb{Z}_6$. Though $\mathbb{Z}_2$ and H_1 are not identical, they can't be too different: both have only one non-identity element. We shall see that simply having the same number of elements is not enough to consider two groups equivalent: both must behave in similar fashions under their group operations. If there is such a correspondence, then we can think of one group as a copy of the other, with its elements renamed. As far as their mechanics are concerned—how the groups operate—they are identical.

This kind of correspondence has a name: we say that groups G_1 and G_2 are *isomorphic* if there is a function $f : G_1 \to G_2$ with these properties: (i) f is 1-1 and onto (so the elements of G_1 and G_2 are in 1-1 correspondence; for finite groups this implies they have the same number of elements), (ii) f

preserves the group operations, that is, for all $a, b \in G_1$, $f(a+b) = f(a)+f(b)$, where $+$ denotes the operation in both G_1 and G_2 (a function satisfying (ii) is a *homomorphism*, it "preserves the form" of the groups), and (iii) the inverse function f^{-1} preserves the group operations. For example, the function $f(x) = 3x$ is an isomorphism $f : \mathbb{Z}_2 \to H_1$. We write $\mathbb{Z}_2 \cong H_1$. Note that an isomorphism $f : G_1 \to G_2$ must take the identity element of G_1 to the identity element of G_2.

Here's a simple example of two non-isomorphic groups having the same number of elements, consider

$$\mathbb{Z}_4 = \{0, 1, 2, 3\} \text{ and } \mathbb{Z}_2 \times \mathbb{Z}_2 = \{(0, 0), (1, 0), (0, 1), (1, 1)\},$$

with the group operation of $\mathbb{Z}_2 \times \mathbb{Z}_2$ defined coordinate-wise: $(a, b) + (c, d) = (a + c, b + d)$. The group $\mathbb{Z}_2 \times \mathbb{Z}_2$ is called the *Cartesian product* of $\mathbb{Z}_2$ with itself. Now the element $1 \in \mathbb{Z}_4$ has order 4, yet the elements of $\mathbb{Z}_2 \times \mathbb{Z}_2$ have order 1 or 2. Consequently, $\mathbb{Z}_4$ cannot be isomorphic to $\mathbb{Z}_2 \times \mathbb{Z}_2$.

The last group theory concepts we need are cosets and quotient groups, which we introduce through examples in $\mathbb{Z}_6$.

We have seen that the subgroup $H_1 = \{0, 3\}$ of $\mathbb{Z}_6$ is isomorphic to $\mathbb{Z}_2$. Note that

$$\begin{aligned} 0 + H_1 &= \{0 + 0, 0 + 3\} = \{0, 3\} \\ 1 + H_1 &= \{1 + 0, 1 + 3\} = \{1, 4\} \\ 2 + H_1 &= \{2 + 0, 2 + 3\} = \{2, 5\} \end{aligned}$$

form a partition of $\mathbb{Z}_6$. That is, the union of these sets is all of $\mathbb{Z}_6$ and each set has no element in common with any other set. We say $0 + H_1$, $1 + H_1$, and $2 + H_1$ are the *cosets* of H_1. More precisely, these are called *left cosets* of H_1 because the element is added on the left.

What about $3 + H_1$, $4 + H_1$, and $5 + H_1$? Write down the elements and you'll see that $3 + H_1 = 0 + H_1$, $4 + H_1 = 1 + H_1$, and $5 + H_1 = 2 + H_1$. For example,

$$4 + H_1 = 4 + \{0, 3\} = \{4 + 0, 4 + 3\} = \{4, 1\} = \{1, 4\} = 1 + H_1$$

where we've used the fact that in $\mathbb{Z}_6$, $3 + 4 = 1$, that is, $3 + 4 \equiv 1 \pmod 6$.

This generalizes: for any subgroup H of any group, if $b \in a + H$, then $b + H = a + H$. For this reason, an element of a coset is called a *representative* of that coset.

These cosets, $0 + H_1$, $1 + H_1$, and $2 + H_1$, form a group under the operation

$$(a + H_1) + (b + H_1) = (a + b) + H_1.$$

As an illustration, note that $(1 + H_1) + (1 + H_1) = \{1, 4\} + \{1, 4\}$. Adding all the elements, we obtain

$$\{1, 4\} + \{1, 4\} = \{1 + 1, 1 + 4, 4 + 1, 4 + 4\} = \{2, 5, 8\} = \{2, 5\} = 2 + H_1$$

where for the penultimate equality we've used $8 \equiv 2 \pmod 6$.

This group of cosets is called the *quotient group* $\mathbb{Z}_6/H_1$. The function $a + H_1 \to [a]_6$, the congruence class of $a \pmod{6}$, is an isomorphism $\mathbb{Z}_6/H_1 \cong \mathbb{Z}_3$. Similarly, $\mathbb{Z}_6/H_2 \cong \mathbb{Z}_2$.

A group G is called *abelian* if the group operation is commutative, that is, if for all a and b in G, $ab = ba$. The examples we've seen so far all are abelian groups, so a complication we could ignore is that in order to form a quotient group G/H, the subgroup H must be *normal*, that is, for all $g \in G$ and all $h \in H$, we have $ghg^{-1} \in H$. An important result is the *first isomorphism theorem*. For this we need two more notions. The *kernel* of a homomorphism $f : G \to H$ is $\ker(f) = \{g \in G : f(g) = 1\}$, where 1 is the identity element of H, and the *image* of the homomorphism $\text{im}(f) = \{f(g) : g \in G\}$. Then the first isomorphism theorem is this: for any homomorphism $f : G \to H$,

1. $\ker(f)$ is a normal subgroup of G,
2. $\text{im}(f)$ is a subgroup of H, and
3. $G/\ker(f) \cong \text{im}(f)$.

The intuition of the first isomorphism theorem is a bit like squinting: squint so much that $\ker(f)$ looks like a blob. Then the elements of G appear as blobs, two elements of G belong to the same blob if they differ by an element of $\ker(f)$. And the group structure of the blobs is the same as that of $\text{im}(f)$.

This name suggests there is a second isomorphism theorem. There is, and also a third, but we won't need them for what we'll do here.

Group theory is an immense, complex branch of mathematics. Our Pascal's triangle exclusions illustrate only a few of the most basic concepts.

3.4.4 Procedure

Use the Pascal's triangle template for exercises done by hand; for all others use the Pascal's triangle software of 3.4 in Mathematica or Python codes.

3.4.5 Sample A

By hand exercise. On a Pascal's triangle template shade the squares containing odd numbers. Using the arithmetic facts odd + odd = even, etc. explain why the Sierpinski gasket appears. Hint: explain why the next-to-the left-most column and the first subdiagonal (the diagonal immediately below the hypotenuse of the triangle) alternate even and odd numbers. Describe the first 8 rows, then generalize.

Label the number in the first row of Pascal's triangle $p_{1,1}$, the numbers in the second row $p_{2,1}$ and $p_{2,2}$, ... the numbers in row n by $p_{n,1}, \ldots, p_{n,n}$. In this notation, we've seen the rule for Pascal's triangle is $p_{n+1,i} = p_{n,i-1} + p_{n,i}$.

1								
1	1							
1	2	1						
1	3	3	1					
1	4	6	4	1				
1	5	10	10	5	1			
1	6	15	20	15	6	1		
1	7	21	35	35	21	7	1	
1	8	28	56	70	56	28	8	1

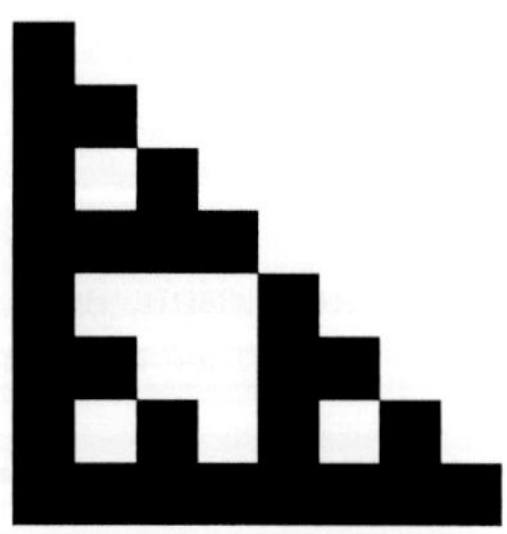

Because each $p_{n,1} = 1$, we see each $p_{n+1,2} = p_{n,1} + p_{n,2} = 1 + p_{n,2}$. That is, the parity (evenness or oddness) of $p_{n+1,2}$ alternates with n.

Similarly, each $p_{n,n} = 1$ and so $p_{n+1,n} = p_{n,n-1} + p_{n,n} = p_{n,n-1} + 1$. That is, the parity of the first subdiagonal alternates with n.

In the first four rows, all squares are shaded except the square containing $p_{3,2}$. In particular, all $p_{4,i}$ are odd, and so $p_{5,2}, \dots, p_{5,4}$ are even. Then these three consecutive even numbers in row 5 are followed by two, aligned right, in row 6 and one in row 7. Also, note $p_{7,2}$ and $p_{7,6}$ are even; all other entries are odd. So row 8 contains all odd numbers.

Because the pattern of shaded boxes depends only on the parity of the $p_{i,j}$, not on the precise values, the pattern in rows 9 through 16 reproduce two copies of the pattern in rows 1 through 8; one copy aligned under $p_{9,1}$, the other under $p_{9,9}$. The numbers $p_{9,2}, \dots, p_{9,8}$ are even. Successive rows see the length of the collection of these even boxes decreasing one by one, till they vanish at row 16, all of whose numbers are odd.

From here the pattern repeats, rows 17 through 32 filled with two copies of the pattern of rows 1 through 16, and so on.

3.4.6 Sample B

Computer exercise. Consider Pascal's triangle (mod 4). Shade the squares congruent to 0 (mod 4), to 1 (mod 4), to 2 (mod 4), to 3 (mod 4), to 1, 2, or 3 (mod 4), to 1 or 3 (mod 4), and to 0 or 2 (mod 4). Interpret these patterns in terms of the group $\mathbb{Z}_4$ and its subgroup $\mathbb{Z}_2$.

First, in Fig. 3.26 we see the patterns generated by shading the Pascal's triangle boxes with numbers congruent to 1, to 2, and to 3 (mod 4). Next, in Fig. 3.27 we see some combinations: shading the boxes of Pascal's triangle with numbers congruent to 0 or 2 (mod 4), to 0 (mod 4), to 1 or 3 (mod 4) and to 1, 2, or 3 (mod 4).

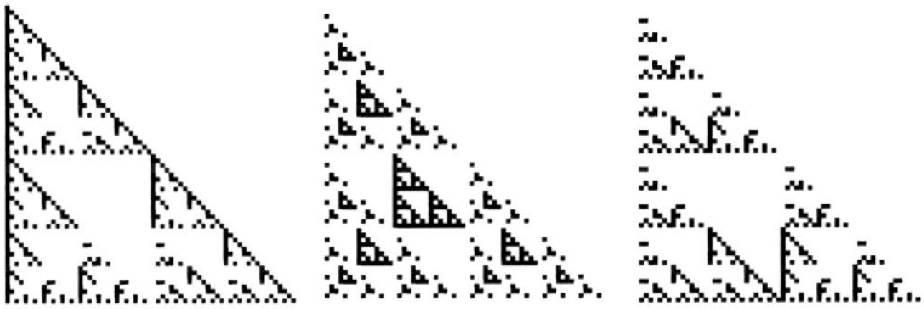

Figure 3.26: Left to right: In Pascal's triangle shade the boxes congruent to 1 (mod 4), to 2 (mod 4), and to 3 (mod 4).

The group $\mathbb{Z}_4$ has two proper subgroups, $H = \{0, 2\} \cong \mathbb{Z}_2$ and $\{0\}$. The function $f : \mathbb{Z}_4 \to \mathbb{Z}_2$ given by $f(x) = x \pmod 2$ is a homomorphism onto $\mathbb{Z}_2$: $f(0) = 0$, $f(1) = 1$, $f(2) = 0$, and $f(3) = 1$. Note that $\ker(f) = H$. In this case, the first isomorphism theorem becomes $\mathbb{Z}_4/H = \{0 + H, 1 + H\} \cong \mathbb{Z}_2$. The coset $0 + H = \{0, 2\}$ corresponds to $0 \in \mathbb{Z}_2$, and the coset $1 + H = \{1, 3\}$ corresponds to $1 \in \mathbb{Z}_2$.

In particular, shading the Pascal's triangle boxes with numbers in the coset $0 + H$ of $\mathbb{Z}_4/H$ gives the same image as shading the boxes with numbers in

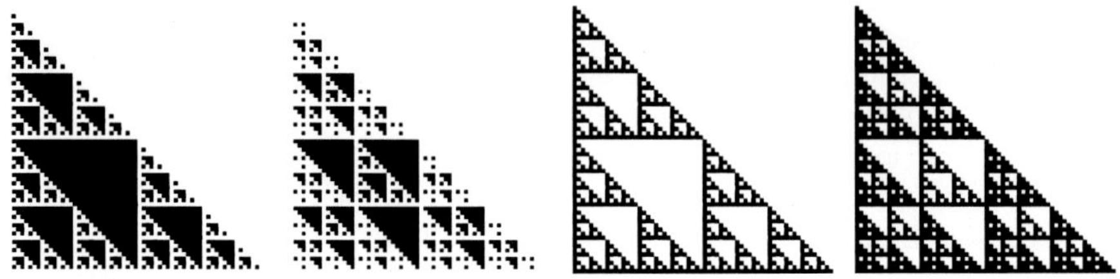

Figure 3.27: Left to right: in Pascal's triangle shade the boxes congruent to 0 or 2 (mod4), to 0 (mod4), to 1 or 3 (mod 4), and to $1, 2,$ or 3 (mod 4).

$0 \in \mathbb{Z}_2$. This is the first image of Fig. 3.27. Similarly shading boxes with numbers in the coset $1 + H$ gives the same image as shading $1 \in \mathbb{Z}_2$. This is the third image of Fig. 3.27. The fourth image is fractal, but not a simple fractal..

3.4.7 Conclusion

Fractal patterns in Pascal's triangle can be used to find subgroups of cyclic groups, and to give a visual illustration of the first isomorphism theorem. These are good examples of how fractals can provide another visual representation of mathematical concepts. The use of IFS to illustrate the geometry of plane transformations is another instance.

The idea of constructing Pascal's triangle for finite groups, in addition to the familiar cyclic groups $\mathbb{Z}_n$, is presented in [82]. The website [83] contains many examples. The idea of constructing Pascal's triangles for other polynomials is presented as an interesting collection of puzzles in [84].

3.4.8 Exercises

Prob 3.4.1 On a Pascal's triangle template shade the squares containing numbers that are not multiples of 3. Use addition (mod 3) to explain why a relative of the Sierpinski gasket appears.

Prob 3.4.2 Consider Pascal's triangle (mod 5). Shade the squares congruent to 0 (mod 5), to 1 (mod 5), to 2 (mod 5), to 3 (mod 5), to 4 (mod 5), and to $1, 2, 3$ or 4 (mod 5). Which shadings produce a fractal pattern? Explain your answer in terms of group theory.

Prob 3.4.3 Recall the Pascal triangle rule $p_{n+1,i} = p_{n,i-1} + p_{n,i}$. Give a geometric characterization for collections of shaded squares to correspond to a subgroup. (Hint: think of the closure requirement.)

Prob 3.4.4 Consider Pascal's triangle (mod 6). Shade the squares congruent to 0 (mod 6). Which combinations of shaded squares give fractal patterns? What other Pascal's triangle patterns do those produce? Interpret these results in terms of the subgroups of $\mathbb{Z}_6$.

Prob 3.4.5 For the given n consider the Pascal's triangle (mod n). Shade the squares that correspond to subgroups of $\mathbb{Z}_n$ and on a separate plot, shade the complements of these. Which correspond to fractal patterns?
(a) $n = 7$,
(b) $n = 8$,
(c) $n = 9$,
(d) $n = 10$, and
(e) $n = 15$.

Prob 3.4.6 Compare the patterns obtained by shading the squares containing numbers congruent to $1, \ldots, n-1 \pmod{n}$ for prime n and for composite n. Under what circumstances do simple fractal patterns arise?

Prob 3.4.7 For prime n compute the dimension of the limiting pattern (rescaling to the same size Pascal's triangle with n^k rows, as $k \to \infty$) obtained by shading the squares containing numbers congruent to $1, \ldots, n-1 \pmod{n}$. Use this observation to generalize the fact that almost all binomial coefficients are even.

The next three problems involve Pascal's triangles of some other polynomials. Successive rows of the Pascal's triangle correspond to successive polynomial exponents n. The entries of each row are the coefficients of powers of x.

Prob 3.4.8 (a) Generate Pascal's triangle for $(1 + x + x^2)^n$. Shade the boxes containing odd numbers. Is this the standard right isosceles Sierpinski gasket?
(b) Now shade the boxes containing numbers congruent to 1 or 2 (mod 3). Compare this with the corresponding picture for the standard Pascal's triangle.

Prob 3.4.9 (a) Generate Pascal's triangle for $(1+x+x^2)\cdot(1+x)^n$. Shade the boxes containing odd numbers. Compare this with the corresponding picture for the standard Pascal's triangle.
(b) Now shade the boxes containing numbers congruent to 1 or 2 (mod 3). Compare this with the corresponding picture for the standard Pascal's triangle.

Prob 3.4.10 (a) Generate Pascal's triangle for $x^n \cdot (1 + x)^n$. Shade the boxes containing odd numbers. Compare this with the corresponding picture for the standard Pascal's triangle.
(b) Now shade the boxes containing numbers congruent to 1 or 2 (mod 3). Compare this with the corresponding picture for the standard Pascal's triangle.

1															
1	1														
1	2	1													
1	3	3	1												
1	4	6	4	1											
1	5	10	10	5	1										
1	6	15	20	15	6	1									
1	7	21	35	35	21	7	1								
1	8	28	56	70	56	28	8	1							
1	9	36	84	126	126	84	36	9	1						
1	10	45	120	210	252	210	120	45	10	1					
1	11	55	165	330	462	462	330	165	55	11	1				
1	12	66	220	495	792	924	792	495	220	66	12	1			
1	13	78	286	715	1287	1716	1716	1287	715	286	78	13	1		
1	14	91	364	1001	2002	3003	3432	3003	2002	1001	364	91	14	1	
1	15	105	455	1365	3003	5005	6435	6435	5005	3003	1365	455	105	15	1

Figure 3.28: The Pascal's triangle template.

3.5 Fractals and Pascal's triangle relatives

In Lab 3.4 we learned to generate fractals from the familiar Pascal's triangle used to generate binomial coefficients. Modifications of this led us to construct relatives of Pascal's triangles for cyclic groups. Examples suggested that the complements of patterns that represent subgroups are fractal, and simple fractals if the order of the subgroup is a prime number.

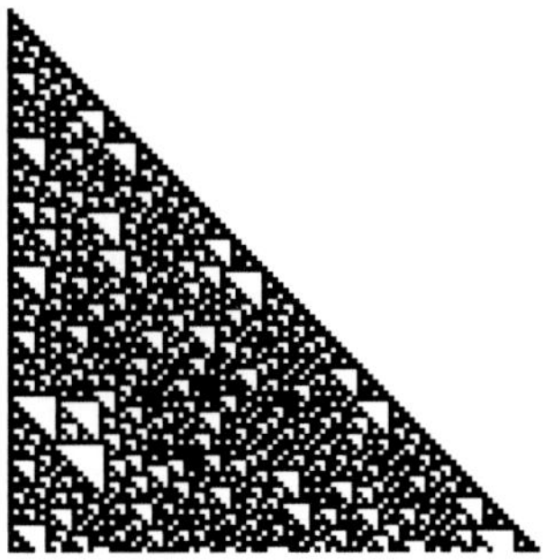

Here we'll extend the Pascal's triangle construction to non-cyclic groups. Will the conjectures of Lab 3.4 be supported on a more general class of groups?

3.5.1 Purpose

We'll extend the Pascal's triangle construction to non-cyclic groups and study the circumstances under which fractal pattens arise from the group structure.

3.5.2 Materials

Copies of the Pascal's triangle template, a pencil, an eraser, and Pascal's triangle software (3.5 of the Mathematica or Python codes).

3.5.3 Background

We'll introduce some more group theory, illustrate some groups with symmetries of regular polygons, and present the group operation as a group table (this is how we'll incorporate a group into the Pascal's triangle software), and end with permutation groups.

Certainly, you should be familiar with Lab 3.4

3.5.3.1 A tiny bit more group theory

Recall the numbers (mod n) cycle under addition: these numbers form the *cyclic group of order n*, denoted $\mathbb{Z}_n$. An equivalent geometric characterization is to think of the rotational symmetries of the regular n-sided polygon.

Of course, regular polygons exhibit symmetries besides rotations. We shall see the collection of all symmetries of a polygon, together with the operation of composition, can be described by a group more complex than cyclic. First we note a way in which the group operation can differ from familiar addition and multiplication.

The order of addition of elements of a cyclic group is irrelevant: $1 + 2 = 2 + 1$, for example. However, for symmetry groups that include rotations and reflections, the order of the operations can matter. For example, label an equilateral triangle's vertices so we can keep track of their positions. Consider the symmetry operations rotate 120° about the center and reflect across the altitude.

Certainly, the order of the transformations has an effect on the outcome. See Fig. 3.29. So in general, we must consider groups in which the order of the operation $\cdot$ can matter: $a \cdot b \neq b \cdot a$.

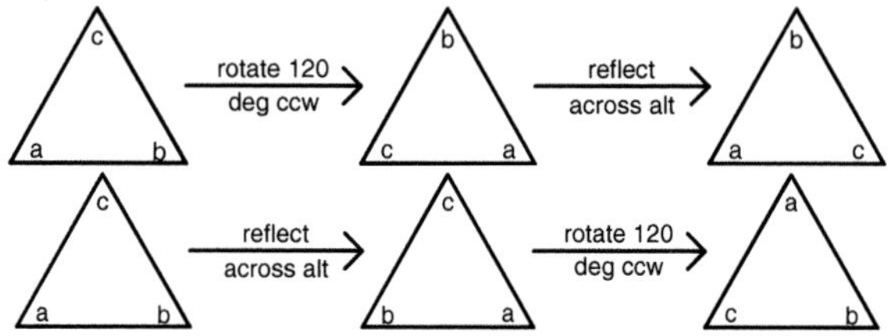

Figure 3.29: Top: rotation, then reflection. Bottom: reflection, then rotation.

With this in mind, in general left cosets and right cosets need not be equal: $gH \neq Hg$. Of course, if $\cdot$ is commutative then $gH = Hg$ for all subgroups H and all g. Even when $\cdot$ is not commutative, $gH = Hg$ can be true for all $g \in G$, for certain subgroups H of G. These are called the *normal* subgroups of G. In Sect. 3.4.3.4 we saw an equivalent characterization: $g \cdot h \cdot g^{-1} \in H$ for all $h \in H$ and all $g \in G$.

If H is a normal subgroup of G, we can define an operation of cosets:

$$(aH) \cdot (bH) = (a \cdot b)H$$

With this operation, the cosets of H form a group, called the *quotient group* G/H.

For example, for the (necessarily normal because $\mathbb{Z}_6$ is abelian) subgroup $H_1 = \{0, 1\}$, consider $\mathbb{Z}_6/H_1$. We have seen the cosets are $0 + H_1, 1 + H_1$, and $2 + H_1$. This table shows the structure of the quotient group.

$+$	$0+H_1$	$1+H_1$	$2+H_1$
$0+H_1$	$0+H_1$	$1+H_1$	$2+H_1$
$1+H_1$	$1+H_1$	$2+H_1$	$0+H_1$
$2+H_1$	$2+H_1$	$0+H_1$	$1+H_1$

The function $a + H_1 \rightarrow a$ is an isomrphism from $\mathbb{Z}_6/H_1$ to $\mathbb{Z}_3$. Because $H_1 \cong \mathbb{Z}_2$, we can write this isomorphism as $\mathbb{Z}_6/\mathbb{Z}_2 \cong \mathbb{Z}_3$.

More interesting examples require groups that are not commutative. We build examples of these in the next section.

3.5.3.2 Symmetries of regular polygons

A regular polygon with n sides has two types of symmetries: rotations about the center and reflections about a line through the center. These symmetries form the *dihedral group* D_n of order $2n$. The order is $2n$ because there are n rotations, by $(m/n)360°$ for $0 \leq m \leq n - 1$, and n reflections.

The number of rotations that preserve the figure is evident. A bit more thought is needed to study the reflections. For each of the items 1 through 3 below, inverstigate these with regular pentagon and regular hexagon templates and from these observations try to deduce the general answers before reading the solutions.

1. Find all the reflections that are symmetries of a *regular n-gon*, a regular polygons having n sides. Describe the reflections in terms of sides or angles of the polygons. Different descriptions may be needed for even n and for odd n.

2. Certainly, all the rotations of a regular n-gon can be generated by compositions of rotation by $(1/n)360°$. Perhaps less obviously, all reflections are compositions of any given reflection and a rotation. Illustrate this by showing reflection across a diagonal of a square is the same as a rotation followed by reflection across the perpendicular bisector of two opposite sides. Prove this for all regular n-gons.

Consequently, the dihedral group D_n is generated by rotation a through $(1/n)360°$, and a reflection b. Denote by 1 the identity symmetry, and observe that $a^n = 1$ and $b^2 = 1$. Are there other relations?

3. In the notation of item 2, show $bab = a^{-1}$. Show this relation is equivalent to $(ab)^2 = 1$.

4. Show $a^n = 1$, $b^2 = 1$, and $(ab)^2 = 1$ are the only relations. That is, if some other combination w of a and b satisfies $w = 1$, then the three relations can be used to show w is equivalent to one of the relations.

Solutions of 1–4

1. If the number of sides is odd, there is a reflection across the perpendicular bisector of each side, so n reflections. If the number of sides is even, the perpendicular bisector of each side is also the perpendicular bisector of the opposite side, so there are only $n/2$ reflections across perpendicular bisectors of sides. But there are also reflections across the bisectors of each vertex. The bisector of each vertex also bisects the opposite vertex, so these account for $n/2$ reflections. (If n is odd, the perpendicular bisector of each side also bisects the opposite vertex.)

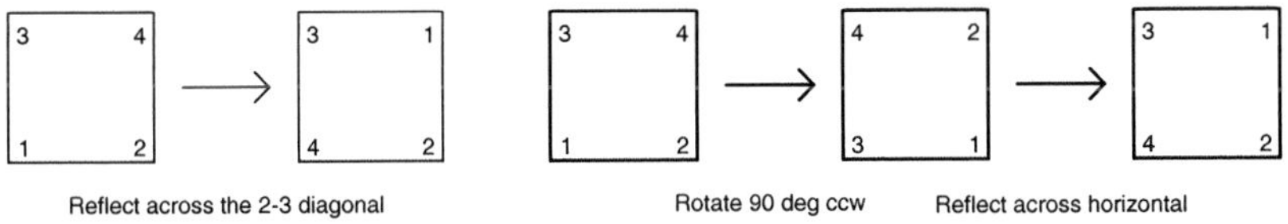

Figure 3.30: Left: Reflection of a square across a diagonal. Right: Rotation followed by reflection.

2. First, the left image of Fig. 3.30 shows a reflection of the square across a diagonal. Next, the right image of Fig. 3.30 shows a 90° ccw rotation followed by reflection across the horizontal bisector of the square. So reflection across this diagonal is equivalent to rotating 90° ccw and then reflecting across the horizontal bisector of the square.

The symmetries of a regular polygon are reflections or rotations, because symmetries must take consecutive vertices to consecutive vertices; the orientation-preserving symmetries are rotations, the orientation-reversing symmetries are reflections. Pick a reflection c and suppose d is any other reflection. Then cd is orientation-preserving, so $cd = r$, a rotation, and $d = c^2d = c(cd) = cr$ because c^2 = identity, for any reflection c. ("I am the mirror image of my mirror image.")

3. Here is an example using a pentagonal template. Reflect across a vertical line, rotate $72° = 360/5°$ ccw, then reflect again across a vertical line. Fig. 3.31 shows this is equivalent to a rotation by $72°$ cw.

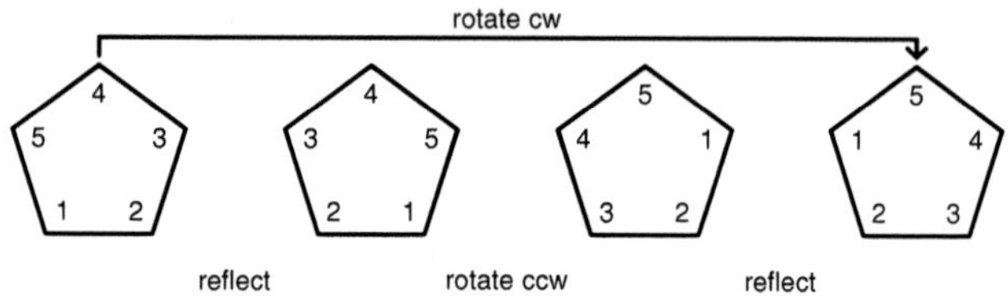

Figure 3.31: A relation between reflections and rotations.

The hexagonal template example is similar.

From $bab = a^{-1}$ we see $(ab)^2 = (ab)(ab) = a(bab) = a(a^{-1}) = 1$.

4. First, we'll say the 2n words (strings of group elements) of the form

$$1, a, a^2, \ldots, a^{n-1}, b, ba, ba^2, \ldots, ba^{n-1}$$

are the *canonical words.*

Now we show that any word can be reduced to one of these canonical words by applications of the relations $a^n = 1, b^2 = 1$, and $(ab)^2 = 1$.

• Using $b^2 = 1$ we can replace every every occurrence of b^k in w by b (if k is odd) or 1 (if k is even).

• Next, $abab = 1$ is equivalent to $bab = a^{-1}$, and because $b = b^{-1}$ this in turn is equivalent to $ab = ba^{-1}$. With this observation, all occurrences of b can be moved to the left. For example,

$$a^2b = a(ab) = a(ba^{-1}) = (ab)a^{-1} = (ba^{-1})a^{-1} = ba^{-2}$$

• With this, w is equivalent to b^ia^j. Finally, i can be reduced to 0 or 1, and j to $0, 1, \ldots, n-1$.

In abstract terms, the dihedral group is written

$$D_n = \langle a, b : a^n = b^2 = (ab)^2 = 1\rangle$$

3.5.3.3 Group tables

Finite groups can be described completely by listing the products of all possible combinations of group elements. These can be organized into a *group table*: arrange the elements $g_1, \ldots, g_N$ along the top and left side of a matrix. The entry in the ith row and j^{th} column of the matrix is the product g_ig_j. For example, here is the table for the cyclic group $\{0, 1, 2, 3\}$ of order 4. (This group is abelian, so the operation usually is written $+$.)

+	0	1	2	3
0	0	1	2	3
1	1	2	3	0
2	2	3	0	1
3	3	0	1	2

No surprises there. More interesting are the group tables of the dihedral groups, D_3 for example. To fill in the table entries, for example to show $b \cdot a = a^2b$, multiply $1 = abab$ on the left by a^2 and on the right by b to obtain

$$a^2b = (a^2)abab(b) = (a^3)ba(b^2) = ba$$

where we've used $a^3 = 1$ and $b^2 = 1$.

The rest of the table for D_3 is filled in a similar fashion. Some shortcuts reduce the amount of work. The first row and column are easy, representing products with the identity. Also, every row and every column of the table must contain each element of the group. Consequently, the last entry of each row can be filled in without any calculation, and the entire last row can be filled in by completing each column.

$\cdot$	1	b	a	a^2	ab	a^2b
1	1	b	a	a^2	ab	a^2b
b	b	1	a^2b	ab	a^2	a
a	a	ab	a^2	1	a^2b	b
a^2	a^2	a^2b	1	a	b	ab
ab	ab	a	b	a^2b	1	a^2
a^2b	a^2b	a^2	ab	b	a	1

For later reference we show the the subgroup $H = \{1, a, a^2\}$ of D_3 is normal, and the subgroup $G = \{1, b\}$ is not.

To show H is normal, we show $bH = Hb, abH = Hab$, and $a^2bH = Ha^2b$. What about $aH = Ha$ and $a^2H = Ha^2$? Because $a \in H$ and $a^2 \in H$ and because subgroups are closed, we see $aH = H = Ha$ and $a^2H = H = Ha^2$.

$$\begin{aligned}
bH &= \{b, ba, ba^2\} = \{b, a^2b, ab\} \\
Hb &= \{b, ab, a^2b\} \\
abH &= \{ab, aba, aba^2\} = \{ab, b, a^2b\} \\
Hab &= \{ab, a^2b, a^2ab\} = \{ab, a^2b, b\} \\
a^2bH &= \{a^2b, a^2ba, a^2ba^2\} = \{a^2b, ab, b\} \\
Ha^2b &= \{a^2b, aa^2b, a^2a^2b\} = \{a^2b, b, ab\}
\end{aligned}$$

so H is a normal subgroup of the dihedral group D_3.

On the other hand, note $aG = \{a, ab\}$ and $Ga = \{a, ba\} = \{a, a^2b\}$, so G is not a normal subgroup of D_3.

We emphasize this useful property of group tables:: every element of the group occurs exactly once in every row and in every column of the table.

3.5.3.4 Permutations

A standard way to denote a permutation σ of n elements is as a $2 \times n$ matrix with first row the numbers 1 through n, and second row $\sigma(1)$ through $\sigma(n)$. The six permutations of $\{1, 2, 3\}$ are listed below. The first is the identity permutation $1 \to 1$, $2 \to 2$, and $3 \to 3$. The second transposes 1 and 2: $1 \to 2$, $2 \to 1$, and $3 \to 3$. And so on.

$$\begin{pmatrix} 1 & 2 & 3 \\ 1 & 2 & 3 \end{pmatrix} \begin{pmatrix} 1 & 2 & 3 \\ 2 & 1 & 3 \end{pmatrix} \begin{pmatrix} 1 & 2 & 3 \\ 1 & 3 & 2 \end{pmatrix} \begin{pmatrix} 1 & 2 & 3 \\ 3 & 2 & 1 \end{pmatrix} \begin{pmatrix} 1 & 2 & 3 \\ 3 & 1 & 2 \end{pmatrix} \begin{pmatrix} 1 & 2 & 3 \\ 2 & 3 & 1 \end{pmatrix}$$

In the order listed, call these elements $e, \alpha, \beta, \gamma, \sigma$, σ^2. Each of α, β, and γ is a *transposition*, interchanging two numbers.

The group operation is composition. For example,

$$\alpha\gamma = \begin{pmatrix} 1 & 2 & 3 \\ 2 & 1 & 3 \end{pmatrix} \begin{pmatrix} 1 & 2 & 3 \\ 3 & 2 & 1 \end{pmatrix} = \begin{pmatrix} 1 & 2 & 3 \\ 2 & 3 & 1 \end{pmatrix} = \sigma^2$$

To see this, note for example that α takes 1 to 2, and γ takes 2 to 2, so $\alpha\gamma$ takes 1 to 2.

The group of permutations of $\{1, 2, \ldots, n\}$ is called the *symmetric group* on n elements, denoted $\mathcal{S}_n$. If we work out the group table of $\mathcal{S}_3$, we see that $\mathcal{S}_3 \cong D_3$. Dihedral and symmetric groups are isomorphic only for $n = 3$. The easiest way to see that $\mathcal{S}_n \not\cong D_n$ for $n \neq 3$ is to note that the order of D_n is $2n$ and the order of $\mathcal{S}_n$ is $n!$. Isometric (finite) groups must have the same number of elements.

Permutations are familiar, probably quite a bit more familiar than groups. One reason we've mentioned permutation groups is to offer a concrete approach to every problem about group theory. This is through *Cayley's theorem*: every (finite) group is isomorphic to a subgroup of a permutation group.

3.5.4 Procedure

Use the Pascal's triangle template for exercises done by hand; for all others use the Pascal's triangle software from 3.5 of the Mathematica or Python codes.

To adapt Pascal's triangle to a finite group, we specify which group elements go down the left edge of the triangle and which elements go down the right edge. Denote by $p_{i,j}$ the group element in row i and column j of the triangle. Then the element in $p_{i,j}$ is the product $p_{i-1,j-1} \cdot p_{i-1,j}$. That is, for each triangle box not on the left or right edge, the group element in that box is the product of the group element in the box directly above and the group element in the box above and to the left. Because order can matter in group multiplication, we multiply these elements in the order they appear in the triangle, left on the left times right on the right.

a	b				
a	ab	b			
a	a^{2}b	a	b		
a	b	ab	ab	b	
a	ab	a^2	1	a	b

Figure 3.32: Dihedral group Pascal's triangle.

For example, in Fig. 3.32 we use the group D_3 to fill in Pascal's triangle. Put the element a down the left side and the element b down the right side. Then fill in the other triangle entries with products from the group table.

3.5.5 Sample A

By hand exercise. (a) Write out the group table for $\mathbb{Z}_2 \times \mathbb{Z}_2$. Verify that every element has order 1 or 2.

(b) Place $(1, 0)$ in all the boxes on the left side of the triangle template (the first 10 rows suffice). Place $(0, 1)$ in the boxes along the right side, starting in row two. Shade the boxes that contain $(0, 0)$.

(a) Elements of $\mathbb{Z}_2 \times \mathbb{Z}_2$ are ordered pairs of elements of $\mathbb{Z}_2$. Here is the group table.

$+$	$(0,0)$	$(1,0)$	$(0,1)$	$(1,1)$
$(0,0)$	$(0,0)$	$(1,0)$	$(0,1)$	$(1,1)$
$(1,0)$	$(1,0)$	$(0,0)$	$(1,1)$	$(0,1)$
$(0,1)$	$(0,1)$	$(1,1)$	$(0,0)$	$(1,0)$
$(1,1)$	$(1,1)$	$(0,1)$	$(1,0)$	$(0,0)$

The diagonal entries of a group table are the squares of the group elements. So if a diagonal entry in the row and column of a group element g is the identity element 1, this says $g^2 = 1$. The presence of the identity

element $(0,0)$ of $\mathbb{Z}_2 \times \mathbb{Z}_2$ down the diagonal of the group table shows every element has order at most 2; only the identity element has order 1.
(b) The left side of Fig. 3.33 shows the first ten rows of Pascal's tiangle filled using the group table for $\mathbb{Z}_2 \times \mathbb{Z}_2$. On the right side those boxes that contain $(0,0)$ are shaded. We see the start of the upside down triangles that form the holes in a gasket.

(1,0)										
(1,0)	(0,1)									
(1,0)	(1,1)	(0,1)								
(1,0)	(0,1)	(1,0)	(0,1)							
(1,0)	(1,1)	(1,1)	(1,1)	(0,1)						
(1,0)	(0,1)	(0,0)	(0,0)	(1,0)	(0,1)					
(1,0)	(1,1)	(0,1)	(0,0)	(1,0)	(1,1)	(0,1)				
(1,0)	(0,1)	(1,0)	(0,1)	(1,0)	(0,1)	(1,0)	(0,1)			
(1,0)	(1,1)	(1,1)	(1,1)	(1,1)	(1,1)	(1,1)	(1,1)	(0,1)		
(1,0)	(0,1)	(0,0)	(0,0)	(0,0)	(0,0)	(0,0)	(0,0)	(1,0)	(0,1)	
(1,0)	(1,1)	(0,1)	(0,0)	(0,0)	(0,0)	(0,0)	(0,0)	(1,0)	(1,1)	(0,1)

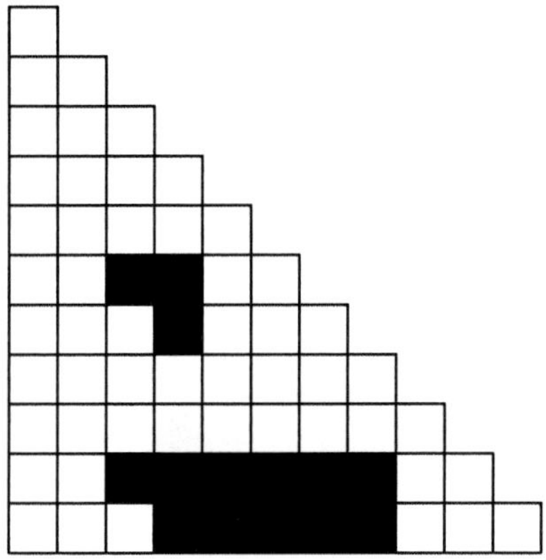

Figure 3.33: Left: The first ten rows of Pascal's triangle using the group $\mathbb{Z}_2 \times \mathbb{Z}_2$. Right: The squares that contain $(0,0)$ are shaded.

3.5.6 Sample B

Computer exercise. For the group $\mathbb{Z}_2 \times \mathbb{Z}_2$, build a Pascal's triangle with $(1,0)$ down the left edge, and $(0,1)$ down the right edge.
(a) Plot the first 100 rows, shade those boxes that contain $(0,0)$. In a separate plot, shade those boxes that contain $(1,0)$, $(0,1)$, or $(1,1)$.
(b) The group $\mathbb{Z}_2 \times \mathbb{Z}_2$ has three subgroups, each isomorphic to $\mathbb{Z}_2$. They are $\{(0,0),(1,0)\}$, $\{(0,0),(0,1)\}$, and $\{(0,0),(1,1)\}$. Recall the conjecture in Lab 3.4 about simple fractal patterns in Pascal's triangle and subgroups of prime order. Investigate the patterns formed by shading pairs of elements and identify these subgroups. Alternately, deduce which elements constitute these subgroups and verify that the observation of Lab 3.4 extends to non-cyclic groups.

(a) The left image is the Pascal's triangle for $\mathbb{Z}_2 \times \mathbb{Z}_2$ with $(1,0)$ down the left edge, and $(0,1)$ down the right edge. The squares that contain $(0,0)$ are shaded. These correspond to a subgroup, the trivial subgroup that consists just of the identity element. Note the filled-in triangles.

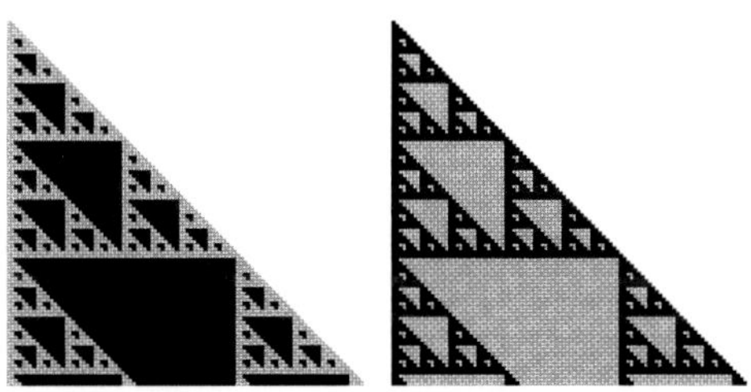

In the right image we have shaded the complement, those squares that contain $(0, 1)$, $(1, 0)$, or $(1, 1)$. This will give a simple fractal. In both, the background grid is to identify individual cells in Pascal's triangle.

Figure 3.34: Top: Shade squares that contain elements of the subgroups $\{(0,0),(0,1)\}$ (left), $\{(0,0),(1,0)\}$ (middle), and $\{(0,0),(1,1)\}$ (right). Botton: the complements of the images above.

In the top row for Fig. 3.34 we shade the Pascal's triangle squares that correspond to the three order-2 subgroups of $\mathbb{Z}_2 \times \mathbb{Z}_2$. Because every subgroup contains the identity group element, $(0, 0)$ for $\mathbb{Z}_2 \times \mathbb{Z}_2$, the shaded squares for every subgroup will contain those shaded for the trivial subgroup. Consequently, for every subgroup the squares shaded for the subgroup complement will be contained in the squares shaded for the complement of $\{(0, 0)\}$. So far, we have seen no obvious reason why all of these should be simple fractals.

Nevertheless, all three complements are simple fractals, so the observation of Lab 3.4 extends to at least this non-cyclic group. We'll explore other extensions in the exercises.

3.5.7 Conclusion

Pascal's triangle provides visual representations of subgroups and quotient groups for non-cyclic groups as well as for the cyclic groups of Lab 3.4.

3.5.8 Exercises

For exercises 3.5.1 and 3.5.2 we use Pascal's triangle for the di-

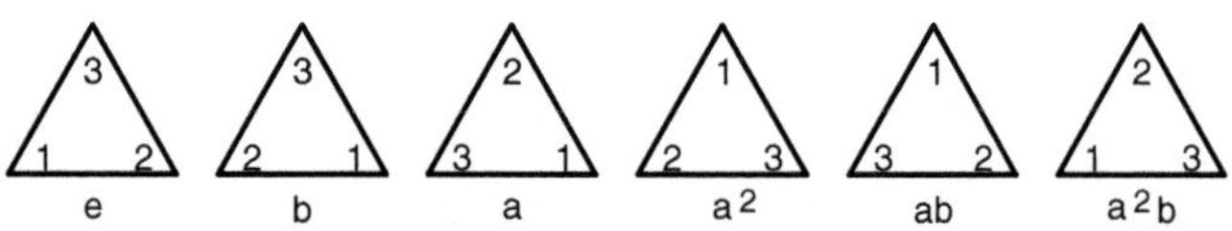

hedral group D_3 of symmetries of the equilateral triangle. In this representation, a and a^2 are counterclockwise rotations about the center of the triangle by 120° and 240°, and b is reflection across the perpendicular bisector of the base of the triangle. Note that ab and a^2b are reflections across the perpendicular bisectors of the other two sides.

Prob 3.5.1 Generate Pascal's triangle by placing b along the left side and a along the right side of the triangle. The product of two rotations is another rotation, so the rotations form a subgroup of D_3. Shade the boxes for that subgroup and for the complement of that subgroup. Does a simple fractal pattern appear?

Prob 3.5.2 Generate Pascal's triangle by placing b along the left side and a along the right side of the triangle. Shade the boxes for the subgroup $\{1, b\}$ generated by the reflection b, and in a separate plot for the complement of that subgroup. Does a simple fractal pattern appear?

For exercises 3.5.3, 3.5.4, and 3.5.5 we use Pascal's triangle for the dihedral group D_4 of symmetries of the square. Here r_1, r_2, and r_3 are counterclockwise rotations about the center of the square by 90°, 180°, and 270°, v is reflection across the vertical line that bisects the square, h reflection across the horizontal line that bisects the square, d_+ reflection across the diagonal of slope $+1$, and d_- reflection across the diagonal of slope -1.

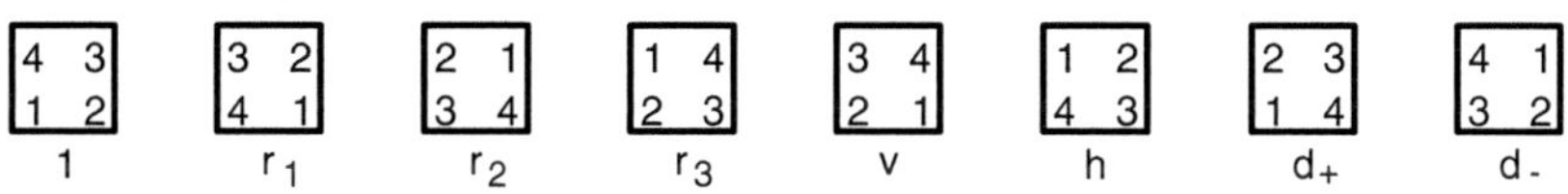

Figure 3.35: Symmetries of the square.

These are functions, so the group operation is composition, read right to left. For example, $r_1d_+ = v$ and $d_+r_1 = h$.

Prob 3.5.3 Write the group table for D_4.

Prob 3.5.4 Generate Pascal's triangle by placing r_1 along the left side and v along the right side.
(a) Shade the boxes containing elements of the rotation subgroup $H = \{1, r_1, r_2, r_3\}$. In a separate plot, shade the complement.
(b) Show H is a normal subgroup of D_4.
(c) How does (a) change if r_3 is placed along the right side of the triangle?

Prob 3.5.5 Generate Pascal's triangle by placing r_1 along the left side and d_+ along the right side.
(a) Shade the boxes containing elements of the reflection subgroup $G = \{1, d_+\}$. In a separate plot, shade the complement.
(b) Is G a normal subgroup of D_4?

Prob 3.5.6 For every $n \geq 2$ find a subgroup of the symmetric group S_n isomorphic to the cyclic group $\mathbb{Z}_n$.

Prob 3.5.7 *Project.* Investigate Pascal's triangle for subgroups of other groups, characterize those subgroups with complements that are simple fractals, more complex fractals, and apparently chaotic patterns. The rotation subgroup of D_3 is an example of the first, reflection subgroups in D_4 are examples of the second, reflection subgroups in D_3 are examples of the third. How does the choice of which elements are placed on the right and left sides of the triangle affect the pattern produced?

3.6 Mandelbrot sets and Julia sets

The Mandelbrot set is one of the most evocative images of fractal geometry, in all of math. This image first reached popular attention when a part of the boundary of Mandelbrot set was the cover image of the August, 1985 issue of *Scientific American.* The accompanying article, "A computer microscope zooms in for a look at the most complex object in mathematics" was that month's *Computer Recreations* column, written by A. K. Dewdney. Some images of the Mandelbrot set and of Julia sets appeared in Chapter 19 of [1], but the color image on the cover of *Scientific American* reached a large audience immediately.

The images are alluring, visually complex but with underlying patterns that repeated some general features, each with variations. For mathematicians and for coders the true surprise was in the text that accompanied the pictures. All the images—really all, and infinitely many others besides—are described by a single simple equation, and are generated by a few straightforward lines of computer code. Two loops and a graphics command are all you need. Thousands began to explore the Mandelbrot set. These were heady times, wonderful times to be a math teacher.

While it was difficult to imagine this book without a lab about the Mandelbrot set, by now these pictures are so familiar that we wondered what we could do that was more substantial than generating deeper zooms into the Mandelbrot set. A solution, the one we'll use here, is to apply the Mandelbrot set and Julia set constructions to functions different from the usual $z^2 + c$ that generates the Mandelbrot set and Julia sets. Will anything be different? Will everything be different? Let's see.

3.6.1 Purpose

We'll investigate the analogs of the Mandelbrot set and Julia sets for the functions $z^3 + c$ (simple cubic), $z^4 + c$ (simple quartic), $z^5 + c$ (simple quintic), for $z^3/3 + z^2/2 + c$ (cubic, two critical points), and for $z^3 + c_1 z + c_2$ (general cubic).

3.6.2 Materials

Use the Mandelbrot set and Julia set software, 3.6 (a)–(d) of the Mathematica or Python codes.

3.6.3 Background

Even from their initial appearance as crude images drawn for Benoit with a pen plotter (consult Google if you don't know what a pen plotter is), the Mandelbrot set was recognized as having considerable visual complexity. Advances in processor speed and in graphical displays continued to reveal details that vary without end. In the late 1980s when we began to investigate the Mandelbrot set, the generation of a decent image would tie up our computers overnight. You will not encounter that bother nowadays.

Here we see a picture of the familiar Mandelbrot set, along with a magnification of a tiny part of the set. The definition of the Mandelbrot set is based on the family of functions

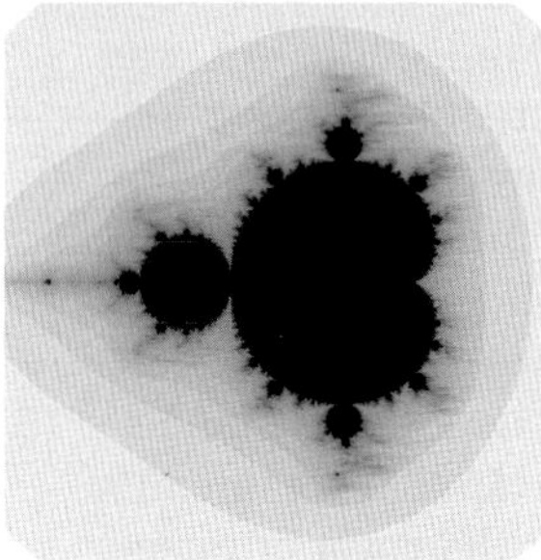

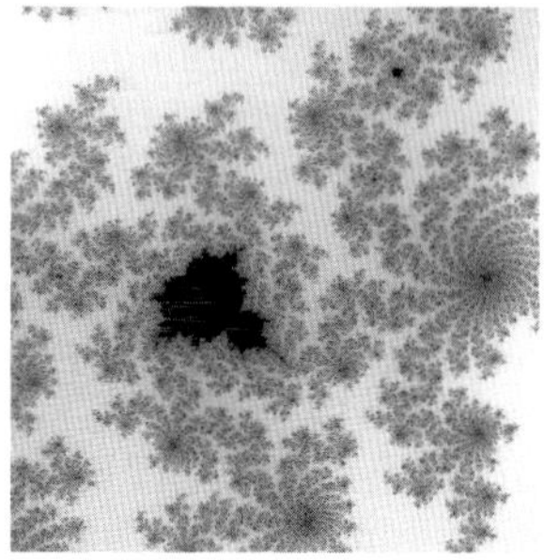

$$f_c(z) = z^2 + c$$

Then the Mandelbrot set M is the set of all those c for which the iterates

$$z_0 = 0, \ z_1 = f_c(z_0), \ z_2 = f_c(z_1), \ \ldots \tag{3.3}$$

do not run away to ∞.

This seems simple enough, but a moment's thought reveals four questions:
(1) The complex plane contains uncountably infinitely many points. How can we possibly test (3.3) for each c?
(2) Surely we can't continue to generate terms of (3.3) until the value ∞ appears. How can we tell if the sequence (3.3) runs away to ∞?
(3) A continuation of (2): how many iterates must we do before we decide that the sequence (3.3) never will run away to ∞?
(4) Why do we care if these iterates run away to ∞?

The first question is easy to answer because we'll generate images on a screen made of pixels. We need to carry out the iterations (3.3) only for those complex numbers $c = a + ib$ where (a, b) are the coordinates of the center of each pixel.

For (2) we recall the *modulus* of a complex number $z_n = x_n + iy_n$ is $\|z_n\| = \sqrt{x_n^2 + y_n^2}$. In Appendix B.10.1 we'll establish the *escape criterion*, that if for some n, $\|z_n\| > 2$, then later iterates z_{n+k} run away to ∞. Replacing ∞ with > 2 is an improvement.

Question (3) is the only place where we must make a choice. Suppose for a particular c we observe $\|z_n\| \leq 2$ for $n = 1, \ldots, 100$. How do we know that $\|z_n\| \leq 2$ for all n? The answer is that we don't know. For points close enough to, but still outside of, the Mandelbrot set, the first hundred, or the first thousand, or the first million, iterates may stay within 2 of the origin, while later iterates will run away to ∞. So we make a choice. Say the *dwell* is the maximum number of iterates we'll check. If for some c, $\|z_n\| \leq 2$ for $n \leq$ *dwell*, then we'll assert that $\|z_n\| \leq 2$ for all n and so the point c, and every point in the pixel centered at c, belongs to the Mandelbrot set. Then we'll color that pixel black. Occasionally this will result in a black pixel that doesn't belong to the Mandelbrot set. The trade-off is that a higher dwell gives a more accurate image, but takes more time to produce.

If for some $n \le dwell$, we find $\|z_n\| > 2$ then paint the pixel centered on c a color assigned to the number n, so the colors of the region outside the Mandelbrot set encode how quickly points in that region escape to ∞. Certainly, striking color images were responsible for much of the early interest in the Mandelbrot set. Rather than colors, we'll encode the escape speed by gray levels. We think some images just look better in black, white, and gray. Would you want to watch *Casablanca* or *Citizen Kane* or *City Lights* or *Duck Soup* or *Dr. Strangelove* in color? We wouldn't.

This brings us to question (4). Here we'll need a brief detour. For each complex number c, the *Julia set* J_c is the collection of all those complex numbers z_0 for which the iterates

$$z_0,\ z_1 = f_c(z_0),\ z_2 = f_c(z_1),\ \ldots \tag{3.4}$$

do not run away to ∞. (Really this is the *filled-in Julia set*; the Julia set is the boundary of the filled-in Julia set.) The points we made about the sequence (3.3) also apply to (3.4). The centers of pixels are the complex numbers z_0 at which the iteration sequence (3.4) starts. If for some n, $\|z_n\| > 2$, then later iterates diverge to ∞. And as with the Mandelbrot set iterates, we must choose a dwell at which to stop every iteration process that has not diverged.

Here we see two Julia sets. The picture at the start of this lab is a magnification of the middle of the Julia set in the right image.

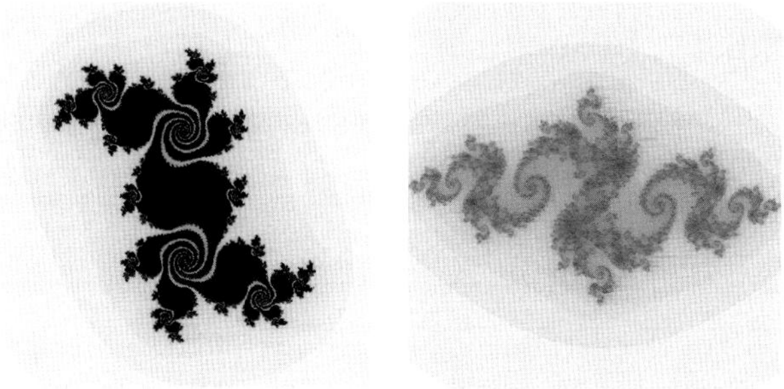

A difference between the Mandelbrot set and Julia sets is the planes they inhabit. The Mandelbrot set lives in the plane of c values, while Julia sets live in the plane of z_0 values. Also, the Mandelbrot iteration (3.3) always begins at $z_0 = 0$. To see the reason for this, and to answer question (4), we'll need a theorem of Pierre Fatou and Gaston Julia. First, replace f_c in (3.4) with any polynomial f to obtain the sequence

$$z_0,\ z_1 = f(z_0),\ z_2 = f(z_1),\ \ldots \tag{3.5}$$

If the sequence (3.5) diverges when z_0 takes the value of each critical point of f, then the Julia set is a Cantor set; if the sequence (3.5) remains bounded when z_0 takes the value of each critical point, then the Julia set is connected.

The polynomial $f_c = z^2 + c$ has only one critical point, $z_0 = 0$, so the Julia sets of f_c are either Cantor sets or connected, and they are connected if and only if the iterates of $z_0 = 0$ remain bounded for all n. Knowing this dichotomy of the Julia set topology, Benoit wondered about the set of all c for which the Julia set of f_c is connected. That is, the set of all c for which (3.3) remains bounded, a perfectly natural question. This is the Mandelbrot set.

We'll explore variations of this construction for polynomials other than $f_c(z) = z^2 + c$. Only two points require some care: the escape to ∞ criterion

$\|z\| = 2$ may be replaced by other bounds. We'll address this in Appendix B.10.1. Second, if the polynomial has more than one critical point, we must chose which critical point (or points) to iterate.

Many of the geometrical properties of the Mandelbrot set have been found by now. Some are sketched in Appendix B.10.2.

The polynomials $z^3 + c$, $z^4 + c$, and $z^5 + c$ each have only one critical point, $z_0 = 0$, so we expect many similarities between the Mandelbrot sets and Julia sets of these functions and those of $z^2 + c$. The polynomial $z^3/3 + z^2/2 + c$ has two critical points, $z = 0$ and $z = -1$, so we may find Julia sets that have topological types different from the two types, connected and Cantor sets, that we observe for $z^2 + c$.

Perhaps you're wondering why the Mandelbrot set is based on iteration of $z^2 + c$, instead of the more general $z^2 + bz + c$? In Appendix B.10.3 we'll show that by a linear change of variables we can rescale any polynomial to remove the next-to-the highest power. Then any quadratic can be written as $z^2 + c$ and any cubic as $z^3 + c_1 z + c_2$. But remember, the coefficients are complex numbers, so the general cubic Mandelbrot set lives in two complex dimensions, equivalent to four real dimensions. We won't be able to see the whole set, but we will explore some cross-sections.

3.6.4 Procedure

Make appropriate modifications to the programs from 3.6 of the Mathematica or Python codes.

3.6.5 Sample A

For the iteration $z_{n+1} = z_n^3 + c$, plot the Mandelbrot set, magnify a small copy of this Mandelbrot set on a branch that runs from the main body of the set. For the whole set, take the window to be $-1.5 \le x \le 1.5$ and $-1.5 \le y \le 1.5$. Plot a Julia set of a c in this Mandelbrot set, and another for a c outside this set. What do these images tell you?

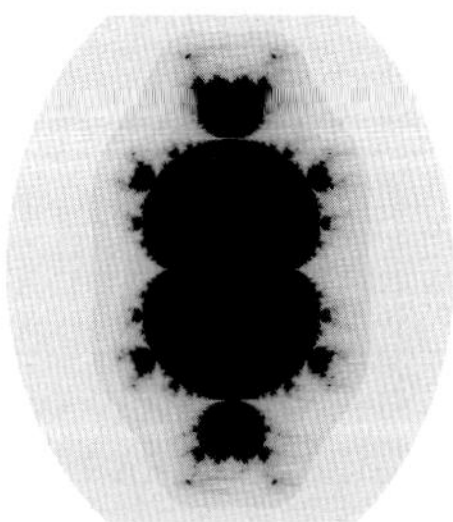

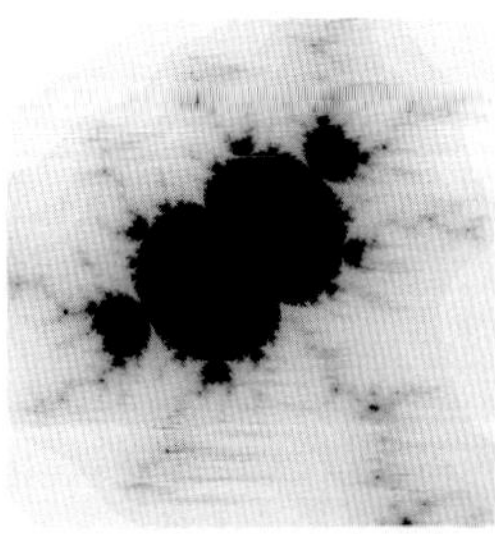

The left image is the Mandelbrot set for $z^3 + c$, the right image is a magnification of the region $-0.293 \le x \le -0.233$ and $1.228 \le y \le 1.288$. While the $z^2 + c$ Mandelbrot set is symmetric about the real axis, this cubic Mandelbrot set appears to be symmetric about both the real axis and the imaginary axis. The magnified copy does look similar to the whole set, but is a bit distorted. In the small copy the line that corresponds to the imaginary axis in the whole shape is rotated about $-45°$ (alone this isn't distortion, it's a rigid motion), but more significantly, both top and bottom of this rotated axis are bent toward the northwest.

In Fig. 3.36 we see three Julia sets for the $z^3 + c$ iteration. The first is for $c = 0.539 + 0.527i$. This is in the Mandelbrot set, and we see that the Julia set is connected. The second is for $c = 0.683 + 0.616i$, a point outside the Mandelbrot set, but near the previous c. As expected, this Julia set is a Cantor set. Is the evident (approximate) 3-fold symmetry a consequence of the iteration's being $z^3 + c$? Let's see.

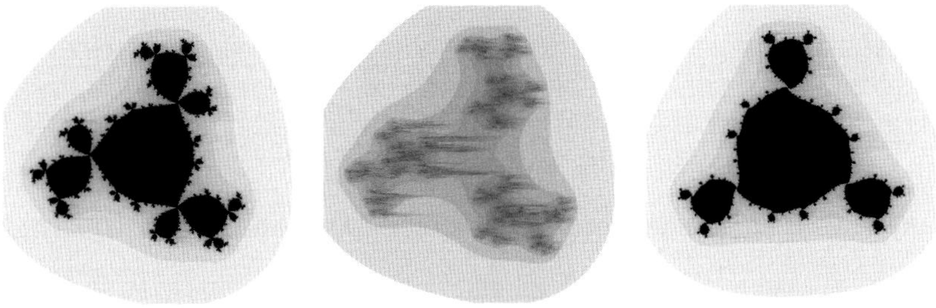

Figure 3.36: Three Julia sets for $z^3 + c$.

The third Julia set of Fig. 3.36 is for $c = -0.06 + 0.95i$, also in the Mandelbrot set. This c is in the large component at the top of the Mandelbrot set, while the c of the first Julia set image is from the largest component in the northeast direction around the top half of this Mandelbrot set. The principal sequence of disks attached to the main cardioid of the quadratic Mandelbrot set described in Appendix B.10.2 also occurs around this Mandelbrot set. In addition, it appears that the number of Julia set lobes that meet at a branch point is the cycle number of the Mandelbrot set component in which the c lies. Can you think of a way to test this conjecture?

3.6.6 Sample B

Now we'll iterate the function $f_c(z) = z^3/3 + z^2/2 + c$. The derivative is $f'(z) = z^2 + z$, so the critical points are $z_a = -1$ and $z_b = 0$. Because this polynomial has two critical points, we can define a Mandelbrot set for this function in three ways:

- M_a is those c for which $f_c(z_a)$, $f_c^2(z_a)$, $f_c^3(z_a)$, ... remains bounded.
- M_b is those c for which $f_c(z_b)$, $f_c^2(z_b)$, $f_c^3(z_b)$, ... remains bounded.
- $M_{a,b}$ is those c for which both $f_c(z_b)$, $f_c^2(z_b)$, $f_c^3(z_b)$, ... and $f_c(z_a)$, $f_c^2(z_a)$, $f_c^3(z_a)$, ... remain bounded.

Plot each of these in the window $-2.5 \leq x \leq 2$ and $-2.25 \leq y \leq 2.25$. Magnify a region that contains a small part of $M_{a,b}$. Does it look like a scaled, perhaps distorted, copy of $M_{a,b}$?

Recall the theorem of Fatou and Julia mentioned in Sect. 3.6.3. If the iterates of all critical points remain bounded, the Julia set is connected; if the iterates of all critical points escape to ∞, then the Julia set is a Cantor set. For the iteration of functions with a single critical point, the Fatou-Julia theorem implies the *Dichotomy theorem*: Julia sets are either connected or Cantor sets. For this cubic function f_c, we have the possibility that the iterates of one critical point remain bounded while the iterates of the other escape to ∞.

Find a point c for which that happens. Is the Julia set connected? Is it a Cantor set? Is it something else?

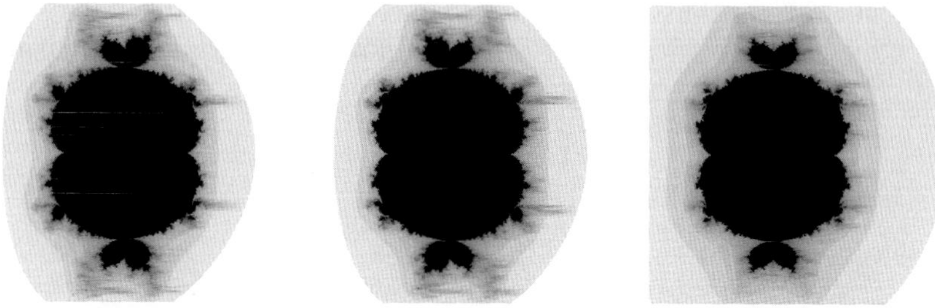

Figure 3.37: The Mandelbrot sets M_a (left), M_b (middle), and $M_{a,b}$ (right).

The three Mandelbrot sets M_a, M_b, and $M_{a,b}$ are shown in Fig. 3.37. All three look similar, but neither M_a nor M_b is left-right symmetric. Do you think that $M_{a,b}$ is left-right symmetric? How would you test this?

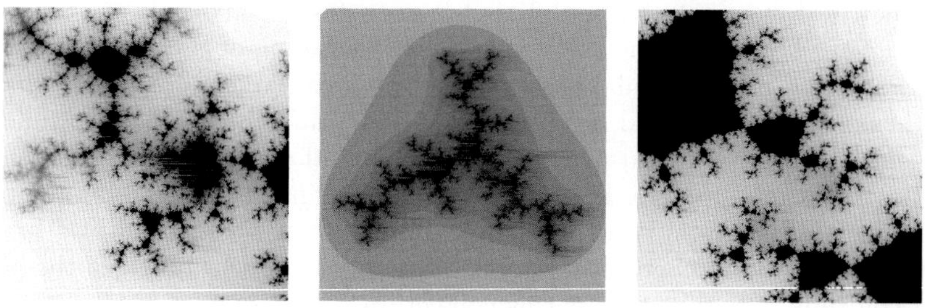

Figure 3.38: A magnification of a region in $M_{a,b}$ (left), the Julia set for $c = -0.25 + 1.75i$ (middle), and a magnification of a region of that Julia set (right).

The left image of Fig. 3.38 is a magnification of $M_{a,b}$, specifically the region $-0.95 \leq x \leq -0.91$ and $1.74 \leq y \leq 1.78$. This magnification contains a complex array of shapes, but the appearance of a (distorted) quadratic Mandelbrot set is unexpected. The reason it appears here is tied up in the universality of the Mandelbrot set mentioned in Appendix B.10.2. Look around a bit. Can you find other copies of the quadratic Mandelbrot set in $M_{a,b}$? What about in M_a or in M_b?

The middle image of Fig. 3.38 is the Julia set for $c = -0.25 + 1.75i$. This Julia set does not appear to be connected, as shown more clearly by the right image, a magnification of the region $-0.5 \leq x \leq 0.4$ and $-0.3 \leq y \leq 0.6$. On the other hand, it is not a Cantor set because it contains filled-in regions and Cantor sets are dusts, they are totally disconnected. For this c value, with $z_0 = -1$ we find $z_{100} = -0.930578 + 0.0355649i$. If we continue, the sequence of iterates appears to converge to a 4-cycle

$$\begin{aligned} &-0.0850869 + 1.74769i, && -1.539 - 0.165438i, \\ &-0.232869 + 1.6228i, && -0.930578 + 0.035565i \end{aligned}$$

and so the iterates of $z_0 = -1$ remain forever bounded. On the other hand, with $z_0 = 0$ we find $z_{22} = 8648.94 - 20851.9i$ and it's clear that the iterates of $z_0 = 0$ escape to ∞.

So with two critical points we have the possibility of a different topological type of Julia set. Would three critical points open up still other possibilities? Suppose the iterates of two critical points remain bounded while the iterates of the third escape? Or two escape while the iterates of the third critical point remain bounded? Would these reveal new shapes for Julia sets, or have we by now seen all possible topological types?

3.6.7 Sample C

Now we'll explore the Mandelbrot set and Julia sets for the cubic $f_{c_1,c_2}(z) = z^3 + c_1 z + c_2$ of Appendix B.10.3. The critical points are the roots of $0 = f'_{c_1,c_2}(z) = 3z^2 + c_1$, that is, $z_\pm = \pm\sqrt{-c_1/3}$. If you're worried about taking the square root of a complex number, the polar representation is the key. A complex number $z = a + bi$ can be written in polar form as $z = r(\cos(\theta) + \sin(\theta)i)$, where $r = \sqrt{a^2 + b^2}$ is the modulus of z and $\theta = \arctan(b/a)$ is the argument of z. (If $a = 0$ then $z = bi$ and $\theta = \pi/2$ if $b > 0$, $3\pi/2$ if $b < 0$.) The branch of the arctan depends on the signs of a and b. Complex number multiplication has a simple expression in polar form: multiply the moduli and add the arguments. Then to square a complex number we square the modulus and double the argument, so the square root is given by

$$\sqrt{z} = \sqrt{r}\big(\cos(\theta/2) + \sin(\theta/2)i\big)$$

For example, $z = 1 + \sqrt{3}i$ has modulus $r = 2$ and argument $\theta = \pi/3$, so

$$\sqrt{z} = \sqrt{2}\big(\cos(\pi/6) + \sin(\pi/6)i\big) = \sqrt{2}\left(\frac{\sqrt{3}}{2} + \frac{1}{2}i\right) = \frac{\sqrt{6}}{2} + \frac{\sqrt{2}}{2}i$$

This Mandelbrot set depends on two complex parameters, c_1 and c_2, and consequently four real parameters, a, b, c, and d, where $c_1 = a + bi$ and $c_2 = c + di$. Because of the complexity of these images, we'll plot planar cross-sections and make whatever deductions we can. From what we learned in Sample 3.6.6 we'll study the Mandelbrot set that consists of those c_1 and c_2 values for which the iterates of both critical points $\pm\sqrt{-c_1/3}$ remain bounded.

This is a pretty symmetric polynomial. Do you think it's possible for the iterates of one critical point to escape to ∞ while the other remains bounded? If this isn't possible, we can cut in half the number of iterations the program must calculate. This would roughly double the speed of image generation. See Exercise 3.6.8 before you shorten the code.

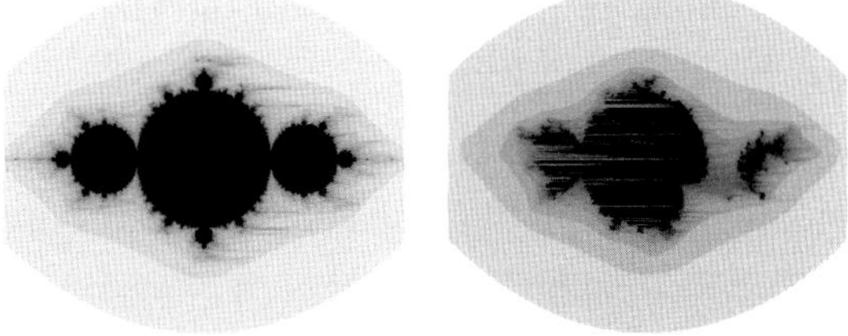

Figure 3.39: Cross-sections: $c = d = 0$ (left) and $c = d = 0.15$ (right).

In Fig. 3.39 we plot the Mandelbrot set cross-section for $c = d = 0$ (the left image) and for $c = d = 0.15$ (the right image). For both, use the window $-3 \leq x \leq 3$ and $-3 \leq y \leq 3$. The right image is not connected. From this can we deduce that the full Mandelbrot set is not connected? No, because the full Mandelbrot set is 4-dimensional. Disconnected pieces in a 2-dimensional cross-section may be connected by a region outside this plane. In fact, the cross-sections in the planes along the path $c = d$ from $c = d = 0.15$ to $c = d = 0$ gradually merge together and connect. The interpretation of 2-dimensional cross-sections of a 4-dimensional object can require some care, and usually additional images.

3.6.8 Conclusion

Mandelbrot and Julia sets for cubic and higher-order polynomials provide some variations on the patterns familiar for the quadratic Mandelbrot and Julia sets. In particular, if the polynomial has several critical points, we may encounter Julia sets that are neither connected nor Cantor sets.

3.6.9 Exercises

Prob 3.6.1 In the window $-2 \leq x \leq 1$ and $-1.5 \leq y \leq 1.5$ plot the Mandelbrot set for iteration of $z^4 + c$. Identify a region that contains a small bit of this Mandelbrot set and magnify that region. Does it look like the whole Mandelbrot set?

Prob 3.6.2 In the main body of the Mandelbrot set of Exercise 3.6.1 identify a point in each of what looks like it should be a 2-cycle disk and a 3-cycle disk attached to the main shape and plot their Julia sets. Identify a point in the main body of the small Mandelbrot set copy generated in Exercise 3.6.1. Plot the Julia set for that point. What do these images tell you?

Prob 3.6.3 In the window $-2 \leq x \leq 1$ and $-1.5 \leq y \leq 1.5$ plot the Mandelbrot set for iteration of $z^5 + c$. Speculate on a relation between the exponent in the iteration function and the number of largest lobes attached to the main body. Identify a region that contains a small bit of this Mandelbrot set and magnify that region. Does it look like the whole Mandelbrot set?

Prob 3.6.4 In the main body of the Mandelbrot set of Exercise 3.6.3 identify a point in each of what looks like it should be a 2-cycle disk a 3-cycle disk, and a 4-cycle disk attached to the main shape. Plot the Julia sets of these points. Identify a point in the main body of the small Mandelbrot set copy generated in Exercise 3.6.3. Plot the Julia set for that point. What do these images tell you?

Prob 3.6.5 For the c value of each of the Julia sets of Exercise 3.6.2 and 3.6.4, determine the length of the cycle to which the iterates of the critical point converge. Interpret this in terms of the number of Julia set lobes that meet at each branch point.

Prob 3.6.6 For the polynomials of Sample 3.6.6, find a point c_1 that belongs to M_a but not to M_b, and find a point c_2 that belongs to M_b but not to M_a. Plot the Julia set for c_1 and the Julia set for c_2. Can you identify some feature of these Julia sets that determine to which Mandelbot set the Julia set c-value belongs?

Prob 3.6.7 Pick a point c that belongs to a 3-cycle component of M_a, that is, the iterates of the critical point $z_a = -1$ converge to a 3-cycle. Also, take this point c not in M_b. Plot the Julia set for this c. Can these conditions explain some of the geometry of this Julia set?

Prob 3.6.8 For $f_{c_1,c_2}(z) = z^3 + c_1 z + c_2$, recall the critical points are $z_1 = \sqrt{-c_1/3}$ and $z_2 = -\sqrt{-c_1/3}$. Take $c_1 = -1$ and $c_2 = 1$. Do the iterates of both critical points remain bounded, or both escape to ∞, or one remain bounded and one escape to ∞?

Prob 3.6.9 For the Mandelbrot set of f_{c_1,c_2} with $c_1 = a + bi$ and $c_2 = c + di$, (a) hold c and d constant ($c = d = 0$ is one choice, but not the only choice) and plot the cross-section of this Mandelbrot set in this ab-plane.
(b) Hold a and b constant and plot the cross-section of this Mandelbrot set in this cd-plane.

Prob 3.6.10 For the Mandelbrot set of f_{c_1,c_2} with $c_1 = a + bi$ and $c_2 = c + di$, (a) Hold a and c constant and plot the cross-section of this Mandelbrot set in this bd-plane.
(b) Hold b and d constant and plot the cross-section of this Mandelbrot set in this ac-plane.

3.7 Circle inversion fractals

In this lab we'll introduce, or review, the geometry of inversion in a circle. Then we'll replace the familiar transformations (each determined by r, s, θ, φ, e, and f) of IFS with inversion in a collection of circles. Because inversion in circles is nonlinear, among other new fractal types we'll produce nonlinear gaskets.

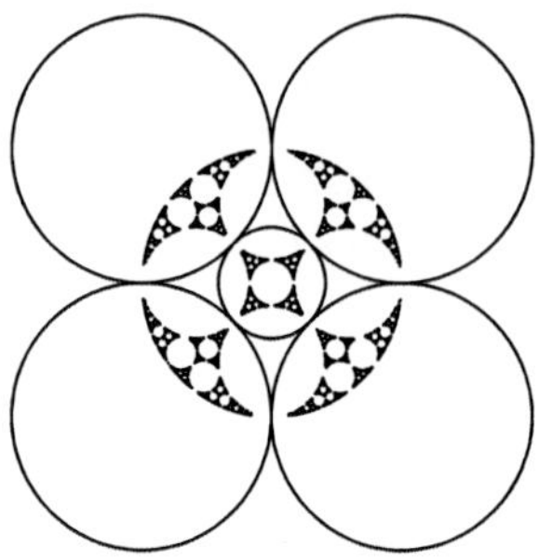

3.7.1 Purpose

Our goals are to learn some of the geometry of circle inversion, and to modify the IFS algorithms to generate fractals based on circle inversion. Extension of IFS algorithms to circle inversions offers several complications. When applied twice, inversion in any circle is the identity transformation. If T is a transformation of an IFS for any of the Labs of Chapter 1, then for any initial point (x_0, y_0), $T^n(x, y)$ converges to the fixed point of the T. But long sequences of inversion in a circle alternates between the identity transformation and a single inversion. Then, too, inversion in a circle C does not have a single fixed point, rather, every point of C is fixed. Unlike our earlier IFS transformations, inversion is not a contraction everywhere. Inversion in C is a contraction for regions outside of C, but is an expansion for regions inside of C. Finally, the contraction factor of inversion in C gets closer to 1 (that is, no contraction) as points approach C. Attempts to adapt the random IFS algorithm to circle inversion suffers from extraordinarily slow convergence near C.

All these issues mean we'll need to be clever in our adaptation of IFS to circle inversions. Achieving this adaptation is the goal of this lab.

3.7.2 Materials

For the manual experiments, a compass (the circle-drawing kind of compass, not the lost-in-the-woods compass), ruler, paper, and pencil. For the computer labs, circle inversion software (3.7 of the Mathematica or Python codes).

3.7.3 Background

If you are familiar with inversion in circles, you can omit the first three sections and jump straight to Sect. 3.7.3.3.

3.7.3.1 Circle inversion definition

Inversion in a circle was introduced by Apollonius of Perga (262 BC - 190 BC) in his last book *Plane Loci.* Coordinate geometry would not be developed for over a millenium after Apollonius, so naturally his approach was synthetic.

• Given a point P outside the circle C, draw the segment PQ tangent to C. From Q drop the perpendicular to OP intersecting at P'. The point P' is the *inverse* of P.
• Conversely, given a point P' inside the circle C, draw the perpendicular to OP', intersecting C at Q. The tangent to C at Q intersects the line extending OP' at P. The point P is the inverse of P'.

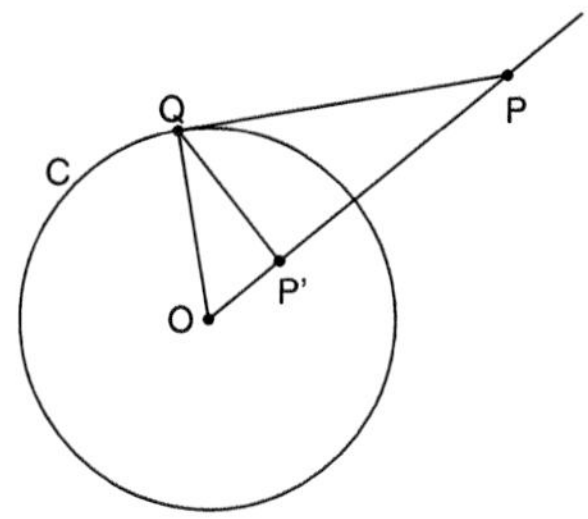

Figure 3.40: The point P' is the inverse of P in C.

Equivalently, P is the inverse of P' if
(i) P and P' lie on the same ray from O, and
(ii) $|OP| \cdot |OP'| = |OQ|^2$.

The equivalence of these two characterizations of inversion follows from this similarity $\Delta OP'Q \sim \Delta OQP$, which implies $OP'/OQ = OQ/OP$.

To find an analytic expression for the inverse of a point, note that the requirement that (x', y') lies on the ray from (a, b) to (x, y) means

$$(x', y') = (a, b) + t(x - a, y - b) \text{ for some } t.$$

To find t, use the distance condition

$$\sqrt{(x-a)^2 + (y-b)^2} \cdot \sqrt{(x'-a)^2 + (y'-b)^2} = r^2$$

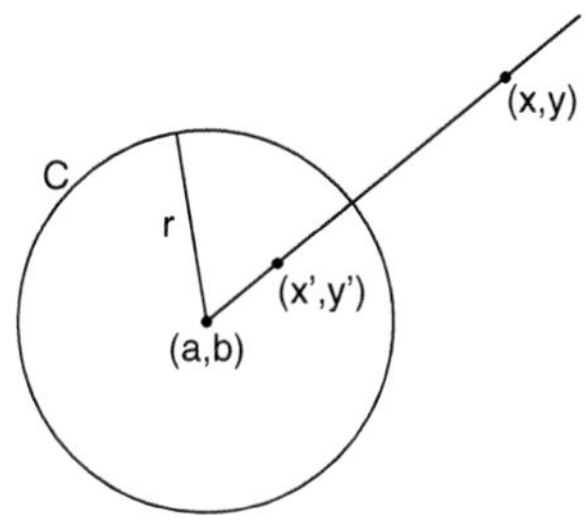

and obtain $t = \dfrac{r^2}{(x-a)^2 + (y-b)^2}$. That is, the inverse of the point (x, y) in the circle with radius r and center (a, b) is

$$(x', y') = (a, b) + \frac{r^2}{(x-a)^2 + (y-b)^2}(x - a, y - b) \tag{3.6}$$

Note that if the center of the circle is $(a, b) = (0, 0)$ and the radius is $r = 1$, the formula (3.6) takes a very simple appearance:

$$(x', y') = \frac{1}{x^2 + y^2}(x, y) \tag{3.7}$$

An easy application of Eq. (3.7) is to show that inversion in the circle of radius 1 and center $(0, 0)$ has a variable contraction factor. The segment of the x-axis between $x = 2$ and $x = 3$ inverts to the segment between $1/2$ and $1/3$. That is, this segment of length 1 inverts to a segment of length $1/6$. And the segment between $x = 4$ and $x = 5$ inverts to the segment between $x = 1/4$ and $x = 1/5$, which has length $1/20$.

3.7.3.2 Circle inversion properties

The basic properties of inversion in a circle C with center O can be derived easily from the characterization "same ray from the center, product of distances equals square of radius" characterization of I_C, inversion in C. Think of the first three of these as preliminary exercises. The others we'll describe in more detail in Appendix B.11.1.

1. Every point on the circle C is left fixed by I_C.

2. Inversion I_C interchanges the part of the plane outside of C with the part of the plane inside C.

3. If the point A is the inverse of the point B, then B is the inverse of A. That is, I_C is an *involution*: I_C^2 is the identity transformation.

What happens to the center of the circle C? In Appendix B.11.1 we'll see a way, called stereographic projection, to make sense of the answer that intuition gives to this question: the center of the circle inverts to a point infinitely far away.

4. Inversion I_C is a contraction on sets that lie outside of the disk D bounded by C, and an expansion on sets that lie inside D.

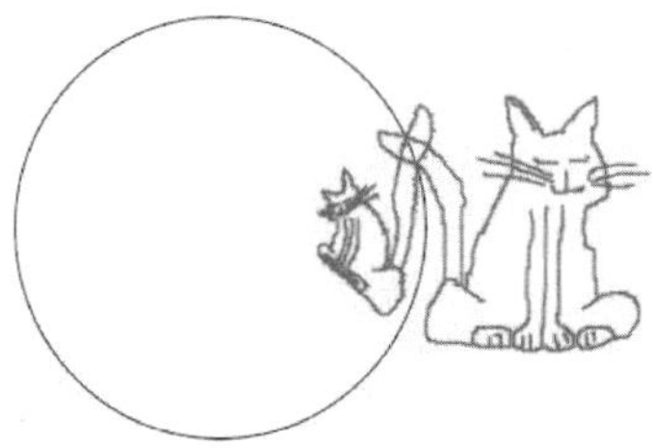

Instead of a calculation, we'll illustrate this point by inverting a sketch of a cat. The portion of the cat outside the circle—all of the cat except the tip of its tail—inverts inside the circle. The tip of the tail inside the circle inverts outside the circle, and the points where the tail crosses the circle do not move. As a preview of some other properties, note that the cat's front legs, approximately straight lines, invert to curves inside the circle.

5. Inversion I_C takes to itself every circle that intersects C orthogonally.

6. Inversion I_C takes every circle that does not pass through O to a circle that does not pass through O.

7. Inversion I_C takes every circle passing through O to a straight line.

These last three properties can be derived analytically from the formula for inverting a circle in another circle. We'll derive the formula in Appendix B.11.1.

3.7.3.3 Limit sets

Suppose X is a set of points in the plane. A point q is a *limit point* of X if for every distance $\delta > 0$, there is a point $w \in X$ with $0 < \text{dist}(q, w) < \delta$. Here $\text{dist}(q, w)$ denotes the Euclidean distance between $q = (q_1, q_2)$ and $w = (w_1, w_2)$, that is, $\text{dist}(q, w) = \sqrt{(q_1 - w_1)^2 + (q_2 - w_2)^2}$. Note $0 < \text{dist}(q, w)$ means that $q \neq w$.

The *limit set* of X, denoted $\Lambda(X)$, or just Λ if X is clear, is the set of all limit points of X.

Recall we have encountered limit sets in Labs 1.2 and 1.6. There we commented that if we start the random IFS algorithm with a point of the fractal, the random algorithm will produce a clean fractal image; while if we start with a point outside the fractal, the random algorithm will give a scattering of points that converge to the fractal. To reconcile this apparent difference, we point out that the fractal generated by the random IFS algorithm is the limit set of the iterates of the starting point, not the iterates themselves.

To help build intuition for limit sets in Appendix B.11.2 we'll give details of a few examples.

3.7.3.4 Circle inversion limit sets

Limit sets of circle inversions are generated by an adaptation of the random IFS algorithm.

For the moment consider circles $C_1, \dots, C_N$ bounding discs $D_1, \dots, D_N$ having disjoint interiors. Start with a point z_0 lying outside all the D_i. Pick a circle C_{i_1} randomly and let z_1 be the inverse of z_0 in C_{i_1}. Now pick another circle C_{i_2} randomly, with the restriction that $C_{i_2} \neq C_{i_1}$, because by property (3) of inversion inverting twice successively in C_{i_1} would produce the sequence $z_0 \to z_1 \to z_2 = z_0$. With this restriction, z_2 is the inverse of z_1 in C_{i_2}. Continuing in this way generates a sequence of points $z_0, z_1, z_2, \dots$. In a moment we'll see that the limit set Λ of this sequence does not depend on the choice of starting point z_0 or on the sequence in which the inverting circles are chosen. For this reason, Λ is called the limit set of inversion in the circles $C_1, \dots, C_N$.

In Appendices B.1.3 and B.2.2 to show the convergence of the random IFS algorithm we first show the deterministic IFS algorithm converges (this is a consequence of contractivity, viewed appropriately), and then show the random and deterministic IFS algorithms have the same attractor (this uses addresses). Here we take a similar approach. First we'll investigate the equivalent of the deterministic IFS algorithm for circle inversions.

To illustrate this, we consider an example with four circles, C_1, C_2, C_3, C_4, all with radius 0.9 and having centers $(1, 1)$, $(-1, 1)$, $(-1, -1)$, and $(1, -1)$. The limit set Λ of inversion in these four circles lies in the discs D_i bounded by these circles. Recall the limit set is invariant (as a set, not point by point) under inversion in all four circles. We can be much more precise about where in these discs Λ lies.

For instance, the portion of Λ lying in D_1 must lie in the inverses in C_1 of D_2, D_3, and D_4. These and the corresponding inverses in C_2, C_3, and C_4 are shown in Fig. 3.41 (a).

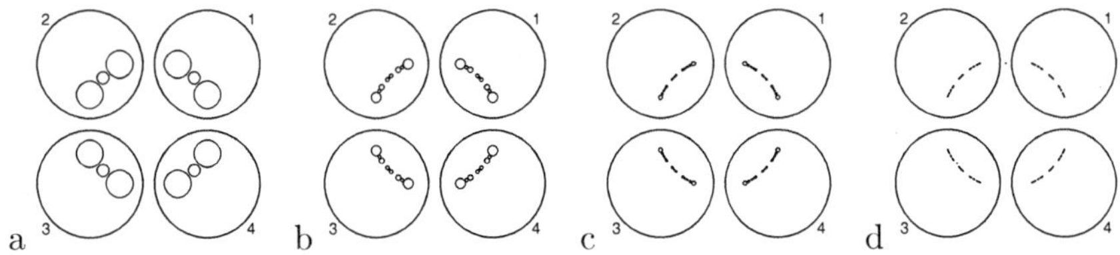

Figure 3.41: An illustration of convergence to a circle inversion limit set.

Call C_{12}, C_{13}, C_{14} the inverses in C_1 of C_2, C_3, and C_4, and so on. Similar reasoning shows that the part of Λ in C_1 lies not just in C_{12}, C_{13}, C_{14}, but in the inverses in C_1 of C_{21}, C_{23}, C_{24}, C_{31}, C_{32}, C_{34}, C_{41}, C_{42}, and C_{43}. That is, in C_{121}, C_{123}, C_{124}, C_{131}, C_{132}, C_{134}, C_{141}, C_{142}, and C_{143}. These and the corresponding inversions in C_2, C_3, and C_4 are pictured in Fig. 3.41 (b). Continuing, Fig. 3.41 (c) shows the 27 length-4 address circles lying in C_1, in each of the other C_i. Because inverting from outside to inside is a contraction, iterating this process gives a sequence of circles converging to the limit set, shown in Fig. 3.41 (d).

The subscripts of these collections of inverted circles provide addresses for the discs they bound. To complete the proof of convergence of the random circle inversion algorthm, note that the randomness of selecting the sequence of inverting circles guarantees every finite address is visited by the iterates of any starting point z_0.

This shows the independence of Λ from the choice of z_0 and from the particular random sequence in which the inverting circles are selected.

Another characterization of the limit set is the intersection of all closed sets left invariant by inversion in all the inverting circles.

3.7.3.5 Some circle inversion limit set examples

Take C_1, C_2, C_3, and C_4 to have radius 1 and centers $(1, 1)$, $(-1, 1)$, $(-1, -1)$, and $(1, -1)$, C_5 the circle with radius 1 and center $(1+\sqrt{3}, 0)$, and C_6 the circle with radius $\sqrt{2} - 1$ and center $(0, 0)$. The radii of C_5 and C_6 guarantee that all intersections occur at points of tangency.

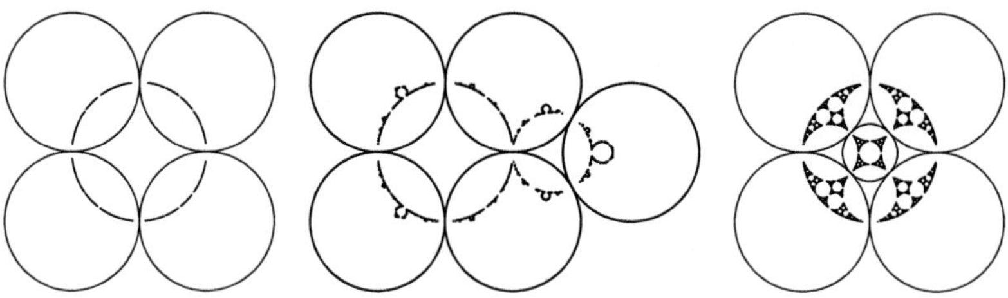

Figure 3.42: Circle inversion limit sets.

In Fig. 3.42 we see limit sets—well, approximations of limit sets—for inversions in circles C_1, C_2, C_3, and C_4 (left); in circles C_1, C_2, C_3, C_4, and C_5 (middle); and in circles C_1, C_2, C_3, C_4, and C_6 (right).

Why did we say "approximations of"? Let's start with the left image. The circle S with center $(0, 0)$ and radius 1 is orthogonal to each of C_1, C_2, C_3, and C_4, so S is invariant under inversion in each of these four circles. With a bit more work, we can see that no proper subset of S is invariant under inversion in all four circles, so S is the limit set. The random circle inversion algorithm doesn't fill in S completely because for points near the inverting circle, inversion moves the points only slightly from one side to the other. Even for a relatively large number of points, 20,000 in this image, the random algorithm leaves gaps near the points of tangency. These gaps invert to other gaps in the fill of the portion of S in each C_i. A consequence of the nonlinearity of inversion is that regions of the attractor fill in at different rates, some very slowly.

The limit set for left image of Fig. 3.42 is S: for the middle image the limit set includes S and the circle with center $(1 + \sqrt{3}/3, 0)$ and radius $\sqrt{3}/3$, and for every little bit that looks like points on a circle, that whole circle is part of the limit set.

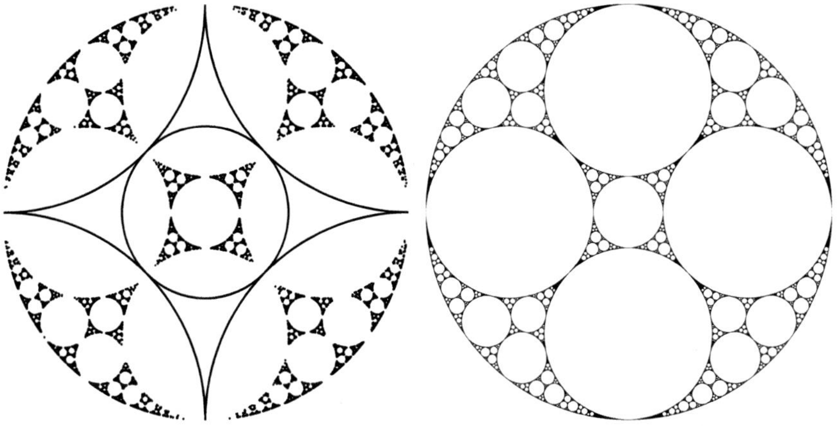

Figure 3.43: Left: a magnification of the right image of Fig. 3.42. Right: the full limit set.

The left image of Fig. 3.43 is a magnification of the right image of Fig. 3.42. The complexity of the figure is apparent, as are the gaps that are the result of slow convergence near the inverting circles. The right image is a much netter approximation of the full limit set. Do you see why the portion of the limit set in each of C_1, C_2, C_3, and C_4 could be called a “nonlinear gasket”?

How did we generate the right image of Fig. 3.43? Did we run the random circle inversion program for billions of iterations? No, of course not. In 1983 Benoit published a different way to generate the limit sets of some collections of circle inversions [102]. In broadest terms, rather than build up an image by generating points in the limit set, this approach removes areas not in the limit set. For this example, it's fairly easy to show that the limit set must be contained in the disk D bounded by the circle S. The empty circles are the inverses, in each of the five circles, of the region outside D. And the inverses of these inverses, and so on. Fairly soon the inverted circles become smaller than a pixel and so disappear on the computer screen. A bit more information can be found in Sect. 2.7 and Appendix A.9 of [17]; a lot more information, and many more pictures, are in [103].

3.7.4 Procedure

With the circle inversion software in 3.7 of the Mathematica or Python codes, investigate how the choice of inverting circle center and radius influences the limit set.

3.7.5 Sample A

By always inverting from outside a circle to inside the circle, we guarantee that every application of circle inversion is a contraction. We need this condition to be sure that the points generated converge to the limit set. So long as the open disks bounded by, but not containing, the inverting circles are disjoint, this is easy. Start with a point outside one of the circles, invert in that circle, and note which circle was used. To find the next point, pick a circle randomly, but forbid the circle just used. This guarantees that every inversion acts as a contraction, and that the same inversion is not applied twice in succession, which would have the effect of undoing both inversions.

What happens if we relax this condition that the circles do not overlap? If C_i and C_j overlap, inversion in C_i can produce a point inside both C_i and C_j. We can't immediately invert again in C_i, but we can invert in C_j and that inversion is an expansion. Instead of a clean fractal attractor, it could give points that execute a complicated messy dance with no appearance of convergence to an attractor.

On the left we see (an approximation of) the attractor of inversion in the circles C_1, C_2, C_3, and C_4 of Sect. 3.7.3.5, and the circle with center $(2,0)$ and radius 1. This last circle overlaps with C_1 and C_4, yet the limit set appears to be well-behaved: the circle S with smaller circles attached, and still smaller circles attached to those, and so on.

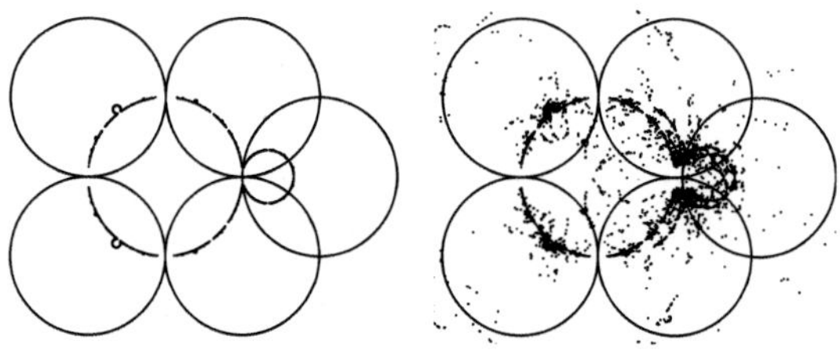

Figure 3.44: Approximations of the attractors of Sample 3.7.5.

For the right image the center of the fifth circle has been moved from $(2,0)$ to $(2.1,0)$. This image is much less clear. Although most points concentrate near the attractor of the left image, other points scatter around the circles. One goal of the exercises is to discover conditions on circle overlap that guarantees the attractor is a clean fractal.

3.7.6 Conclusion

Circle inversions can generate new types of fractals. The nonlinearity of inversion makes these shapes visually interesting. That inversion is a contraction only on the outside of the inverting circle can lead to some small changes in the circles yielding immense changes in the limit set.

3.7.7 Exercises

In Exercises 3.7.1–3.7.8, C_1, C_2, C_3, and C_4 are the circles with radius 1 and centers $(1,1)$, $(-1,1)$, $(-1,-1)$, and $(1,-1)$.

Prob 3.7.1 To the circles C_1, C_2, C_3, and C_4 add circles with centers $(\pm 2, 0)$ and radius 1. Is the limit set a clean fractal?

Prob 3.7.2 To the circles C_1, C_2, C_3, and C_4 add circles with centers $(\pm 2.1, 0)$ and radius 1. Is the limit set a clean fractal?

Prob 3.7.3 (a) To the circles C_1, C_2, C_3, and C_4 add a circle C_5 with center $(2, 1)$ and radius 1. Is the limit set a clean fractal?
(b) Find the smaller angle between the tangents of C_1 and C_5 at their points of intersection.

Prob 3.7.4 (a) Find the center $(a, 1), a > 1$ of the circle C_5 of radius 1 so the smaller angle between the tangents of C_1 and C_5 at their points of intersection is $\pi/4$.
(b) Is the limit set of inversion in $C_1, \dots, C_5$ a clean fractal?

Prob 3.7.5 Repeat Exercise 3.7.4 for the smaller angle between the tangents equal to $\pi/6$.

Prob 3.7.6 From the results of Exercises 3.7.1–3.7.5 suggest a condition about overlapping inverting circles that guarantees a clean fractal limit set. State a conjecture about this angle and design an experiment to test your conjecture.

Prob 3.7.7 (a) To the circles C_1, C_2, C_3, and C_4 add a circle C_5 with center $(1.5, 1.0)$ and radius $1/2$. Plot 50,000 points in the window $-2.1 \leq x \leq 2.1$, $-2.1 \leq y \leq 2.1$.
(b) Expand the plot window to $-2.1 \leq x \leq 4.1$. Explain why the attractor extends beyond the region bounded by these circles.

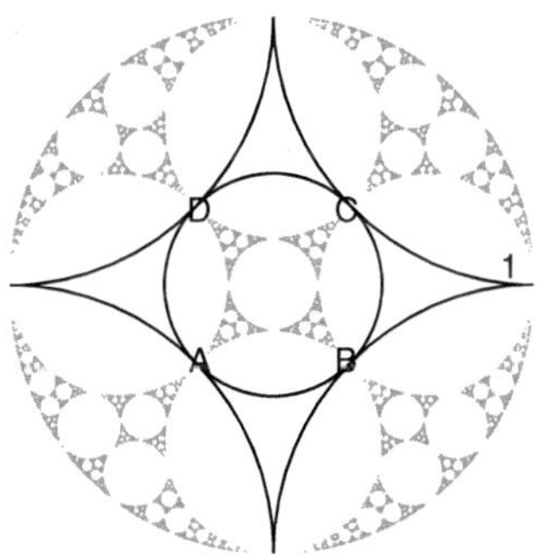

Prob 3.7.8 In this limit set, indicate the images of points A, B, C, and D in circle C_1, one quarter of which is shown in the upper right of the figure. Explain your reasoning.

These last two are just to show you some pretty fractals that are different from others we have generated. Perhaps these will suggest some aesthetic experiments.

Prob 3.7.9 Draw the limit set for inversion in the circles of radius 1 and with centers $(-2, 0)$, $(0, 0)$, $(2, 0)$, $(-2, 2)$, $(0, 2)$, $(2, 2)$, $(-2, -2)$, $(0, -2)$, and $(2, -2)$.

Prob 3.7.10 Draw the limit set for inversion in the circles of radius 1 and with centers $(-2, 0)$, $(2, 0)$, $(-2, 2)$, $(0, 2)$, $(2, 2)$, $(1, -1)$, $(-1, -1)$, $(1, 1)$, $(-1, 1)$, $(-2, -2)$, $(0, -2)$, and $(2, -2)$.

3.8 Fractal tiles

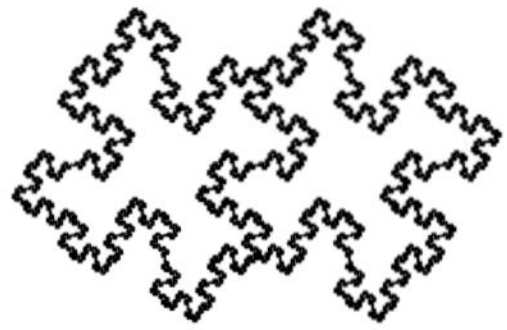

By a *tiling* of the plane we mean a way to cover the entire plane with exact copies of a shape (or perhaps several shapes) so the shapes touch only along edges. The squares of a chessboard could continue on forever. So could the hexagons of a honeycomb. Among the regular polygons, only the equilateral triangle, square, and regular hexagon tile the plane. This is easy to see: at each vertex of the tiling some number $n \geq 3$ of tiles must meet. In order for the tiles to not overlap or leave a gap at that vertex, $n\theta = 360°$, where θ is the vertex angle of the tile. The only vertex angles that evenly divide 360° are 60° (triangle), 90° (square), and 120° (hexagon).

As a topic of study, tilings is immense. Extensions to three dimensions, relations to crystallography, applications in art and design, lots of applications in pure math, Roger Penrose's aperiodic tilings, to the discovery by Dan Shechtman that these aperiodic tilings occur in nature and now are called quasicrystals. Entire books could be written about tilings. In fact, entire books have been written about tilings. Our goal is much more limited. We'll learn ways to construct tilings where the tiles have fractal perimeters.

3.8.1 Purpose

We'll develop two methods to produce tiles with fractal boundaries. The first is a simple geometric processes, the second requires some work with matrices. Specifically, we'll use matrix multiplication, the determinant and trace of a matrix, and the inverse of a matrix.

3.8.2 Materials

We'll use paper, pencil, ruler, protractor, and lots of erasers; for the second method use the IFS software (1.1 of the Mathematica or Python codes) and matrix commands (3.8 of the codes, or work them out by hand).

3.8.3 Background

Here we'll describe two methods to convert a tiling of the plane by squares into a tiling by other shapes, including tiles with fractal perimeters.

3.8.3.1 Opposite side modification

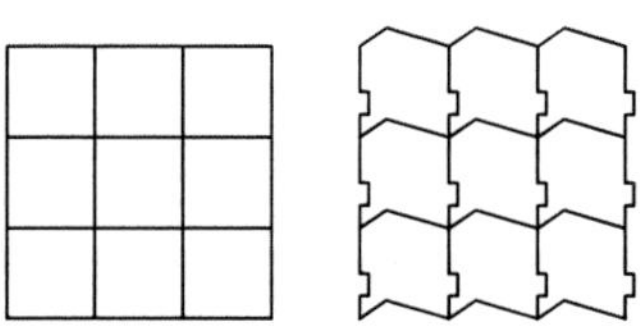

A good reference on how to construct tilings is Doris Schattschneider's exposition [104]. Here we'll begin with a tiling of the plane by squares. Opposite side modification consists of making changes to one side of the square and making the complementary changes to the opposite side of the square. That is, modify one side of the square and translate the modified side to the opposite. This will produce a new shape that still tiles the plane, so long as we're careful that the modifications do not involve changes so large that they cause the new sides to intersect each other.

Of course we can replace each side by something much more complicated, a fractal curve, for example. On the left we see the first iteration (thin lines) of an IFS applied to a horizontal line segment (thick gray line). On the right we see a later iterate of this IFS. Certainly, the limiting shape is a fractal curve. Do you see why the similarity dimension of this curve is $d = 3/2$?

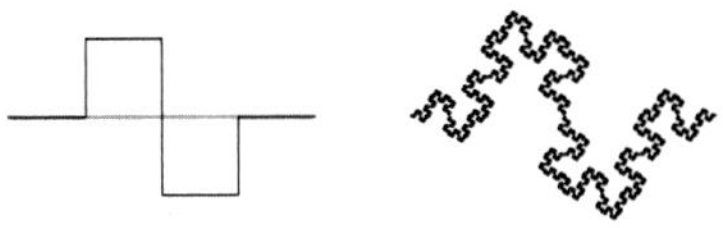

Here is the IFS to generate this curve.

r	s	θ	φ	e	f
0.25	0.25	0	0	0	0
0.25	0.25	90	90	0.25	0
0.25	0.25	0	0	0.25	0.25
0.25	0.25	−90	−90	0.5	0.25
0.25	0.25	−90	−90	0.5	0
0.25	0.25	0	0	0.5	−0.25
0.25	0.25	90	90	0.75	−0.25
0.25	0.25	0	0	0.75	0

Make four copies. Leave one in its orientation, rotate one by 90°, one by 180°, and one by 270°. Then translate them so the appropriate copy goes along each edge of the unit square. This is how we generate the images of Fig. 3.45.

Figure 3.45: Left to right: initial square and the first iterate, first and second iterates, initial square and the fourth iterate.

As complicated as this shape appears, it is generated by opposite side modification and so we know that this shape tiles the whole plane.

3.8.3.2 Some matrix arithmetic

For the applications to fractal tiles we'll use 2 × 2 matrices, but all the concepts we'll review here have versions for matrices of other sizes. We'll use matrix multiplication, the inverse of a matrix, and the eigenvalues, trace, and determinant of the matrix. Background on all these topics, except the trace, is given in Appendix B.8.

There we discuss how to compute eigenvalues of 4×4 matrices and derive the characteristic equation $\det(M - \lambda I) = 0$, for which the eigenvalues λ of a matrix M are the solutions. For 2×2 matrices the characteristic equation takes on a simple form. For this we need one more term. The *trace* $\text{tr}(M)$ of a matrix M is the sum of the elements that lie on the diagonal of M. We can write the characteristic equation of a 2×2 matrix M as

$$\begin{aligned} 0 &= \det \begin{bmatrix} a-\lambda & b \\ c & d-\lambda \end{bmatrix} = (a-\lambda)(d-\lambda) - bc = \lambda^2 - (a+d)\lambda + ad - bc \\ &= \lambda^2 - \text{tr}(M)\lambda + \det(M) \end{aligned}$$

Then the eigenvalues of M are the solutions of this quadratic equation,

$$\lambda_{\pm} = \frac{\text{tr}(M) \pm \sqrt{(\text{tr}(M))^2 - 4\det(M)}}{2} \tag{3.8}$$

3.8.3.3 Matrices and fractal tiles

Another method to construct fractal tiles is presented by Christoph Bandt in [105]. A good explanation and excellent examples are given in [106]. To describe Bandt's approach, in Sample 3.8.5 we'll see that a fractal tile constructed from the example of the last section can be decomposed into smaller copies of itself. (Really, can this be a surprise for a fractal tile?) Bandt's method uses a matrix $M = \begin{bmatrix} a & b \\ c & d \end{bmatrix}$ with integer entries to expand one of these component tiles to the size of the original tile. Then the inverse matrix M^{-1} is a contraction that takes the larger tile to one of these pieces. Consequently, M^{-1} determines the IFS parameters r, s, θ, and φ. Because

$$M^{-1} = \frac{1}{\det(M)} \begin{bmatrix} d & -b \\ -c & a \end{bmatrix}$$

we see from Eq. (1.1) of Lab 1.1 that

$$\begin{aligned} r \cdot \cos(\theta) &= \frac{d}{\det(M)}, & s \cdot \sin(\varphi) &= \frac{b}{\det(M)}, \\ r \cdot \sin(\theta) &= \frac{-c}{\det(M)}, & s \cdot \cos(\varphi) &= \frac{a}{\det(M)} \end{aligned}$$

In Sect. 1.3.3.5 of Lab 1.3 we work through examples of how to convert matrix entries to r, s, θ, and φ.

All that remains is to find the translation parameters, the e and f, that give each subtile.

To transform a subtile into the whole tile, the matrix M must be an expansion, that is, $|\lambda_{\pm}| > 1$. (If the eigenvalues are complex, recall $|x + iy| = \sqrt{x^2 + y^2}$.) Suppose S is the unit square with vertices $(0, 0)$, $(1, 0)$, $(0, 1)$, and $(1, 1)$. Take $\vec{e}_1$ to be the *edge vector* from $(0, 0)$ to $(1, 0)$, and $\vec{e}_2$ from $(0, 0)$

to $(0, 1)$. Then M transforms S into the parallelogram $M(S)$ with two edges given by the vectors $M\vec{e}_1$ and $M\vec{e}_2$,

$$M\vec{e}_1 = \begin{bmatrix} a & b \\ c & d \end{bmatrix} \begin{bmatrix} 1 \\ 0 \end{bmatrix} = \begin{bmatrix} a \\ c \end{bmatrix} \qquad M\vec{e}_2 = \begin{bmatrix} a & b \\ c & d \end{bmatrix} \begin{bmatrix} 0 \\ 1 \end{bmatrix} = \begin{bmatrix} b \\ d \end{bmatrix}$$

The area of this parallelogram is $|\det(M)|$. There's a bit of algebra between this and the familiar base $\times$ altitude formula for the area of the parallelogram. Here's an intermediate step: the base of the parallelogram has length $\sqrt{a^2 + c^2}$ and altitude $\sqrt{(bc - ad)^2/(a^2 + c^2)}$. Because $\det(M) = \lambda_+ \cdot \lambda_-$ (Multiply the expressions of Eq. (3.8)), and because $|\lambda_\pm| > 1$, we see $|\det(M)| > 1$. Consequently, M^{-1} contracts areas by a factor of $1/\det(M)$.

If a fractal tile is composed of N smaller copies of itself, each copy has area $1/N$ times the area of the original tile. Consequently, $\det(M)$ is the number of copies into which the tile is decomposed.

To find the translations, observe that the parallelogram $M(S)$ has vertices $(0, 0)$, (a, c), (b, d), and $(a + b, c + d)$. These are *lattice points*, points whose coordinates are integers. Pick's theorem (Google is your friend) states that

$$\text{Area}(M(S)) = I(M(S)) + B(M(S))/2 - 1$$

where $I(M(S))$ is the number of lattice points in the interior of $M(S)$ and $B(M(S))$is the number of lattice points that lie on the boundary of $M(S)$.

How many lattice points can lie on the boundary of $M(S)$? Certainly, the four vertices of $M(S)$ are lattice points. Also, if an edge of $M(S)$ contains a lattice point other than its endpoints, so will the parallel edge. Call these points *interior edge points*, by which we mean lattice points that lie on an edge, but not at the endpoints of an edge. Then we see that

$$B(M(S)) = 4 + 2 \cdot (\text{number of interior edge points on } M\vec{e}_1 \text{ and } M\vec{e}_2)$$

and consequently

$$\text{Area}(M(S)) = I(M(S)) + \text{number of interior edge points on } M\vec{e}_1 \text{ and } M\vec{e}_2 + 1$$

Recall that the number of IFS transformations is $|\det(M)| = \text{Area}(M(S))$, and there is a 1-to-1 correspondence between the IFS translations and the lattice points of $M(S)$. One set of translations are those that take the origin to

- the lattice points in the interior of $M(S)$,
- the interior edge points on $M\vec{e}_1$ and $M\vec{e}_2$, and
- the origin.

Other choices of translations are possible. Recall that translates of $M(S)$ tile the plane. Bandt shows that the IFS translations can be taken to be the coordinates of any collection of $|\det(M)|$ lattice points in translates of $M(S)$, provided that for each of the lattice points listed above, one of the new points occupies the same relative position in its translate of $M(S)$. These collections of points are called a *residue systems* for M. Residue classes may be easier to understand by the examples of 3.8.6 and 3.8.7 than from the general argument.

3.8.4 Procedure

With opposite side modification (Sect. 3.8.3.1) we can change a square into a fractal tile. The only remaining question is how to assemble these tiles to cover the plane. We'll use an early stage of the tile construction of Sect. 3.8.3.1 to show the way.

On the left we see the unit square and the first iteration of this fractal tile construction, subdivided into squares. For reference, one of these squares is darkened. On the right we show the next iterate, along with the unit square to locate the pieces. Each square of the left image is replaced by a scaled copy of the whole tile. This suggests that the sizes and placements of the pieces determine an IFS with attractor the fractal tile. In Sample 3.8.5 we'll see how to do this, though it is not so difficult to figure out.

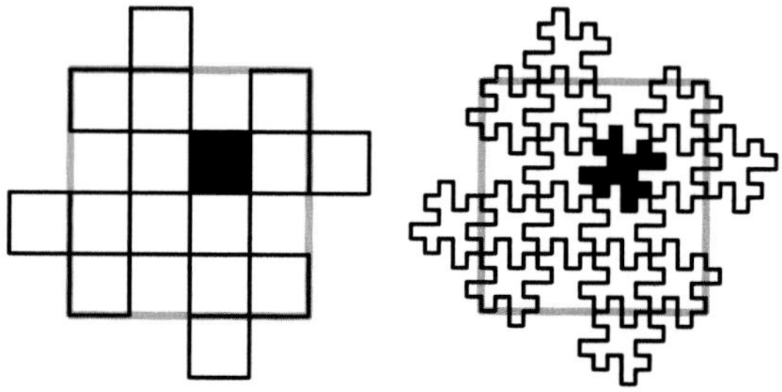

For Bandt's method we begin with an integer 2×2 matrix M with eigenvalues $\lambda_{\pm}$ that satisfy $|\lambda_{\pm}| > 1$. Use the matrix entries to find the IFS parameters r, s, θ, and φ. The number of IFS transformations is $|\det(M)|$. To find the translation parameters e and f, find the parallelogram $M(S)$ where S is the unit square. Within this parallelogram and its translates identify a residue system of lattice points. These are the e and f values for the IFS. The IFS software from 1.1 of the Mathematica or Python codes can generate a picture of the tile. Different choices of residue system produce different tiles. We'll work through examples in Samples 3.8.6 and 3.8.7.

3.8.5 Sample A

Here we see the unit square (outlined in gray) is subdivided into 16 $1/4 \times 1/4$ subsquares. Consequently, the IFS to generate the tile has 16 transformations. Each piece can be taken to be in the same orientation as the whole, so $r = s = 1/4$ and $\theta = \varphi = 0$ for each transformation. To find the translations, place the origin at the lower left corner of the gray square. Then the coordinates of the lower left corner of each of the 16 subsquares are the corresponding e and f values. For example, $(e_1, f_1) = (1/2, -1/4)$, $(e_2, f_2) = (0, 0)$, $(e_3, f_3) = (1/2, 0)$, ..., and $(e_{16}, f_{16}) = (1/4, 1)$.

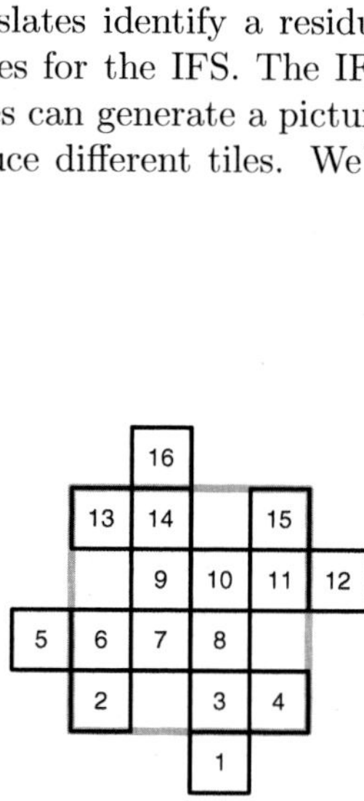

In Fig. 3.46 we start with the unit square and apply the IFS four times. The right image consists of 65,536 little squares, certainly enough to get a good idea of the appearance of the tile. Note that the shape generated by each iterate can be obtained from the unit square by opposite side modification, so each iterate tiles the plane. The IFS attractor also is produced from the unit

Figure 3.46: The first four iterates of the IFS to generate the fractal tile.

square by opposite side modification, but now the modifications are fractal curves, not polygonal curves.

3.8.6 Sample B

Take the matrix $M = \begin{bmatrix} -1 & -1 \\ 1 & -1 \end{bmatrix}$. The eigenvalues are $\lambda_{\pm} = -1 \pm i$ so the matrix is an expansion. Because $\det(M) = 2$, the tile is generated by an IFS with 2 transformations. To find the r, s, θ, and φ values, note

$$M^{-1} = \begin{bmatrix} -1/2 & 1/2 \\ -1/2 & -1/2 \end{bmatrix} = \begin{bmatrix} r\cdot\cos(\theta) & -s\cdot\sin(\varphi) \\ r\cdot\sin(\theta) & s\cdot\cos(\varphi) \end{bmatrix}$$

From this we find

$$r^2 = r^2\cos^2(\theta) + r^2\sin^2(\theta) = (-1/2)^2 + (-1/2)^2 = 1/2$$
$$s^2 = s^2\sin^2(\varphi) + s^2\cos^2(\varphi) = (1/2)^2 + (-1/2)^2 = 1/2$$

so $r = \pm 1/\sqrt{2}$ and $s = \pm 1/\sqrt{2}$. We have some choices here. Let's take $r = s = 1/\sqrt{2}$. Then to match $r\sin(\theta) = -1/2$ we have $\theta = -45°$, and to match $-s\sin(\varphi) = 1/2$ we have $\varphi = -45°$.

To find the translations, first find the parallelogram $M(S)$ where S is the unit square. Check the vertices,

$$M\begin{bmatrix}0\\0\end{bmatrix} = \begin{bmatrix}0\\0\end{bmatrix}, \quad M\begin{bmatrix}1\\0\end{bmatrix} = \begin{bmatrix}-1\\1\end{bmatrix}, \quad M\begin{bmatrix}0\\1\end{bmatrix} = \begin{bmatrix}-1\\-1\end{bmatrix}, \quad M\begin{bmatrix}1\\1\end{bmatrix} = \begin{bmatrix}-2\\0\end{bmatrix}$$

This shows that $M(S)$ is the left plot of Fig. 3.47. With this we'll find two residue systems. Recall that these are the lattice points in the interior of $M(S)$, the lattice points on $M\vec{e}_1$ and on $M\vec{e}_2$ that are not endpoints of these edges (the interior edge points), and the origin.

The only interior lattice point in $M(S)$ is $(-1, 0)$, marked with a small circle in the left figure. There are no interior edge points. The origin, also marked with a small circle, is the other point of this residue system, which we'll call residue system 1.

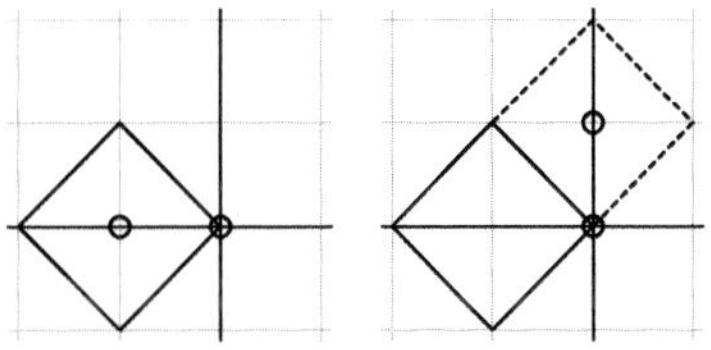

Figure 3.47: The parallelogram $M(S)$ and a translate.

In the right figure we plot $M(S)$ and its translate $M(S) + (1, 1)$. Residue system 2 consists of the origin, along with the point $(0,1)$, the translate of the point $(-1,0)$. With these residue systems we have the translations to complete two IFS to generate fractal tiles based on this matrix.

Here is the IFS for residue system 1.

r	s	θ	φ	e	f
0.707	0.707	-45	-45	0	0
0.707	0.707	-45	-45	-1	0

Here is the IFS for residue system 2.

r	s	θ	φ	e	f
0.707	0.707	-45	-45	0	0
0.707	0.707	-45	-45	0	1

Here we see the tiles generated by these IFS. To indicate the relative positions of these tiles, in both images the coordinate axes have been included. The right tile appears to be the left tile rotated 90°, yet the IFS tables differ only in their e_2 and f_2 values from $e_2 = -1, f_2 = 0$ to $e_2 = 0, f_2 = 1$. What would happen if we set $f_2 = 3$? Will the tile disconnect? No, it simply gets larger. What about $e_2 = 2, f_2 = 3$? Larger still. Do you see why? If this isn't clear, in our original gasket IFS change the $e = 0.5$ and $f = 0.5$ to $e = 1$ and $f = 1$.

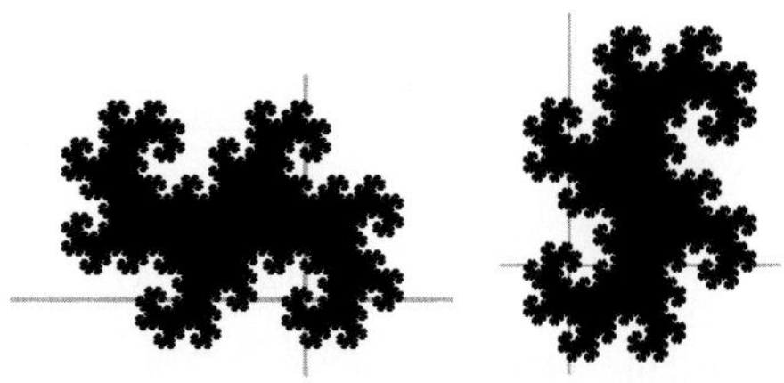

Figure 3.48: Tile for residue system 1 (left) and 2 (right).

3.8.7 Sample C

Take the matrix M to be $M = \begin{bmatrix} 2 & -2 \\ 2 & 0 \end{bmatrix}$. The eigenvalues are $\lambda_{\pm} = 1 \pm \sqrt{3}i$ so the matrix is an expansion. Because $\det(M) = 4$, an IFS to generate a tile for this matrix will have 4 transformations. To find the r, s, θ, and φ parameters, we use

$$M^{-1} = \begin{bmatrix} 0 & 1/2 \\ -1/2 & 1/2 \end{bmatrix} = \begin{bmatrix} r \cdot \cos(\theta) & -s \cdot \sin(\varphi) \\ r \cdot \sin(\theta) & s \cdot \cos(\varphi) \end{bmatrix}$$

First, $r\cos(\theta) = 0$ tells us that $\theta = \pm 90°$. Then for $r\sin(\theta) = -1/2$ we have two choices, $r = -1/2$ and $\theta = 90°$, or $r = 1/2$ and $\theta = -90°$. We'll take the former. Finally, $-s\sin(\varphi) = 1/2$ and $s\cos(\varphi) = 1/2$ are satisfied for $s = 1/\sqrt{2}$ and $\varphi = -45°$.

To find the translations, we use the parallelogram $M(S)$. Its vertices are

$$M\begin{bmatrix} 0 \\ 0 \end{bmatrix} = \begin{bmatrix} 0 \\ 0 \end{bmatrix}, \quad M\begin{bmatrix} 1 \\ 0 \end{bmatrix} = \begin{bmatrix} 2 \\ 2 \end{bmatrix}, \quad M\begin{bmatrix} 0 \\ 1 \end{bmatrix} = \begin{bmatrix} -2 \\ 0 \end{bmatrix}, \quad M\begin{bmatrix} 1 \\ 1 \end{bmatrix} = \begin{bmatrix} 0 \\ 2 \end{bmatrix}$$

Again we'll find two residue systems, one that uses lattice points in $M(S)$, the other uses lattice points of $M(S)$ and a translation of $M(S)$.

In the interior of $M(S)$ we find one lattice point, $(0, 1)$, two interior edge points, $(-1, 0)$ and $(1, 1)$, and the origin $(0, 0)$. These points are residue system 1, indicated on the left. On the right we see residue system 2, in which the interior edge point $(-1, 0)$ is replaced by the corresponding point $(1, 0)$ in the translate $M(S) + (2, 0)$.

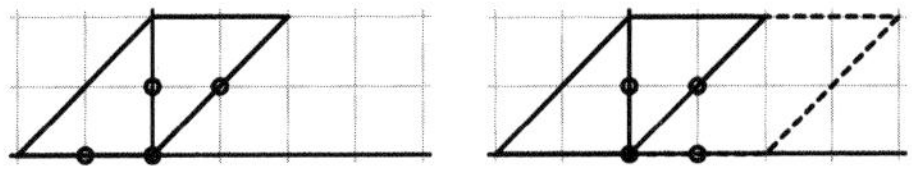

Figure 3.49: The parallelogram $M(S)$ and a translate.

Here is the IFS for residue system 1.

r	s	θ	φ	e	f
0.5	0.707	−90	−45	0	0
0.5	0.707	−90	−45	−1	0
0.5	0.707	−90	−45	0	1
0.5	0.707	−90	−45	1	1

Here is the IFS for residue system 2.

r	s	θ	φ	e	f
0.5	0.707	−90	−45	0	0
0.5	0.707	−90	−45	1	0
0.5	0.707	−90	−45	0	1
0.5	0.707	−90	−45	1	1

Here we see the tiles generated by these IFS. More symmetric choices are constructed in [106]. A relation between this construction and complex base representations is given in [107]. The two tiles presented in Sample 3.8.6 are related by a simple rotation. Clearly this is not the case for the tiles of Sample 3.8.7.

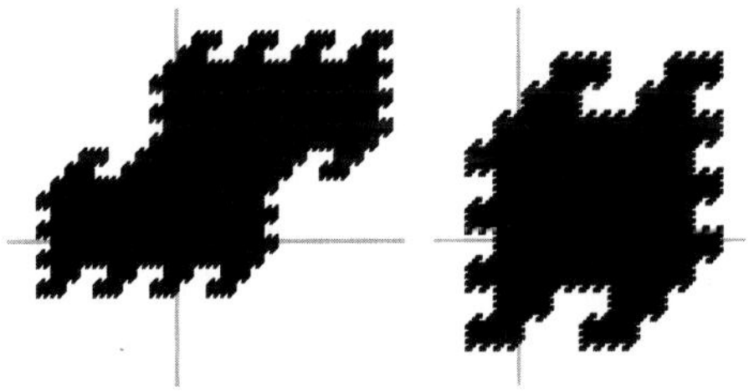

Figure 3.50: Tile for residue system 1 (left) and 2 (right).

In Sample 3.8.6 we saw that some choices of residue systems just produce larger tiles of the same shape as smaller tiles. Could the number of points in a residue system be related to the number of distinct tile shapes? How would you explore this?

3.8.8 Conclusion

Tilings are an immense, and visually appealing, part of mathematics. The weight of the Branko Grünbaum and G. Shephard's *Tilings and Patterns* [108] should convince you of the breadth of this field; for visual appeal of tilings, probably the most familiar examples are the works of M. C. Escher. We wonder what Escher would have done with fractal tilings. Others have thought about this. See [109, 110], for example.

Another direction is to tile the interior of a region with a fractal perimeter. Examples include those of Robert Fathauer [111, 112] as well as work by Peter Raedschelders.

Here we presented two methods to construct a variety of fractal tiles. These methods allow easy modification to fit artistic experimentation. With some patience for iteration, the first method can be implemented with a simple graphics program: scale, copy, translate, and group. Repeat. And these methods give another class of fractals we can generate with IFS.

But really, we hope you'll generate tiles to use as substructure, think about Escher, then exercise your visual and artistic imagination.

3.8.9 Exercises

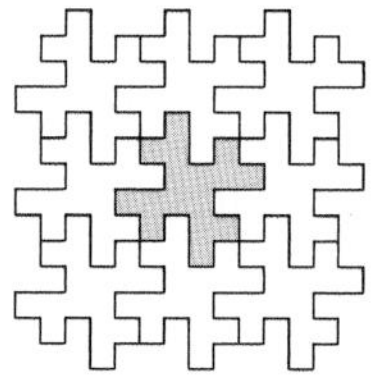

Prob 3.8.1 Here is a variation of the placement of the tiles of Sample 3.8.5. The first tile is made by opposite side modification applied to a square, so we know it tiles the plane. This 3×3 array can be viewed as constructed by opposite side modification applied to a square, so this array also tiles the plane. Now iterate this process: group nine of these, placed in a 3×3 array, then scale them by a factor of $1/3$. Repeat this process. What (familiar) shape is produced in the limit?

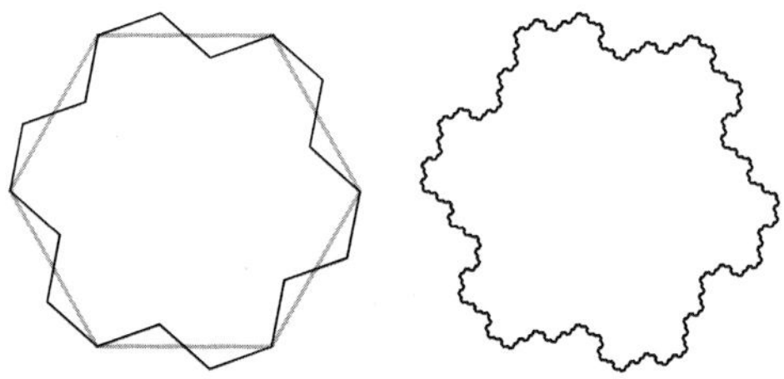

Prob 3.8.2 Opposite side modification can be applied to tilings by hexagons, too. On the left we replace each edge (say of length L) with three segments of length $L/\sqrt{7}$. An approximation of the limiting tile shape, called a *Gosper island*, is shown on the right.
(a) Find the dimension of the boundary curve of the Gosper island.
(b) Find the IFS to generate a segment of the Gosper curve (one piece of the boundary of the Gosper island with endpoints two adjacent endpoints of the original hexagon) from the segment between $(0,0)$ to $(1,0)$.

$$M_1 = \begin{bmatrix} 1 & 2 \\ -1 & 1 \end{bmatrix} \quad M_2 = \begin{bmatrix} 2 & 0 \\ 1 & 2 \end{bmatrix} \quad M_3 = \begin{bmatrix} 2 & -1 \\ 1 & 2 \end{bmatrix} \quad M_4 = \begin{bmatrix} 1 & 3 \\ -1 & 1 \end{bmatrix} \qquad (3.9)$$

In Exercises 3.8.3–3.8.6, for the specified matrix M from the list (3.9),
(a) sketch the parallelogram $M(S)$ and find the residue system that consists of points in $M(S)$.
(b) Find the IFS that generates the corresponding tile.
(c) Sketch the tile with the deterministic IFS algorithm from 1.1 of the Mathematica or Python codes. To get a more filled-in image, take the initial square to have vertices $(\pm 1, \pm 1)$.
(d) Repeat steps (b) and (c) with a different residue system.

Prob 3.8.3 Take $M = M_1$.

Prob 3.8.4 Take $M = M_2$.

Prob 3.8.5 Take $M = M_3$.

Prob 3.8.6 Take $M = M_4$.

Prob 3.8.7 Why can't the matrix $\begin{bmatrix} 1 & 1 \\ 2 & 3 \end{bmatrix}$ generate a tiling of the plane? Give two reasons that involve characteristics of the matrix.

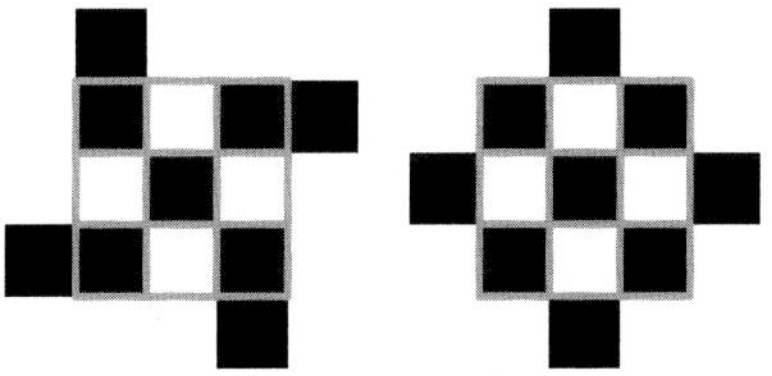

Prob 3.8.8 Divide the unit square into a 3×3 grid.
(a) Use the placement of these nine subsquares in the left image to define an IFS. Is this an opposite side modification? Call the left image the first iterate. Plot the second, third, and fourth iterates. Does the limiting shape tile the plane?
(b) Use the placement of these nine subsquares in the right image to define an IFS. Is this an opposite side modification? Call the right image the first iterate. Plot the second, third, and fourth iterates. Does the limiting shape tile the plane?

Prob 3.8.9 Design your own tiles by opposite side modification. Find the corresponding IFS and generate the second and third iterates of the IFS attractor.

Prob 3.8.10 Find a 2×2 integer matrix M with eigenvalues that satisfy $|\lambda_{\pm}| > 1$. Because $\lambda_+ \cdot \lambda_- = \det(M)$ (This follows from Eq. (3.8)), this matrix is an expansion. Consequently, M^{-1} and a residue system determine an IFS that generates a tile of the plane. Plot the third iterate of the IFS attractor. Experiment a bit. Find something pretty.

Chapter 4

Labs in the Studio and in the Kitchen

Now we'll leave the abstract world we entered through computer screens and keyboards, and focus on experiments in the physical world. To ease the transition between the computer code and mathematical reasoning of Chapters 1, 2, and 3 and the electronics and chemistry experiments of Chapter 5, in this chapter we experiment in the artist's studio and in the kitchen.

In Labs 4.1 and 4.2 we show how paint and two pieces of paper or two pieces of plexiglass can produce very intricate patterns, ramified branching reminiscent of trees and ferns. Called decalcomania, this technique has been known to artists for decades, and it remains a rich ground for fractal experimentation. We also experiment with patterns made by the flow of paint in Lab 4.3, patterns made by mixing paints in Lab 4.4, and, with every possible apology to Jackson Pollack, patterns made by dripping paint in Lab 4.5.

In Lab 4.6 we investigate variations on origami that give another avenue to build physical intuition about scaling. Alone among our labs, these experiments had an impact on holiday cards. Students modified the cut and fold models to produce (mostly) Valentine's day cards. Well, quite a few people pointed out the xkcd Sierpinski Valentine https://xkcd.com/543/ which we happily admit is more elegant than our origami constructions. Still, this lab allowed considerable room for creative exploration.

In Labs 4.8 and 4.9 we move into the real kitchen. We'll discover fractals in many foods, cauliflower and broccoli are obvious choices, but we'll find others. Finally, we'll learn to cook fractal crepes. Many other fractals can be produced in the kitchen. Maybe you'll find your own recipes.

A word about materials: before you rush out and acquire all the materials listed, you should read through the entire lab to see which materials you will want to use. You may not need to assemble the whole list if you plan to do only some of the experiments.

4.1 Fractal painting: decalcomania 1

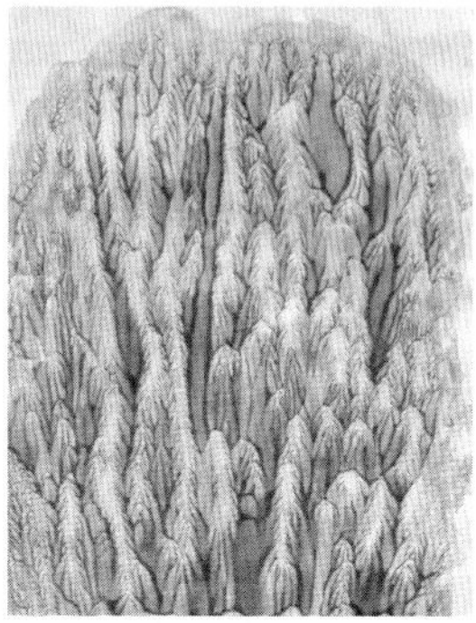

Most of our experiences with paints begin with finger paints in kindergarten or in elementary school. Drop some paint, maybe several colors of paint, on a piece of paper, then push your fingers around in the paint, smearing it over the paper. Most often this just produces blobs, sometimes with swirls of colors, sometimes a more-or-less uniform mud. Only rarely do kids get images like the one on the left. Subtle, wispy branches off of other branches, the illusion of peering into a misty forest. But these images are not so difficult to make, and offer many opportunities for experiments with fractal patterns.

4.1.1 Purpose

A simple modification of childhood finger paint explorations can produce intricate fractal branching patterns. The main idea is this: flatten the paint between two pieces of paper, then pull the papers apart. In this lab we'll investigate how variations on this process, called *decalcomania*, can alter the fractals that result.

4.1.2 Materials

We'll use acrylic paints (finger paints and watercolors can work well, too), stiff paper cut into 3 inch squares, dividers (a compass is a workable substitute), a mm scale ruler, a fine-point marking pen, a magnifying glass, and a good light source.

4.1.3 Background

Here we'll present a tiny bit of the background of decalcomania and sketch an idea of its mechanics. Then we'll describe four ways to quantify a branching pattern: the box-counting dimension, scaling of the number of branches with the branch level, the number of branch levels, and the ratio of the lengths of successive branches.

4.1.3.1 Decalcomania

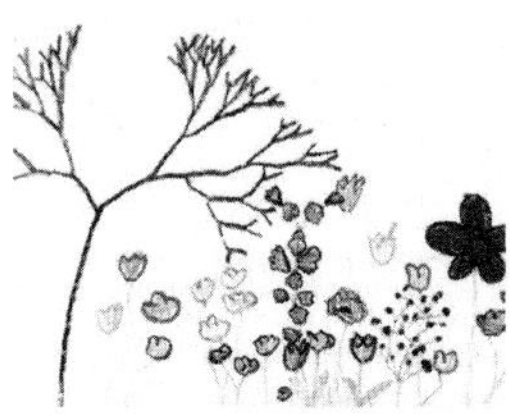

Most views of the natural world are visually complex, though elementary school art classes do not always embrace that complexity. In first grade art classes students are instructed to draw trees as green lollipops with brown trunks. A glance out the window in late March does not show trees at all like that. One of us (MF) produced this image and was criticized by his teacher for not following directions. (Did

I save that picture all these years? No, I didn't. When my mother died, my sister found it among the school papers our mother had saved.)

How can artists reproduce, or at least hint at, the complexity of nature? Impressionists squint, photorealists spend months on a single image, folk artists suggest and call on viewers' memories. Decalcomania is another approach: representing the complexity of one part of nature with the complexity of another.

The his masterpiece *History of Art* [113], art historian H. W. Janson points out that Alexander Cozens felt that landscape paintings should not copy particular landscapes, but rather should reflect their poetic and imaginative spirit. He advocated ink blotted on crumpled and flattened paper as landscape models. Also, Janson reports that Leonardo da Vinci advised artists to use stains on old walls as models for painting mountain ranges.

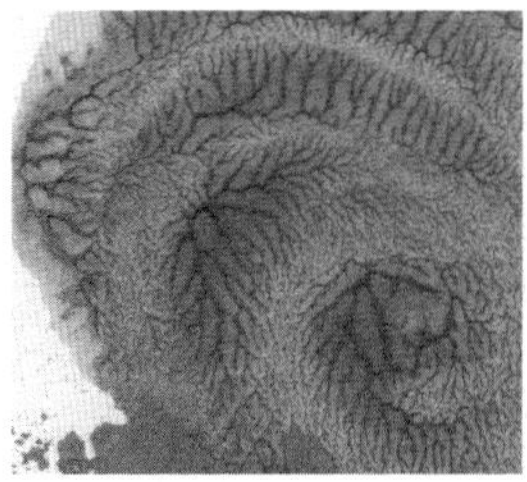

Although they did not have this language, both Cozens and da Vinci reference fractality of nature and its scale-ambiguity: a small natural fractal seen nearby looks like a large natural fractal seen from a great distance. A stain on a wall near my feet can suggest the profile of a distant mountain range. The natural complexity of one source can mimic that of another.

In the 1830s Georges Sand used the technique we call decalcomania, unnamed then, to paint landscapes (the branches of paint suggested the branches of trees) and more abstract compositions. Around 1935 this process was rediscovered and named decalcomania. Óscar Domínguez and Boris Margo used paper; Hans Bellmer, Max Ernst, Marcel Jean, Enrico Donati, and André Masson (the artist, not the economist) used canvas; Max Bucaille used glass.

Other examples, by Natalie Eve Garrett, Tanja Geis, Claire Miller, and Cara Norris, are presented in Sect. 4.9.4 of [17]. Those examples are parts of their course projects for the fractal geometry course one of us (MF) taught at Yale between 1993 and 2016.

4.1.3.2 Branch mechanics

How do the fractal branches arise when the sheets are pulled apart? As the top sheet is pulled away, the paint adheres to both top and bottom, stretching till it breaks. The paint pulls into ridges whose thickness and spacing depends on the paint viscosity and the speed with which the sheets are pulled apart, among other things. As the distance between the sheets increases, the ridges coalesce and produce a branched pattern. This is most clearly seen if the top sheet is pulled away from one side rather than lifted straight up off the bottom sheet.

4.1.3.3 A brief review of box-counting dimension

Recall from Sect. 2.1.3 we defined and computed the box-counting dimension of a shape by three steps: for a sequence of sizes $r_1 > r_2 > \cdots > r_n$,
1. Cover the shape with boxes of side length r_i.
2. Count the number $N(r_i)$ of these boxes that contain any part of the shape.

3. Test for a power law relation $N(r_i) \approx k \cdot (1/r_i)^d$.

In practice, we plot the points $(\log(1/r_i), \log(N(r_i)))$ for $i = 1, \dots, n$. If these points appear to lie on or near a straight line, this supports the existence of the power law relation. Then by Eq. (2.3), $\log(N(r_i)) \approx d \log(1/r_i) + \log(k)$ and we call d the box-counting dimension of the shape.

The greater the range of r_i-values over which the power law relation holds, the better the data support the claim that the shape is fractal. For serious science, the range should cover at least two orders of magnitude, and a wider range is better still. For these labs, we'll be satisfied with more modest ranges.

4.1.3.4 Number of branches

If the mechanical process responsible for branch formation is scale independent (over some range of scales, of course), then we expect branches should grow off branches in more-or-less the same way that branches grow. As the number of branches increases, typically the lengths of the branches decrease. The simplest way to test this is to draw a family of concentric circles centered at the base of a branch. Then for each circle radius r, count the number $N(r)$ of branches in that circle. A power law relation between the $N(r)$ and r would be revealed by a plot of $\log(N(r))$ on the y-axis and $\log(r)$ on the x-axis. If the points lie near a straight line, we've found a power law relation.

4.1.3.5 Number of levels of branching

Another measure of the complexity of these paint images is to count the number of levels of branching. Plausibly the number of levels is related to the thickness of the branches: if successive branches become very thin, we expect to see many more levels of branching than we will for paint producing thicker branches. Thickness of branches might be related to the viscosity of the paint in this way: more viscous paint should make thicker branches and so fewer levels of branching. Before you feel too satisfied with this explanation, note that petroleum jelly is very viscous and yet makes extremely fine branches between overhead transparencies. This problem presents ample room for experimentation.

To compare different samples, appropriate measures include the maximum, minimum, and average number of levels. Should higher complexity be signaled by larger maximum and minimum, by a larger range between maximum and minimum, or by width of the distribution of the number of levels around the average? Should this distribution be a bell curve, or something else? All of these questions, and more besides, can be addressed by experiments provided you have enough paper, paint, and patience.

4.1.3.6 Ratios of lengths of successive branches

One of the most important developments in the current incarnation of chaotic dynamics was Mitchell Feigenbaum's discovery [114] of the asymptotic scaling of the distance between successive period-doubling bifurcation parameters as the logistic map approaches chaos. We expect that the paint branch lengths decrease with successive branchings. How can we test if the branch length distribution exhibits a scaling relation?

Select a branch. With a divider and ruler, measure the length, L_1, from the root to its first split. Select a subbranch, measure the length L_2 from the root of this subbranch (the point that defines one of the ends in the measurement of L_1) to its first split, the second split of the original branch. Continue until you encounter a limit imposed by the branches themselves, by your patience, or by your eyesight. Plot the points $(i, L_i/L_{i+1})$. If these points approach a horizontal line, then the ratio of lengths approaches a limit.

4.1.4 Procedure

The process is simple: affix a piece of paper to a hard, flat surface. A desk or a table are ideal. Put some paint near the middle of the paper, then place another piece of paper on top of the first. Press the top piece of paper onto the bottom piece, then lift off the top piece. You'll see a branching pattern of paint ridges.

Experimental parameters include
- the paint viscosity,
- the paper stiffness and roughness,
- the pressure with which the top paper is pressed onto the bottom paper,
- the uniformity of the pressure: a heavy book, your fingertips, or press a pattern on the top paper with a table knife,
- the direction and speed with with the top paper is lifted off the bottom paper, and
- reprocessing: lift the top paper and place it again on the bottom paper, in the same orientation or after a rotation.

These are a few experiments. Many, many others can be found. Combine paints of different colors, thicken the paint with the addition of corn starch or glycerine or guar gum, thicken only some areas of the paint, add small glass beads ("seed beads") to the paint. Think a bit. You'll find your own variations on these experiments.

4.1.5 Sample A. Decalcomania variations.

Now we'll sketch a few simple experiments and illustrate each with pictures of the experiments. These are meant to be a guide, not in any case the whole story of the family of related experiments.

Viscosity experiments

First we'll experiment with paint viscosity. We won't measure viscosity directly, but will let dilution stand in for viscosity. The left image is prepared with undiluted acrylic, the right with an approximately 1 : 1

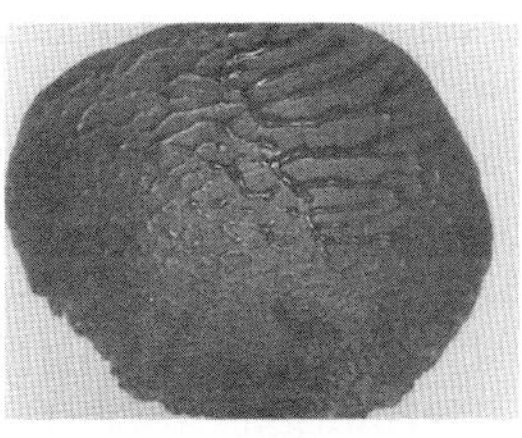

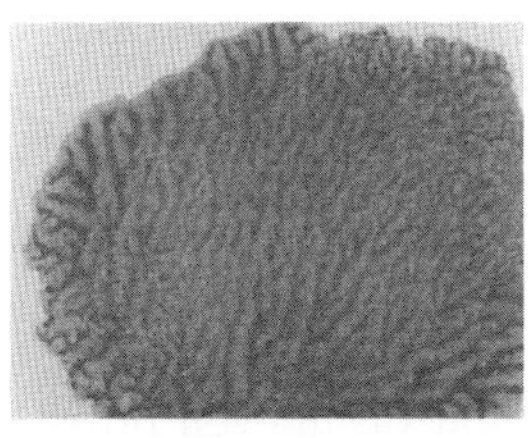

dilution by volume. The undiluted paint produces thicker branches (not a surprise), and at least in the region of thicker branches we see fewer branches. A proper experiment would control more precisely the volume of paint used and the pressure applied to the top sheet. Here our point is to illustrate that dilution the paint can alter the branch population.

Reprocessing experiments

By *reprocessing* we mean after the top sheet is lifted off the bottom sheet, it is replaced, flattened, and lifted off again. The replacement can be in the same orientation as the original, or can be rotated. How does reprocessing change the branch pattern?

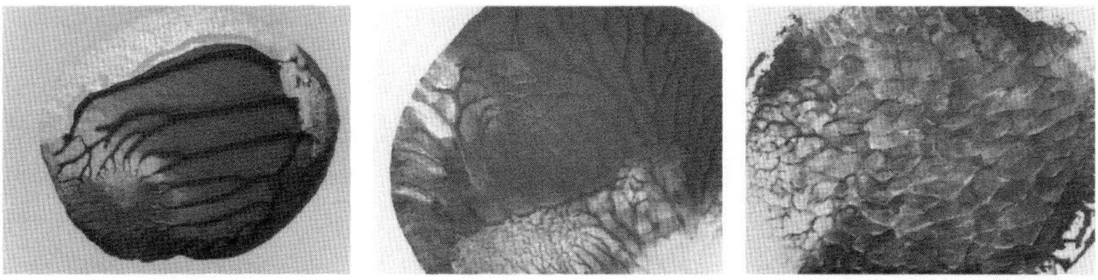

Figure 4.1: Left to right: no reprocessing, five in the same orientation, and five, each rotated 90° relative to the previous.

The left image of Fig. 4.1 is our original decalcomania blob, zero reprocessing. The middle image is the decalcomania blob after being reprocessed five times, each with the top sheet replaced in the same orientation as that of the left image. The right image also has been reprocessed five times. In each, before it is replaced the top sheet rotated 90° relative to its placement in the previous reprocessing. What do we see?

Careful branch counts could reveal more, but for now we are interested only in what we can see in the pictures. Reprocessing in the same direction appears to reduce the lengths of the longest branches and to increase the density of branches. We cannot replace the top sheet in *exactly* the same orientation as its original placement, so this relatively minor change of structure could result from reprocessing slightly misaligned branched.

In the right image of Fig. 4.1 we see that all of the long branches have been broken into smaller bits. In fact, the dendritic structures revealed in the left and middle images have been replaced by something closer to a net. When the rotated top sheet is placed on the bottom sheet, the branches of one cross the branches of the other approximately perpendicularly. This breaks up the longest branches and forms the web of ridges.

So 0° rotation does not disrupt long branches, while 90° rotation does disrupt them. Is there a critical rotation angle below which long branches are not disrupted? Does this angle depend on the number of times the image is reprocessed? Fresh experiments are all around.

Pressure experiments

How we flatten the top sheet on the bottom sheet can have a significant impact on the decalcomania pattern. The neutral choice is to place a heavy book on the top sheet. This distributes the pressure approximately evenly. How do uneven pressure distributions alter the pattern?

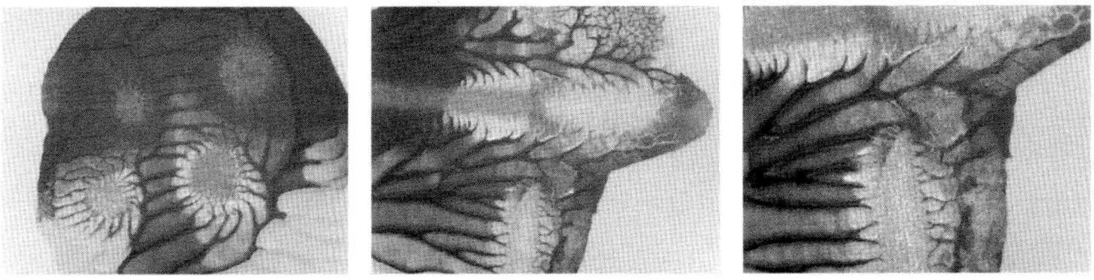

Figure 4.2: Left to right: fingertip pressure, straight-edge pressure, a magnification.

In the left image of Fig. 4.2 we see a pattern produced by application of fingertip pressure to the top sheet. In fact, four fingertips. Not surprisingly, at these sites the paint is thinner. But notice that in these regions all the branches bend inward, toward the region of highest pressure.

In the middle image, pressure is applied with the dull side of a pair of scissors, in two perpendicular directions. The right image is a magnification of the middle. Here again we see that the branches bend inward toward the highest pressure region, and approach it perpendicularly. So far as we know, this effect was first noticed by Tanja Geis in her project for the autumn, 2001, fractal geometry course at Yale.

4.1.6 Sample B. Measurement experiments.

Number of levels of branching

Consider the two pictures of the viscosity experiments in Sample 4.1.5. A quick inspection shows many branches with at least ten levels of branching for the diluted paint, while for the undiluted paint (at least the upper half of the picture, where the count is relatively easy), we see few branches with more than five levels.

Ratios of lengths of successive branches

Another relatively simple quantification is the ratios of lengths of successive branches. To make an effective measurement, we should select branches with many levels. For the purpose of illustration, a simple example suffices. Here we see a decalcomania pattern with 5 successive branches indicated. The lengths are $L_1 = 1.4$, $L_2 = 1.1$, $L_3 = 0.4$, $L_4 = 0.2$, $L_5 = 0.1$, and $L_6 = 0.1$. Then the ratios L_i/L_{i+1} are

$$1.27,\ 2.75,\ 2.0,\ 2.0,\ 1.0$$

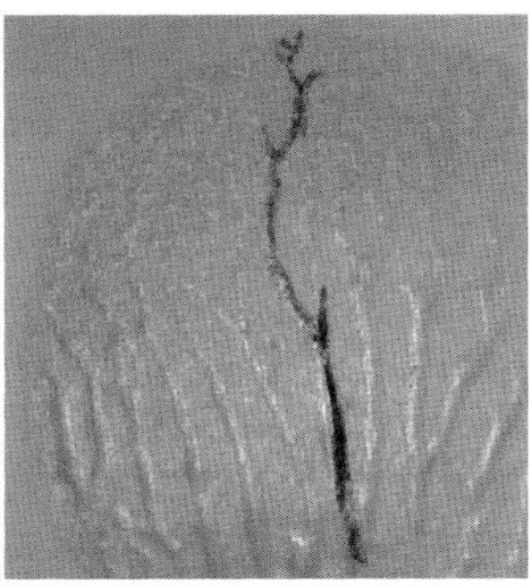

In order to take these calculations seriously, we must measure the lengths much more accurately, and for that matter, find ridges with more branching levels. Until the last measured length, the sequence looked promising. But we haven't nearly enough data or accurate enough measurements to believe that any ratio we compute might be close to a limit. This example is just to illustrate how to do the calculations.

Number of branches

A quantification placing substantially more strain on the eyes is testing for a power law relation in the number of subbranches as a function of distance from the start of the branch.

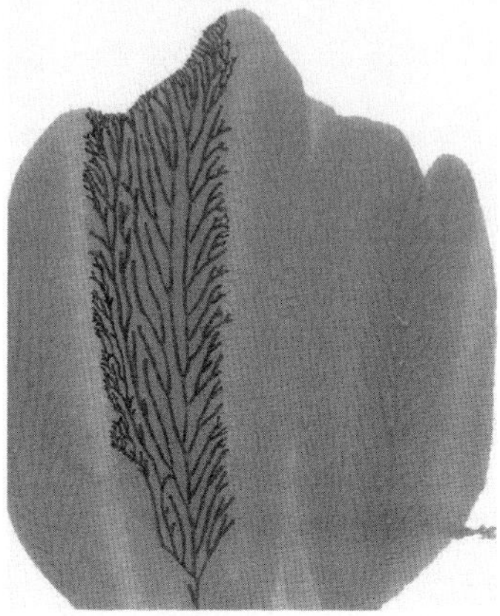

Here is an example, counting the subbranches of the highlighted branch as a function of the distance from the the base of the branch. One way to make this count is to draw concentric circles on an overhead transparency, so the placement of the circles can be adjusted over the decalcomania pattern. When you find a position for the transparency that places the circles where you want them, immobilize the transparency and decalcomania sheet with tape or paper clips. The number $N(r)$ of branches in a circle of radius r we counted are

r		2	3	4	5	6	7	8	9	10	11	12
$N(r)$	2	9	26	66	89	107	128	143	164	185	217	239

The points of the log-log plot shown here appear to fall along two lines. The first four points lie near a line with slope 2.5, the next six points lie near a line of slope 1.1. The last two points we'll discuss in a moment.

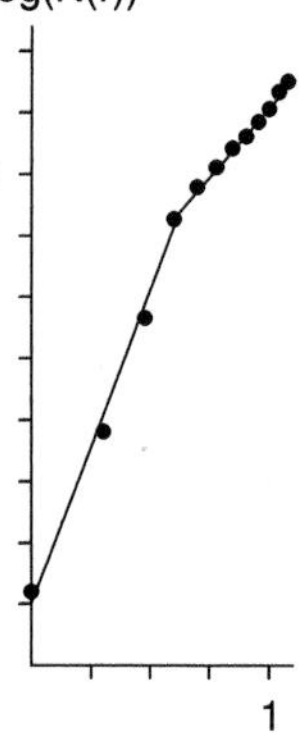

The fourth circle crosses the highlighted branches about where they collide with the left wall, the effect of hand pressure. Up to that point, at least part of the cluster was exhibiting its natural, unfettered growth, though the power law exponent, 2.5, does not represent a physical dimension because we are counting the number of branch points, not something like mass as a function of distance.

The branches of the part of the pattern within the next six circles are growing but are bounded left and right by pressure ridges. We see greater complexity toward the sides, but each segment between successive arcs is about a translation of any of the others. The growth of the number of branch points is approximately linear, $N(r) \approx k \cdot r^d$ for $d \approx 1$.

The last two circles cut through the top of the branches, so do not represent anything intrinsic to the cluster.

Box-counting dimension

Most demanding is the box-counting dimension. For the number of branches sample, here are the counts of the number $N(r)$ of boxes of side length r needed to cover the branching pattern.

r	1	1/2	1/4	1/8	1/16
$N(r)$	33	113	381	1077	2342

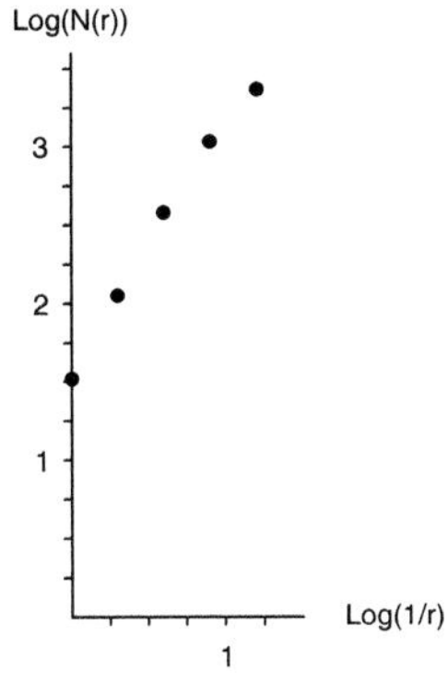

The log-log plot shows a clear down-turn for smaller r. The best-fitting line from $r = 1$ through $r = 1/8$ has slope $d = 1.68$. On the other hand, the line determined by $r = 1/8$ and $r = 1/16$ has slope $d = 1.12$. Note we are *not* asserting that dimension can be deduced from a two point log-log plot. Two data points are fit exactly by a line, and consequently provide no evidence of linear trend, hence of a power law.

One source of this apparent change of slope is that the branches are so close together that for larger r the occupied boxes come close to filling in an area, so these counts suggest a dimension closer to 2. Only for the smallest of boxes in this example do many of the pattern gaps become visible. Better still would be to take even smaller grids, but to illustrate the concept, and point out some possible issues, these suffice.

4.1.7 Conclusion

Paint experiments are a rich environment for generating natural fractals. Experimental parameters abound. With several relatively simple quantification schemes we can measure how the pattern complexity varies with these parameters.

One final comment: an artist's eyes and sensitivity to color and light, and familiarity with the mechanics of paint, can greatly improve the impact of the project. We are not artists, as is clear from our examples presented here. Natalie Eve Garrett, Tanja Geis, Claire Miller, Brianna Murratti, and Cara Norris are artists. Their work exhibits subtlety at a level altogether different from the one we inhabit. For example, the first decalcomania image of this lab was part of the fractal geometry class project of Claire Miller and Cara Norris.

This lab is a good opportunity for students with some experience in painting to use their skills to make fractals.

4.1.8 Exercises

Prob 4.1.1 Get some paint and papers. Experiment. Test paint viscosity, reprocessing, pressure. What about the roughness of the paper? What about the paper stiffness or porosity? Will glossy printer paper produce patterns different from those on an overhead transparency?

Prob 4.1.2 How reproducible are these patterns? Prepare another arrangement of paper and paint, as nearly identical as you can to an experiment you performed in Exercise 4.1.1. How similar are the two patterns?

Prob 4.1.3 How does reprocessing alter the image? Prepare paper and paint as in Exercise 4.1.1. After lifting off the top paper, photograph the paint pattern, replace the top paper and apply mild pressure, lift the top paper, and photograph the paint pattern. Repeat.

What changes if you rotate the top paper 90° before you return it to the bottom paper?

Prob 4.1.4 (a) Quantify one of the pictures of Exrecise 4.1.1. Measure the number of levels of branching, the ratios of the lengths of successive branches, the number of subbranches as a function of the distance from the start of the branch, or the box-counting dimension.
(b) Repeat (a) for a reprocessed pattern of Exercise 4.1.3.

Prob 4.1.5 Do your own pressure experiments similar to those of Fig. 4.2. Do the decalcomania ridges always approach the pressure hollows?

Prob 4.1.6 Suppose the experiment is performed on a painting that already has dried. Or on a piece of thrift store art, for example, as shown here. The body of the monster is a decalcomania watercolor blob to which legs, wings, and ears have been added. Do the decalcomania branches look like the monster's hair, or maybe something more frightening?

These are the possibilities we thought of in a few minutes. Use your imagination. What will you find?

4.2 Fractal painting: decalcomania 2

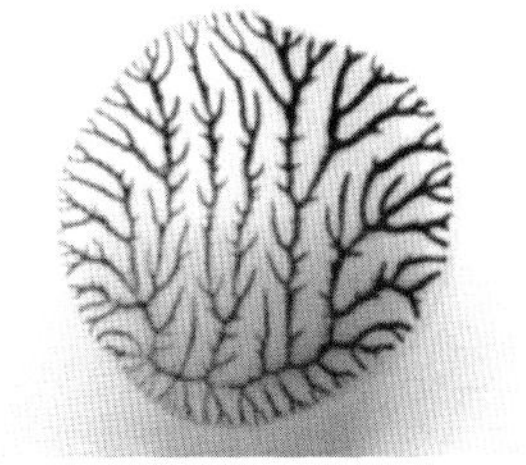

In Lab 4.1 we explored patterns made when two pieces of paper with acrylic paint between them are pressed together and then pulled apart. Usually the bottom sheet of paper is attached to a hard surface, a table or a desk, and the top sheet is pulled away. The flexibility of the top sheet of paper may contribute some aspects of the dendritic patterns we observed. One way to test this is to keep both surfaces rigid when they are pulled apart. In Sect. 4.1.3.1 we mentioned that Max Bucaille constructed decalcomania patterns between sheets of glass; in this lab we'll use squares of plexiglass. The website [115] gives useful video instructions on this process.

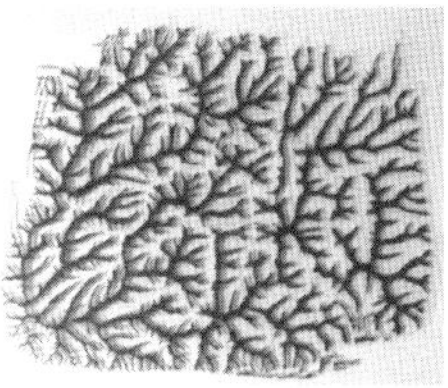

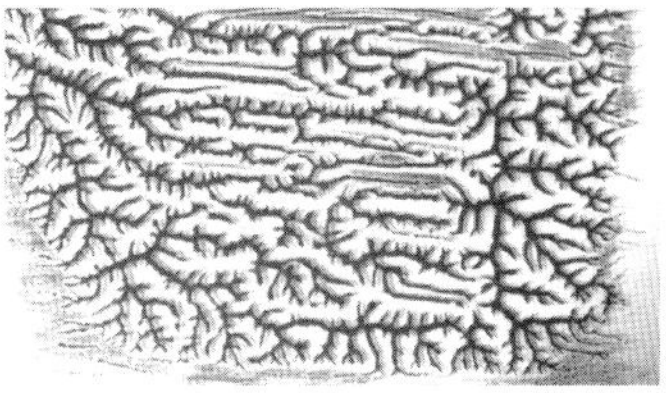

Here are two other examples. Try to find common features in these images. This will inform how you look at the other images in this lab, and at your own experiments. Eventually, these will help us investigate any effects the rigid upper surface has on the dendrite structure.

4.2.1 Purpose

Here we'll explore a way to make fractal branching patterns with acrylic paint and plexiglass or glass plates. With these examples we will investigate how these patterns form.

4.2.2 Materials

To do all the experiments of this lab you will use acrylic paint; two approximately $8'' \times 10''$ plexiglass or glass panels, at least $0.1''$ thick so they won't bend or break when pried apart; a $2''$ paint brush and a thin putty knife; plain copier paper; a camera; old rags, a non-abrasive household cleaner, and running water.

4.2.3 Background

The patterns made with plexiglass or glass plates are distinct from those made with paper because now both surfaces are rigid, and also because acrylic paint adheres to paper differently than it adheres to plexiglass. In the Exercises we'll suggest experiments to assess the relative contributions of these two effects. Now we'll focus on the effects of adhesion.

The paint sticks to paper much more than it sticks to glass which gives the paint branches on glass a better-defined appearance than that of paint branches on paper. As the plates are pulled apart, air intrudes and pushes paint out of the way. Suppose the plates are pulled apart from the top to the bottom. At first, very little paint is near the top edge and the air intrudes to form many small pockets. The paint along the top is pushed out of these pockets to form small ridges between the pockets. As we continue to separate the plates, small ridges coalesce to form (perhaps only slightly) larger branches. When the plates are pulled apart, we have a forest of dendrites. Specific details of how these branches form are influenced by differential adhesion of paint to different regions of the glass. Different methods of applying the paint to the glass can give a variety of branching structures.

All the analysis methods of Lab 4.1 can be applied here to investigate the patterns you get. Count the number of levels of branching, find the ratios of lengths of successive branchings, seek a power law relation between the number of subbranches and the distance to the branch start, and compute the box-counting dimension if your sample has enough levels of branching.

4.2.4 Procedure

Apply acrylic paint to one of the two glass or plexiglass plates placed flat on a suitable surface. Place the second plate on top of the first plate, press the top plate down slightly, then separate the plates. The resulting pattern then can be photographed or transferred to paper.

Here are the details of the procedure.

• Paint can be applied as a thick blob to be spread out by applying pressure to the plates or by brushing.

• If the paint is to be spread by applying pressure to the plates, little or much pressure can be applied to plexiglass, but be careful with pressure applied to glass plates, to be sure the glass doesn't shatter.

• The plates can be difficult to separate. If this happens, a very thin flexible putty knife can be used to start the separation. In addition, the start of plate separation can be made easier if the top plate is offset slightly from the bottom plate, rather than aligning exactly the edges of the top and bottom plates. Insert the putty knife between the plates, near a corner may be easiest, and slowly twist the knife. For glass plates, be very careful with this.

• When the plates are separated, you can photograph them, with or without flash, or transfer the pattern to paper. Transfer to paper must be done quickly because acrylic paint dries rapidly and finer details will not transfer from dry paint. Use a fairly stiff copy paper and press very lightly to make the transfer. The paint pattern will show the depth of the branches, but this detail is lost on a paper transfer.

• Under running water the paint will wash off glass plates easily. Plexiglass plates may have a residue pattern, like an etching, left by the paint. This can be removed by a non-abrasive household cleaner and some effort that one of us (NN) found annoying.

4.2.5 Sample A. Spreading a blob of paint by pressure.

Place one of the plates on a flat protected surface and apply a blob of paint near the middle of the plate. Put the other plate on top of the first and apply pressure to the top plate. If you use glass plates, be careful to apply the pressure evenly. Follow the steps of the procedure to separate the plates and produce your dendrite designs. Experiments include
• vary the amount of paint,
• vary the viscosity of the paint (thin with water, thicken with vasoline),
• vary the pressure,
• apply the paint in something other than a blob, and
• reprocess: put the two plates together again, apply some pressure, then separate.
Fig. 4.3 shows two paper impressions and a photograph of paint blobs spread by pressure.

Figure 4.3: Two paper impressions (left and right) and a photograph (middle) of pressure spreading experiments.

The pictures of Fig. 4.3 were chosen for a reason. While the design we are usually interested in and focused on is the dark branching of the paint, it is easy to see the paths established by the air that flows through the paint in these pictures. It seems the air paths compete as much or more for your attention here as do the paint branches. The top plate for the left picture was pulled down from the upper left corner, the middle picture from the top middle of the plate, and the right picture from the top right corner. Notice how the paint is pushed and piles up as you travel to the bottom of the plate. This shows up especially with the transfer to paper that flattens out the paint.

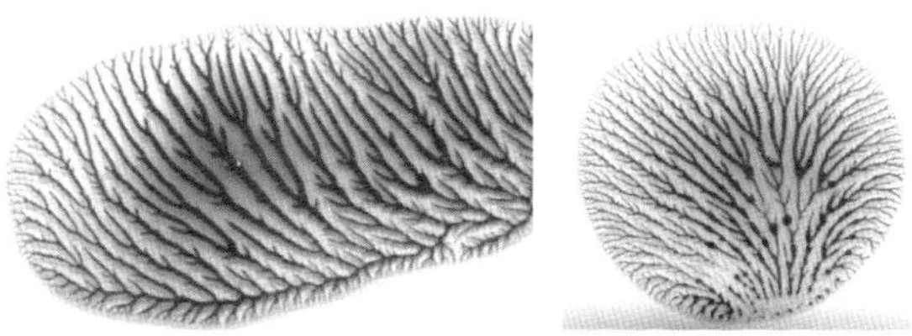

On the right are some designs where the air paths are not so obvious. Here you need to take a page from Douglas Hofstadter and switch the foreground and background. Try to concentrate on the absence of paint and not the paint. This is not so easy at first, but with a little practice you can clearly see the air pathways. Once the airways pop out at you, you will have a hard time seeing the paint branches.

4.2.6 Sample B. Applying paint with a brush.

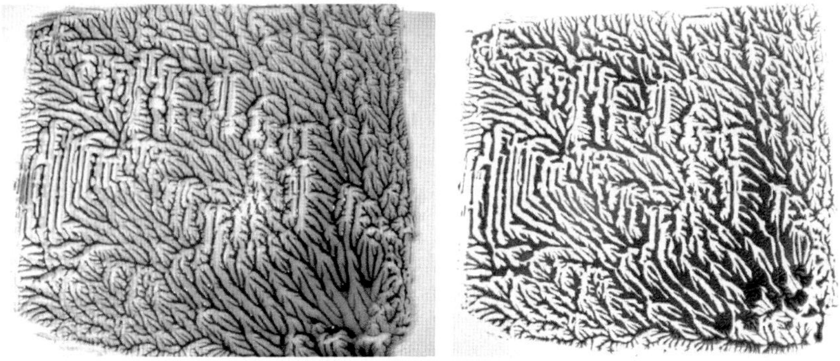

Figure 4.4: A photograph (left) and paper impression (right) of a brush spreading experiment.

You might not think that how you apply the paint would make that much of a difference, but it does. Once again there is a competition for territory between the air and the paint but here with the use of a brush the paint can be put on in a thinner layer than before. As a result the air has less resistance and gets to push the paint around more. This results in more branching and designs distinct from those of Sample 4.1.6. So, put some paint on one of the plates, spread it around with a brush and see what you get. In Fig. 4.4 we see a photograph (left) and a paper impression (right) of dendrites photographed on the left. In the photo on the left we see that the plates were separated from the upper left corner by the pile-up of the paint that was pushed to the lower right corner. The branches in the lower right corner have more depth than those in the upper left.

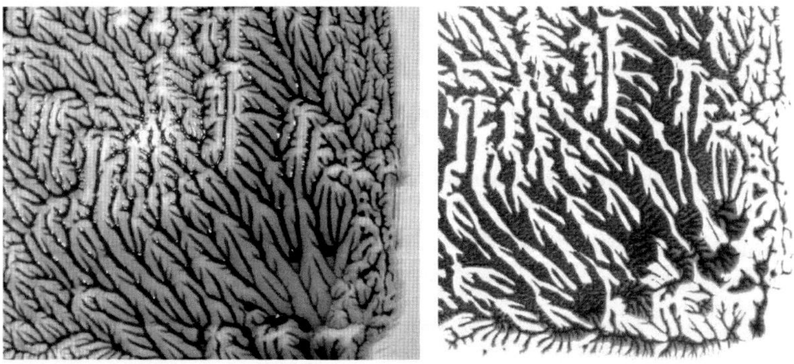

Figure 4.5: Magnifications of the lower right corners of the images of Fig. 4.4.

The images of Fig. 4.5 are close-ups of the lower right corner of the pictures of Fig. 4.4. Notice the accumulation of paint and the depth of the branches of the left picture; it really is a 3D design. On the right, we lose depth and detail and flatten the branches when making a paper transfer but we gain another set of fractal branching when the paper is pulled from the design. Look carefully and you will see this in the flattened branches. We also see the air pathways more clearly in the paper transfer than in the photo.

4.2.7 Conclusion

Air invasion into paint flattened between rigid plates offers another avenue to explore decalcomania. Probably much more can be done with this by people who are more artistic than we are (NN designed and executed this lab; we've already seen that MF can claim no artistic vision), but we hope this lab has given you some ideas and motivated you to experiment. The difference between the patterns using paper or glass and the method of application of the paint surprised us. What else can be done that we have not considered? Go discover.

4.2.8 Exercises

Prob 4.2.1 Generate some decalcomania patterns by the method of Sample 4.1.6. Record the experimental parameters: amount of paint, pressure applied to the plates, direction of lifting the top plate, and so on.

Prob 4.2.2 Generate some decalcomania patterns by the method of Sample 4.2.6. Record the experimental parameters: amount of paint, direction of brush strokes, direction of lifting the top plate, and so on.

Prob 4.2.3 With three equal amounts of paint, generate three pressure decalcomanias, one between two sheets of paper, one between two glass or plexiglass plates, and one between two overhead transparencies. Try to use identical experimental parameters. To apply the same amount of pressure uniformly, rather than press on the top plate or sheet, place a book on the top plate or sheet. Because the weight of a sheet of paper or an overhead transparency is negligible, but that of a sheet of glass or plexiglass is not, put the book alone on the top glass or plexiglass plate, but put the book and a glass or plexiglass plate on the top sheet of paper and on the top overhead transparency.

By comparing these dendrites, can you speculate on which features of the pattern formed on the glass or plexiglass plates are due to the rigidity of the top plate, and which are due to the plate smoothness?

4.3 Fractal painting: bleeds

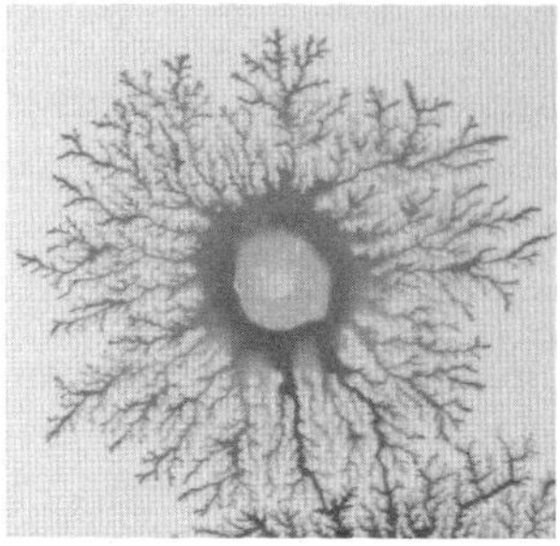

Here we'll see what can be done when isopropyl alcohol (mixed with ink) bleeds into a high-viscosity paint. Early art lessons emphasize the need to allow time for a layer of one color to dry before the next color is applied. The reason is to avoid bleed of one color into the other. As we'll see, encouraging bleed can produce some interesting patterns, many of them fractal.

While the artist can influence many aspects of these bleeds—colors, the contour along which the bleed occurs, the density of the branches and rate of their formation—the fine structure is determined entirely by the physics and chemistry of the interaction of the fluids with one another and with the surface on which they are applied.

YouTube has several videos that illustrate fractal painting. The methods used here were inspired by two videos at Myriam's Nature [116] and [117]. The videos are really worth watching to get a feel for what follows. Thank you Myriam.

4.3.1 Purpose

Fractal painting can produce images that appear to be closely related to computer generated trees, decalcomania patterns, diffusion-limited aggregation (as we'll see in Lab 5.4), and some patterns in nature, for example, lightning and river branching. We will experiment with different methods of making images that branch by using paint bleeding techniques.

4.3.2 Materials

The images you get in these experiments are very delicately related to the materials used. Please read through the entire lab, including the exercises, to determine what you need to collect. In order to make the experiments robust enough that everyone will obtain reliable outcomes, the materials in the samples are standardized by specifying them in detail. You should use the particular materials listed to insure the best results at least for the sample experiments. We really do not have stock in the corporations manufacturing them and are not looking to make our fortunes from the sale of these products. In the exercises you will be encouraged to experiment with several different materials. When doing the exercises keep in mind what Bob Ross would always say, "There are no mistakes, only happy accidents." (Episodes of Bob Ross' *Joy of Painting* television series can be found on YouTube.)

We used white acrylic paint, a well palette, a ball stylus, acrylic ink, isopropyl alcohol, ceramic glazed tiles, overhead transparencies, a comb, and a razor blade scraper.

Suggested exercise materials: you may want to use these same materials in the exercises, but do not be limited by our suggestions. Read the exercises and determine what materials you want to use.

4.3.3 Background

We will take a quick look at binary branching to see how fractal trees are formed. We can generate a binary fractal tree recursively by symmetric binary branching. A trunk of length 1 splits into two branches of length $r < 1$, each making an angle θ with the direction of the trunk. Both of these branches divide into two branches of length r^2, each making an angle θ with the direction of its parent branch. This iteration process is continued to form, as Benoit would say, "a multiplicative cascade" of branches. Fig. 4.6 illustrates this process.

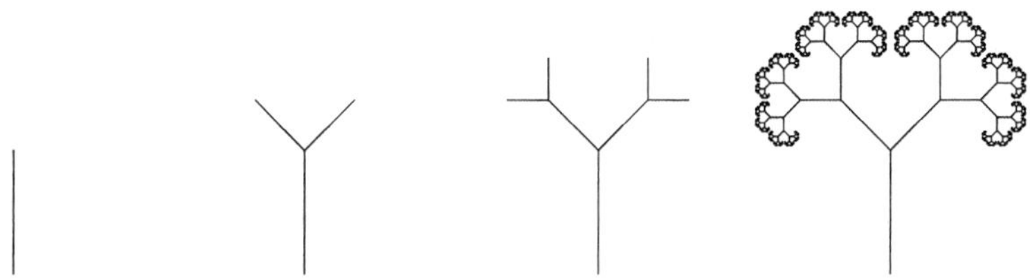

Figure 4.6: The initial stage and iterates 1, 2, and 10.

The whole tree is not a fractal. For example, the trunk is not made of smaller copies of the whole tree. However, the limit points of the ends of the banches, that is, the set of *branch tips*, is a fractal. It consists of two copies, one on the left and one on the right, that are scaled copies of the whole. Both these copies are further divided into two smaller copies, and so on.

It is easy to see that the tree contains scaled copies of itself. Break off the left branch at the first branch point. It is a scaled version of the whole tree. Now break off the right branch at the first branch point of the branch that was just separated from the trunk. It also is a scaled version of the whole tree. As this procedure is continued each branch broken off will be a scaled version of the entire tree. See Fig. 4.7.

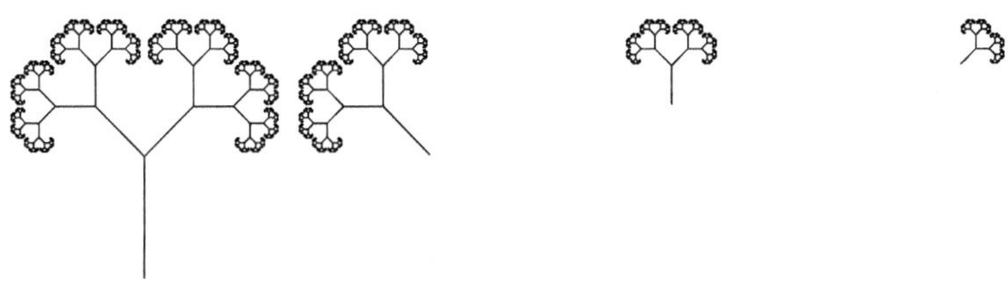

Figure 4.7: The original tree, left branch removed from the trunk, right branch removed from the previous branch, right branch removed from previous branch.

The appearance of computer generated binary trees is determined by the angle of branching and the scale ratio of successive stages of branching. These conditions were thoroughly investigated in [118] and other places. Depending on the angle of branching and the scale size of successive branches, three things can occur. The branches and branch tips can fail to touch anywhere, they can just touch so that some tips of the left branch contact some tips of the right branch but no tip contacts any non-tip part of the tree, or the left branch can overlap the right branch at points that are not branch tips. We say the tree is *self-avoiding*, *self-contacting*, or *self-overlapping*, respectively. An example of each is shown in Fig. 4.8. To make the comparisons clearer, in each tree we have taken $\theta = 45°$. Other angles can be used, of course.

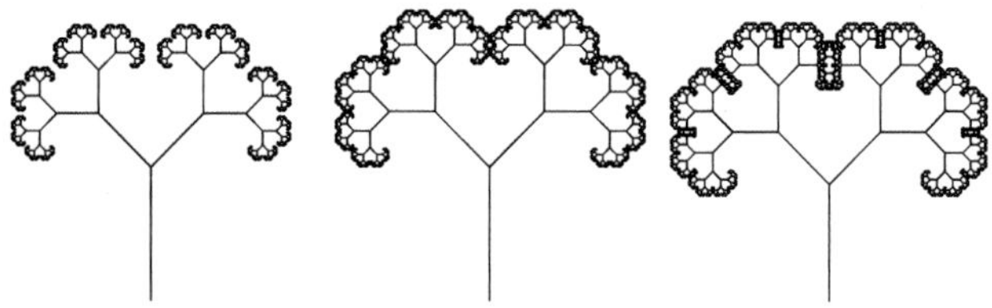

Figure 4.8: Left to right: self-avoiding, self-contacting, and overlapping binary trees.

If the branches are self-avoiding, the branch tips are a fractal dust. The canopy of a tree is the set of branch tips that can be reached from a place distant from the tree without crossing any other part of the tree. When the branches are self-contacting the canopy is a self-similar continuous curve. The canopy by itself is a fractal as is the set of branch tips.

Often branching structures in nature do not exhibit many levels of branching. The forces that produce branching are the dominant forces over only a limited range of scales. This holds for the paint bleeds of this lab. Establishing a power law relationship or calculating a dimension (Lab 2.1) for the bleeds is very challenging. We'll be content to study the density and number of levels of the branches.

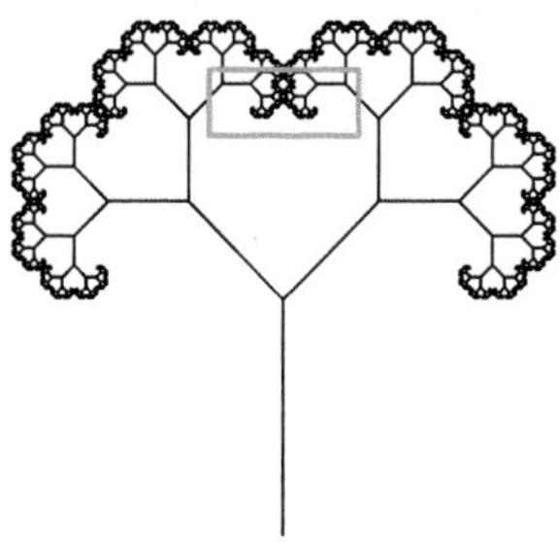

The part of the tree in the box is not part of the canopy nor are any of the other parts of the tree that correspond to it. Look closely at the part of the canopy directly above the box. Does it remind you of any other fractal?

Some of the images we generate will look very much like those from Lab 5.4 on diffusion-limited aggregation, but they are formed in a different way. The dendrites we get in Lab 5.4 are particles coming from a distant source sticking when they reach some part of the growing cluster. Fractal painting dendrites such as decalcomania branching are examples of a less viscous medium, air in this case, intruding into the more viscous paint. In the fractal bleeds of this lab isopropyl alcohol is the less viscous medium and thinned paint is the more viscous. The ink is used only to make the alcohol visible. The paint bleed images clearly are different from decalcomania. As we will see, the dendrites you get depend delicately on the base paint medium, the medium doing the bleeding, the surface the base paint is applied to, and the size of the stylus used to apply the bleeding mixture.

4.3.4 Procedure

First a mixture of white paint and a flow medium is spread on a glazed tile. Then a mixture of ink and alcohol is applied to the paint. Fractal branches should radiate out from the ink into the paint. The resulting image can be photographed and if you wait overnight, the dried paint can be scraped off the tile. Experiments can be done with all of the materials being mixed together and how the ink is applied to the paint. Some of this will be explored in the exercises. In samples 4.3.5 and 4.3.6 you should stick strictly to the materials and methods used to insure reasonable results.

4.3.5 Sample A. Central acrylic bleeds

In this sample we will produce a branching pattern from a central point that radiates out into the paint. You should actually do this sample and sample 4.3.6 in order to have success with the exercises. Start by having a clear flat workspace with some protective covering on top. First mix enough acrylic paint and Floetrol in an appropriate container to cover the surface of the tile you are working with. The paint mixture does not have to go all the way to the edges of the tile, close is good enough. Mix one part acrylic paint to four parts Floetrol. Make sure to mix these together well. Pour the paint mixture from the container onto the tile and spread it out with the comb. Something like a comb works best to spread the paint as opposed to a flat spatula. The idea is the get the paint surface to be as smooth and even as possible; this does not work too well with a spatula.

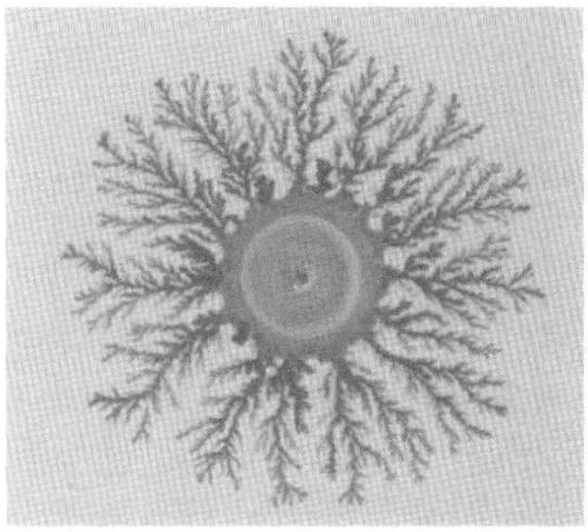

Figure 4.9: Radial branching of Sample 4.3.5.

Put one drop of Dr. Ph Martin's India ink into one of the wells of the palette and add 3 or 4 drops of the 70% alcohol using the dropper. The ratio of ink to alcohol is not that important. The more alcohol the faster and further the branching will occur, but the lighter the color will be. Choose the larger end of the stylus and mix the ink and alcohol with the stylus. Dip the stylus into the ink mixture and touch the stylus to the paint somewhere near the middle of the tile. You may not see much happen at this time. You will have to repeatedly add more ink to increase the volume of ink in the paint. So continue to dip the stylus into the ink mixture and try to touch the stylus to the paint at the same place as before. As you do this, the volume of ink builds up, the ink bleeds out into the paint and forms branches. It will take some time to build up the alcohol mixture in the center and as you do this the alcohol may evaporate from the mixture in the palette. If the branching appears to stop just add some more alcohol to the ink. Do not be impatient and dump a lot of ink onto the paint all at once, you will just get a big central blob with small branches.

If everything has gone well you should have a pattern such as the one in Fig. 4.9. You can photograph the pattern and also leave it to dry overnight. The next day you should be able to scrape the paint off in one piece from the tile. Use the razor blade scraper and loosen the edges of the paint first. After the edges are released from the tile the remaining part should scrape off easily.

4.3.6 Sample B. Edge bleeds

The set-up for this sample is the same as above but this time be sure to make the paint mixture goes right to one of the edges of the tile. When you add the ink mixture, do the same as before but add it in several distinct places along the edge of the tile. Instead of a radial pattern you should get branches that are roughly parallel.

4.3.7 Sample C. Watercolor bleeds

Watercolors also can produce interesting bleed patterns. Here we see the pattern when a layer of light watercolor paint, slightly diluted 3 : 1 with water, is painted on a piece of paper and then, before it dries, a region of darker watercolor, diluted in the same way, is applied on the lighter layer.

The bleed begins as small wobbles along the boundary between the paints. The wobbles are unstable against small perturbations, so branches grow side branches, which grow smaller side branches. But unlike acrylic bleeds, here the border between paints is a bit wispy. Still, watercolor bleeds provide another way to explore the interaction between artistic intent and physical processes.

4.3.8 Conclusion

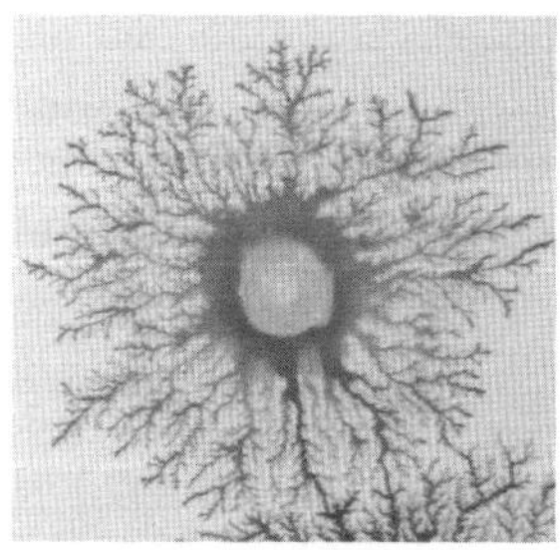

Paint bleeds are another way to generate fractal branching patterns. In Fig. 4.10 we see some other examples of dendrites. Top left is a piece of limestone in which a manganese solution has diffused through cracks in the stone and crystalized out. Top center is a bacteria colony [119] grown on a 1.75% agar concentration with 0.01 g/L peptone, a low concentration for bacteria growth. Top right is a decalcomania example from Lab 4.2. Bottom left (Lab 5.4) was grown in a 0.5 M zinc sulfate solution with a 3 V current. Bottom center is a Lichtenberg figure from 6.6.1. Bottom right is a viscous fingering pattern in a guar gum solution.

The mechanisms for these processes differ from one another. Why does branching occur so often? Can we quantify these comparisons? Do similar

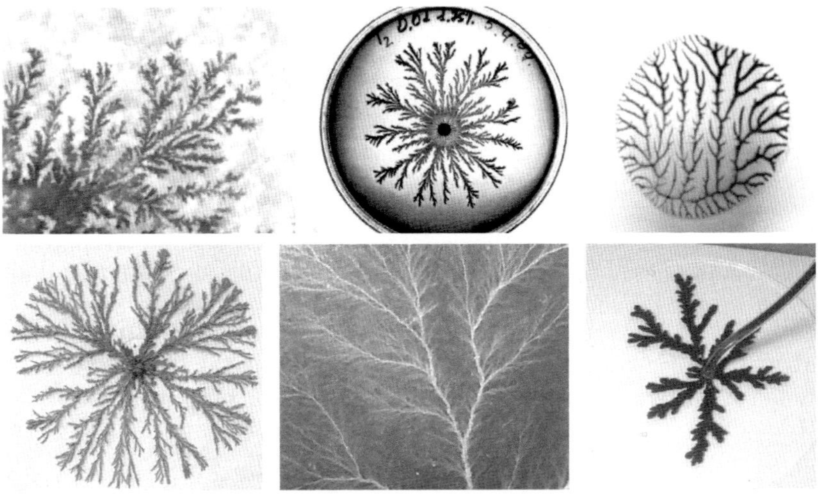

Figure 4.10: Dendrites. Top row, left to right: manganese deposits in limestone, $\mathbb{T}$ morphotype of *Bacillus subtilis* 168 (Fig. 1(d) of [119]), and a decalcomania example from Lab 4.2. Bottom row, left to right: a zinc electrodeposit from Lab 5.4, a Lichtenberg figure from 6.6.1, and a viscous fingering pattern from Lab 5.5.

measurements reinforce similar appearances? More questions than answers.

4.3.9 Exercises

Prob 4.3.1 *Base paint.* I (NN) was unfamiliar with Floetrol until I did these experiments. It is a commercial latex paint flow medium readily available in paint stores. It is added to latex paint in the suggested ratio of 8 oz. to one gallon to increase the flow properties of the paint in order to not leave brush marks or roller dimples. So I thought why not try latex paint as a base instead of acrylic paint. I found a can of old Kilz latex paint and tried it with the Floetrol. See if you have some light colored latex paint lying around and try it with the Floetrol and India ink alcohol mixture. How did you do? Does it produce any fractal branching?

Prob 4.3.2 *Flow medium.* I used two flow additives, Floetrol and Decoart Pouring Medium for acrylic paints. They produced distinctively different patterns. Also the ratio of the flow additive to the paint made different images. If you have two different flow mediums try each of the suggestions below for each medium, if not do them for the one flow medium you have.
(a) Use just paint with no flow medium.
(b) Try it with a ratio of 1 flow medium to 1 paint, 2 to 1, 3 to 1, and 4 to 1. When the ratio gets up to 4 to 1 the base mixture is mostly flow medium. So what about this:
(c) Try it with just flow medium and no paint.
(d) Use a combination of only Floetrol and Decoart at the same time.

Prob 4.3.3 *Penetration medium.* We have been using isopropyl alcohol as the penetrating medium, but what about other liquids which are compatible with India ink? You might try paint thinner, denatured alcohol, naphtha, mineral spirits, nail polish remover, etc. These are flammable materials. **Please use them carefully.** You might also try acrylic ink instead of India ink to color the penetration medium. Give it a try and see what you get.

Prob 4.3.4 *Stylus size.* Vary the size of the stylus used to apply the ink/alcohol mix. Applying larger amounts of ink will not necessarily give larger designs faster. You could even try a small dropper. Find what you like best.

Prob 4.3.5 *Edge patterns.* Instead of creating radial patterns try to get a more linear pattern by introducing the ink in several consecutive places along the edge of the paint. The paint will not bleed onto the tile and so it will be forced to go in one direction away from the edge of the paint.

Prob 4.3.6 *Where's the fractal?* After the bleed has dried, place an overhead transparency over the bleed and gently mark the tip of each branch. Place the transparency on a blank piece of paper so you can focus on the branch tips. Do you see small (approximate) copies of larger parts of the branch tip pattern?

4.4 Fractal painting: mixing

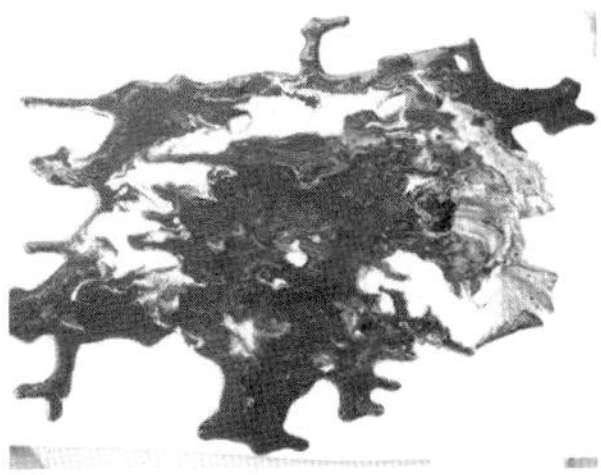

The method of Labs 4.1 and 4.2 certainly allows ample opportunity for artistic expression, but much of the patterns of each example results from the physics and mechanics of paint, paper, and process, a less viscous fluid (air) intruding into a more viscous fluid (paint). In Lab 4.3 much of the final result depends on the diffusion of different density fluids rather than on the deliberate actions of the artist. In this lab we'll investigate patterns produced by the mixing of paint under gravity-driven flow. The initial distribution of paints, and when and how to change the tilt of the surface, are the avenues of artistic expression. In Lab 4.5 we'll study drip painting which affords much greater input of the artist. We'll find that Jackson Pollock is *not* easy to mimic.

4.4.1 Purpose

We'll test paints of various viscosities and densities as they flow on a tilted surface. Our goal is to find conditions that give rise to intricate patterns, swirls within swirls. Visual investigation of these structures is how we'll assess the fractality of the paintings.

4.4.2 Materials

We'll use at least two colors of high-flow acrylic paints (mixtures of acrylic paint and pouring medium), paper, disposable plastic cups for mixing paint and pouring medium, plastic or wooden sticks to mix paint and pouring medium, a tilt table (a large box top and a few books to prop up one side is a substitute of limited utility, but still can give interesting results), and a painter's drop cloth or an old shower curtain, or a decommissioned cat litter box, to keep your workspace free of paint spills.

Figure 4.11: Left: a universal ball joint. Right: a tilt table.

For the tilt table, with wood screws attach one plate of a universal ball joint (the left image of Fig. 4.11) to a base, a block of wood will do. Attach the other plate to a flat surface with a raised edge. You can build this surface with a sheet of plywood ($9'' \times 12''$ or a bit larger) and four strips of moulding for the edges; or use a small, old cookie sheet that you won't want to use again for baking; or the top of a tub of cat litter. Drill holes in this surface to match several holes in the top plate of the universal joint and attach the surface to the top plate with nuts and bolts. See the right image of Fig. 4.11. Unless you are unbothered by a messy workspace, use a tilt table top with a lip, or place the tilt table on a drop cloth or shower curtain, or in an old cat litter box.

4.4.3 Background

In the spring semester of 1993 at Yale one of us (MF) introduced that university's first fractal geometry course, an elaboration of the course developed with Dave Peak at Union College. That semester, for his project a student, whose name has vanished in several office moves, produced a remarkable painting with different colors swirling together on smaller and smaller scales. How had he done this? He's built what he called a "tilt table" with which he could change the angle of the table that supported the canvas. Then he experimented with different densities of paints. Almost 30 years later, the beauty and intricacy of this painting has stayed with me.

We won't try to quantify any aspect of these pictures; we'll just look for levels within levels of patterns. This is visual processing, not mathematical. On the right we see a magnification of a portion of one of our experiments. Note the wiggles upon wiggles. These pictures are relevant to to our study of fractals and chaos because this sort of mixing is central to the "stretch and fold" paradigm of chaotic dynamics, described in Sect. 10.4 of [18]. Sensitivity to initial conditions, one of the characteristics of chaos, implies that nearby trajectories diverge, the "stretch" part of stretch and fold. But how is this possible if the trajectories converge to an attractor in a bounded region of phase space? This is where "fold" comes in. For systems described by continuous behavior, fluid flow is an example, we may find complicated, many-layered patterns. This is described beautifully, in great detail and with lovely pictures, by the chemical engineer Julio Ottino in his book [120].

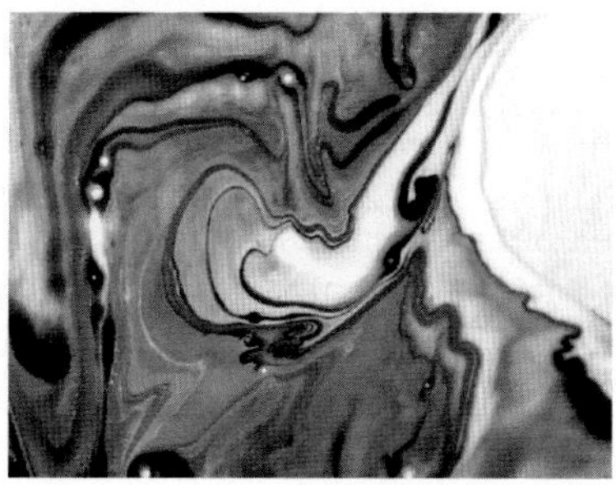

4.4.4 Procedure

In the plastic cups, one for each color of paint, put a volume of pouring medium appropriate for the paint-pouring medium mixtures, then add the corresponding volume of acrylic paint to each cup and mix. In our experiments we used the same ratio for each paint color. Specifically, we took pouring medium:acrylic paint ratios 1:1, 2:1, 4:1, and 8:1. With tape secure a piece of paper to the tilt table. Place the tilt table in the horizontal orientation and one at a time pour the paint onto the paper.

One approach is to pour the paints, one at a time, in extended blobs near the center of the paper; another is to pour one paint and spread it across all of the paper, then pour the remaining paints, one at a time, in extended blobs near the center of the paper.

Now tilt the table and let the paint flow. You may want to keep the paint on the paper, but this isn't necessary because of the lip around the edge of the tilt table. The drop cloth is a back up if the paint overflows the table lip. Then pick another direction, 90° from the original tilt direction is a natural choice, tilt the table in that direction, and let the paint flow. Then pick another direction, tilt the table, and watch the paint flow. Continue until the paint dries so much that the flow is greatly reduced, or until you get tired.

The greater the angle of the table tilt from horizontal, the faster the paint will flow. Because the paint dries as it flows, this effect is a bit more complex that it might appear.

Photographing the pattern before each change in tilt direction may reveal interesting patterns.

4.4.5 Sample A

We prepared 4:1 mixtures of pouring medium to acrylic paint for white, blue, and yellow paint. For conversion to these grayscale images, yellow wasn't the best color choice. In Fig. 4.12 we see the initial pour and flow that results from the first tilt. The right image is flipped vertically from the orientation of the left image.

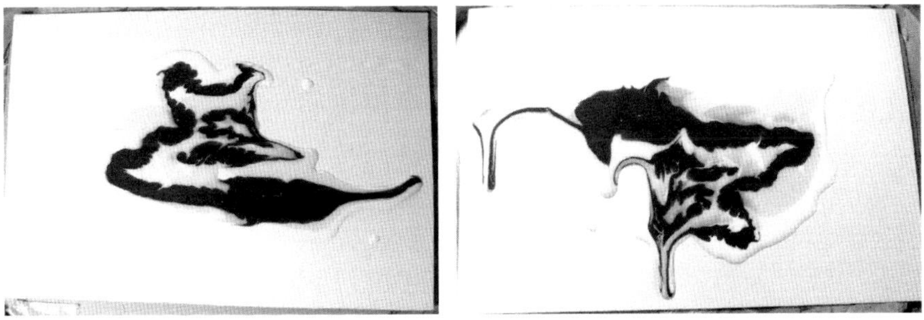

Figure 4.12: Initial pour and first tilt for Sample A.

In Fig. 4.13 we see the flows that result from the second and third tilts. The left image of Fig. 4.13 is in the same orientation as the right image of Fig. 4.12; the right image of Fig. 4.13 is flipped vertically from this orientation.

Notice that in addition to mixing along the branches, the patterns in the large central region becomes increasingly complicated with each additional tilt of the table.

In Fig. 4.14 we see the flow after the eighth tilt, and a magnification of part of that image.

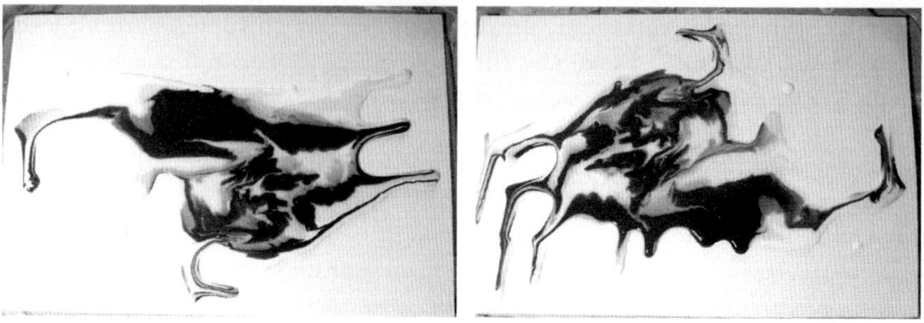

Figure 4.13: Second and third tilt for Sample A.

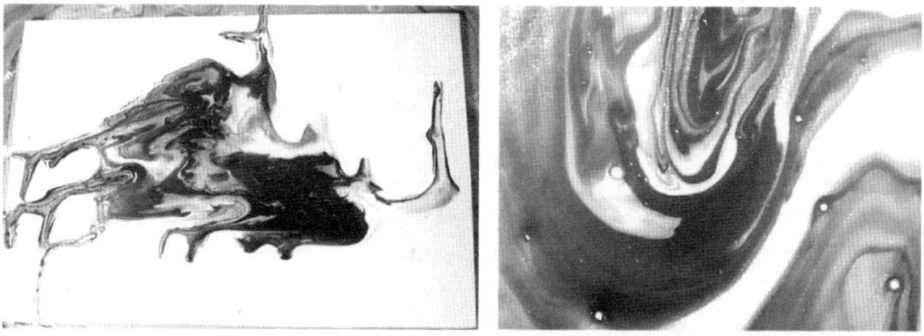

Figure 4.14: Eighth tilt and a magnification for Sample A.

We see that a small number of tilts suffice to produce complex folded structures on many levels. In principle, we imagine we could continue this mixing down to bands only a few dozen molecules wide. But the continued drying of the paint limits the number of tilts that produce visible changes in the paint pattern.

4.4.6 Sample B

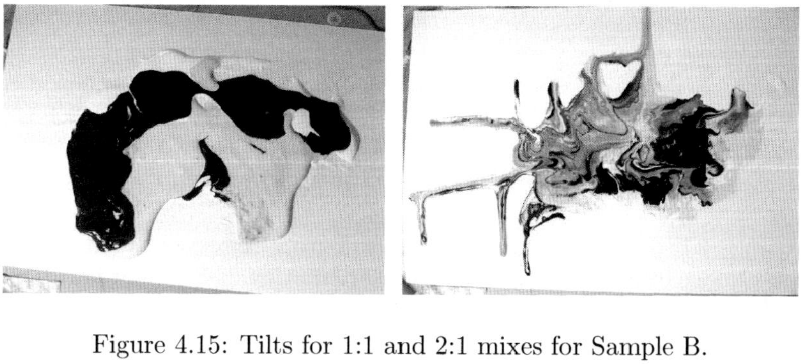

Figure 4.15: Tilts for 1:1 and 2:1 mixes for Sample B.

In Fig. 4.15 we see tilt flows for 1:1 (left) and 2:1 (right) mixtures of pouring medium and paint. The left mixture is so viscous that each tilt required between 10 and 15 minutes of flow to produce a noticeable change. By the end of the flow after the third tilt, the paint has dried so much that additional tilts produced no discernable difference.

For the right mixture the viscosity has been reduced adequately to allow several tilts before visible flow stops. The result of five tilts is seen.

In Fig. 4.16 we see the flow after the seventh tilt for an 8:1 mix of pouring medium and paint. While we see some fine structures, mostly what has happened is that the blue and yellow paint have mixed to form green. To get an interesting mix, we appear to need ratios greater than 1:1 and less than 8:1.

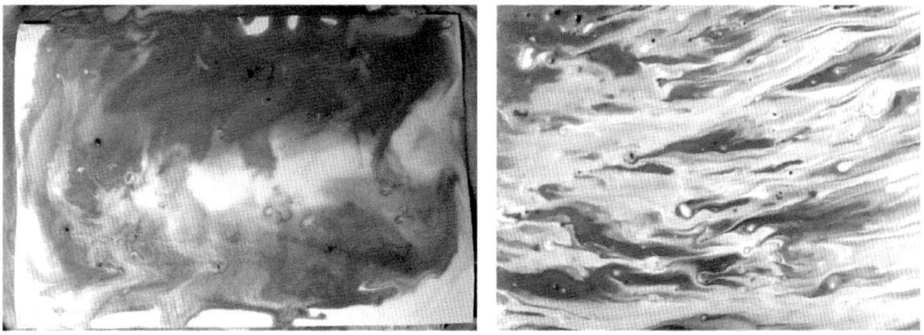

Figure 4.16: The seventh tilt for an 8:1 mix (left) and a magnification (right), for Sample B.

4.4.7 Conclusion

When the viscosity of acrylic paint is reduced by the addition of pouring medium, tilting the painted surface induces paint flows that mix bands of similar colors. A sequence of tilts in different directions can give rise to intricately interwoven bands. This offers many avenues for artistic exploration.

4.4.8 Exercises

Prob 4.4.1 (a) Prepare pouring medium and paint mixtures in the ratio 2:1. Tilt the table four times, rotating the tilt direction 90° between successive tilts.
(b) Repeat (a) for a 4:1 ratio
(c) Repeat (a) for a 6:1 ratio.
(d) Compare the patterns obtained after the final tilt of (a), (b), and (c). Do you notice significant differences?

Prob 4.4.2 Select two colors of paint and prepare pouring medium and paint mixtures in the ratio 2:1 for one color (this is the more dense mixture), 4:1 for the other color (the less dense mixture).
(a) Pour the less dense paint first, then the more dense paint on top of that. Tilt the table four times, rotating the tilt direction 90° between successive tilts.
(b) Pour the more dense paint first, then the less dense paint on top of that. Tilt the table four times, rotating the tilt direction 90° between successive tilts.

(c) Compare the patterns obtained after the final tilt of (a) and (b). Do you notice significant differences?

Prob 4.4.3 Repeat Exercise 4.4.2 for one color paint mixed with pouring medium in the ratio 2:1 (the 2 is the pouring medium volume), and the other color mixed in the ratio 6:1.

Prob 4.4.4 (a) Prepare two paint mixtures, both in the ratio 2:1. Pour part of one, part of another, more of the first, more of the second. Tilt the table four times with a 90° rotation of tilt direction between tilts.
(b) Prepare two paint mixtures, one in the ratio 2:1, the other in the ratio 4:1. Repeat the steps of part (a).
(c) Compare the results of (a) and (b).

Prob 4.4.5 (a) Prepare two paint mixtures, both in the ratio 4:1. Now change the tilt directions only along one axis: tilt to the left, then to the right, then to the left, etc.
(b) Describe how this result differs from that of Exercise 4.4.1 (b).

Prob 4.4.6 (a) Prepare two paint mixtures, both in the ratio 4:1. Toss two coins to randomize the tilt direction. For example, HH means tilt in the positive x-direction, HT in the positive y-direction, TH in the negative x-direction, and TT in the negative y-direction. Tilt the table four times.
(b) Compare this result with those of Exercises 4.4.1 (b) and 4.4.5 (a).

Prob 4.4.7 Adapt the method of Exercise 4.3.6 to look for (approximate, statistical) self-similarity in a paint flow pattern. What is a good candidate to replace the branch points of that exercise?

Prob 4.4.8 Prepare two paint mixtures, both in the ratio 4:1. When the paint is poured, push a stick through the paint in a complicated path. Then tilt the table several times. By comparison with previous exercises, how does this disturbance of the initial paint distribution alter the flow patterns?

Prob 4.4.9 Combine the technique of this Lab with that of Labs 4.1 and 4.2. That is, prepare two samples with the tilt table and photograph them. Before they have dried, push them together, wet surface in contact with wet surface, then pull them apart. Compare the resulting patterns with the photographs.

Prob 4.4.10 Make something pretty. Don't limit yourself to the variations we suggest. Use your imagination.

4.5 Fractal painting: dripping

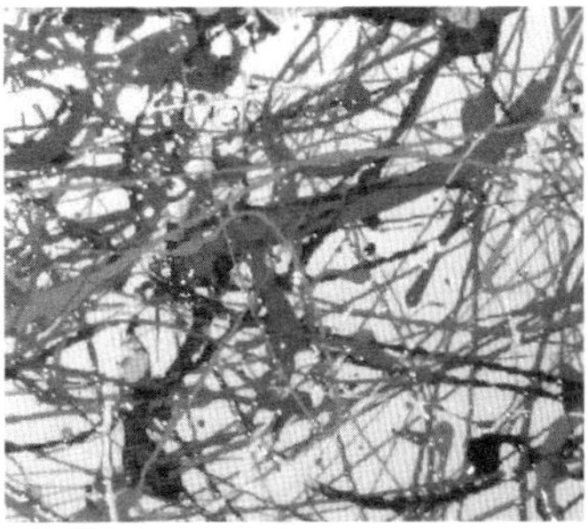

In his far too short life, Jackson Pollock (1912–1956) developed a powerful technique called *drip painting*. The picture here is not a Pollock, but a sample done by the one of us (NN) with more artistic talent. A good sketch of Pollock's life and work is B. H. Friedman's biography [123]. Rather than focus on the substantial artistic importance of Pollock's work, here we'll study some fractal aspects of his drip paintings.

In his *Scientific American* article "Order in Pollock's chaos," [129] Richard Taylor reports that he noticed general similarities between Pollock's drip paintings and automatic paintings he produced as tree branches blown by the wind moved cans of paint and dripped paint onto a canvas. Taylor wrote that Pollock "must have adopted nature's rhythms when he painted."

Here we'll investigate ways to generate our own drip paintings, not to try to mimic Pollock's masterpieces, but to seek a physical understanding of how movement on several scales can produce structure across scales.

4.5.1 Purpose

Inspired by Pollock's drip paintings we will describe how we went about making one of our own and how to illustrate the scale ambiguity and angle ambiguity of such a painting. We will also discuss what we think were the sources of some of Pollock's motivation.

4.5.2 Materials

You need a large floor area on which to spread out a protective coating. I (NN) worked in the basement and used an old shower curtain. You should use a good-sized canvas, not too small and not too big, unless you are very ambitious. I used one $18'' \times 24''$ bought at a local craft store. I used several sample size cans of latex paint in the bright colors of white, black, blue, yellow, and red. Many other colors can be gotten by combining some of these colors. Several wood stirring sticks and a digital camera. Caution: wear old clothes and shoes.

4.5.3 Background

Often in nature self-similarity provides evidence that the same forces have acted across many scales: viewed from nearby, little bits of coastlines, mountain ranges, clouds, and trees look like large bits of coastlines, mountain ranges, clouds, and trees viewed from far away. Much of nature exhibits scale ambiguity. On page 120 of [129], Taylor demonstrates scale ambiguity in Pollock's drip paintings. A natural question is to what extent Pollock was motivated by the appearance or the dynamics of nature. Two of Pollock's most famous quotes are "I am nature" (page 65 of [123]) and "My concern is with the rhythms of nature. I work inside out, like nature" (page 48 of [121]).

In pictures of bare trees in New Hampshire we'll find visual echoes of some of Pollock's drip paintings. Was he inspired by thickets of young trees? B. H. Friedman wrote, "Walking back to the house Jackson and Lee are struck once again by the look of bare trees traced against the winter sky" (page 97 of [123]) and in an interview with Lee Krasner he stated, "Several writers have connected some black-and-white paintings—and some colored ones too—with the feel of the East Hampton landscape, particularly in winter: the look of bare trees against the sky and flat land moving out toward sea."

Pollock did not often write about his art, but much that he wrote is memorable. For example (page 140 of [123]) Friedman recounts, "When he [Pollock] described his method to them[Dr. and Mrs. Seixas], the wife asked, 'But Mr. Pollock, how do you know when you're finished?' He replies, 'How do you know when you're finished making love?' "

The view some had is that randomness was the real artist in Pollock's paintings. To this, Pollock countered, "I can control the flow of paint; there is no accident" (page 54 of [122]).

In the Appendix B.12 we present some of the details of Taylor's work to apply box-counting to quantify the fractality of Pollock's drip paintings. We should note that this approach has some critics.

We'll approach fractality visually, through scale ambiguity.

In [138] Taylor observes another geometric aspect of Pollock's paintings. Referring to *Blue Poles*, Taylor writes that "The painting doesn't have an up or a down, a left or right, or even a center of focus!" That is, some of Pollock's paintings exhibit ambiguity of view angle and viewpoint. Viewpoint ambiguity means any point in the picture can be used as a center for scaling and view angle, and the scaling and view angle will be ambiguous. (All sections of the painting are scale and view angle ambiguous.) Scale, view angle, and viewpoint ambiguity we'll group together and refer to collectively as *visual ambiguity*. We'll investigate examples of this, too.

4.5.4 Procedure

Lay out the protective covering on the floor in a well-lit area with sufficient space for you to move around. Place the canvas near the center of the covering. Mix any variations of colors you may want in addition to the purchased colors. You may not want to use all of the colors listed above and that is fine. Pick a color and a paint stick and holding the paint can in one hand and the stick in the other, dip the stick into the paint and proceed to drip paint on the canvas following the method described below. If the procedure is difficult to follow, it is easy to find internet videos of Pollock painting.

All the while you are dripping paint on the canvas keep in mind what you are trying to achieve. You want to cover the canvas with paint so that all areas are equally complex and all areas have approximately the same distribution of thin, medium, and thick lines. The number of thick, medium, and thin lines do not have to be the same but their distribution over the whole canvas should be the same so that any one part of the canvas is equally as complex as any other part no matter the size of the regions. This is going to necessitate that the final painting is very dense in lines.

The way you apply the paint is important. Your shoulder, arm, and wrist movement should be loose and flowing as if you were painting in the air above the canvas. Imagine making flowing infinity symbols in the air or perhaps waving a baton at the canvas to conduct a symphony orchestra. Don't make straight lines, but curves and arcs. Go off the canvas onto the floor covering with the paint. Don't stand in one place, but walk around the canvas. I found I was not coordinated enough to paint and step around as I painted so I would paint in one spot, step to another place and paint, etc. The height you hold the stick above the canvas and the amount of paint on the stick determines how thick the paint drips are. Experiment with different heights and amounts of paint. If you snap your wrist you can produce splatters of paint to go with the curves.

Once you have laid down some paint with one color switch to another color and repeat the process above. I did not let the paint dry between layers of paint and found that there was minimal bleeding between colors. What little bleeding there was I liked. Do not feel that you are finished with a color after you have done a layer with it. You can apply multiple layers of the same color. I used between eight and ten layers of paint on my canvases.

Look at what you have. See if you like it. If not you can always add more layers of paint at any time even after the original paint has dried. Keep in mind that as you add more layers of paint not only are you adding structure but you are also subtracting by obscuring some of the previous parts. Drip painting is both an additive and subtractive process.

4.5.5 Sample A

In Fig. 4.17 are three sections from a painting. Without looking at the original painting can you arrange them from the largest section to the smallest section? For each section can you determine if the camera was held horizontally, vertically, or at an angle?

Figure 4.17: Left to right: section # 1, # 2, and # 3.

The original painting with the three sections indicated by rectangles is in Fig. 4.18. How did you do?

The section's size and angle should be obvious. Section # 1 is the largest, shot at an angle, # 2 is the smallest, shot horizontally, and #3 is between the largest and smallest shot vertically. The camera picture size is the same for all the photos and as a result the camera automatically scales smaller regions up and larger regions down. This scaling becomes obvious in the photos because

Figure 4.18: The original painting, with the sections of Fig. 4.17 outlined.

we are looking at objects we are familiar with and can determine their sizes relative to each other. Angles are also easy to determine because the images are of familiar objects whose orientation to each other is easy to see. Paintings like this are not scale or angle ambiguous, nor are they intended to be.

4.5.6 Sample B

Figure 4.19: A drip painting.

In Fig. 4.19 we see an example of a drip painting made by the method described in Sect. 4.5.4. Your results may vary.

4.5.7 Conclusion

Using Pollock's technique, drip paintings can be made that are visually ambiguous, but not all drip paintings are visually ambiguous. You should have a plan and work at it. Any self-similar fractal is scale ambiguous, but not all scale ambiguous objects are fractal. (Imagine a flat, featureless plane, a plain plane. Are you looking at a little bit from close by or a large area from far away? A plane is scale ambiguous, but no one would call a plane fractal.) Pollock surely did not know about fractals and not formally about scale, angle, and viewpoint ambiguity, but did he have a plan from nature that he employed to produce pictures that are visually ambiguous. "My paintings do not have a center," he says, "but depend on the same amount of interest throughout." [124] "There's no central point of focus, no hierarchy of elements in this allover composition in which every bit of the surface is equally significant." [125]

In [126] Taylor wrote

> So, when you stare out at nature, you are staring at repeating fractal patterns. The same is true of Pollock paintings. Art critic Clement Greenberg acknowledged this repetition, observing that the artist's drips and splashes are 'knit together of a multiplicity of identical or similar elements.' This repetition has important visual consequences. In particular, if patterns at different magnifications look similar to each other, it becomes very hard to tell which magnification you are looking at. Because of this, you can't tell if you are looking at a Pollock from the back of the gallery or up close. As journalist Alfred Frankenstein noted: 'Pollock is as strong from a distance as he is close to.'

Such descriptions lead us to believe he was trying to achieve what we call visual ambiguity. Where did he get his ideas? This we can never be certain of but there are some clues given to us by people who knew him. J. Potter was a neighbor of Pollock and got to know him well. "The art world changed forever in 1945, the year that Jackson Pollock moved from downtown Manhattan to Springs, a quiet country town at the tip of Long Island, New York. Friends recall the many hours that Pollock spent on the back porch of his new house, staring out at the countryside as if assimilating the natural shapes surrounding him." [127] What inspired him in nature? Here is a photo of dead pine tree branches taken while on a walk through Bedrock Gardens in New Hampshire.

What did Pollock see? Maybe more than we have thought. Like his art or not, he was unique and made a lasting impression on the art world. I don't think he was trying to copy nature but rather extend it. Pollock was not just throwing paint on a canvas. Consciously or unconsciously, he had a plan to produce paintings inspired by nature.

4.5.8 Exercises

Prob 4.5.1 In Fig. 4.20 are photos of four sections from a drip painting that were all taken with the camera held horizontally illustrating only scale ambiguity. See if you can arrange them in order from what you think is the largest section of the painting to what you think is the smallest section.

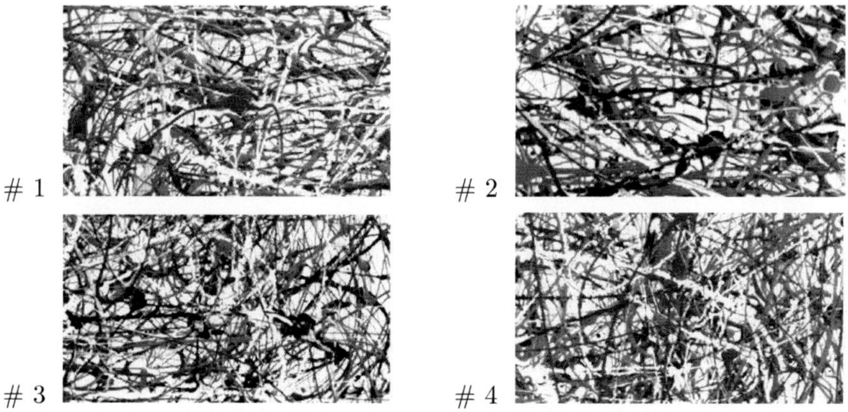

Figure 4.20: Images for Exercise 4.5.1.

Prob 4.5.2 Make a drip painting of your own and photograph different sized sections of it. When you photograph the different sized sections take care to know which photos are of which sections because if your painting is successfully scale ambiguous and you do not do this you will never identify the sections of the painting your photos came from. Does your panting support scale ambiguity? If not can you add more layers of paint to make it scale ambiguous?

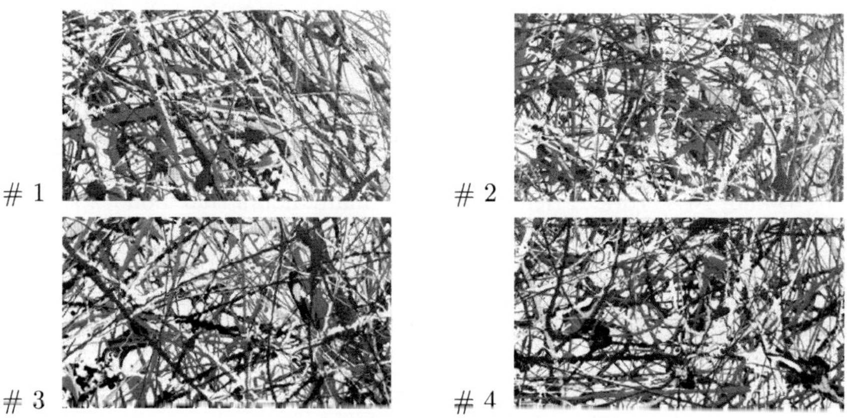

Figure 4.21: Images for Exercise 4.5.3.

Prob 4.5.3 The photos in Fig. 4.21 represent different sized regions of a drip painting. In this case the camera may have been held horizontally, vertically, or at some other angle to illustrate angle ambiguity. Can you identify the largest region and the smallest region and also identify each photo as having been taken horizontally, vertically, or at some other angle?

Prob 4.5.4 Go outside and look at some things in nature. Can you find examples of things in nature that might have inspired Pollock? Tree branches silhouetted against the sky are a good bet. Take some photos and see if they look close to what Pollock did. You might want to enhance the photos on a computer if you have the capability.

4.6 Fractal paper folds

Paper folding originated centuries ago, probably independently in Asia and in Europe. The Japanese senbazuru, "1000 cranes", is practiced around the world. Folding a thousand cranes is a wish for health, a wish for good fortune, a memorial for a devastating loss. Delicate paper cranes are left exposed to the elements and release their wish as weather destroys them. The story of Sadako Sasaki did much to popularize this practice.

In addition to cranes, other animals, human faces, and abstract geometrical shapes are folded. In origami paper is folded; in kirigami paper is folded and cut. Kirigami can be used to make remarkable pop-up cards. Little surprise, then, that eventually fractal kirigami appeared. We'll use these to add another level of intuition of physical scaling. And we'll see examples of fractal pop-up cards.

4.6.1 Purpose

Here we'll make paper sculptures of shapes that approximate fractals in the sense that their construction follows physical implementations of several iterations of processes that generate mathematical fractals. These constructions are cuts and folds of paper. People familiar with origami know that the thickness of paper limits the number of folds upon folds, so our paper sculptures will not exhibit similar shapes over a two-decade scaling range. Rather, these are meant to provide insight into the physicality of scaling.

4.6.2 Materials

Origami paper, scissors or an x-acto knife, ruler, and a pencil. For larger projects, and those that do not require so many successive folds, printer paper or construction paper can be used.

4.6.3 Background

The paper models made in this lab have been inspired by the book *Fractal Cuts: Exploring the Magic of Fractals with Pop-Up Designs* by Diego Uribe [139] and the article [140] by Elaine Simmt and Brent Davis. In addition, appropriate cuts of folded paper invite exercise of our imaginations.

Background research revealed several interesting results. One was Britney Gallivan's formula for the length of paper needed to fold the paper in half n times. We discuss this in Sample 4.6.6. Another was the *fold and cut theorem*: Any shape with straight edges can be made from a single (perhaps extraordinarily thin) sheet of paper by folding it and making a single straight cut. This result was proved in [141]; some history is provided in [142] and [143]. We must mention that examples of single cut shapes were published by the magician Harry Houdini [144].

To prepare for some of the computational exercises, and to reinforce the point, made in Sample 2.1.6, about what happens when we measure a shape in the wrong dimension, we'll compute the area and the length of a Sierpinski gasket. We'll use the equilateral Sierpinski gasket. The first few steps of its construction are sketched in Fig. 4.22.

Figure 4.22: The first few stages of the equilateral gasket construction.

The equilateral triangle that is the left image of Fig. 4.22 has side length 1, so base 1, altitude $\sqrt{3}/2$, and area $A_0 = \sqrt{3}/4$.

In the first stage of the construction of the gasket, the second image of Fig. 4.22, we remove a triangle of 1/2 the height and base of the original, hence of area $(1/4)A_0$.

In the second stage (third image of Fig. 4.22) we remove 3 triangles of area $(1/4)^2 A_0$. In the third stage we remove $9 = 3^2$ triangles of area $(1/4)^3 A_0$. Continue in this way. We see the areas of the removed triangles sum to

$$A_0\left(\frac{1}{4} + \frac{3}{4^2} + \frac{3^2}{4^3} + \cdots\right) = \frac{A_0}{4}\left(1 + \frac{3}{4} + \frac{2^2}{3^2} + \cdots\right) = \frac{A_0}{4}\frac{1}{1 - 3/4} = A_0$$

Consequently, the area that remains, the area of the gasket, is 0.

Next, to compute the length of the gasket, we'll find the sum of the perimeters of the orignal triangle and of all the triangles removed. Because we remove *open* triangles, each of these perimeters is part of the gasket. The perimeter of the original equilateral triangle is $P_0 - 3$. The first removed triangle has sides of length 1/2, so has perimeter $(1/2)P_0$. Each of the three second removed triangles has perimeter $(1/4)P_0$. Because we remove three of these, the second stage contributes length $(3/4)P_0 = (3/2^2)P_0$. Continue this way and we find that the length of the gasket is at least

$$P_0 + \frac{1}{2}P_0 + \frac{3}{2^2}P_0 + \frac{3^2}{2^3}P_0 + \cdots = P_0 + \frac{P_0}{2}\left(1 + \frac{3}{2} + \frac{3^2}{2^2} + \cdots\right)$$

That is, the gasket has infinite length. Yet again, we see that when measured in a dimension too low (length is a 1-dimensional measure and $1 < \log(3)/\log(2)$, the dimension of the gasket) the result is ∞. And when measured in a dimension too high (area is a 2-dimensional measure and $2 > \log(3)/\log(2)$) the result is 0.

4.6.4 Procedure

The mantra for this lab is "Careful measurement, straight lines, clean cuts, crisp folds." Specifics are determined by the particular construction. Samples 4.6.5 and 4.6.6 provide ample illustration of the procedure.

4.6.5 Sample A

Here we'll build a representation of the third stage of the construction of the Cantor middle-thirds set. The process consists of three successive folds and cuts, then refold in the opposite direction, and finally reverse two levels of the folds.

First fold and cut. Begin with a piece of origami paper. Fold the sheet in half, folding parallel to the long side if the sheet isn't square. Mark the folded sheet into three equal sections along the crease and at these marks make two cuts halfway to the opposite edge. In Fig. 4.23 we see a photograph, and a line drawing to clarify the positions of the cuts.

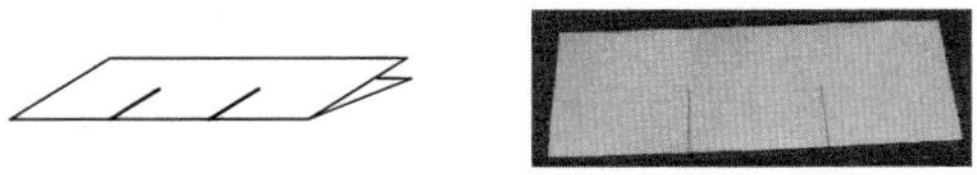

Figure 4.23: First fold and cut.

Second fold. Now fold over the two end pieces made by the cuts and leave the middle piece horizontal. Fold against a ruler to guarantee a straight crease.

Fold both ways several times to insure a good crease and then unfold so the paper sheet is flat. See Fig. 4.24. In the top line drawing on the left of the figure, the dashes represent the folds we made.

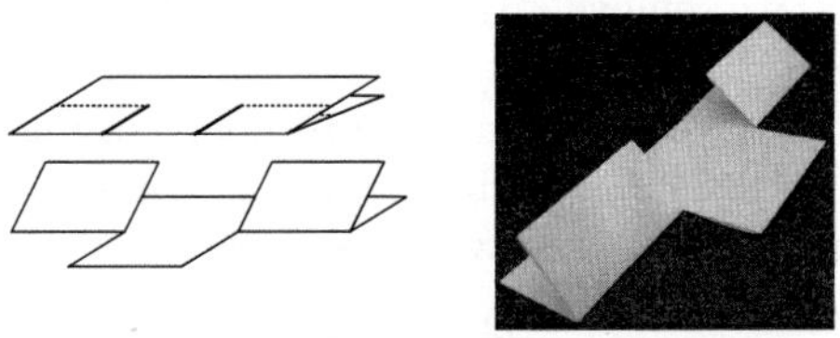

Figure 4.24: Second fold.

Second cut and third fold. Divide each of the two end pieces into thirds along the crease of the large sheet, and make cuts at these marks halfway to the newly formed folds.

Fold the two middle pieces, again bending both ways several times, and unfold. See Fig. 4.25.

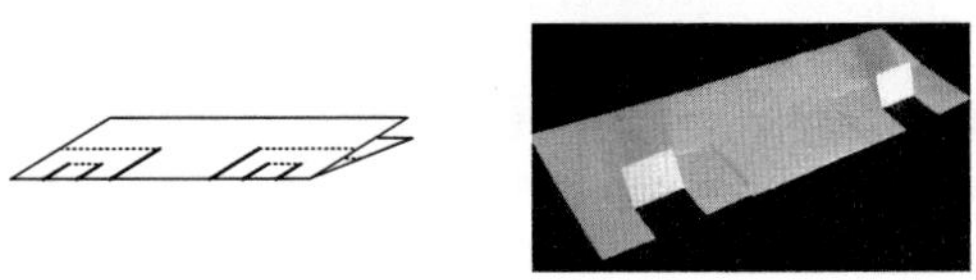

Figure 4.25: Second cut and third fold.

Last cut and fold. Continue this process one more time, to obtain four new folded middle pieces. See figure Fig. 4.26. We'll stop the cuts and folds here, but of course more levels better illustrate scaling and fractality.

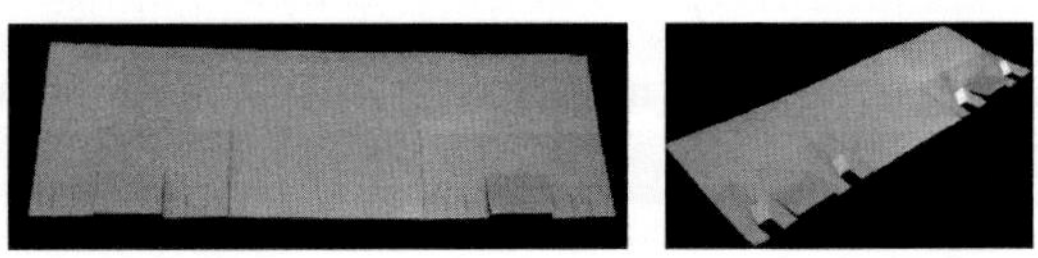

Figure 4.26: Last cut and fold.

Fold in the opposite direction. Without changing the large middle section, fold the two largest outside pieces in, the opposite of their current position, so that they point in between the halves of the large sheet. This is illustrated in Fig. 4.27.

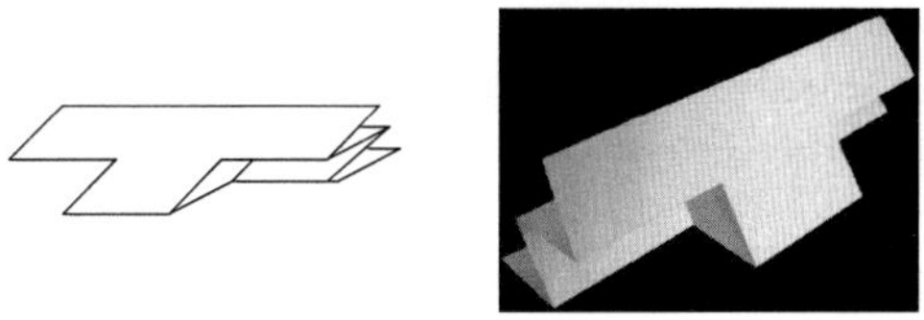

Figure 4.27: Fold in the opposite direction.

Now open up the large sheet and turn it as illustrated in Fig. 4.28. This leaves the two large end pieces reverse folded. These two large end pieces represent the two intervals that are the first stage in the construction of the Cantor middle-thirds set. The middle piece folded in the other direction represents the removed middle interval in this first stage.

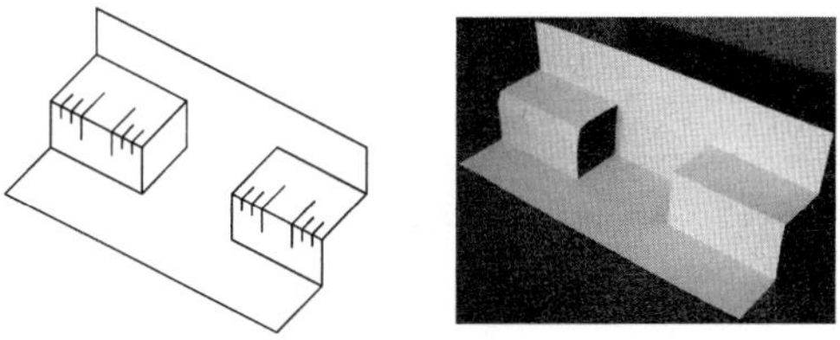

Figure 4.28: The first reverse fold.

The last two reverse folds. Finally, push in each of the six middle sections, to reverse the direction of their folds. The two larger folds are reversed on the left side of Fig. 4.29, the four smaller on the right side. The right side of Fig. 4.29 is the completed first three stages of a paperfolding representation of the Cantor middle thirds set.

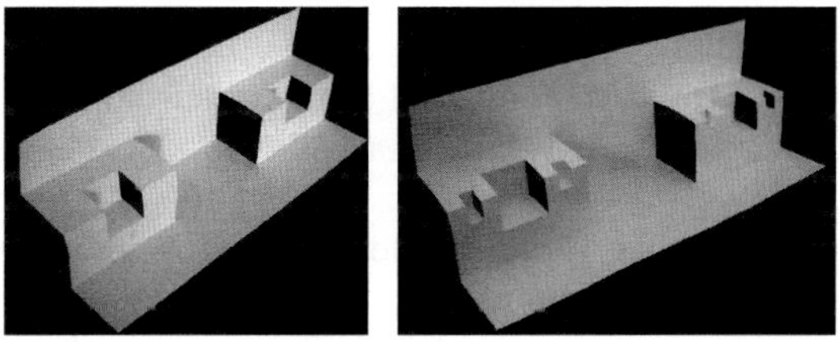

Figure 4.29: The last two reverse folds.

4.6.6 Sample B

Here we'll build a model of the fourth stage of the construction of a Sierpinski gasket.

First cut and fold. Start with a piece of paper with length twice its height. If you use square origami paper, tape two pieces together by placing them flat side-by-side and taping across that edge. In any case, fold the paper perpendicularly to the long side. This produces a square piece

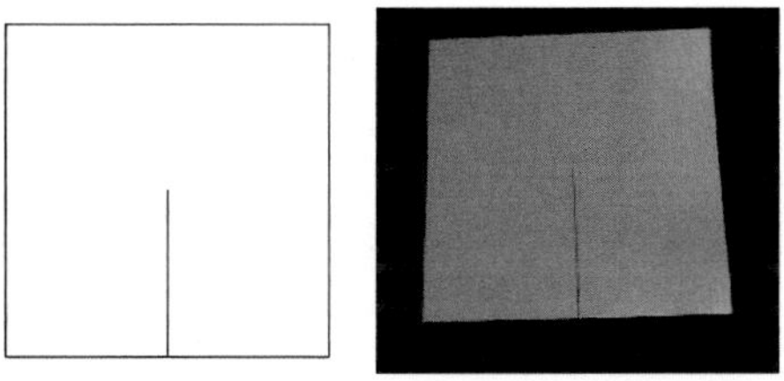

Figure 4.30: Folded sheet and first cut.

of (folded) paper.

Mark the middle of the crease and make a cut from the crease half way to the opposite side. See Fig. 4.30.

Fold the piece immediately to the left of the first cut to bring the crease even with the opposite side. See Fig. 4.31. Fold this piece back and forth a few times to make sure the fold moves easily in both directions.

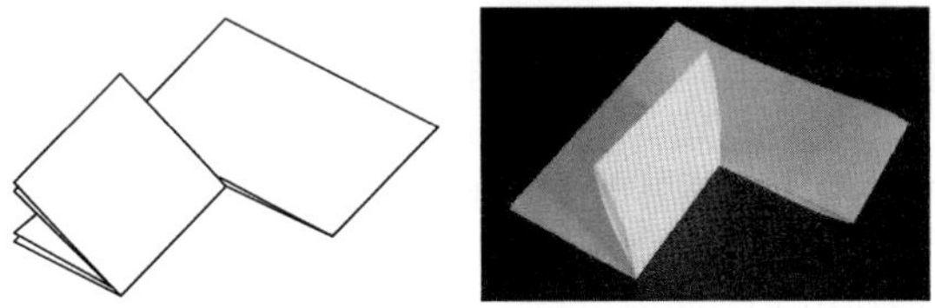

Figure 4.31: The first fold.

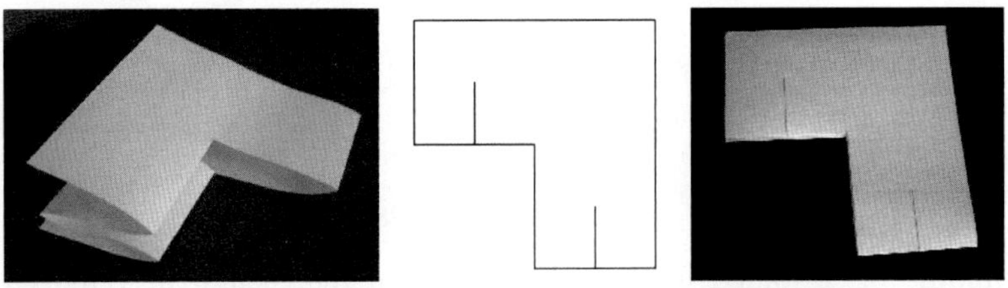

Figure 4.32: First fold refolded and the second cuts.

First fold refolded and the second cuts. Refold inward so the piece on the left goes between the halves of the large sheet. See the left image of Fig. 4.32. Now mark the middle of each half on either side of the first cut and make a cut half the length of the first cut. See the middle and right images of Fig. 4.32.

With each successive fold, the next cut is through twice as many layers of paper. Paper folding folklore holds that the maximum number of times a piece of paper can be folded in half is seven. However, in 2002 Britney Gallivan [145] demonstrated twelve successive folds, with a strip of toilet paper 1.2 km long. In addition to this remarkable physical demonstration, which involved walking about 3.6 km, she derived a formula $L = ((\pi t)/6)\big(2^n + 4\big)\big(2^n - 1\big)$ for the length L of paper needed to make n folds (all in the same direction) of a paper of thickness t.

We do need paper stronger than toilet paper, and we don't want to use a kilometer of origami paper, so we'll stick to a maximum of four successive folds.

Second fold refolded and the third cuts. Now we have four pieces. Fold pieces one and three, counting from the left, to the ends of the cuts. Then fold these pieces inward as before. See Fig. 4.33.

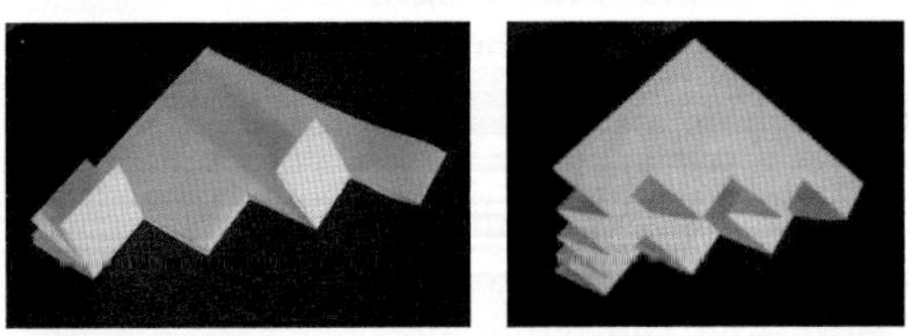

Figure 4.33: The second cuts folded and refolded.

Mark the middle of each of the horizontal folds and cut half the length of the previous cuts. Fold and refold inward as before. See Fig. 4.34. These will be the third fold. Next we will do one more fold, the last step to build our model of a Sierpinski gasket.

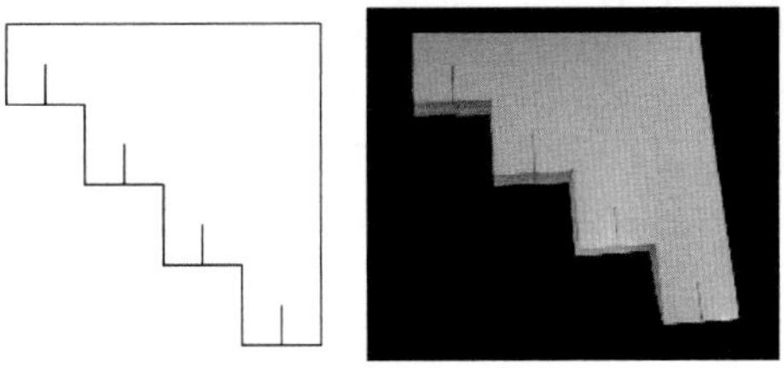

Figure 4.34: The third cuts.

Third fold refolded and the fourth cuts. In Fig. 4.35 From the folklore about paper folds, we could do three more, for a total of 7 (12 if we fold toilet paper). You are welcome to try another one or two if you wish, but four gives a reasonable model, and crisp folds after four require some special efforts.

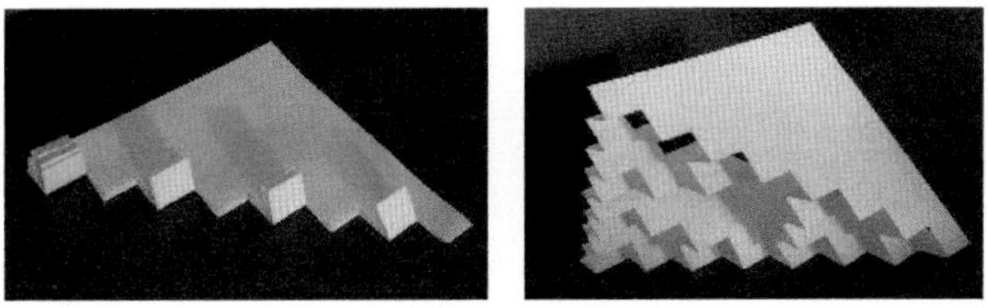

Figure 4.35: The third cuts folded and refolded inward.

Repeat the entire process one last time. Mark centers, cut half the length of the previous cuts, fold the left side of each cut and refold inward. See Fig. 4.36.

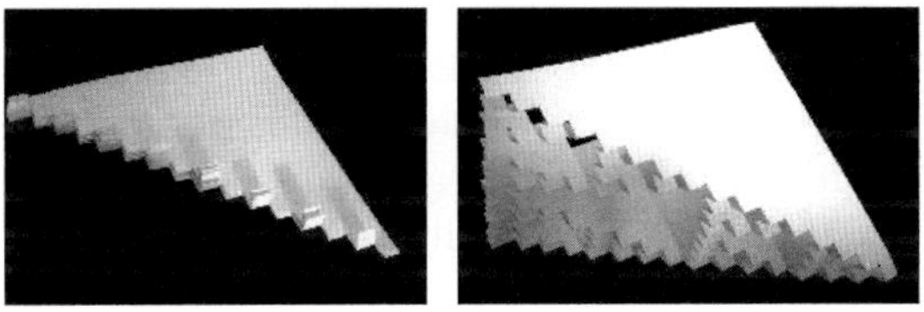

Figure 4.36: The fourth cuts folded and refolded inward.

In all these steps, the final model is improved by more accurate measurements, crisper folds, and cleaner cuts. After several folds, scissors may produce some sliding during the cuts. An x-acto knife may be a better choice.

Now all that remains is to open the two sides out to a right angle, as in Fig. 4.37. To see how this suggests a gasket, you must pick a good view point. A bit above and off-center seems to work well.

To make more visible the gaps formed by the cuts and folds, which correspond to the holes in the gasket, you might make a backing with paper of a contrasting color. If you do this, you've made a Sierpinski gasket pop-up card.

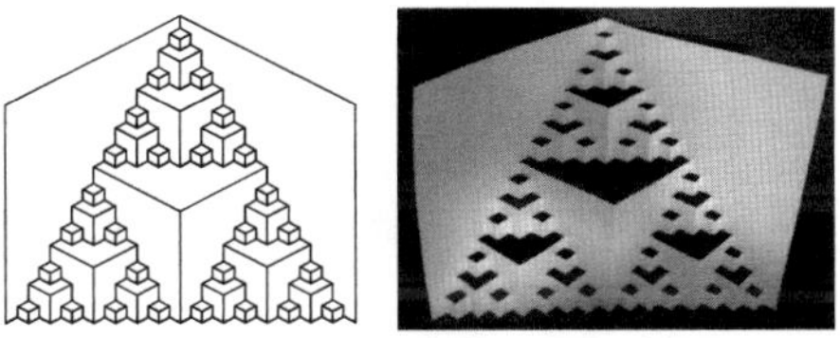

Figure 4.37: Gasket model, schematic and actual.

4.6.7 Gallery of examples

In Fig. 4.38 we see some images of the fractal pop-up cards made by Tova Feldmanstern and Jennifer Michelstein as part of their fractal geometry project at Yale. Some of these were inspired by designs in [139]. In addition to the construction of these models, their project included calculations of the type outlined in Exercises 4.6.1, 4.6.2, and 4.6.3.

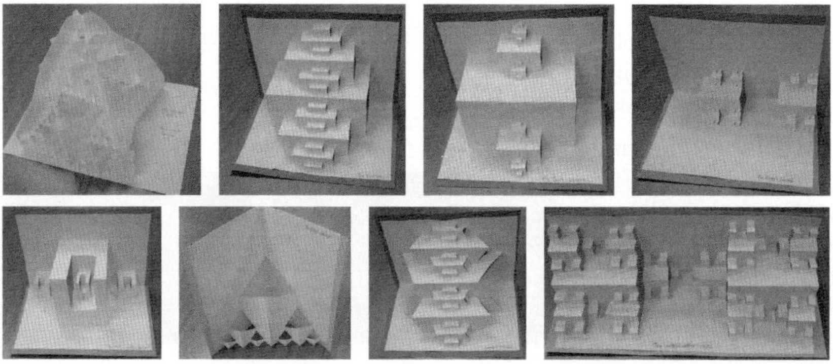

Figure 4.38: Some fractal paper folds by Tova Feldmanstern and Jennifer Michelstein.

In Fig. 4.39 we see some images of fractal paper-cut models made by Stephen Smith's Enfield High School math class. Classes at Enfield High School participate in a holiday classroom door decoration contest. Stephen's students covered their classroom door with fractal cut-outs and won the first prize, bagels for the whole class, an unexpectedly tasty benefit of the study of fractal geometry.

On the right we see a picture of the award-winning classroom door of Stephen Smith's class. Fractal decorations on the tree, fractal stars in the sky.

Figure 4.39: Some fractal paper folds by Stephen Smith's class.

4.6.8 Conclusion

Only our imaginations, and Gallivan's calculation, limit the variety of sculptures that can be made in this fashion. Some represent fractals, others do not. Mathematical questions can be asked about some; all can be enjoyed visually.

4.6.9 Exercises

These questions can be asked for the limit of the construction for the Cantor set of Sample 4.6.5 and of the Sierpinsi gasket of Sample 4.6.6. Solutions are given for the Sierpinski gasket.

Prob 4.6.1 (a) Compute the lengths of the cuts, assuming the process is continued for infinitely many steps to produce the limiting shape.
(b) Compute the length of the edges of the limiting shape. (Both cuts and folds create edges.)

Prob 4.6.2 Folding creates families of triangles in the limiting shape. The left image of Fig. 4.40 shows the three largest triangles, and the three groups of three next-to-the-largest triangles. Find the sum of the areas of these triangles.

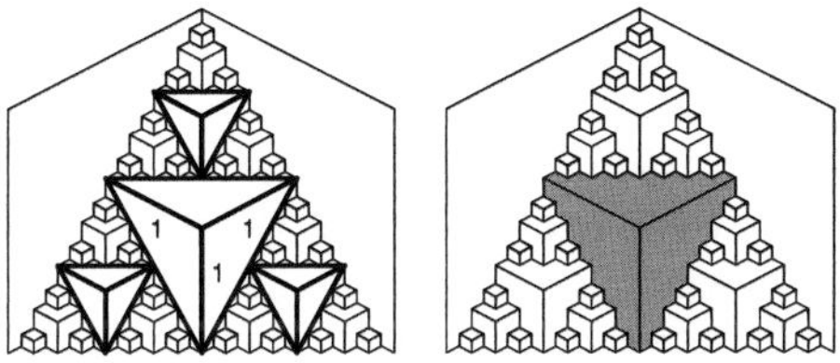

Figure 4.40: Sketches for 4.6.2 and 4.6.3.

Prob 4.6.3 If we put a back on this gasket model to make a pop-up card, we can view each fold as forming the two front faces of a cube. The back forms the two back faces. Suppose we add the remaining cube faces, top and bottom, of each cube. The part of the largest cube that is visible from outside the shape is shaded in the right image of Fig. 4.40. Find the volume of this collection of cubes.

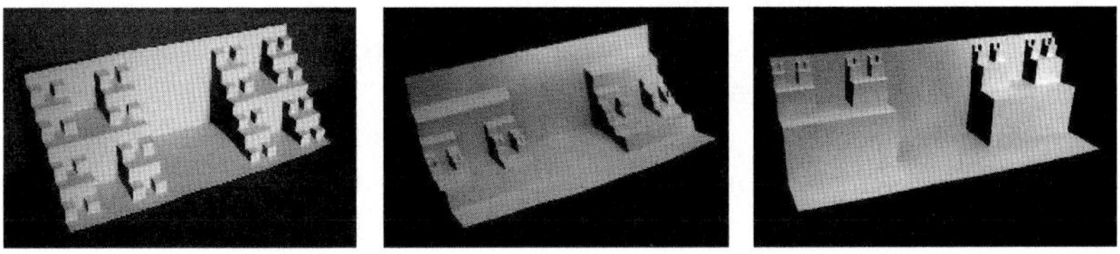

Figure 4.41: Three Cantor set variations.

Prob 4.6.4 In Fig. 4.41 we see three variations of the Cantor set representation. Find a construction procedure for each example.

Prob 4.6.5 Make something beautiful.

4.7 A closer look at leaves

Trees and their leaves have been around for a long time, about 385 million years, but how many of us have looked really closely at the patterns in leaves. If you are like me (NN) you may have noticed the outline shape of leaves and the pretty colors some of them make, especially in the fall, but have not examined their structure up close. An easy way to do this is to take a leaf, find a clean window with the sun shining through it and press the leaf up to the window. See Fig. 4.42. This should clearly show the veins in the leaf. The patterns of the veins in a leaf vary greatly from plant to plant. We will investigate a particular vein pattern and see if fractal geometry can help us understand how to make sense of it.

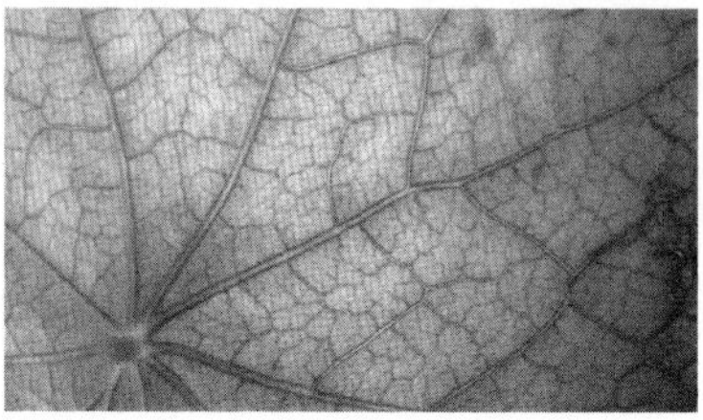

Figure 4.42: Vein patterns in a nasturtium.

4.7.1 Purpose

In this lab we will investigate the structure of leaves with reticulate venation (defined in Sect. 4.7.3) for the purpose of finding a geometrical pattern for the venation using ideas from fractal geometry.

4.7.2 Materials

Several leaves from dicotyledonous Angiosperms that exhibit reticulated venation. Some suggestions are: dahlia, foxglove, lantana, nasturtium, ox eye sunflower, and rhododendron. Availability will vary depending on where you live. A magnifying glass is useful. If you do not intend to take photos, a clean sunny window will do; if you want photos then you will need the following: two $8'' \times 10''$ panes of clear glass or Plexiglas, something to raise the glass such as two $2'' \times 4''$ pieces of wood, some light source to slide under the glass (such as a press on stick up LED light), a digital camera capable of macro photography, and a white piece of paper to put under the light source. Fig. 4.43 shows the set-up used.

Figure 4.43: A simple light box.

4.7.3 Background

The name given to the arrangement of veins in a leaf is *venation* [146, 147]. We are interested in a type of venation called reticulate venation that occurs when the veins form a net-like pattern. This occurs in dicotyledonous Angiosperms, flowering plants whose embryos have two seed leaves. Leaf veins consist of xylem, which carries water to the leaf, and phloem which carries sugars from the leaf to the rest of the plant. The xylem and phloem always occur together to make up the vein no matter how large or small the vein is. To be effective the veins need to be close to the cells of the leaf. Reticulate venation employs veins of at least three orders of size that branch out to form a network that covers the body of the leaf. There are large primary veins (there may be only one) from which branch several smaller secondary veins that in turn have even smaller third order veins and perhaps some other higher order veins that make a dense net-like pattern.

The reticulate pattern of venation has sometimes been called random and indeed it does look that way at first glance. But perhaps if we look at it with eyes accustomed to fractal patterns we can see something more than randomness. In Fig. 4.44 we see a venation pattern that shows the primary vein, some secondary veins and many third order veins. The pattern may not form a true fractal but there might be a hierarchical distribution of regions similar to the distribution of different sized regions in fractals. Let us see.

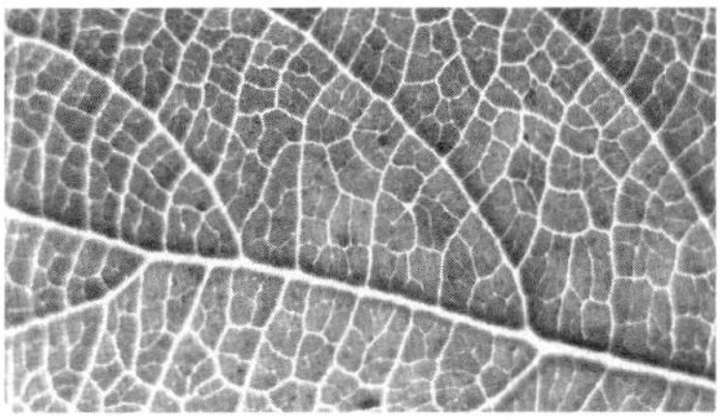

Figure 4.44: Reticulate venation in a lantana leaf.

4.7.4 Procedure

First collect some leaves that have reticulate venation. Next you want to observe the venation of the leaves. You can see a lot with your eyes and a magnifying glass. To get a good look at finer detail using a sunny window helps a lot. By taking backlit photos you will be able to see a lot more detail and the photos are very impressive. The apparatus described in the materials section was used with a hand held camera. If the leaf does not flatten out, a second pane of glass placed on top may help flatten it. A tripod was not used but the camera was simply moved in and out from the leaf. This produced very good in-focus pictures.

4.7.5 Sample

How are we to make any sense of the venation of the foxglove leaf of Fig. 4.45? This leaf is approximately 8″ long by 3.5″ wide and shows reticular venation. Even from a distance the veins are seen to have varying diameters. This suggests that we could use veins of the same diameter to serve as boundaries for regions of the leaf. We will proceed in stages from the largest vein to the smallest to look at the regions these veins bound.

Figure 4.45: A foxglove leaf.

As shown on the left side of Fig. 4.46, in stage 1 the primary vein divides

Figure 4.46: Left: stage 1, the primary vein. Right: stage 2, secondary veins in the bottom half of the leaf.

the leaf into two sections, a top and bottom. Secondary veins branch directly off the primary vein and run to the edge of this leaf and further divide the leaf into sections. The right side of Fig. 4.46 shows secondary veins coming off the primary vein for the bottom part of the leaf. You can also see some secondary veins for the top part of the leaf.

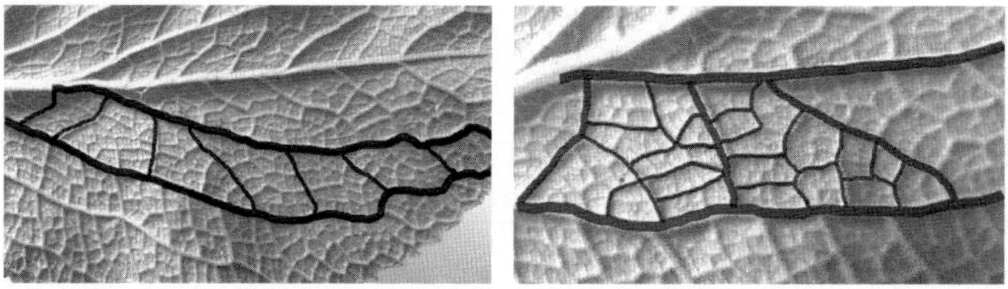

Figure 4.47: Left: some stage 3 veins going between two stage 2 veins. Right: some stage 4 veins going between stage 3 and 4 veins.

Next we zoom in and consider some stage 3 veins running between two stage 2 veins. See the left side of Fig. 4.47. Going in even closer, on the right side of Fig. 4.47 we now find some smaller stage 4 veins between stage 3 veins. Now a magnifying glass could be very helpful because it is difficult to distinguish the different sized veins at this stage.

There are some smaller stage 5 veins visible in the right side of Fig. 4.47, but they are difficult to see. At this point holding the leaf against a sunny window should clearly show more detail. If you want photos you should use the apparatus described in the materials section to backlight the leaf, as shown in Fig. 4.48.

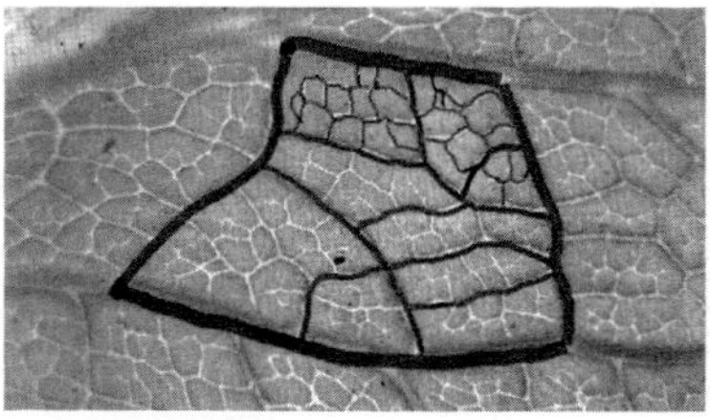

Figure 4.48: Some stage 5 veins.

4.7.6 Conclusion

The circulatory and pulmonary networks of mammals are visibly fractal because both consist of iterated branching. Leaves with reticulate venation exhibit not only iterated branching, but also a complex network of interconnections. Any fractal characteristics leaves might possess are better hidden that those of mammal lungs. Of course, plants evolved earlier than mammals, so perhaps when plants appeared evolution had not yet discovered the efficiencies of fractal geometry.

But leaf veins do need to get close to the cells they serve and to do so they must get smaller and smaller in diameter and more numerous. In doing this they form a complicated pattern resembling space filling curves. In contrast to the pattern of squares on a checkerboard the distinct regions of the leaf are bounded by veins of different diameters. This creates a hierarchical geometric system with the number of regions increasing as the diameter of the veins decreases. Fractals such as the Sierpinski gasket also do this with the number of triangles increasing as the length of the sides of the triangles decreases. Being familiar with the structure of fractals helps to see the hierarchical growth pattern of the leaves.

4.7.7 Exercises

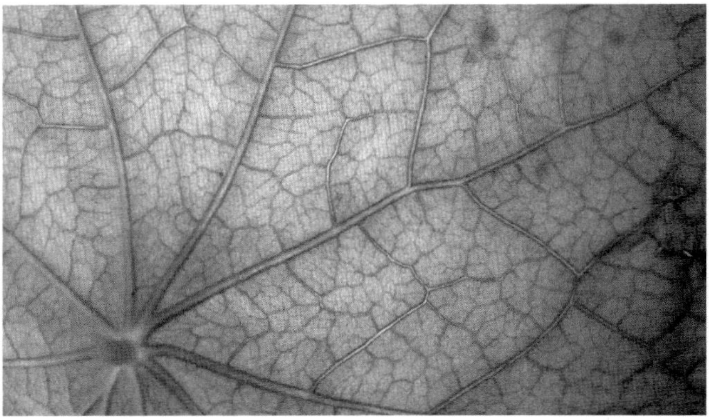

Figure 4.49: The nasturtium leaf image for Exercise 4.7.1.

Prob 4.7.1 (a) How many primary veins are there in the nasturtium leaf shown in Fig. 4.49?
(b) Find some secondary veins.
(c) Find some third order veins. The diameters of these veins may decrease as their lengths increase.

Prob 4.7.2 How many orders of veins can you find for the ox eye sunflower leaf shown in Fig. 4.50? Viewing the picture from a distance rather than close up may help distinguish the smaller veins.

Prob 4.7.3 Using the rhododendron leaf pictured in Figs. 4.51 and 4.52, identify some different sized veins and the regions they bound. Once again the diameter of the veins may decrease as the length of the veins increase and the different sized veins may best be distinguished at a distance from the picture. Note how the number of regions bounded by the veins increases as the diameter of the veins decreases.

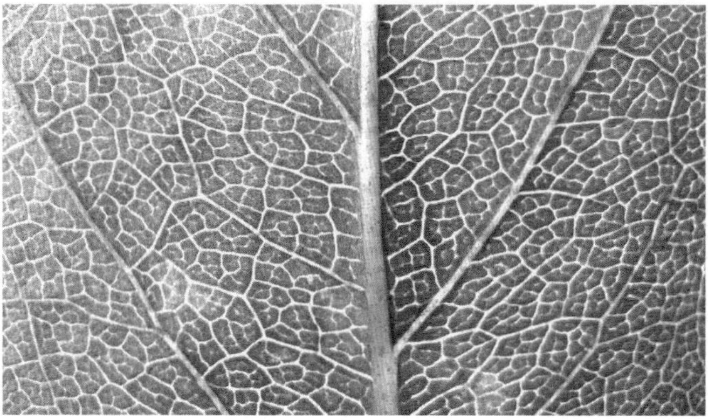

Figure 4.50: The sunflower leaf image for Exercise 4.7.2.

Figure 4.51: The rhododendron leaf image for Exercise 4.7.3.

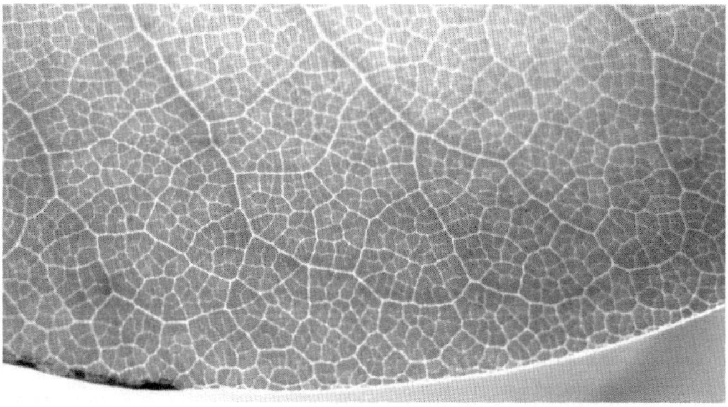

Figure 4.52: A magnification of a portion of the leaf image of Fig. 4.51.

Prob 4.7.4 Collect some of your own leaves and observe and measure their venation.

Prob 4.7.5 This exercise os for people with better eyesight than your authors

have.
(a) Count the number of veins of stage 1, of stage 2, of stage 3, of stage 4, and of stage 5.
(b) Count the number of regions enclosed by stage 1 veins (and possibly the leaf edge), the number of regions enclosed by stage 2 veins (and possibly stage 1 veins and the leaf edge), and so on.
(c) Do either of these data sets exhibit power law scaling? If so, you may have found a way that leaf veins use fractal geometry.

4.8 Structures of vegetables

So far, over 390,000 species of plants have been discovered. While they exhibit a wide variety of forms, some themes repeat. Branching, for example. The top of the yarrow plant on the left looks like just tightly clustered blossoms, but from underneath (right) we see recursive branching.

We think similar patterns appear in many instances because evolution discovers and reuses, or rediscovers, the laws of geometry, including fractal geometry. Let's see what we can find.

4.8.1 Purpose

Here our goal is to find fractal patterns in vegetables. We look for similar patterns in different vegetables, and also for different patterns. Can we find fractals other than branching?

4.8.2 Materials

Use what you find in your kitchen, in your garden, or in the produce section of the local grocery store.

4.8.3 Background

Many patterns are fractal. We'll focus on three archetypes, pictured here.

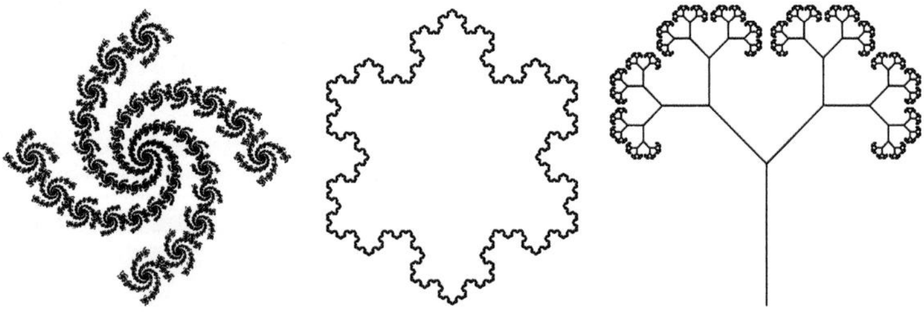

Figure 4.53: Left to right: spiral, rough edge, and branching fractals.

4.8.4 Procedure

This is a visual experiment. We aren't looking for measurements and power laws. Just use your eyes to look for fractal patterns.

4.8.5 Sample A. Broccoli Romanesco

Perhaps the clearest example of a fractal structure in vegetables is that seen in broccoli Romanesco. A search of Google images gives many pictures; here we see photographs taken of a head of broccoli Romanesco purchased at a local market in Connecticut. The right image is a magnification of the left. Note the spiral made of spirals made of spirals. This vegetable is clearly composed of smaller and smaller copies of itself. Are the copies exactly self-similar? No, close inspection will show imperfections in the copies but you have to look very carefully to see them. Do the copies repeat an infinite number of times? Again, no. We would not expect natural objects to be able to reproduce themselves an infinite number of times or to be exact copies of themselves on smaller scales. This is reserved for mathematical fractals, but the patterns found in nature can truly suggest a fractal structure.

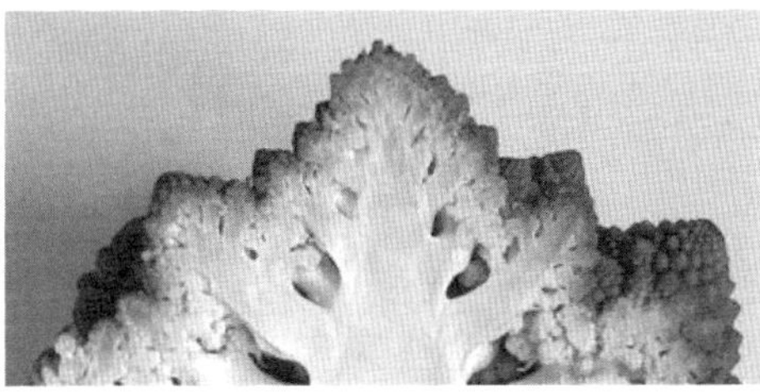

Is this canopy of self-similar spirals generated by branching from underneath? The slice shown on the right exhibits branching but it is rather coarse and we do not think it could be entirely responsible for the refined spirals seen on the surface. Any ideas? Look at the spirals in sunflowers and in pine cones for a hint.

4.8.6 Sample B. Electrical tape and kale

Yes, yes, yes, we know electrical tape is not even a plant let alone a vegetable, but it illustrates a point. If you cut plastic electrical tape with a scissors it leaves a nice straight edge. But I (NN) can remember my father never had a scissors handy when he needed to cut a piece so he would pull it until it stretched enough to break. (My father did the same thing, and I do, too. MF) When it broke, the edge would always be wrinkled as you can see in the comparison photos on the right.

What makes this happen? Stretching the tape makes it thinner until it breaks. Doing this does not cause any tape to be lost so there is just as much tape between A and B in the stretched tape as in the cut tape. Then there has to be as much thin tape between A and B as there was before so it has to occupy more space than before. To accomplish this the tape bulges out causing a wrinkled edge.

In kale the growth pattern of the kale leaves is consistent as the kale grows away from the veins until it gets close to the edge of the leaves as seen in the photo. Then the growth pattern changes and the leaves produce

more matter than in previous generations. The leaves do not get thinner as the tape did, but there is more material between two points than in previous generations. To accommodate this excess of material the leaf bulges out as the tape did.

In kale the growth is such that even the bulges have bulges suggesting something similar to a randomized Koch curve (right) if viewed straight on as in the right image of Fig. 4.54. Looking at kale straight on its edge we see a very complicated pattern of folds. Isolating a small section we can see bulges on bulges. Does any part remind you of a Koch curve?

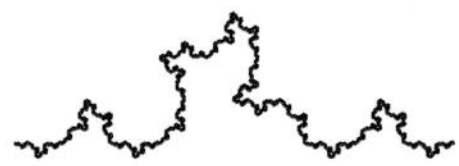

Figure 4.54: Left: an edge view of kale. Right: an edge view from straight on.

4.8.7 Sample C. Sedum

No sedum is not a vegetable, please don't try to eat it, but it is a plant that very nicely illustrates branching. Let's look at some photos. In

the left photo we see obvious branching from the main stem as well as branching from the secondary stems. The right photo shows even more branching from the secondary stems and branching from these. How many levels of branching can you identify?

Figure 4.55: Views of the sedum canopy.

The left photo of Fig. 4.55 shows the canopy of the sedum from above, the result of the branching. In the right photo the clusters of the secondary branching have been separated apart. Within these cluster are smaller clusters resulting from branching from the secondary branches and even more still smaller clusters. These are difficult to separate and photograph. You should get some of your own sedum and closely examine it. See how many clusters you can identify.

4.8.8 Conclusion

Plants give us examples that closely resemble fractal structures. While not composed of exact scale copies of themselves as (nonrandom) mathematical fractals are, they never the less give us physical examples of naturally occurring structures that grow with an iterative pattern of smaller and smaller approximate copies of themselves. Perhaps nature has used this principle to simplify the DNA coding for generating these structures. Of course, as with all naturally occurring processes, there is a limit to how many levels of iteration can occur because of the physical size constraints. We hope this lab will inspire you to collect your own vegetables and look at them closely. Anyway your mother was right: in the end you always should eat your vegetables.

4.8.9 Exercises

Prob 4.8.1 Look at the top surface of some heads of ordinary broccoli or cauliflower.
(a) Can you find clusters resembling the whole head?
(b) Do these clusters have clusters?
(c) How many levels of clusters can you find?
(d) How closely are these clusters scale copies of the whole head and of each other?
(e) Carefully slice the whole head or one of the larger clusters vertically. Look at the branching that is present. Do you think the branching accounts for the canopy that is formed?

Prob 4.8.2 Kale was a good vegetable to consider for illustrating wavy edges. In the realm of flowers, daffodils also work nicely. Find a vegetable or flower with wavy edges and see how many levels of bulges you can find.

Prob 4.8.3 Can you find a vegetable or flower with branching similar to sedum? Observe the levels of branching and how they relate to the number of clusters in the canopy.

Prob 4.8.4 Can you find a vegetable with cross-sections that resemble the Sierpinski carpet, shown on the right? Such a rigid distribution of holes is unlikely to occur in natural objects, so instead look for cross-sections that have a small number of large holes, a moderate number of middle-size holes, and a large number of small holes.

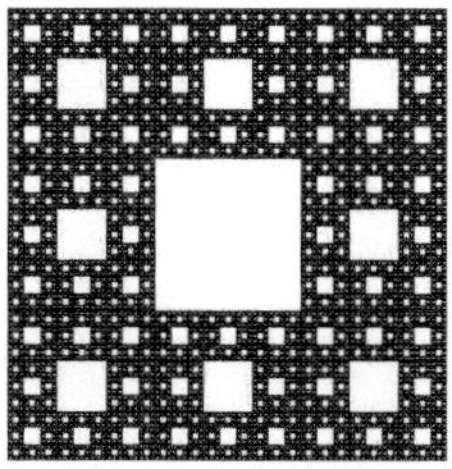

Prob 4.8.5 Pictured here are some blossoms of a celosia plant, a genus of the amaranth family. The name is derived from the Greek word for flame, referencing the flame shape of the blossoms. The blossoms are roughly conical, with roughly conical side branches, which in turn have roughly conical side sub-branches. Note that the side branches resemble shrunken copies of the entire blossom. Can you find other plants that exhibit a similar sort of fractality?

Prob 4.8.6 Can you find parts of our bodies that have branching similar to sedum, or rough edges similar to kale? How might evolution have discovered and exploited similar structures in plants and in animals? Surely we are not very closely related to sedum or kale. Perhaps evolution has discovered and uses the laws of geometry.

4.9 Cooking fractals

That we find fractals among many vegetable shapes in the kitchen is no surprise. After all, simple branching architecture is common to many plant body, root system, and leaf vein networks. But can we find fractals in things we cook or bake? Why should we? Difficult to imagine branches in anything we cook. If it's early evening, you have a clear view of the eastern horizon, and the near-full moon is visible in a cloudless sky, what do you see? You see the moon, with a few large darker areas (the maria), and many craters, more smaller than larger, over a wide range of sizes. In [148] we find

Figure 4.56: The moon

> As a specific example consider the frequency-mass distribution of asteroids. Direct measurements give a fractal distribution. Since asteroids are responsible for for the impact craters on the moon, it is not surprising that the frequency-area distribution of lunar craters is also fractal.

Benoit wrote about the distribution of craters in Chapter 33 of [1].

Can we cook something that looks like the moon? Yes, and you may have done this for years. Let's see how.

4.9.1 Purpose

If you search Google images for fractal cookies you'll find cookies in the shapes of Sierpinski gaskets, Koch snowflakes, Sierpinski carpets, Julia sets, Apollonian gaskets (a kind of circle inversion fractal), and even the Mandelbrot set. Probably other fractal shapes, too. While these look interesting, and no doubt are tasty, we have in mind something else. Rather than design fractal shapes, we'll let the the cooking process produce fractals.

4.9.2 Materials

For this we need a stove, a decent non-stick pan, a mixing bowl, measuring cups and spoons, a spatula, an egg beater, flour, salt, butter, eggs, and milk, and a crepe recipe. We use the recipe in *The Vegetarian Epicure*[150].

4.9.3 Background

The distribution of craters on the near side of the moon (the far side looks considerably different, the NASA website has plenty of photographs) is revealed through a power law scaling. Suppose $N(r)$ is the number of craters with radius about r. We see more small craters than large craters, so a power law scaling would have the form

$$N(r) = k\left(\frac{1}{r}\right)^d \quad \text{which gives} \quad \log(N(r)) = d\log(1/r) + \log(k)$$

when we take the log of both sides and simplify. Now we need to address an issue about measurement. Most likely, no pair of these circles have the same radius, so for each measured radius r_i we'd get $N(r_i) = 1$, not useful at all. One approach to this problem is to partition the radius values into intervals and count the number of circles whose radii fall into each interval. Another approach is to define $N_+(r)$ to be the number of circles with radius $\geq r$. If $N(r)$ exhibits a power law scaling, must $N_+(r)$ also exhibit a power law scaling? If both have power law scalings, must they have the same scaling exponent? See Appendix B.13.

A crepe exhibits approximate circles of many radii, the browned edges of bubbles trapped under the batter as it solidifies.

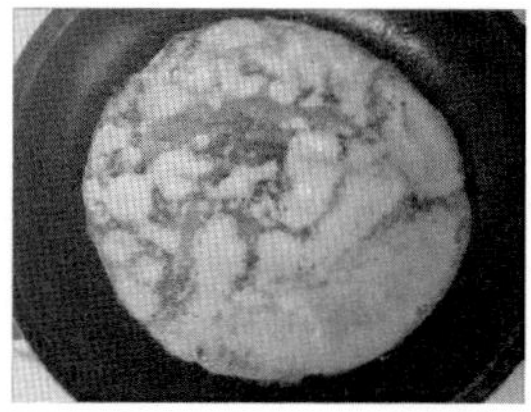

To prepare us to recognize power law distributions, we see three examples in Fig. 4.57. The circles in these pictures were generated by placing the centers uniformly randomly in a square in the plane, with the number of circles of each radius r_i determined by rounding $(1/r_i)^d$ where $d = 1.4$ for (a), $d = 1.6$ for (b), and $d = 1.9$ for (c).

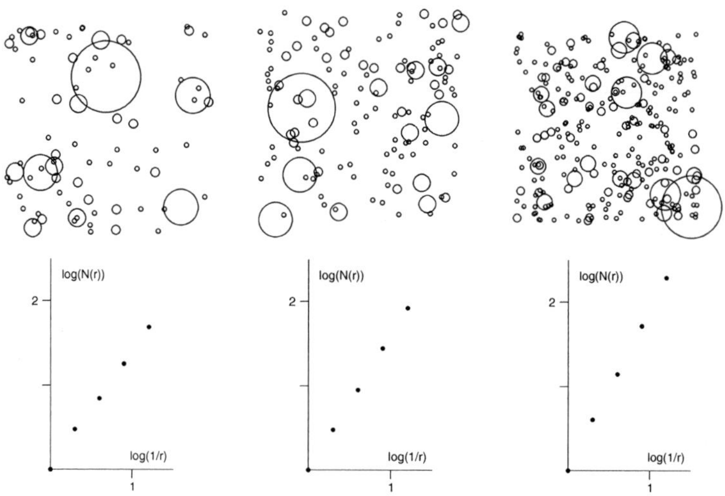

Figure 4.57: Some distributions of circles and their log-log plots.

Here are the numbers of circles of each radius for the three examples of Fig. 4.57, (a) represents the circles of the left image, (b) of the middle, and (c) of the right. Because of how the collections of circles were generated, it's no surprise that the points of the log-log plot of (a) fall along a line of slope 1.4, those of (b) along a line of slope 1.6, and those of (c) along a line of slope 1.9.

r_i	1	1/2	1/4	1/8	1/16
(a) $N(r_i)$	1	3	7	18	49
(b) $N(r_i)$	1	3	9	28	84
(c) $N(r_i)$	1	4	14	52	194

4.9.4 Procedure

Assemble the ingredients and implements and cook some crepes. If cooking were our only goal, this would be a cookbook instead of a lab manual. When you've prepared a crepe, you'll measure the distribution of circle sizes on the crepe and look for a power law relation between r and the number of circles with radius $\geq r$.

There is an analog way to do this count, but it is delicate and can be messy. Gently place an overhead transparency on the crepe and immobilize crepe and transparency. With a marking pen, on the transparency outline all the circles you see. Use a light touch: too much pressure can deform the crepe and the circles on it.

Better is to take a digital photograph of the crepe and import the photo into a graphics program. The programs should have a circle function and an undo function. This works much better with two people: one to cover crepe circles, the other to record the sizes of the circles. Covering the crepe circles with graphics circles first and then trying to count them is not any fun at all. If you lose your place and need to start again, counting these circles can become quite annoying. Also, finding crepe circles has something in common with scraping paint from a wall: the more paint you scrape, the more paint you find. When you think you've got them all, you'll find still more, smaller, circles. Patience is an important tool.

4.9.5 Sample A. Crepes

In Fig. 4.58 we see some crepes. The top left image is the crepe just before it has been flipped. The others have been flipped and the other side cooks. The process of counting the circles is a test of patience, and of eyesight.

Figure 4.58: Crepes: Top left is before the crepe is flipped, the bottom right is plated and ready to fill. The others have been flipped and the underside cooks.

One reason to cook several crepes is that some practice is needed to get the a sense of when and how to flip the crepe. If this is your first time cooking crepes, expect the first few to wind up in the garbage. After you can reliably cook crepes, several examples could allow several circle counts and give a sense of the reproducibility of the power law exponent measurements. Or, after one count, you may prefer to spread a bit of black raspberry or lingonberry jam, roll up the crepes, and enjoy the reward for the eyestrain of the count.

Here we see a crepe image with circles superimposed. For the reference radii are $r_1 = 0.5$, $r_2 = 0.25$, $r_3 = 0.125$, $r_4 = 0.0625$, and $r_5 = 0.03125$. The number of circles with $r \geq r_1$ is 1, with $r_1 > r \geq r_2$ is 5, with $r_2 > r \geq r_3$ is 37, with $r_3 > r \geq r_4$ is 121, and with $r_4 > r \geq r_5$ is 356. From this we deduce

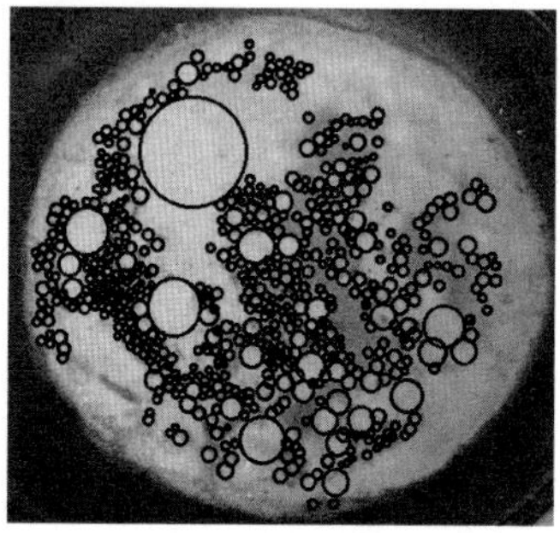

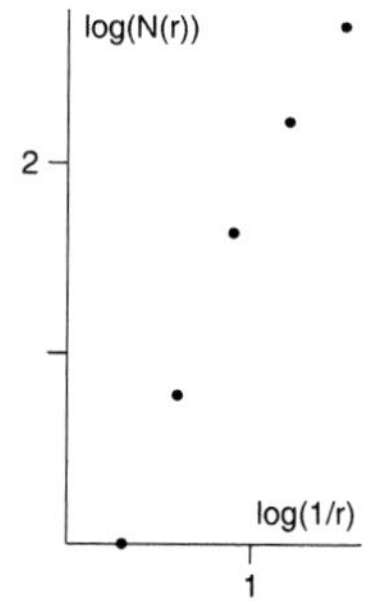

$$N_+(r_1) = 1,\ N_+(r_2) = 1 + 5 = 6,\ N_+(r_3) = 6 + 37 = 43,$$
$$N_+(r_4) = 43 + 121 = 164,\ N_+(r_5) = 164 + 356 = 520$$

Points of the plot of $(\log(1/r_i), \log(N_+(r_i))$ appear to approach a straight line, though to be sure we do not have enough data points to support this suspicion. The slope of the line determined by the last two points is about 1.66. If we continued the circle count to even smaller radii, and if the additional points of the log-log plot do appear to fall along a straight line, we expect the slope of that line may be a bit less than 1.66, but not a lot less.

4.9.6 Sample B. Bread and butter, and beer

Crepes are not the only fractals produced by cooking. And no, we don't mean something as simple as pancakes instead of crepes. If it isn't kneaded too much, bread reveals a fractal pattern of holes when sliced. In Fig. 4.59 we see (left) a piece of bread that shows a distribution of gaps of different sizes, (center) prints made by inking the bread surface and pressing it gently onto paper, and (right) a section of bread that does not reveal a fractal pattern.

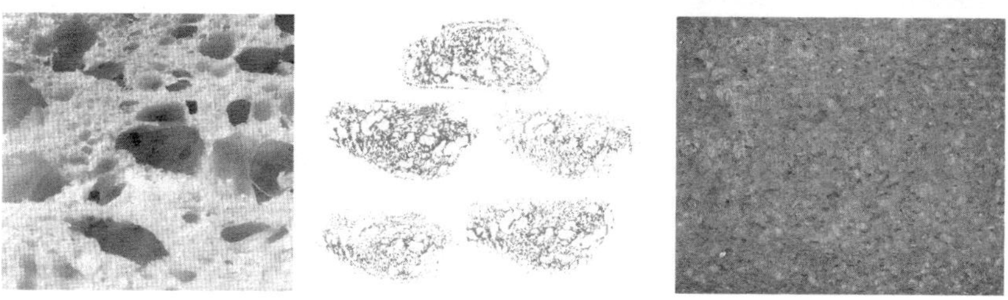

Figure 4.59: Left to right: fractal bread, bread prints, nonfractal bread.

In college one of us (NN) worked in a commercial bakery where the bread dough was kneaded so much that carbon dioxide bubbles produced by yeast were broken into small, roughly uniform sizes. This bread is not fractal. Less kneading allows the bubbles to grow to different sizes, producing bread that may be fractal. One way to test this is to make a print of the bread surface, as shown in the middle image of Fig. 4.59. Then as with the crepes, we could tally the number of holes with diameters in a given range and test if the data reveal a power law scaling.

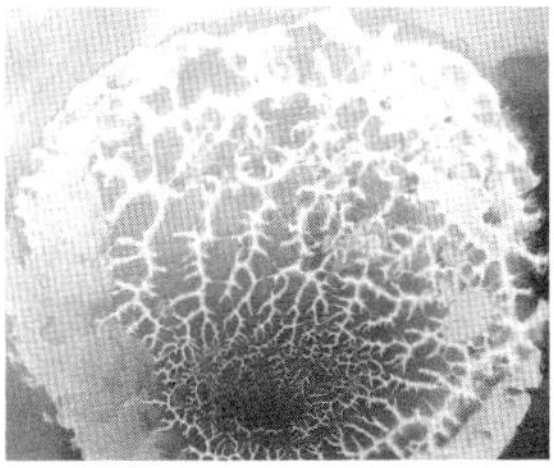
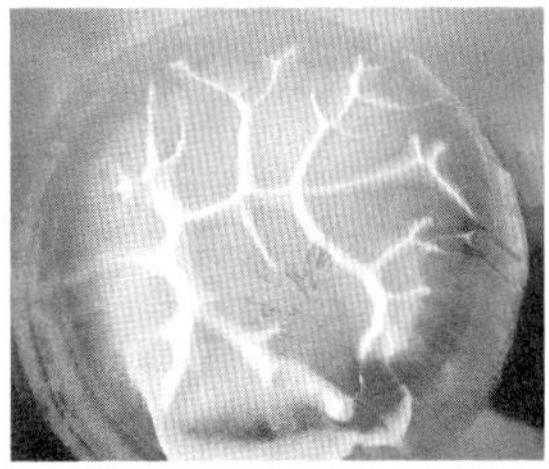

Because we've mentioned bread, we may as well think about butter. We'll follow the idea of Lab 4.2. Here we put butter between two rigid plates, apply pressure to flatten the butter, then pull the plates apart. But because butter softens at room temperature, now temperature is a parameter we can vary to some effect. In the left image above the butter was cold, warm in the right image.

In volume 2 of [151], on page 362 we find a letter Sir Isaac Newton wrote to Dr. Thos. Burnett. The letter includes this.

> As I am writing, another illustration of ye generation of hills, proposed above, comes into my mind. Milk is as uniform a liquor as the chaos was. If beer be poured into it, and ye mixture let stand till it be dry, the surface of ye curdled substance will appear as rugged and mountainous as the earth in any place.

Certainly, Newton was a careful observer. When he wrote "rugged and mountainous" he must have had in mind what we now call the fractality of mountains, a cascade of ever smaller peaks off of larger peaks, but here the peaks are curdled milk. Sadly, we have been unable to reproduce Newton's result. Perhaps contemporary pasteurization techniques impede the curdling.

4.9.7 Conclusion

Obviously, the distribution of lunar craters is not caused by the mechanism that produces circles on crepes. Asteroids do not fly through your kitchen and hit the crepe batter in your frying pan. Rather, we just look for power law scalings. Why do power laws occur so often in the world? That's a really, really good question.

4.9.8 Exercises

Prob 4.9.1 First, make your own crepes and count the circles. Do you find a power law in the radius scaling? Is your power law exponent close to the one we got in Sample 4.9.5?

Prob 4.9.2 Measure the power law exponent of two of your crepes. Do you find the same exponent?

Prob 4.9.3 Modify the amount of liquid in the recipe and see what effect this has on the distribution of circles. We might think more fluid would make less viscous batter and so more smaller bubbles. But does it?

Prob 4.9.4 Following Sample 4.9.6, make a print of a slice of bread with holes of many sizes. Make a log-log plot of the hole sizes. Do you find evidence that supports a power law distribution?

Prob 4.9.5 Following Sample 4.9.6, perform decalcomania experiments with butter of different temperatures. What about margarine? Or peanut butter?

Prob 4.9.6 Following Sample 4.9.6, perform a milk curdling experiment. Can you reproduce Newton's results? Maybe substitute vinegar for beer.

Prob 4.9.7 Look around your kitchen. Can you find a way to make fractals that is different from those we've described in this lab?

Chapter 5

Labs in the Lab

Now we move into a real lab, though only a few of these experiments involve much equipment: a video camera with video out cable, a DC power supply, and an oscilloscope are as complicated as we need.

In Lab 5.1 we'll build a magnetic pendulum and explore the complex structure of the map of initial points that land, eventually, over each magnet. Lab 5.2 we'll construct an optical gasket with mirrored spheres and illustrate fractal basin boundaries with a laser pointer. Both these labs illustrate a puzzling phenomenon called Wada basins.

We'll build analog version of IFS with a video camera, monitor, and two mirrors in Lab 5.3.

In Lab 5.4 we'll grow fractal clusters by electrodeposition. In Lab 5.5 we'll grow fractal branching patterns when we mix fluids of different viscosities. This is a variation of Lab 4.4. Here we'll use a DC power supply, zinc sulfate powder, a laboratory balance, and a magnetic stirrer.

Recall that in Lab 2.2 we estimated the mass dimension of crumpled paper balls. In Lab 5.6 we study how the number of creases of a crumpled, uncrumpled, and flattened paper ball grow as the paper is repeatedly recrumpled.

Labs 5.7 and 5.9 involve fractal electrical networks. We'll look for scaling behavior in the voltage across a fractal network of resistors. For this we need a collection of resistors, a multimeter, and some breadboards.

In Lab 5.8 we'll study the field around a fractal network of magnets.

In Lab 5.9 we build a Sierpinski gasket network of oscillators and explore circumstances under which individual oscillators synchronize. This is our most equipment-intensive lab: an oscilloscope, breadboards, and a collection of capacitors, inductors, and resistors.

So many familiar experiments can be modified by a modest amount, the focus redirected just a bit, to grow or discover fractals. The labs presented here are just a start. Look around. Perhaps you can find others that are just as interesting.

5.1 Magnetic pendulum

A feature of chaos, in fact the most familiar feature, is sensitivity to initial conditions. Trajectories that begin from two nearby points will trace paths that are nearby for a while, but then diverge and exhibit behaviors that appear unrelated to one another. By now, this is familiar to many people.

In this lab and the next, Lab 5.2, we'll investigate another form of sensitivity, final state sensitivity. Here the experimental set-up is a magnetic pendulum free to swing in any direction over three immobilized magnets. Eventually the pendulum will come to rest over one of these magnets, but over which one can depend very delicately on the initial position of the pendulum. While small uncertainties in some initial positions do not alter the eventual behavior, for other initial positions small changes can alter the outcome drastically. For these initial positions, the pendulum has no long-term predictability. Because this system includes friction, we know that every trajectory will wind up over one of the fixed points. But over which one?

5.1.1 Purpose

We'll build a magnetic pendulum and observe some examples of final state sensitivity. Lack of both patience and sufficiently steady hands prevent us from mapping out the basins of attraction of each magnet. For this, we'll turn to the software from 5.1 of the Mathematica or Python codes.

5.1.2 Materials

To build the pendulum we'll use an overhead projector, seven disk-shaped magnets, an overhead transparency, and either (1) strong thread, tape, and a suction cup, or (2) a straight pin, a button with a hole near its center, glue, two balsa wood rods, a thumb tack, a fishing swivel, and a carpenter's clamp.

To test final state sensitivity we'll use overhead transparency pens of three different colors (red, blue, and green for example). If you want to try to map out the basins of attraction of the magnets, use one of the grids from Sect. 2.1.10. Probably the largest-box grids will suffice, and any smaller grids likely will exhaust your patience.

For the computer experiments, use the software of 5.1 (a) and (b) of the Mathematica or Python codes. The differential equations used in these codes are derived in Appendix B.14.

5.1.3 Background

The intuition behind the term "basin of attraction" is pretty clear. To be more precise, and to show some of the complications that can arise, we'll illustrate it with graphical iteration (See Sect. 3.1.3.1 if you need a refresher) for the function $f(x) = -x^3/2 + 3x/2$. We'll use Newton's method (introduced in Sect. 5.1.3.2) to illustrate basin boundaries, fractal basin boundaries and the Wada property. While the meanings of basin boundaries and fractal basin boundaries are easy enough to guess, we'll spend a bit of time with the Wada property, definition and examples.

5.1.3.1 Basins of attraction

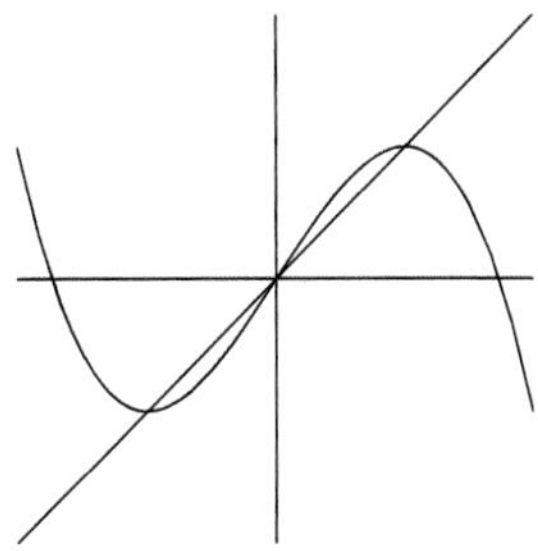

To illustrate the notion of basin of attraction we'll use the function $f(x) = -x^3/2 + 3x/2$. In a moment we'll use the locations of the zeros of f: $f(x) = 0$ for $x = \pm\sqrt{3}, 0$. The fixed points are the solutions of $f(x) = x$; geometrically they are the points of intersection of the curve $y = f(x)$ and the line $y = x$. This function has three fixed points, $x = -1$, $x = 0$, and $x = +1$. A fixed point z is stable (all sufficiently nearby points iterate toward the fixed point) if $|(df/dx)|_z| < 1$ and is unstable (some points arbitrarily near to the fixed point iterate away from it) if $|(df/dx)|_z| > 1$. From the slopes of the tangent lines to $y = f(x)$ at the fixed points, we see that $x = 0$ is unstable, while $x = \pm 1$ are stable. Let's explore a some aspects of the basin of attraction of $x = 1$.

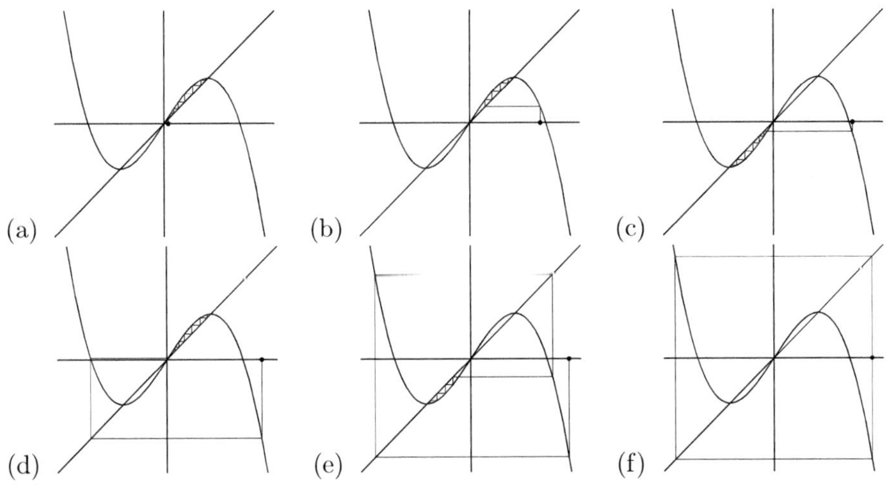

Figure 5.1: Graphical iteration of f, with $x_0 = 0.1, 1.6, 1.8, 2.15, 2.225$ and $\sqrt{5}$.

In Fig. 5.1 we see graphical iteration from six initial points x_0. We see that the iterates of $x_0 = 0.1$ and $x_0 = 1.6$ converge to $x = 1$. In fact, for any x_0 in $(0, \sqrt{3})$, the iterates of x_0 converge to $x = 1$. That is, the basin of attraction of $x = 1$ includes the interval $(0, \sqrt{3})$. Is this the entire basin? No, because while $x_0 = 1.8$ iterates to $x = -1$, we see that $x_0 = 2.15$ iterates to $x = 1$, and $x_0 = 2.225$ iterates to $x = -1$. The basin of attraction of $x = 1$ consists of $(0, \sqrt{3})$ together with an infinite sequence of ever-smaller intervals that approach $\sqrt{5}$ and $-\sqrt{5}$. The points $x = \sqrt{5}$ and $x = -\sqrt{5}$ are a 2-cycle under graphical iteration. What about $|x_0| > \sqrt{5}$? These iterates escape to

∞. For example, with $x_0 = 2.24$ (note $\sqrt{5} \approx 2.23607$) the tenth iterate is about 5×10^{651}. Even for simple functions, the basins of attraction can be a bit complicated.

5.1.3.2 Newton's method

Newton's method is an iterative procedure to approximate the zeros of a differentiable function f. Start from an initial guess x_0, draw the tangent line T_0 to the graph of f at the point $(x_0, f(x_0))$. If this line is not horizontal, it intersects the x-axis at some point x_1. (If this tangent line is horizontal, Newton's method fails. Pick another initial guess and try again.) Now repeat the process: denote by T_1 the tangent line to the graph of f at the point $(x_1, f(x_1))$. If T_1 isn't horizontal, it will intersect the x-axis at some point x_2. Continue. This generates a sequence of points $x_3, x_4, \ldots$ that (usually) converges to a zero x_* of f.

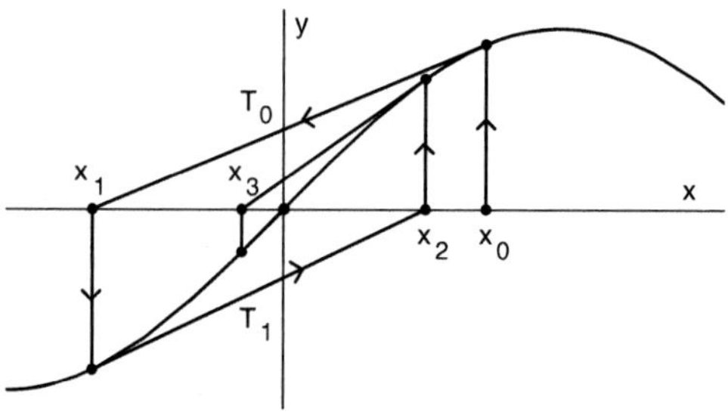

The tangent line T_0 has slope $f'(x_0)$ and passes through the points $(x_0, f(x_0))$ and $(x_1, 0)$. From this we obtain

$$f'(x_0) = \frac{f(x_0) - 0}{x_0 - x_1} \quad \text{that is, } x_1 = x_0 - \frac{f(x_0)}{f'(x_0)}$$

If we argue in the same way for other points in this sequence, we arrive at the general formula for Newton's method

$$x_{n+1} = x_n - \frac{f(x_n)}{f'(x_n)} = N_f(x_n) \tag{5.1}$$

For example, suppose we want to find a numerical approximation to $\sqrt{5}$. This is a zero of $f(x) = x^2 - 5$, and we know $\sqrt{5}$. With this f the Newton function is $N_f(x) = x - f(x)/f'(x) = (x^2 + 5)/2x$. Take $x_0 = 2$. Carried to 10 digits, the first five iterates are

2.2500000000, 2.2361111111, 2.2360679779, 2.2360679775, 2.2360679775

The fifth iterate of Newton's method agrees with $\sqrt{5}$ to at least ten digits. Typically Newton's method converges rapidly. This is one reason it is used so frequently in computations.

In 1879, Arthur Cayley [152] extended Newton's method to complex functions and by a clever change of variables showed that for $f(z) = z^2 - 1$, the basin of attraction of $z = 1$ consists of all complex numbers with positive real part, and the basin of attraction of $z = -1$ consists of all complex numbers with negative real part. What about the imaginary axis?

5.1.3.3 Basin boundaries

This question leads to the notion of a basin boundary. We've used intuition about boundaries several times in these labs. Now we'll be more precise. To make visualization easier, we'll restrict our attention to subsets of the plane, but these ideas can be applied in vastly more general settings.

A point p belongs to the boundary of a set A if every disk with center p contains some points that belong to A and also some points that do not belong to A. For example, if $A = \{(x, y) : x^2 + y^2 < 1\}$, then the boundary of A, denoted ∂A, is $\{(x, y) : x^2 + y^2 = 1\}$. Do you see that the set $B = \{(x, y) : x^2 + y^2 \leq 1\}$ has $\partial B = \partial A$?

In case you think, "Yes, of course, this is all obvious," what about these examples: the set $C = \{(x, 0) : x \in \mathcal{C}\}$ where $\mathcal{C}$ is the Cantor middle-thirds set, and $D = \{(x, y) : \text{both } x \text{ and } y \text{ are rational}\}$? What are their boundaries? Do you see why $\partial\mathcal{C} = \mathcal{C}$ and $\partial D = \mathbb{R}^2$?

In Newton's method for $z^2 - 1$, Cayley showed that the boundary of the basin of attraction of $z = -1$ and of the basin of attraction of $z = 1$ both are the imaginary axis. For z_0 on the imaginary axis, Newton's method does not converge. Rather, its dynamics are chaotic. This is not obvious, but requires a bit of work.

5.1.3.4 Fractal basin boundaries

Cayley anticipated a similar analysis for $f(z) = z^3 - 1$, but the problem was more intricate than he had at first anticipated. This cubic polynomial has three zeros, 1, $-1/2 + i\sqrt{3}/2$, and $-1/2 - i\sqrt{3}/2$, so the plane should be divided into three basins. By the computation of some points on the boundary of these basins, Cayley knew that they were not just three 120° wedges. Almost a century later, John Hubbard produced a plot of these basins, similar to this picture. Is it a surprise that Cayley, working entirely by hand, did not produce even a glimpse of these basins?

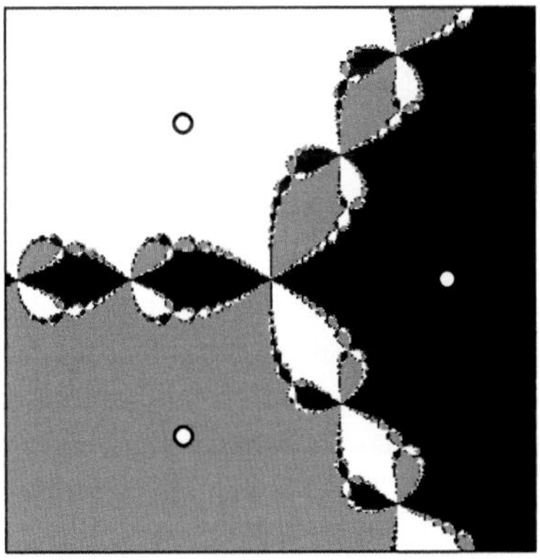

Look at the loops off of loops off of loops. The boundaries of these basins are fractal.

But there's something strange about this picture. For example, between the big gray region and the big white region, we see smaller black regions. And between these black regions and the large white region we see even smaller gray regions. Can this continue to to all magnifications? In the next section we'll see that it can.

Of course, this is a question about a mathematical object. For any physical object, fractality can extend over only a limited range of sizes, the scaling range, over which the same forces dominate. The more extensive the scaling range, the better the claim of fractality. But occasionally we must go back to examples in geometry, to help us interpret physical constructions.

5.1.3.5 The Wada property

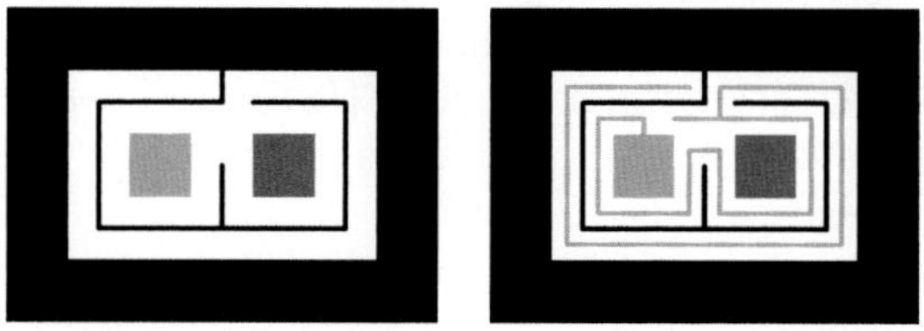

In his 1917 paper Kunizô Yoneyama [153] describes a puzzling construction that he attributes to Takeo Wada. Here's a slight variation of the original construction. Imagine an island, a white rectangle, in a dark sea. On the island are two lakes, here represented by darker and lighter squares. First, a canal is dug from the surrounding sea so every point of the island is no more than 1 km from the canal, and hence from water access to the sea. Next, a canal is dug from the left lake so every point of the island is no more than 1/2 km from that canal, and so from water access to the left lake. Now dig a canal from the right lake so every point of the island is no more than 1/4 km from that canal. But now the people who want easy access to the sea dig a fourth canal so every point of the island is no more than 1/8 km from that canal. Continue forever. To avoid worry about canal sizes, let's suppose that every canal has no thickness; they are just made up of line segments. In the limit every point of the island is arbitrarily close to canals that lead to each lake and to the sea. This construction is called the *Lakes of Wada.*

Think of the sea and all the canals that lead to is as the basin of attraction of the sea. Each lake and all the canals that lead to it are the basin of attraction of that lake. Because every point of the island is arbitrarily close to canals that lead to each lake and to the sea, we see that on the line segment between points of any two basins of attraction, we find a point of the third basin. In fact, on this segment we find infinitely many points of the third basin. And infinitely many of the other two basins, as well.

Any collection of sets intertwined in this way exhibits the *Wada property.* The basins of attraction of the three zeros of $f(z) = z^3 - 1$ have the Wada property. In this lab and the next (Lab 5.2) we'll see physical examples that approximate the Wada property.

5.1.4 Procedure

First we'll describe the experimental set-up.

To immobilize the magnets, attach them in pairs on opposite sides of a transparency. See the first image of Fig. 5.2.

The pendulum can be constructed in (at least) two ways.

(1) Tape a piece of thread to a magnet so the magnet hangs approximately parallel to the projector surface. Attach the other end of the string to the overhead projector head (tape, a suction cup, or a knot work well). See the second image of Fig. 5.2.

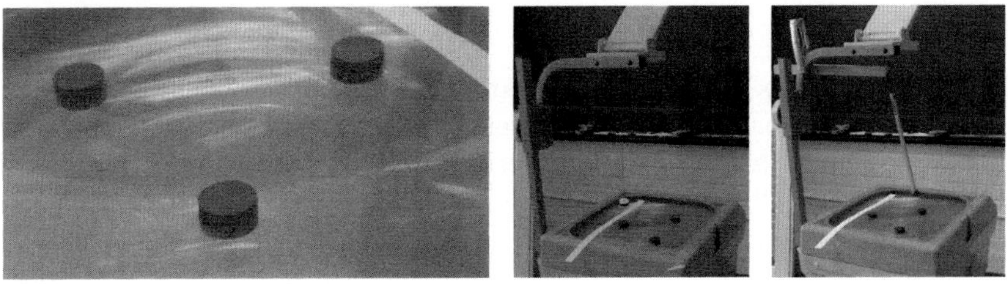

Figure 5.2: The magnetic pendulum set-up.

(2) Glue a straight pin to a button, glue the button to the seventh magnet, and stick the pin into a balsa wood rod. Attach this balsa wood rod to another balsa wood rod with a fishing swivel. Use a carpenter's clamp to attach the second basa wood rod to the overhead projector. See the third image of Fig. 5.2.

Make sure the pendulum and the stationary magnets are arranged so the magnets on the transparency attract the magnet on the pendulum. Failing to do this can result in a demonstration very amusing for the audience, though not so much for the presenters.

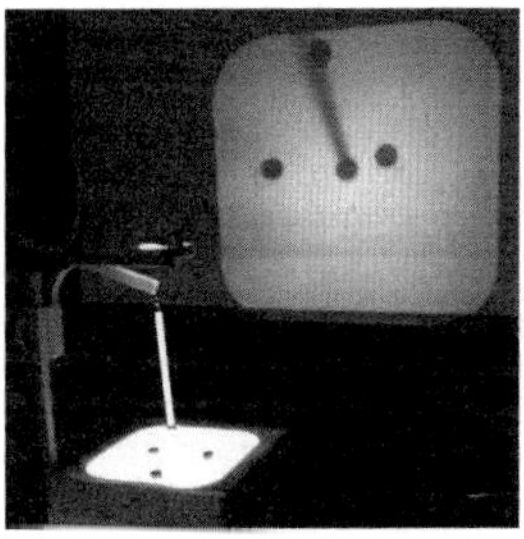

Place the stationary magnets on the overhead projector surface and attach the pendulum to the projector head. Before settling over one of the magnets, the pendulum can undergo very complicated motion, the details of which depends delicately on the point from which it starts.

Starting the pendulum directly over one of the stationary magnets results in the pendulum staying over that stationary magnet. That is, the stationary magnets locate fixed points of the pendulum dynamics. Moreover, from all nearby starting points the pendulum returns to the same stationary magnet, so these fixed points are stable.

The purpose of this lab is to explore what happens between these stable fixed points. We'll take two approaches.

1. *Wada property.* Suppose that if the pendulum starts over a point P it comes to rest over one magnet, and if the pendulum starts over a point Q it comes to rest over a different magnet. Find a point R on the segment between P and Q with this property: if the pendulum starts over the point R, it comes to rest over the third magnet.

2. *Basins of attraction.* Map out the basin of attraction of each magnet. Replace the transparency between the magnets with a grid transparency. Color the squares immediately around one magnet red, the squares immediately around another magnet blue, and the squares immediately around the third magnet green. For each other grid square, start the pendulum from over the center of the square and color that square with the color associated with the magnet over which the pendulum comes to rest.

5.1.5 Sample A

The Wada property. The specific dynamics of the physical pendulum depend on many physical parameters—the strength of the magnets, the length of the pendulum, the distance from the lowest point of the pendulum's path to the transparency, the distance between the magnets on the transparency. We'll leave physical experiments to you. Here we'll illustrate the Wada property of magnetic pendulum basins of attraction with the software from 5.1 of the Mathematica or Python codes.

For this example, we'll use the *Wada Property* software. The magnets, represented by black dots, are located at the points $(-1/2, \sqrt{3}/2)$, $(-1/2, -\sqrt{3}/2)$, and $(1, 0)$, points evenly spaced around the unit circle. The first figure shows trajectories that begin at three nearby points, enclosed by a small box. The second figure is a magnification of the beginning of the three paths, to show just how closely together they start.

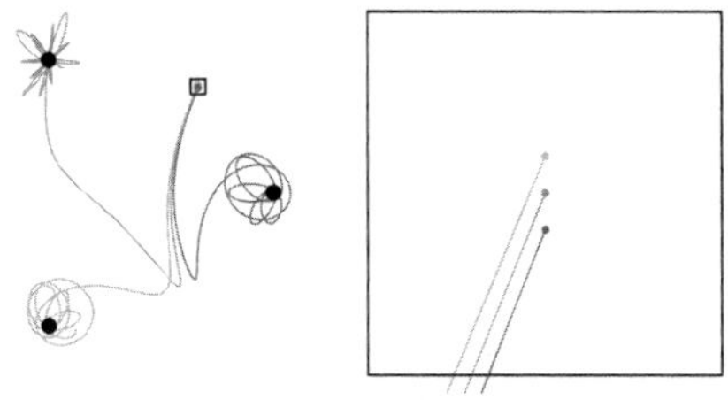

Note that these trajectories do not simply spiral in to the fixed points. Two follow paths that remind us of precessing planetary orbits, the third like a multiple-leaved rose with shrinking petals. Certainly, differential equations can generate trajectories that spiral in to a fixed point. These paths appear more complex than that.

To continue this investigation of the Wada property, we'd show that on the line between each pair of these initial points we can find an initial point that leads to a magnet different from those of the trajectories from the initial points on either side. As far as we can tell, this continues for arbitrarily high magnifications.

5.1.6 Sample B

The basins of attraction. As with Sample A, here we'll leave the physical experiments for you. The project is to produce a rough map of the basins of attraction of the three magnets. We'll use the *Basin of attraction* software.

Here we see the basins of attraction in the window $-2 \le x \le 2$ and $-2 \le y \le 2$. In the left image the boxes have side length 0.2, 0.1 in the right image. For each box take the initial point of the trajectory to be the center of the box. Paint the whole box the color determined by the magnet to which this trajectory leads.

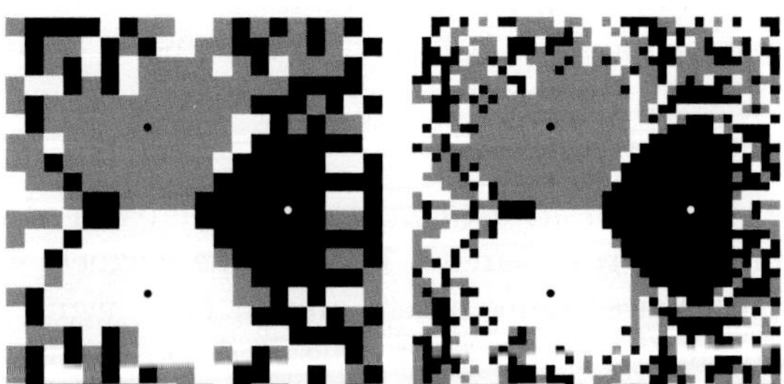

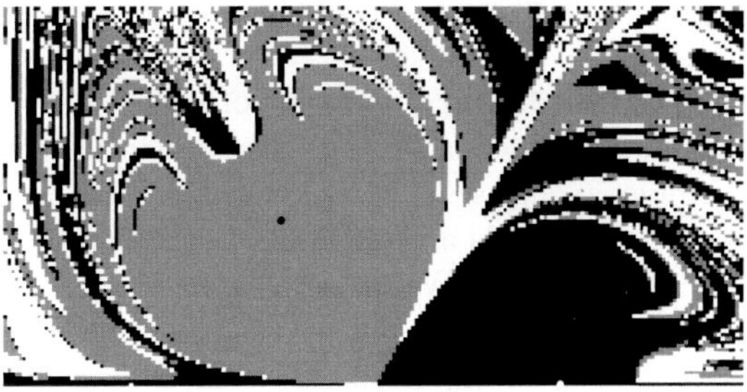

Here we magnify again, now to boxes of side length 0.02. This image is for $-2 \leq x \leq 2$ and $0 \leq y \leq 2$. Certainly, some regions of the attractor are solid, filled-in shapes. On the other hand, near the boundary between two regions the picture is much more complex. The basins are mixed together on ever-finer details, always with bits of the third between the other two. These basins appear to exhibit the Wada property.

5.1.7 Conclusion

Even though eventually the pendulum will come to rest over one of the stationary magnets, the path it takes can be quite complicated and in some regions, tiny changes in the starting position can alter over which stationary magnet the pendulum eventually comes to rest. This is another form of sensitivity to initial conditions, final state sensitivity realized with a pendulum and magnets.

5.1.8 Exercises

Prob 5.1.1 Locate the unstable fixed point for this array of magnets. Friction should allow you to place the pendulum magnet very close to this fixed point so that it stays there. To demonstrate this, one of us (NN) moved the pendulum magnet to this point. To illustrate the fixed point instability, the pendulum was pushed a bit to the side, but rather than swing over to one of the magnets, the pendulum returned to (what we'd thought was) the unstable fixed point. Confused murmurs from the teachers, puzzled looks exchanged between us. Then we realized the transparency had been placed upside down on the projector, so these magnets repelled the pendulum magnet.

Prob 5.1.2 Change the positions of the magnets on the projector. For example, suppose two are close together and the third is far away. Do you think the Wada property still holds? Speculate on how the basins of attraction change.

Prob 5.1.3 Modify the magnetic pendulum programs to investigate your magnet placement in Exercise 5.1.2. In fact, this is straightforward: just change the coordinates of $\{p1, q1\}$, $\{p2, q2\}$, and $\{p3, q3\}$ near the top of the program.

Prob 5.1.4 Add a fourth pair of magnets to the transparency. First place them on the corners of a square, then try other placements. Investigate the Wada property and the shapes of the basins of attraction.

Prob 5.1.5 Modify the programs to include a fourth magnet. This is a bit more work that just adding the coordinates of $\{p4, q4\}$ at the top of the program. You must also modify the functions $fx[x\_, y\_]$ and $fy[x\_, y\_]$, and add another level to the If statement.

5.2 Optical gasket

A tetrahedral arrangement of spherical mirrors can produce reflected light patterns that resemble a curved Sierpinski gasket. In addition, they provide an example of a fractal basin boundary and the Wada property.

If you've sat in a barber shop between two walls, you are familiar with iterated reflection: ever-smaller copies of you and the barber shrink away to nothing. When we replace flat mirrors with spherical mirrors, the geometry of reflection becomes much more interesting, and can generate fractals.

5.2.1 Purpose

We use spherical mirrors and lights to produce a nonlinear version of a Sierpinski gasket. With a laser pointer we'll illustrate multiple basins of attraction and fractal boundaries of these basins.

5.2.2 Materials

Use four reflecting spheres of the same size. Silver Christmas balls work well. Use three rolls of masking tape to support three of the spheres. Many open-top cylinders with common height and diameter will work as supports. Empty and rinsed cat food cans, for example. Several good light sources, three folders or pieces of construction paper of different colors (one red, one yellow, and one blue, for example), and a laser pointer.

Some of the exercises require more spheres, a second laser pointer, and more sophisticated mechanisms to immobilize the spheres.

5.2.3 Background

The background for this lab is identical to that of Lab 5.1. In particular, the concepts of Sects. 5.1.3.4 and 5.1.3.5 are central to understanding the conclusions of this lab. The initial images of the three different colored papers are reflected on the four spheres and are reflected again and again in reduced size on the four spheres. This size reduction and reflection is iterated until no longer physically visible. By shining a laser pointer on one of the colored regions of a sphere we create an inverse process whereby the light from the laser pointer is reflected from colored region to colored region and back to the original colored paper. Some additional information, and beautiful photographs, can be found in [154].

5.2.4 Procedure

Here we'll describe the experimental set-up. The samples illustrate some of the experiments that we can do.

Place a piece of brightly colored construction paper on a flat surface. Place three rolls of masking tape in an equilateral triangular array on the construction paper. These rolls of tape serve to immobilize three of the plastic spheres. Place a plastic sphere on each of the rolls of tape, adjusting the locations of the rolls of tape so the three spheres just touch, as shown on the right above.

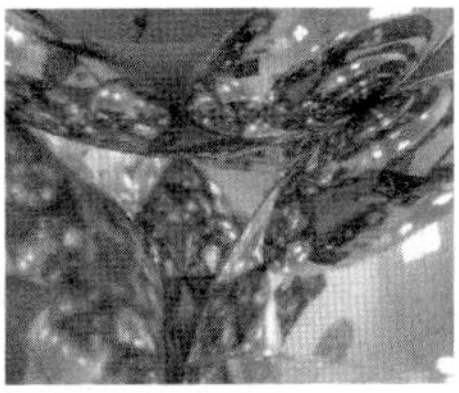

Place the fourth plastic sphere in top of the other three, making a tetrahedron of plastic spheres. On the right we see a closer view looking into one of the spaces between the spheres. The fractal nature of the picture, as well as the reason for calling this an optical gasket lab, should be apparent.

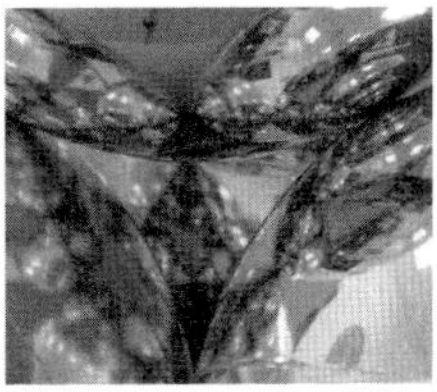

To distinguish the different openings through which light enters the tetrahedron, stand different colored folders in front of two of the other openings. Leave the last opening clear to see inside the tetrahedron. The right image is a magnification of the view of the front of the tetrahedron. This is the experimental set-up.

5.2.5 Sample A

Here we'll interpret regions of like color on the reflecting spheres as a basin of attraction. We'll observe that the three basins are mixed together in a complicated way.

A note about safety: be careful that the laser light exiting the spheres doesn't shine directly into anyone's eyes.

Note that if a laser pointer is shined at any regions of a particular color, then the beam exits the tetrahedron through the opening in front of the folder of that color and leaves a spot of laser light on that folder. We can think of the union of like-color regions on these reflective spheres as the basin of attraction for light that hits the folder of that color.

This is much easier to see in physical space, rather than as a small flat picture on a book page. Build it yourself and look around. To help with the interpretation of this picture, the first spot of laser light on the reflecting spheres is in the white circle on the right and the spot where the laser light hits the folder is in the white circle on the left. Between these, the light bounces off the spheres several more times.

5.2.6 Sample B

The observation of Sample 5.2.5 leads naturally to the the main surprise of this construction: generally smaller regions have been reflected more times, so as we shine the laser on tiny regions the beam bounces around many times between the spheres before it finally leaves the tetrahedron and hits a folder.

On the other hand, if the laser is shined on what appears to be the boundary between two regions, the light exits the tetrahedron in many places. In every experiment we tried, the light hits all three folders. That is, it appears we have an optical expression of the Wada property. Because the light has been split into so many pieces, it is difficult to see the light on the folders, unless the room is darkened. Even then, photographing the light off the folders is difficult. (And again, build it and look.) Instead of this, here we see evidence that the light bounces around the boundaries of the three regions. With spheres of higher reflectivity, a brighter laser, and better camera, we'd see that the boundary between these basins of attraction can be called a fractal basin boundary.

Another way to see the basin boundary is to photograph the spheres in a darkened room and use the flash on your camera. The pattern of reflected flashes is an approximation of the basin boundary.

5.2.7 Conclusion

Flat mirrors are reasonably easy to understand. Really? Why do flat mirrors reverse left and right, but not up and down? An answer is that flat mirrors reverse front and back. The left-right reversal is a consequence of the front-back reversal, which does not reverse up and down.

Curved mirrors, by contrast, have some surprises. Reflection off spheres is similar to inversion in circles. This provides some of the reason for the richness of these images. Do the circle inversion fractals of Lab 3.7 suggest arrangements of the reflecting spheres that might produce interesting images?

5.2.8 Exercises

Prob 5.2.1 Get a second laser pointer, of a different color from the first. Both red and green laser pointers are readily available. Aim both laser beams at nearby spots in the middle of a large single-color region on one of the spheres. Do they exit at nearby spots on a folder? Hold one laser in place and move the other slightly. Describe how the spot of laser light moves on the folder.

Prob 5.2.2 Move the three base spheres slightly apart (about 1 cm), and suspend the fourth so it is close to, but does not touch, the three base spheres. Does this arrangement exhibit the Wada property?

Prob 5.2.3 Replace the tetrahedron with a pyramd: four reflecting spheres placed on the corners of a square, and a fifth placed on the opening between these four. Does this arrangement exhibit the Wada property?

Prob 5.2.4 Place a smaller reflecting sphere in the space between the four spheres of the tetrahedron, or between the five spheres of the pyramid. Does this additional sphere alter the Wada property?

Prob 5.2.5 Arrange eight spheres on the corners of a cube so each sphere touches three others. Does this arrangement exhibit the Wada property? Now move these spheres so nearest neighbors are about 1 cm apart. Does this arrangement exhibit the Wada property?

5.3 Video feedback fractals

Video feedback is a variation on what you see when you are in a barber's chair between parallel mirrors. Replace one mirror with a video monitor and the other with a video camera, feed the output of the camera into the input of the monitor, point the camera at the monitor, and away we go. When the camera zoom is adjusted so the monitor shows a reduced picture of the monitor, we see a sequence of ever-smaller images of monitors. Rotate the camera and the images from a spiral. One of the earliest illustrations of this is Fig. 81, pages 490–491 of Douglas Hofstadter's wonderful *Gödel, Escher, Bach* [155]. Change the zoom so the monitor shows only a portion of the monitor and more complicated patterns, some suggesting chaos, can appear.

Adjust the zoom back so the monitor shows a reduced picture of the monitor, add a mirror or two, and video feedback can generate fractals.

5.3.1 Purpose

We'll learn to construct video feedback images by placing one or two mirrors beside the video monitor. In addition, we'll learn to recognize video feedback with mirrors as a form of IFS or of IFS with memory.

5.3.2 Materials

We'll use a videocamera with a video out jack, a monitor with a video in jack, cables, two mirrors (One square foot wall tile mirrors are available in the tile section of buliding supply stores) and some method to support them (about 10 ft of $1'' \times 2''$ pine, for example and some wood screws), pieces of paper tape, a ruler and protractor, a random IFS program (1.2 of the Mathematica or Python codes), software for determining affine transformations (1.3 of the Mathematica or Python codes) from the images of points, and random IFS with memory software (5.3 of the Mathematica or Python codes).

5.3.3 Background

Here we sketch some of the new material that underlies this lab, and point out techniques from previous labs that we'll use here.

First, in Fig. 5.3 we see some examples of dynamical patterns generated by pointing the videocamera at the monitor and adjusting the zoom and the axial angle. No component of the set-up moves, but the image on the monitor swirls and rotates. Experiments have revealed four classes of dynamical behaviors.

(1) The screen goes blank
(2) A spot of light, stationary or pulsating, appears.
(3) Spots of light move in an apparently unorganized, non-repeating fashion.
(4) Spots of light appear to evolve in an organized fashion.

These correspond to the four classes of cellular automata behavior, the *Wolfram classes*, put forward by Stephen Wolfram [156]. In particular, (3) corresponds to chaotic behavior and (4) to complex behavior.

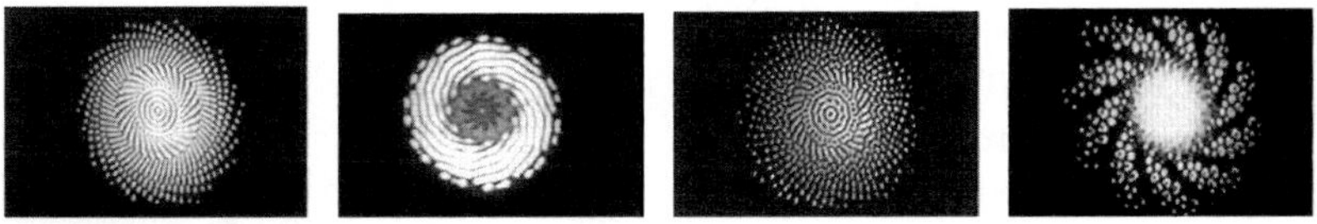

Figure 5.3: Four examples of dynamical video feedback patterns.

This is fertile ground for experimentation. For example, fix everything except zoom and axial angle. Then the space of parameters is 2-dimensional. Explore regions of this space and map the dynamical class as a function of the coordinates of the corresponding point in the parameter plane.

In [157] James Crutchfield presented models for several types of video feedback. Reminiscent of cellular automata, Crutchfield represents how the light intensity at one point and time depends on the light intensities of nearby points at previous times. Crutchfield's technique was able to reproduce some of the simpler video feedback patterns. One of the variations Crutchfield mentioned was the use of mirrors to expand the range of accessible patterns. This is the direction we'll take in this lab, with the restriction that we focus on patterns that are static, not dynamic.

Johannes Courtial and coworkers [158, 159] observed *static* fractal video feedback patterns by adjusting the zoom so the camera sees only a portion of the monitor. The magnified pixels mask portions of the pattern, and iteration generates a pixel mask that reproduces this pattern over many scales. In particular, Fig. 1 (c) of [158] shows an intricate fractal spiral produced by rotating the camera around its axis. A fractal spiral without mirrors. This is an interesting direction, but we'll generate fractals with a mirror.

David Hagar generated video feedback fractals by placing a mirror perpendicular to the monitor and pointing the camera at the edge between the mirror and the monitor. Our recognition of the similarity between the Hagar's video feedback fractals and the canopies of binary fractal trees motivated our approach to use IFS simulations of video feedback.

We extended the class of fractals accessible by video feedback beyond binary trees, and eventually inverted the relation between video monitor and computer screen to think of video feedback with mirrors as an analog version of some IFS.

Some background about binary fractal trees was presented in Sect. 4.3.3 of Lab 4.3, Fractal Painting: Bleeds. The careful study of binary fractal trees began in Chapter 16 of [1] and continued in [160, 118, 161], among others. On the right we see see an example of a self-contacting tree, that is, some descendants of the left branch just touch some of the right branch.

On the left of Fig. 5.4 we see the *canopy*, that is, the branch tips, of a binary fractal tree with $\theta = 40°$ and

$r = 0.565$. On the right we see a video feedback image produced with a single mirror beside the monitor. Except for the mismatch of the orientation, these images are similar. This comparison was our motivation for the development of this lab.

One point to mention is that because the video feedback image is produced with a mirror, the left and right sides are related by reflection. Then why does the IFS of item 4 of Sample 5.3.5 use rotation rather than reflection? The left-right symmetry of the tree is responsible for the fact that reflection and rotation can produce the same result.

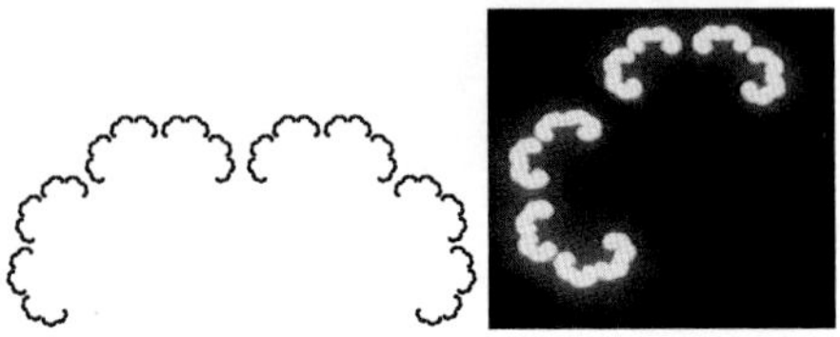

Figure 5.4: Left: binary fractal canopy. Right: one mirror video feedback.

The general formulation of the IFS transformations is presented in Lab 1.1; the random algorithm is described in Lab 1.2.

In this lab we determine IFS parameters by comparing the coordinates of three non-collinear points in the whole fractal with the coordinates of the corresponding points in each piece of the fractal. This method was developed in Lab 1.3.

If part of an image isn't captured in the region of a mirror in the field of the camera, then in order for an IFS to synthesize that image, some combinations of transformations are forbidden. In order to achieve this we'll use the IFS with memory technique of Lab 1.9.

5.3.4 Procedure

First a general comment about monitor settings. Dynamical video feedback experiments depend very delicately on monitor and camera hardware, ambient room light, and perhaps other system parameters. While the precise form of the static patterns produced by our approach also may be delicate, their general forms are not. Nevertheless, with our hardware setting the monitor brightness very low improves the image quality.

The basic setup without mirrors is shown here. This is an old video camera and and old monitor; likely you'll use more modern equipment. One cable lead carries video, two carry sound, the left and right channels. We've connected only one lead to the video out of the camera and to the video in of the monitor. In this set-up, the camera is aimed at the center of the monitor, the camera axis almost perpendicular to the monitor. For our experiments the zoom is adjusted so the camera sees the entire monitor and consequently the camera contracts the image of the monitor. Recall that to generate fractals by IFS the scaling factors r and s must satisfy $|r| < 1$ and $|s| < 1$.

With this setup the main experimental parameters are

- camera position, direction, and axial rotation,
- camera zoom,
- monitor brightness, and
- ambient room light.

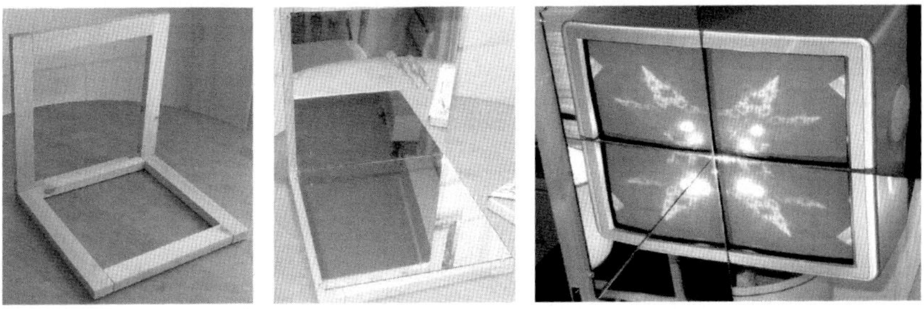

Figure 5.5: Left: A mirror holder. Center: Placement of two mirrors on the holder. Right: How two mirrors produce three copies of the monitor image.

To assemble a mirror holder from about 1″ × 2″ fairly soft wood—pine works well—is straightforward. The left side of Fig. 5.5 is the arrangement we used. The only point requiring care is recalling that the mirrors should be aligned along an edge of the holder, as seen in the center image of Fig. 5.5, so they lie directly along the monitor. Note the placement of the wood screws. The mirrors can be slid in from the right and adjusted to be flush with the right edge of the wooden holder.

The right image of Fig. 5.5 shows how this arrangement of two mirrors can produce three copies of the image on the monitor. Quadrant 1 is the monitor, quadrant 2 is the left-right reflection of the monitor, and quadrant 4 is the up-down reflection of the monitor. Quadrant 3 is a bit more complicated. Part of quadrant 3 is the up-down reflection of part of quadrant 2. Because the composition of a left-right reflection and an up-down reflection is a 180° rotation, part of the image in quadrant 3 is a 180° rotation of a part of the image in quadrant 1. Similarly, the other part of the image in quadrant 3 is a 180° rotation of another part of the image in quadrant 1. With the proper placement of the camera, the image in quadrant 3 is a 180° rotation of the image in quadrant 1.

For the one-mirror arrangement, simply cover the lower mirror with a sheet of dark construction paper or a dark cloth. Any dark opaque, non-reflective cover will do.

5.3.5 Sample A

1. Video feedback without mirrors can produce spirals. See the left side of Fig. 5.6. The tape seeds the spiral. By this we mean that the tape stays at the same place on the screen, so the camera continues to see it and the monitor to project it at the same place.

The angle and zoom of the camera produce a second image of the tape, slightly smaller and rotated. The camera sees the smaller image and produces

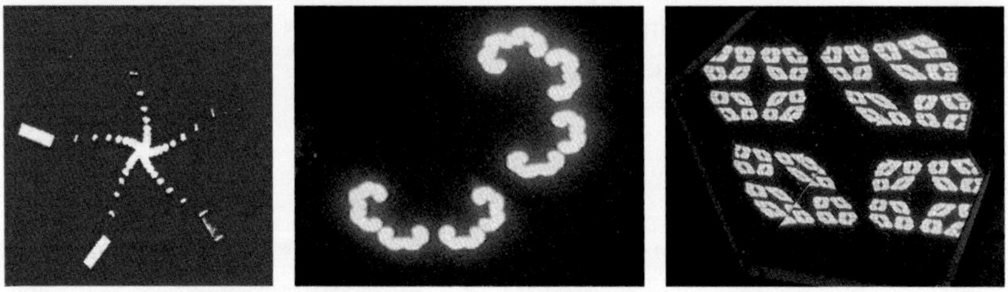

Figure 5.6: Sample video feedback pictures produced with zero (left), one (center), and two (right) mirrors.

a still smaller, and so on, making a spiral. The number of branches of the spiral is determined by the camera angle, as experiments will reveal.

2. Video feedback with one mirror can produce binary trees. See the center of Fig. 5.6. The upper side of the diagonal from upper left to lower right is the monitor, the lower side is the mirror.

The particular structure of the fractal image is determined by the angle of the camera, the position of the camera relative to the monitor, and the angle between the monitor and the mirror. That the image is a binary tree is a result of half the image being a reflection of the other half. In this image the lower side appears foreshortened because the angle between the camera and the mirror is smaller than the angle between the camera and the monitor.

3. Video feedback with two mirrors can produce more complicated fractals. See the right image of Fig. 5.6. The monitor is the upper right quarter of the image; the relation of the other three quarters to upper right is shown in the right image of Fig. 5.5. We presented these results in [162].

The complexity of the image comes from the fact that each mirror reflects the monitor and part of the other mirror. The "part of" is the reason IFS with memory may be needed to synthesize some two mirror video feedback images.

4. IFS synthesis of no mirror video feedback patterns.

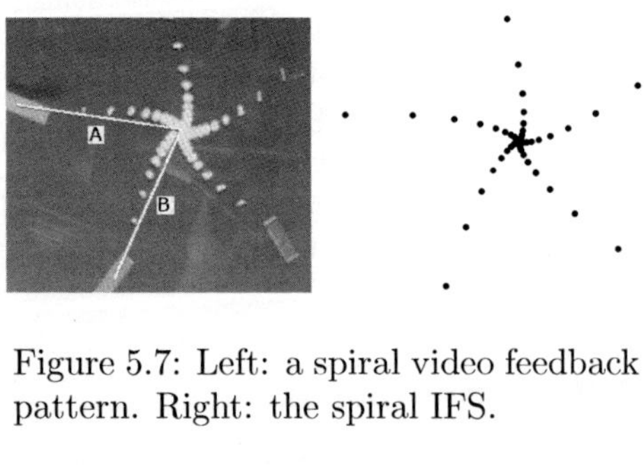

Figure 5.7: Left: a spiral video feedback pattern. Right: the spiral IFS.

Certainly, this video feedback pattern is not fractal. Nevertheless, we will find an IFS to generate this spiral. Here are the steps.

- Place the origin of the coordinate system at the center of the spiral.
- Measure the length of the segment A from the origin to the most distant point of the spiral. For this example we measure 4.5 cm.
- Measure the length of the segment B from the origin to the second-most distant point of the spiral. For this example we measure 4.1 cm.
- The ratio of the lengths B/A gives the scaling factor of one of the transformations. For this example we calculate $r = s = 4.1/4.5 \approx 0.91$.
- The angle between the segments B and A gives the $\theta = \varphi$ value for this transformation. For this example we measure $\theta = \varphi = 71°$.

This information gives the transformation T_1. The IFS that consists of T_1 alone generates a single point, the center of the spiral.

Use a second transformation T_2 to make the point most distant from the center of the spiral. For this, e and f represent the horizontal and vertical displacement of that point. We measure $e = -4.3$ and $f = 0.64$. These two give this IFS, which generates the right image of Fig. 5.7.

	r	s	θ	φ	e	f
T_1	0.91	0.91	71	71	0	0
T_2	0.0	0.0	0	0	-4.3	0.64

5. IFS synthesis of one mirror video feedback patterns.

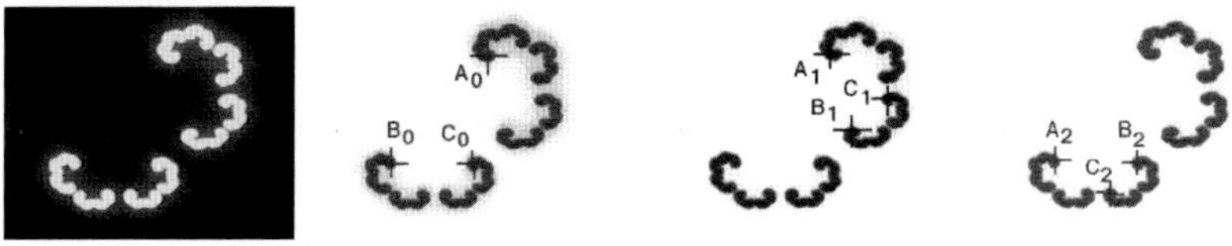

Figure 5.8: Left to right: a one mirror video feedback image, the negative of the left, with initial points indicated, images of the initial points in the right side, images of the initial points in the left side.

On the first image of Fig. 5.8 we see a one mirror video feedback pattern; the second image is its negative (for ease of reading labels), together with the initial points A_0, B_0, and C_0. Noncollinear centers of three prominent spirals in the image are good initial points.

Take the initial point A_0 to be the origin and set up the x-axis to run through B_0. Then measure coordinates $(4.2, 0)$ for B_0 and $(2.6, 1.8)$ for C_0. With this coordinate system, we measure these coordinates for the points $A_1 = A_0$, B_1, and C_1 in the third image of Fig. 5.8, and $A_2 = B_0$, $B_2 = C_0$, and C_2 in the fourth image of Fig. 5.8.

A_1	B_1	C_1	A_2	B_2	C_2
$(0, 0)$	$(1.3, 1.8)$	$(0, 1.8)$	$(4.2, 0)$	$(2.6, 1.8)$	$(3.85, 1.8)$

With these coordinates and the software from 1.3 of the Mathematica or Python codes we obtain these IFS rules.

	r	s	θ	φ	e	f
T_1	0.53	0.59	54	50	0	0
T_2	-0.57	0.52	-48	-43	4.2	0

On the right we see the IFS image generated by these rules. In comparing this image with the left side of Fig. 5.8, note in that figure the x-axis passes through points A_0 and B_0, while in this figure the x-axis is horizontal. Other than this change of orientation, the video feedback image and the IFS attractor look quite similar.

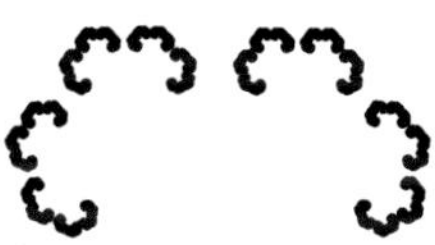

Recall our claim is that when the zoom is adjusted to be a contraction (the image of the monitor is smaller than the monitor), static video feedback patterns can be understood as analog IFS. The five-armed spiral obtained without mirrors was a simple validation of our claim; this one-mirror video feedback example provides stronger support.

6. IFS synthesis of two mirror video feedback patterns.

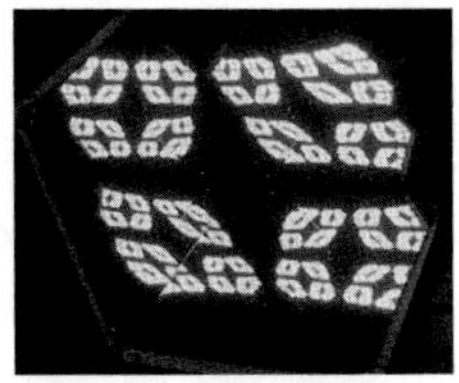

On the right is a two mirror video feedback pattern. The left image of Fig. 5.9 is the negative of the video feedback image, together with the initial points A_0, B_0, and C_0. Take the initial point A_0 to be the origin and set up the coordinate axes running through B_0 and C_0. Then we measure the coordinates $(4, 0)$ for B_0 and $(0, 3.6)$ for C_0.

With this coordinate system, in the right image of Fig. 5.9 we measure these values for the coordinates of the image points.

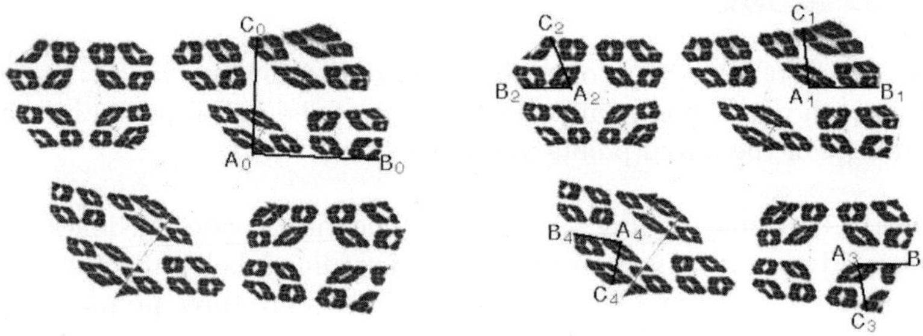

Figure 5.9: Left: the negative of the video feedback image with initial points indicated. Right: the images of the initial points in each of the four pieces of the video feedback fractal.

$$
\begin{array}{lll}
A_1 = (2.4, 2.0) & B_1 = (4.5, 2.1) & C_1 = (2.3, 3.8) \\
A_2 = (-6.0, 1.8) & B_2 = (-7.5, 1.7) & C_2 = (-6.5, 3.2) \\
A_3 = (3.0, -3.2) & B_3 = (4.7, -3.2) & C_3 = (3.4, -4.6) \\
A_4 = (-4.2, -2.8) & B_4 = (-6.0, -2.7) & C_4 = (-4.0, -4.2)
\end{array}
$$

With these coordinates and the software from 1.3 of the Mathematica or Python codes we obtain these IFS rules.

	r	s	θ	φ	e	f
T_1	0.53	0.50	3	3	2.4	2
T_2	-0.38	0.41	4	20	-6	1.8
T_3	-0.43	0.40	180	-164	3	-3.2
T_4	0.45	0.39	176	-171	-4.2	-2.8

Fig. 5.10 shows the corresponding IFS. The match with the original video feedback image is not wonderful. Even a cursory inspection reveals many differences. An instructive experiment is to change the coordinates of an image point by some small amount, rerun the software from 1.3 of the Mathematica or Python codes to obtain a new IFS rule, and then generate the new IFS attractor. Because we measured coordinates on a small image of the video feedback fractal, tiny measurement errors have a proportionately larger effect. With a larger printout of this image you might find an IFS that produces a better match. The first try may not give the bet result.

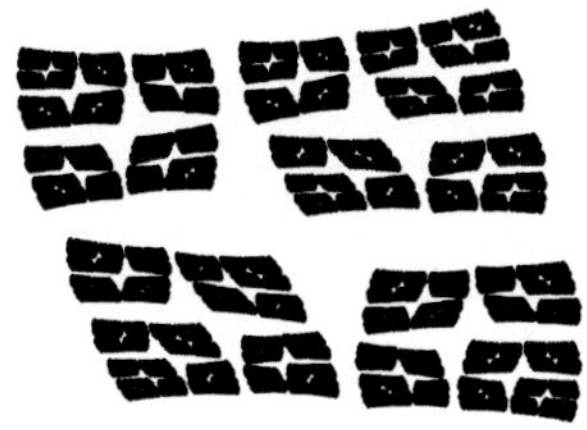

Figure 5.10: 2-mirror IFS attractor.

5.3.6 Sample B

The placement of the camera relative to the monitor and mirrors can cause the truncation of part of the video feedback image. An example is shown on the right. Note that the right portion of the image has only two sub-pieces, while the other three have portions have four sub-pieces. We'll see that while this fractal cannot be generated by IFS, it can be generated by IFS with memory, a technique we explored in Lab 1.9.

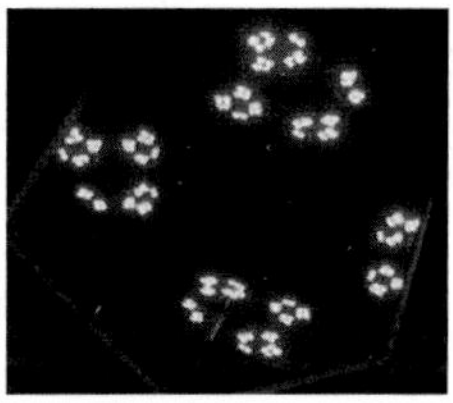

First we'll find the IFS rules to generate a self-similar fractal that contains this attractor as a subset. Then we'll find the transition graph that excludes the regions cut off in the video feedback image. Finally, we'll test whether this IFS with these restrictions generates an attractor close to the video feedback image. Of course, if it doesn't, then we wouldn't include this example, so this final test has no dramatic tension.

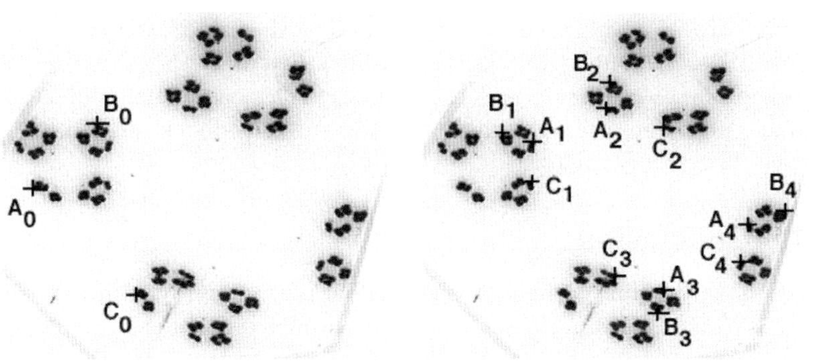

Figure 5.11: Left: the initial points for this video feedback image. Right: the images of the test points in each of the four pieces of the video feedback fractal.

On the left side of Fig. 5.11 we see the negative of the video feedback image with three initial points A_0, B_0, and C_0 indicated. We take $A_0 = (0,0)$ and measure $B_0 = (1.7, 1.7)$ and $C_0 = (2.9, -2.9)$. On the right side of Fig. 5.11 we see four sets of target points A_i, B_i, and C_i.

A_1	B_1	C_1	A_2	B_2	C_2
$(2.1, 1.3)$	$(1.2, 1.5)$	$(2.1, 0.2)$	$(4.3, 2.1)$	$(4.4, 2.8)$	$(5.9, 1.7)$
A_3	B_3	C_3	A_4	B_4	C_4
$(5.9, -2.8)$	$(5.7, -3.4)$	$(4.5, -2.3)$	$(8.2, -1)$	$(9.3, -0.8)$	$(8, -2)$

With these coordinates and the software from 1.3 of the Mathematica or Python codes we obtain these IFS rules.

	r	s	θ	φ	e	f
T_1	-0.30	0.36	26	47	2.1	1.3
T_2	0.33	0.37	24	42	4.3	2.1
T_3	0.31	0.32	-163	-145	5.9	-2.8
T_4	0.31	0.43	-21	-57	8.2	-1

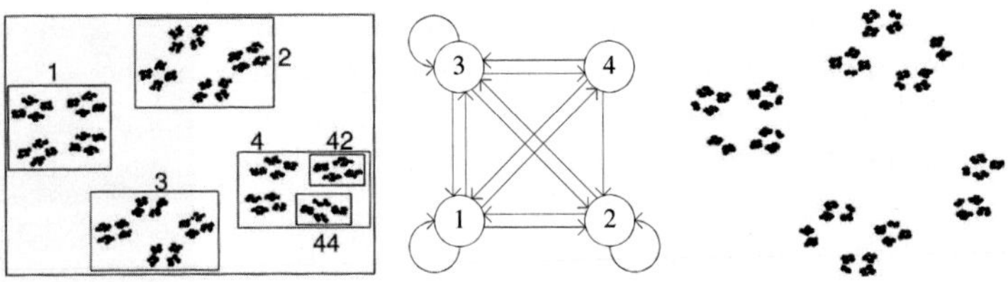

Figure 5.12: Left: IFS attractor and some addresses. Center: the transition graph. Right: the attractor of the random IFS with the restrictions of the transition graph.

In the left image of Fig. 5.12 we see the attractor of the IFS with the length-1 addresses shown and also the two length-2 addresses, 42 and 44, that are empty in the video feedback picture. In the middle we see the transition graph with $2 \to 4$ and $4 \to 4$ forbidden. On the left is the corresponding random IFS with memory (5.3 of the Mathematica or Python code) attractor. The match is good, but not perfect. More careful measurements of the initial and image points in a larger printout of the video feedback image may give a more accurate IFS attractor.

5.3.7 Conclusion

The more familiar form of video feedback involves dynamical behavior, analogous to the Wolfram classes of cellular automaton behavior. Using mirrors we can produce stable fractal video feedback patterns. This form of video feedback can be viewed as analog versions of IFS and of IFS with memory.

5.3.8 Exercises

Prob 5.3.1 Generate a video feedback image without mirrors and find the corresponding IFS. Use one piece of tape to seed the spirals.

Prob 5.3.2 Generate video feedback images without mirrors and find the corresponding IFS. Use two, three, and four pieces of tape to seed the spirals.

Prob 5.3.3 Generate video feedback images with one mirror and find the corresponding IFS.

Prob 5.3.4 Starting with image of Exercise 5.3.3, change the camera angle and generate video feedback images with one mirror. Find the corresponding IFS. How does this IFS differ from that of Exercise 5.3.3?

Prob 5.3.5 Generate video feedback images with two mirrors and find the corresponding IFS.

Prob 5.3.6 Generate video feedback images with two mirrors and adjust the placement of the camera so the corresponding IFS is an IFS with memory. Find the corresponding IFS and the transition graph. Compare the attractor of that IFS with memory with the video feedback image.

Prob 5.3.7 Can you position the camera so that an IFS with 2-step memory (Lab 1.10) is needed to simulate the video feedback image?

Prob 5.3.8 Can you use small opaque stickers to cover parts of the video-camera lens and so generate video feedback patterns that are simulated by IFS with memory in this way?

5.4 Electrodeposition

Electrodeposition is a fairly simple way to grow beautiful dendritic fractals. While the list of materials is longer than that of most labs, once the cell is built (not difficult if you have some practice with tin snips) and the zinc sulfate solution is made, the experiment consists of flipping a switch and waiting a while, maybe several hours, until the aggregate grows to the size you want.

Experiments for growing fractal aggregates by electrodeposition go back at least to 1984, see [163, 164] for example, only five years after the publication of *Fractals: Form, Chance, and Dimension* [165], the first English text on fractal geometry and the precursor of *The Fractal Geometry of Nature* [1]. This is a well-studied process. We'll find a simple way to implement it, and to explore some variations.

5.4.1 Purpose

We'll grow dendritic fractals by aggregation of zinc ions to the cathode of a simple cell constructed from a Petri dish and a strip of zinc. Similarities with some of the vegetable fractals of Lab 4.8 offer the opportunity to think of why, and how, both living and non-living processes often produce similar fractal patterns.

5.4.2 Materials

We'll use zinc sulfate monohydrate, distilled water, a 30 V 3 A DC power supply (for most experiments, two 1.5 V batteries can work, though with less fine voltage control), a 100 ml graduated cylinder, a 1000 ml Erlenmeyer flask and 1000 ml beaker, a mortar and pestle, a magnetic stirrer (can be replaced with a glass stirring rod and some patience), 120 mm × 20 mm Petri dishes, 3 ml transfer pipettes, a digital laboratory scale, 3 1/2″ diameter coffee filters, a 0.02″ × 12″ × 12″ or some other size (8″ × 11″ for example) zinc sheet, tin snips, 12 gauge solid copper wire, and a helping hand alligator clamps and base, and additional aligator clamps.

5.4.3 Background

As zinc ions diffuse through a zinc sulfate solution they move by Brownian motion. Individual solution molecules move under the influence of heat and collide with zinc ions. Zinc ions, which are positively charged, aggregate on the cathode in the cell and grow into the dendrite. For this reason, electrodeposition is an example of *diffusion-limited aggregation*, DLA. This process models many other types of aggregation, for instance, the way snowflakes form.

DLA has been and continues to be an active area of research. In physical experiments it models electrodeposition (this lab), dielectric breakdown (6.6.1), viscous fingering (Lab 5.5), and even the growth of bacteria in Petri dishes with a low level of nutrient in the agar. For this last, see [166, 167, 168], among many other references. We couldn't figure out a way to do this without an autoclave—the oven of your stove is *not* a suitable substitute—and because improper use of an autoclave can hurt you badly, we decided to skip the bacteria-growth lab.

More active still are computer experiments, simulations of DLA. An example is shown here. To generate this image, particles (really, just points) are released from a large circle centered at the origin. Each of these particles walks randomly: the direction of each step is selected uniformly randomly, and the length of the step is selected according to a normal distribution, so the particle makes many more small moves than long moves. When a particle wanders close enough to the center, or to a particle attached to the center, it sticks. That's it. That's how this image was drawn.

These and other simulations have revealed a host of subtle problems. We mention this because once people understand how DLA works, it seems to give a simple explanation for many physical processes. While in broad outline this is true, in detail nature is a bit more complicated.

The spaces between the branches of a DLA cluster are called *fjords*, referencing familiar features of the Norwegian coast. Typically, regions deep inside fjords experience little additional growth. This is because a randomly walking particle is unlikely to reach the cluster deep in a fjord, near the circle, for example. On the other hand, a particle can easily reach regions of the aggregate near the square. Most growth of a DLA cluster occurs near its periphery.

If you think about the geometry of a branch, when a particle is added along the side of the branch, that additional particle can be approached from more directions that can a particle on the branch. So we see that branches tend to grow side branches. When these have grown a bit, they grow their own side branches. This side branching continues for many levels as the dendrite grows; the same rules govern the growth of branches at all levels. This is the source the self-similarity of a DLA cluster.

Zinc ions diffusing through the zinc sulfate solution are modeled by particles walking randomly to form a DLA cluster. This is why the zinc dendrites we grow look a bit like DLA clusters. But only in their rough outlines: much of the fine structure is determined by the physical chemistry of crystallization. Then, too, if the cathode is placed slightly off-center in the Petri dish, aggregation is more rapid on the side closer to the anode. DLA describes the general form; particulars are determined by the details of the experimental set-up and the mechanics of crystal formation.

5.4.4 Procedure

We'll divide the procedure into three categories: prepare the solution, prepare the cell, and run the experiment.

Prepare the solution. We'll make about 500 ml of 1 molar zinc sulfate solution. Put about 400 ml of distilled water in a 1000 ml beaker. We use zinc sulfate monohydrate, so measure 89.75 g of zinc sulfate monohydrate with the digital scale. Details of the calculation are in Appendix B.15. We found the zinc sulfate to be a bit lumpy, so we ground it with the mortar and pestle. Slowly combine the powdered zinc sulfate with the water in the beaker and mix it until all the powder has dissolved. A magnetic mixer is useful, but a glass stirring rod and some time also work. Add distilled water until the solution reaches a volume of 555 ml. Pour the solution into a 1000 ml Erlenmeyer flask and seal the top with plastic wrap and a rubber band to prevent evaporation.

Prepare the cell. A 120 mm Petri dish has circumference close to 15″, so we need an extra step to build the anode. Mark off a 12″ by 1/2″ strip on the zinc sheet. One side of the zinc sheet should be covered with plastic. Mark the uncovered side. Cut along this mark with tin snips. Cut with the tips of the blades. While you cut the zinc, the strip will curve. The direction of the curvature will put the shiny side of the zinc, after the plastic cover is removed, facing into the Petri dish. Mark and cut another strip, 3″ by 1/2″. Bend the strips so both ends touch the table when the strips are placed on a table. Put the strips in a Petri dish, fairly tightly against the side of the Petri dish. With alligator clamps attach the zinc strips at their overlaps. Clip the positive (red) lead of the power supply to the zinc strip. Place a coffee filter on the bottom of the Petri dish. Strip about 1 cm of insulation from both ends of the 12 gauge solid copper wire. With the helping hands alligator clamp, position the piece of wire over the center of the Petri dish, within a few mm of the coffee filter. Attach the negative (black) lead of the power supply to the 12 gauge solid copper wire. Gently pour enough of the zinc sulfate solution into the Petri dish to cover about a mm of the cathode (the wire connected to the negative lead) and cover the bottom of the zinc strip. The cell is ready.

Run the experiment. Turn on the power supply and select a voltage. We used 3 V for most experiments. With 6 V, the solution became quite warm. Let the dendrite grow until it reaches the size you want. Do not shake, bump, or jostle the Petri dish. Be sure to stop the current before the dendrite reaches the zinc strip. After you've disconnected the leads and removed the wire (cathode), use a transfer pipette to remove the remaining solution. Then carefully remove the zinc strips. Set the Petri dish aside and let it dry overnight.

5.4.5 Sample A

In all these experiments we'll set the power supply to 3 V. Place the cathode so only about 1 mm of the bare wire is in the solution. Initially the current is about 0.01 A. As the dendrite grows and more area is available for zinc ions to land, the current increases. Stop the current when the dendrite has reached the size you want.

First grow a dendrite in a 0.5 molar solution of zinc sulfate. To make a 0.5 molar solution, pour 100 ml of 1 molar solution into an Erlenmeyer flask, add 100 ml of distilled water and mix thoroughly. To grow an aggregate with diameter about 8 cm can take a few hours. Several levels of branching still can be seen in smaller clusters if time is an issue. See the left image of Fig. 5.13.

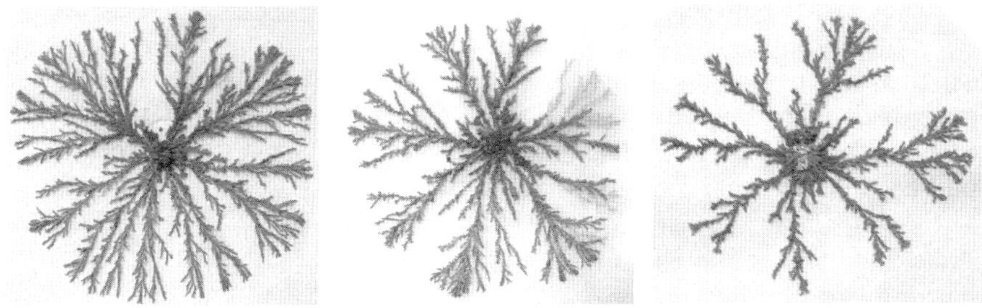

Figure 5.13: Dendrites grown in 0.5 molar (left), 0.25 molar (center), and 0.125 molar (right) solutions of zinc sulfate.

Next, grow a dendrite in a 0.25 molar solution. To make a 0.25 molar solution, pour 50 ml of 1 molar solution into an Erlenmeyer flask, add 150 ml of distilled water and mix thoroughly. In the center picture, at around 3 o'clock and a few other places, you notice several branches that are lighter than most. These are shadows: a bit of a dendrite punctured the filter paper and continued to grow in the thin layer of solution below the filter. See the center image of Fig. 5.13.

Finally, grow a dendrite in a 0.125 molar solution. For this, pour 25 ml of 1 molar solution into an Erlenmeyer flask, add 175 ml of distilled water and mix thoroughly. See the right image of Fig. 5.13. As the zinc sulfate concentration decreases, so does the density of the branches.

Here we see dendrites grown in a 0.125 molar solution for four and a quarter hours. The thin branching of the right image of Fig. 5.13 suggested that a longer growth period could reveal more levels of branching without overlaps. The right image is a magnification of part of the left image.

The longer growing time may make the 0.125 molar cluster appear more like the 0.25 molar cluster with shorter growing time. A possible explanation is that the zinc sulfate concentration influences the rate of branch growth, not the eventual density of branches. All three clusters of Fig. 5.13 grew in about two hours. Is it plausible that if we let a cluster continue to grow, eventually it would fill in something close to a solid disk? A straightforward test of this conjecture won't work, because the longest branches will bridge the cathode to anode gap. Can you find another approach to test this? On the other hand, does fjord screening make this explanation plausible?

5.4.6 Sample B

In a 0.5 molar solution grow dendrites with two cathodes. Place the cathodes about 2 cm apart, near the middle of the Petri dish. For the set-up we used a second piece of 12 gauge solid copper wire, and as we see in the left figure, twisted together the ends of the cathodes that do not go in the solution.

Connect the negative lead from the power supply to the twisted wires. Make sure the ends of the cathodes are very close to the same height, a few mm, above the bottom of the Petri dish. Carefully pour in enough of the zinc sulfate solution to cover the bottoms of the cathodes, and also some of the zinc strip around the inside of the Petri dish. Set the power supply to 3 V.

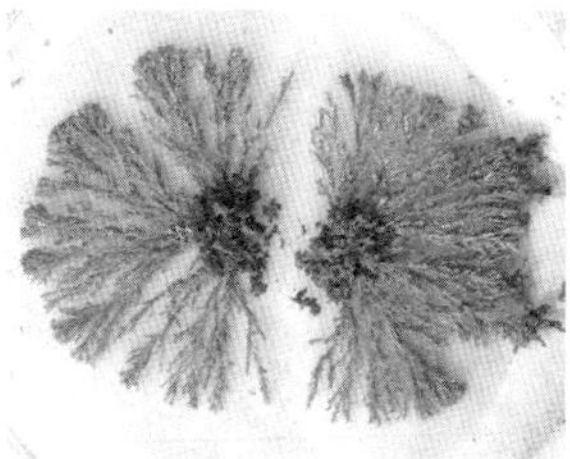

In the second image above we see that each aggregate grows little on the side facing the other aggregate. The image on the right here, after we've turned off the power and stopped the cluster growth, shows left and right halves, with little growth between them. The mostly empty channel between the halves comes about from the mechanism that causes fjords in a DLA cluster. The walls of the fjords likely will intercept a randomly walking particle before it travels deep within the fjord. Similarly, the left and right halves likely will intercept a zinc ion before it travels far along the middle channel.

Cathode geometry can be varied in many other ways. We'll suggest some in Exercises 5.4.6–5.4.10.

5.4.7 Sample C

They're not all pretty. If some of your experiments don't produce such pretty pictures, don't be concerned. Some of ours didn't turn out so well, either. Possible causes include uneven placement of the anode around the Petri dish, so the amount of anode immersed in the solution, and consequently the local ionic current, varies. The cluster grows most rapidly in regions of highest ionic current.

The first two images of Fig. 5.14 were made with a 0.5 molar solution, the other four with a 1 molar solution. All were run with 3 V. None of the dozen dendrites we grew with a 1 molar solution looked roughly symmetrical, like those of Fig. 5.13. We speculate that this concentration is so high that a random extension toward the anode is amplified into a branch that continues to grow more rapidly than all of the cluster except other randomly amplified branches.

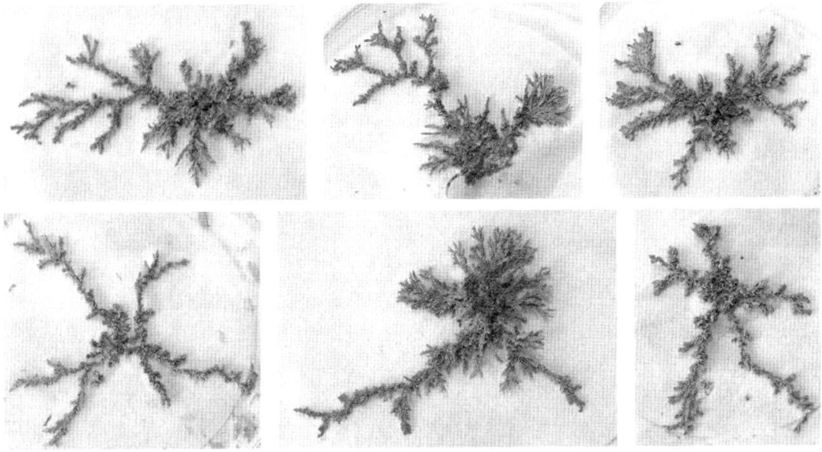

Figure 5.14: Some irregular aggregates.

Generally we avoided all sorts of Rorschach inferences for the many fractal images we've seen. However, it is difficult not to give the name "DLA man" to the right image of the second row.

5.4.8 Conclusion

While the experimental set-up is a bit complicated, and several hours are needed to grow each dendrite, electrodeposition is a very rich ground for experiments in fractal growth. In the Exercises we'll suggest some variations. You may find many other interesting directions to explore.

5.4.9 Exercises

Prob 5.4.1 Repeat the experiments of Sample 5.4.5. Comparing your results with ours may help you refine your technique. Do have a look at Sample 5.4.7 if your results are much different from ours.

Prob 5.4.2 Vary the voltage: for three samples of 0.5 molar solution, grow dendrites with 6 V, 3 V, and 1.5 V.

Prob 5.4.3 Grow a dendrite in 0.5 molar solution with the cathode completely immersed, that is, lower the cathode so it touches the filter at the bottom of the Petri dish. Set the power supply to 3 V.

Prob 5.4.4 Grow a dendrite in a 0.5 molar solution with pulsed voltage. For example, set the voltage to 3 V for the first 10 minutes, then 6 V for the next 10 minutes, and repeat this pattern.

Prob 5.4.5 Repeat the experiments of Sample 5.4.6 but vary the positions of the cathodes.

Pictures of our experiments for Exercises 5.4.6–5.4.10 are given in the solutions. There you'll find only the pictures; any deductions are left to you. But at least the pictures can show you how we constructed the cathodes in each case.

Prob 5.4.6 Grow an aggregate in a 0.5 molar solution at 3 V with a linear cathode. That is, a strip of zinc attached to the negative lead and suspended near the middle of the Petri dish.

Prob 5.4.7 Grow an aggregate in a 0.5 molar solution at 3 V with a circular cathode, a strip of zinc bent into a rough circle—cylinder, actually—attached to the negative lead and suspended near the middle of the Petri dish.

Prob 5.4.8 Grow an aggregate in a 0.5 molar solution at 3 V with a right isosceles triangular cathode suspended near the middle of the Petri dish. Does the angle, 45° or 90°, influence the branch structure?

Prob 5.4.9 Grow an aggregate in a 0.5 molar solution at 3 V with an L-shaped cathode suspended near the middle of the Petri dish. The structure on the outside corner of the L should be similar to that of the right angle in Exercise 5.4.8. What about the inside corner of the L? Does each side shield the other?

Prob 5.4.10 Grow an aggregate in a 0.5 molar solution at 3 V with a Koch snowflake cathode suspended near the middle of the Petri dish.

5.5 Viscous fingering

A Hele-Shaw cell consists of two parallel flat plates, usually glass or plexiglass, separated by a small distance. The dynamics of fluids in three dimensions can be difficult to see, but fluid flow between the two plates can be be studied visually. The interaction of fluids with different densities or viscosities can be studied by adding dye to one fluid. Here we'll inject a low viscosity fluid (water) in the center of a higher viscosity fluid (glycerine or a guar gum solution). Then we'll study the patterns that result. The main variables are the viscosities and the injection speed.

Another parameter we can vary is the size of the opening through which the dyed water is injected into the higher viscosity fluid. Rather than build a cell with a different diameter tubing, we'll cover the tube opening on the top plate with a bit of duct tape and puncture the tape with a pin. Perhaps you can think of other variations.

5.5.1 Purpose

We'll study patterns produced when a less-viscous fluid displaces a more viscous fluid. Although with our glycerine experiments we haven't been able to reproduce the DLA-like patterns some authors report (for example, Figures 3 and 4 of [173]), our guar gum solutions did produce patterns that more closely resemble DLA.

5.5.2 Materials

Glycerine, ink, clear PVC plastic tubing ($1/8''$ inside diameter), glue, two $12'' \times 12'' \times 0.12''$ clear acrylic sheets, four binder clamps, one syringe with tip outside diameter about 4 mm and volume 10 ml, four thin objects with thickness less than 1 mm (we used small zinc strips left over from Lab 5.4), and a drill to make a hole in one of the acrylic sheets. The liquids can flow out from between the acrylic sheets, so to contain the messiness, we performed these experiments in a large cat litter box.

For the second set of experiments, guar gum powder, 1000 ml beaker, a digital laboratory scale, and a magnetic stirrer (or, as mentioned in Lab 5.4, a glass stirring rod and some patience can substitute for the magnetic stirrer).

5.5.3 Background

The Hele-Shaw cell was invented by Henry Selby Hele-Shaw, a British mechanical engineer, probably best known as the co-inventor, with T. E. Beacham, of the variable-pitch propellor, an innovation that contributed to the success of British aircraft in World War II. Hele-Shaw studied the flow of water around obstacles by pumping water through the cell and injecting dye to visualize the streamlines. We'll use the cell in a different way.

When a less-viscous fluid displaces a more viscous fluid, the interface between the fluids is unstable. Smooth curves are perturbed into bumps off of curves, which in turn grow their own bumps, and so on. We've seen a version of this phenomenon in Labs 4.1 and 4.2, where the less-viscous fluid is air and the more viscous fluid is paint, acrylic or watercolors.

Background on viscous fingering is presented in Chapter 4 of *Fractals* [169] by Jens Feder, and in "Viscous fingering in porous media" [170] by George Homsy. The first report of viscous fingering in Hele-Shaw cells was P. Saffman and G. Taylor's "The penetration of a fluid into a medium or Hele-Shaw cell containing a more viscous liquid" [171]. Our experiment is similar to that Jing-Den Chen reported in "Growth of radial viscous fingers in a Hele-Shaw cell" [172] and in "Radial viscous fingering patterns in Hele-Shaw cells," [173].

As a lower-viscosity fluid (dyed water, for example) is injected into a higher-viscosity fluid (glycerine or guar gum derivative), the boundary between the two fluids is unstable: small perturbations cause branches to split. When the high-viscosity and low-viscosity fluids have a high interfacial tension (glycerine and water), the thickness of the fingers increases as the fingers grow. With low interfacial tension (guar gum derivative and water), the branches maintain nearly constant thickness and the viscous fingering pattern more closely resembles the electrodeposition patterns of Lab 5.4.

5.5.4 Procedure

Drill a hole in the middle of one of the acrylic sheets. The hole diameter should be equal to or a bit larger than the outside diameter of the PVC tubing. Cut a piece of tubing about 8 inches long and glue one end of the tube to the sheet, so the tube goes through the sheet and the end of the tube is flush with the other side of the sheet.

For the glycerine experiments, place 15 or 20 ml of glycerine near the center of the lower acrylic sheet, the sheet without the attached tube. Next, draw about 8 ml of water dyed with ink into a syringe and attach the syringe to the tubing. Hold the acrylic sheet with the tubing so the tubing and syringe lie below the sheet and slowly depress the syringe plunger so the dyed water fills the tube. The goal is to push (almost) all of the air out of the tubing without forming a bead of dyed water above the acrylic sheet (in this orientation) where the tubing joins to the sheet. This takes some care, and likely some practice. Air bubbles can perturb the flow pattern when water is injected into the glycerine, while a bead of dyed water can diffuse into the glycerine when the cell is assembled, wetting a path that can guide much of the injected water in a single direction reducing the growth of other branches.

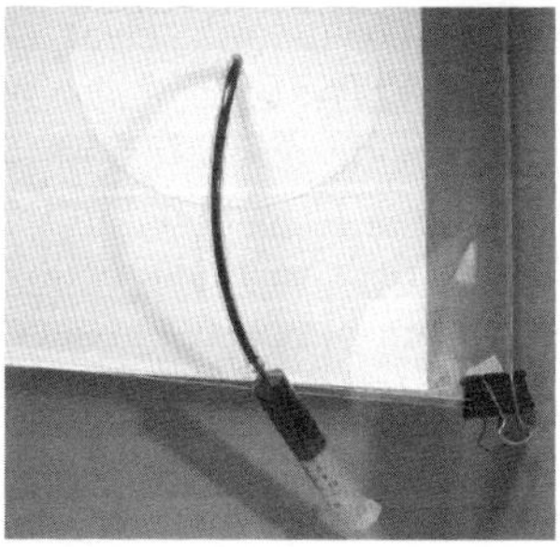

Figure 5.15: Viscous fingering setup, viewed from above.

Place the four spacers, one at each corner of the lower acrylic sheet. Invert the top acrylic sheet so the tubing and syringe are above the sheet and gently place it on top of the lower acrylic sheet. Attach the binder clamps, one at each corner, to secure the two acrylic sheets together. Inject the dyed water into the glycerine, being careful not to let a branch (or "finger") of the injected water extend beyond the edge of the glycerine.

For the guar gum solution experiments, use the same steps as the glycerine experiments, with the guar gum solution replacing the glycerine. Too much guar gum powder and the solution becomes a paste. The molecular weight of guar gum is 535.15 g/mol. We dissolved 1.07 g of guar gum powder in 100 ml of water (the magnetic stirrer is very useful here) to obtain a 0.02 molar solution.

5.5.5 Sample A: Glycerine experiments

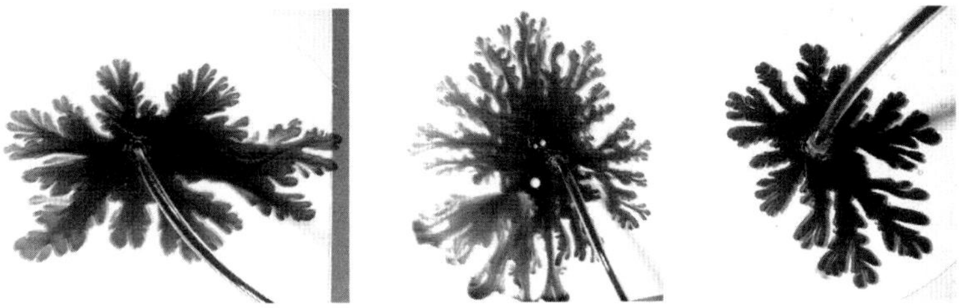

Figure 5.16: Left to right: slow injection, fast injection, and injection through a pin hole.

In Fig. 5.16 we see results of three experiments with dyed water injected into glycerine. For the left image the water was injected slowly, quickly for the middle image. We see that quick injection produces thinner branches and more levels of branching than does slow injection. For the pinhole injection, cover the opening of the tube in the acrylic sheet with a very small piece of duct tape and punch a pin hole through the tape. The number of branching levels appears to be the lowest of the three.

5.5.6 Sample B: Guar gum solution experiments

Figure 5.17: Left to right: first test, second test early, second test late.

In Fig. 5.17 we see results of two experiments where dyed water is injected into about 20 ml of 0.02 molar guar gum solution. For the experiment shown in the left image, as we continued to inject water the top branch stopped growing while the bottom branches continued, producing an asymmetrical pattern. The middle image was taken soon after the start of the second experiment; the right image was taken a bit later. Notice that in contrast with Sample 5.5.5, here the branches are much thinner and resemble more closely the electrodeposition patterns of Lab 5.4. But whereas the zinc electrodeposits were grown by ions diffusing from far away and attaching to the growing cluster, here the water channels grow from the inside out: grow, split, and grow again.

5.5.7 Conclusion

The viscosity difference between the fluids drives the instability that forms branches off branches at the interface of the fluids. Water and glycerine exhibit relatively high interfacial tension, while water and guar gum solution have low interfacial tension. This suggests that branch thickness is inversely related to interfacial tension, but we have too little data to do anything other than suggest. We'll think of viscous fingering as another way to generate fractals, another example in the collection of processes that produce similar fractal patterns.

5.5.8 Exercises

Prob 5.5.1 Do your own versions of the experiments in Sample 5.5.5. Do your results differ from ours?

Prob 5.5.2 Do your own versions of the experiments in Sample 5.5.5 but replace the dyed water with dyed isopropyl alcohol. Do the patterns differ from those of Exercise 5.5.1?

Prob 5.5.3 Do the experiments of Sample 5.5.5 using vegetable oil and dyed water.

Prob 5.5.4 Do the experiments of Sample 5.5.5 using vegetable oil and dyed isopropyl alcohol.

Prob 5.5.5 Do your own versions of the experiments in Sample 5.5.6. Do your results differ from ours?

Prob 5.5.6 Do your own versions of the experiments in Sample 5.5.6 with a 0.01 molar guar gum solution.

Prob 5.5.7 Do your own versions of the experiments in Sample 5.5.6 with a 0.04 molar guar gum solution.

Prob 5.5.8 Repeat Exercise 5.5.5 replacing dyed water with dyed isopropyl alcohol.

Prob 5.5.9 Repeat Exercise 5.5.6 replacing dyed water with dyed isopropyl alcohol.

Prob 5.5.10 Combining the observations of these exercises, what general trends do you see?

5.6 Crumpled paper patterns

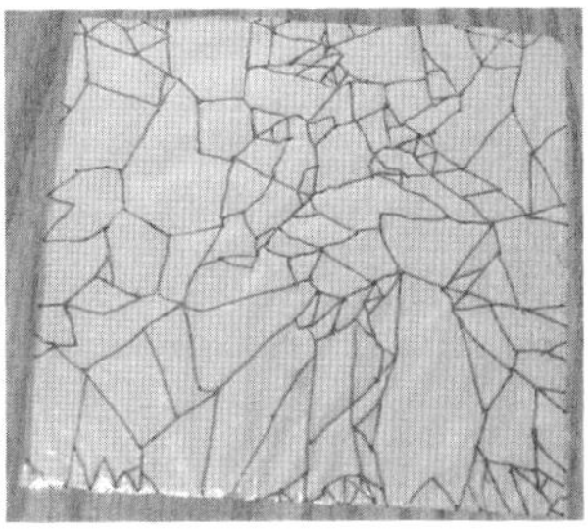

In Lab 2.2 we measured the mass dimension of crumpled paper balls. We had a glimpse of the source of the complexity of these crumpled paper balls when we bisected one and inspected the internal fold structure in Exercise 2.2.6 of that lab.

In this lab we'll take a different approach. We'll crumple the paper, unfold it, carefully mark and measure the creases and ridges, and find the total length of the creases and ridges. We'll explore how this total length grows as the paper is re-crumpled and flattened.

5.6.1 Purpose

We'll measure the total length of creases and ridges in crumpled mylar that has been flattened. In particular, we'll investigate the observation [177] that with successive re-crumpling the total length increases logarithmically. In Exercise 5.6.3 we'll suggest a way to approach a second observation, that the increase from the nth crumpling to the $(n+1)$st does not depend on the history of crumplings before the nth.

5.6.2 Materials

Several 10 cm $\times$10 cm mylar sheets (*very* sharp scissors to cut the mylar), a plexiglass cylinder with inner diameter about 3 cm (we used a cylinder with inner diameter 1 and 1/8 inches) and length between 15 and 20 cm, a plunger that fits snugly into the plexiglass cylinder (we used a 1/4 inch dowel rod, the plastic top of a yogurt container, and tape), an ultra-fine point permanent marking pen, a millimeter scale ruler, a hands-free magnifying glass, and probably some way to treat headache or eyestrain that may accompany marking the crumpled mylar.

5.6.3 Background

Before computer experiments became common in math, college deans loved math departments because they were so inexpensive. A mathematician needed paper, pencils, and a wastebasket. Only philosophy departments were cheaper: they didn't need wastebaskets. In the setting of Lab 2.2, we could argue that mathematicians have made fractals for centuries. Every wrong direction in a proof, every incorrect calculation, was crumpled to a fractal and tossed in the wastebasket.

In this lab, we'll look at crumpling in a different way. We'll take the crumpled paper out of the wastebasket, gently flatten it, and measure the total length of the creases. Why should this be interesting?

(Historically, we'll mention that once Benoit's uncle Szolem retrieved a paper from his wastebasket and handed the paper to Benoit to read on the train ride home. The paper was a review of George Zipf's book [174] on word frequencies. Thinking about this idea was one of Benoit's first excursions into scaling, which eventually led to fractals.)

We know that some statistics of crumpled geometry are repeated across many experiments [175]. In 2018 Omer Gottesman and colleagues [176] reported the results of experiments in which thin mylar sheets were confined in a cylindrical chamber and compressed until they crumpled. An uncrumpled and flattened sheet retains a record of its crumples in the pattern of creases. The sum of the lengths of these creases exhibits some interesting properties:

• under repeated crumpling and flattening, for n crumplings the total length $L(n)$ of creases grows as $\log(n)$, specifically

$$L(n) = a(1 - \Delta) \log\left(1 + \frac{bn}{\Delta}\right) \tag{5.2}$$

where Δ is the compaction ratio, and a and b are constants that are adjusted to fit the data, and

• regardless of their past history of crumpling, for two sheets with about the same total length of creases, under additional crumpling both will accumulate about the same additional damage.

In 2021 Jovana Andrejević [177] and coworkers extended these experiments and constructed a simple geometric model that recovers the two properties observed by Gottesman. Andrejević views crumpling as an application of fragmentation theory, where the sheet is partitioned into flat facets separated by ridges or creases. Then re-crumpling is subdivision of the facets. A central insight is the addition of feedback into the fragmentation process. Smaller facets make the sheet more pliant and reduce the rate of subsequent fragmentation. This makes plausible the logarithmic growth of total crease length; some of the heavy lifting of the paper is the calculation that supports this intuition.

When the mylar sheet is compressed, additional fragmentation follows from the geometric incompatibility of the folded sheet and its confinement in a cylinder. Regardless of the starting configuration, iterated re-crumpling in a cylinder quickly converges to a "crumpled attractor," thus erasing any memory of the initial configuration.

This logarithmic growth is observed in other settings. For example, under repeated impact of dust in vacuum and in microgravity environments, the total compaction of dust scales as the log of the number of impacts [178]. This may have implications for the formation of planets in the solar nebula [179, 180]. Does this hint at an underlying universal mechanism, or is it just a coincidence that both the total length of creases under iterated crumpling and the total compaction of dust under iterated impacts both exhibit logarithmic scaling? We don't know. Here we'll settle for a more modest question: to what extent can we reproduce the logarithmic increase in total crease lengths?

5.6.4 Procedure

Two opposing issues guide our crumpling mechanism:

• If the mylar sheet is crumpled by too little compaction (don't push the plunger in very far), the flattened sheet may not have enough facets to reveal

a pattern.
• If the sheet is crumpled by too much compaction, then it may undergo a sequence of fragmentations across multiple scales, that then would appear to have occurred in a single crumpling iteration.

Following Andrejević, we'll crumple a 10 cm square of mylar to a height of 2.7 cm. That is, we'll use a compaction ratio $\Delta = 2.7/10 = 0.27$.

First, prepare the chamber and plunger. Cut a 15 to 20 cm length of the plexiglass cylinder. Draw a vertical line 10 cm long from the cylinder bottom and make marks at distances 10 cm and 2.7 cm from the cylinder bottom. See the left image of Fig. 5.18.

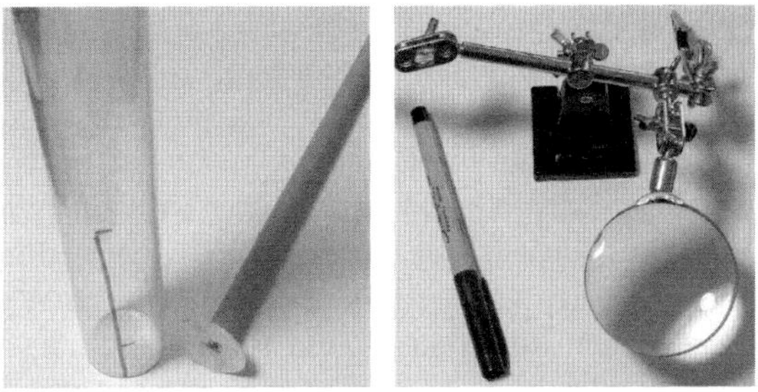

Figure 5.18: Left: marked cylinder and plunger. Right: hands-free magnifier and pen.

The plunger should fit snugly inside the cylinder. Our cylinder has inside diameter 1 and 1/8 inches, so a 1 and 1/8 inch dowel rod should make a good plunger. Unable to easily find this size dowel, we took a different approach. We cut the top of a plastic yogurt container into an approximate disk with diameter a bit less than 1 and 1/8 inches, then used a small tack to fasten the disk to the end of a 1/4 inch dowel. Push the disk into the cylinder and note the locations of the gaps between the disk and the cylinder. Remove the disk from the cylinder, build up the gaps with tape, push the disk into the cylinder, note the remaining gaps. Continue until the disk fits fairly snugly into the cylinder. Not too snugly, because we must compact the mylar with the plunger. The fit must be snug enough that the mylar can't fit between the plunger and the cylinder. See the left image of Fig. 5.18.

Now cut a 10 cm ×10 cm square from the mylar sheet. Mylar tears easily, so use very sharp scissors. Good sewing scissors, or the tiny scissors of a Swiss Army knife, work well; craft scissors not so much. Thin mylar also creases very easily, so carefully roll it up and place it at the bottom of the plexiglass cylinder as shown in the left image of Fig. 5.19.

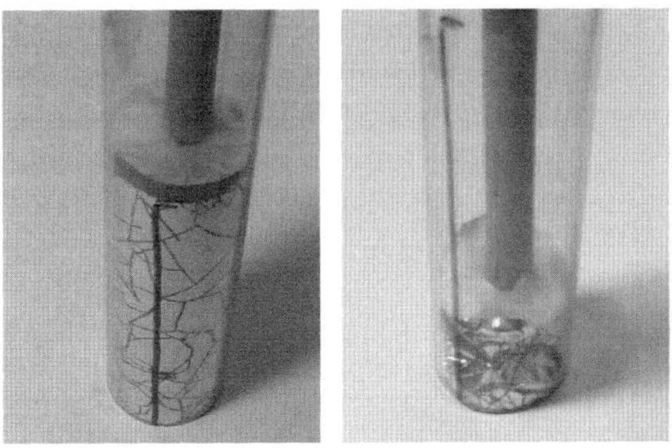

Figure 5.19: Left: rolled mylar sheet in the cylinder. Right: crumpled mylar sheet.

Slowly push the plunger down until it reaches the 2.7cm mark as shown in the right image of Fig. 5.19. Here we picture the second compaction, because without some lines drawn on the mylar sheet, a photograph doesn't show it clearly in the cylinder before compaction.

One way to guarantee that the mylar doesn't slip past the plunger is to roll the mylar from both sides, to form a double scroll the resembles the capital of an Ionic column. This is a more difficult roll to manage without introducing additional crumpling into the mylar sheet, so use appropriate care.

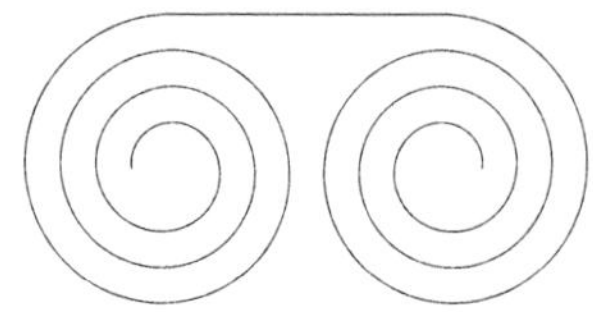

Remove the crumpled sheet from the cylinder. Simplest is to lift the cylinder from the surface on which it rested and slowly push the plunger until the crumpled mylar is ejected. Gently flatten the mylar on a table. Illuminate the sheet with a good light source at a fairly low angle to emphasize shadows. Using a hands-free magnifier and a pen—our choices are shown in the right image of Fig. 5.18—mark each crease. When you've marked a crease, measure its length in millimeters and record the measurement. Use this order: mark, measure, record; repeat for the next crease. If you mark a lot of creases before measuring, keeping track of which you've measured is an unnecessary and unwelcome complication.

Following our own experiences, we need to say a bit about how to approach this "mark and measure" stage of the experiment. When you first look at the crumpled, flattened, and still unmarked piece of mylar, you may may feel overwhelmed. Where to start? That doesn't so much matter: just pick a clearly visible crease, concentrate on it alone, then mark, measure and record. Then pick another clearly visible crease to mark, measure, and record. You needn't stick to one small portion of the mylar sheet. Take care of the most obvious creases first. With these marked, the shorter creases appear clearer. Move the magnifier over the mylar sheet, move your head, gently rotate the mylar sheet to get different perspectives. When you've finished marking, measuring, and recording, sum the crease lengths. This is the first data point.

Gently roll up the marked mylar sheet, place it in the cylinder, compress it, remove and flatten the sheet, then mark, measure, and record. At least four data points are needed to begin to support the logarithmic increase observation [176, 177], and of course more are better. This does require time and patience, and probably eyes younger than ours.

5.6.5 Sample

In Figs. 5.20 and 5.21 we see the results of four successive crumplings of a mylar square. The total measured crease length in the first image is 2138 mm, and the additional crease lengths are 481 mm, 348 mm, and 307 mm.

In Fig. 5.22 we plot the cumulative crease lengths. Because creases never disappear, the total crease length grows with each additional crumpling. We can verify at least that these points lie on a curve of the correct concavity: the line determined by the two left most points—we'll call these the first and second points—has slope greater than that of the line determined by the second and third points, which in turn has slope greater than that determined by the third and fourth points.

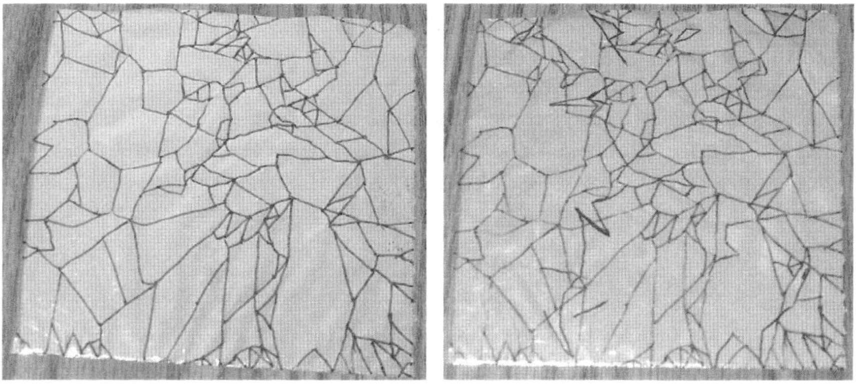

Figure 5.20: The first two crumples, marked

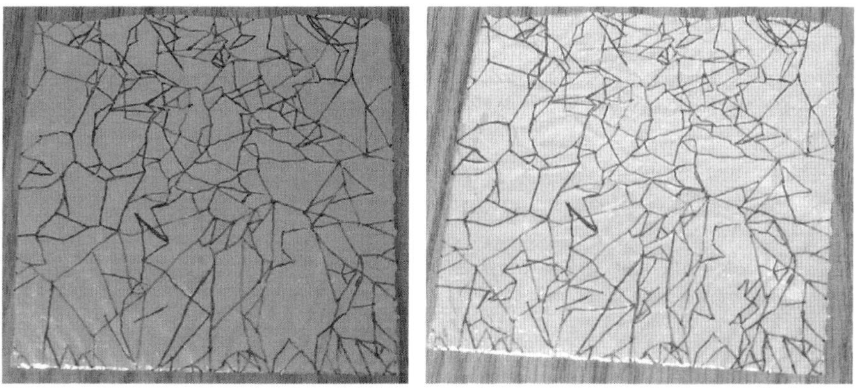

Figure 5.21: The third and fourth crumples, marked

Because Eq. 5.2 has two adjustable parameters, a and b, we don't have enough data points to test the validity of this specific model. We'll leave that to younger eyes, but we have seen that the data points lie on a curve of the right general shape.

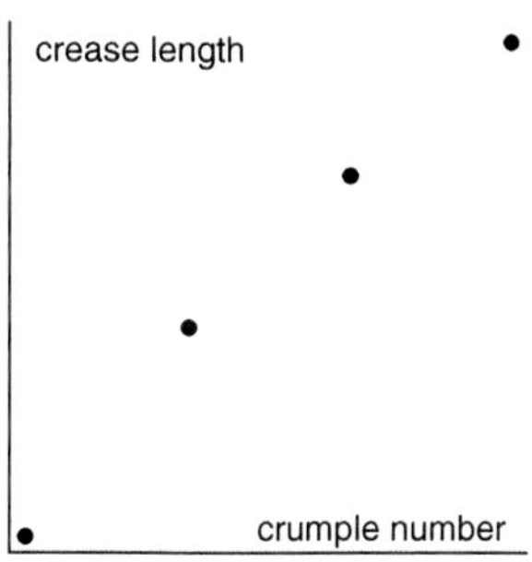

Figure 5.22: Crease length plot

5.6.6 Conclusion

In Lab 2.2 we presented data and observations that support the fractal structure of crumpled paper balls, though we had only a vague hint of why crumpling might produce a power law distribution of gap sizes. While the observations of this lab do not provide a an explanation for the gap size power law, the model of [177] does give a power law for the mean facet area as a function of the number of crumplings. This is an important step, but not the last step. Who knew that crumpling paper is so complicated?

Finally, the comparison between the crease patterns for successive levels of crumpling and the several levels of leaf veins seen in Figs. 4.47 and 4.48. The leaf veins are more ordered than crumpled mylar creases, but recall that the leaves grow more-or-less unconstrained by geometry, while the mylar is crumpled within a cylinder. How would the hierarchy of leaf veins appear if the leaf grew inside a small cylinder? Do you think the comparison is coincidence, or something deeper? How could we test this?

5.6.7 Exercises

Prob 5.6.1 Replicate the experiment of the Sample 5.6.5. How do your total crease length numbers compare with ours?

Prob 5.6.2 Prepare three 10 cm ×10 cm mylar squares. Compress one square to the marked height of 2.7 cm, one to 1.7 cm, and one to 3.7 cm. Compare the total crease lengths of the three squares. Comment on the differences.

Prob 5.6.3 Prepare ten 10 cm ×10 cm mylar squares. Compress each to the marked height of 2.7 cm, flatten, and recrumple. Calculate the total crease length of each square. Take the two whose total crease lengths are closest together and compress these again to the 2.7 cm mark. Compare the additional total crease lengths of these squares. Does this support the claim that additional growth of the total crease length depends only on the current total length and not on earlier history? This is MUCH better done as a group exercise.

Prob 5.6.4 Repeat Exercise 5.6.1 with a 10 cm ×10 cm square of printer paper. Compare these results with those of Exercise 5.6.1.

Prob 5.6.5 Repeat Exercise 5.6.2 with three 10 cm ×10 cm squares of printer paper. Compare these results with those of Exercise 5.6.2.

5.7 Fractal networks of resistors

Familiar rules show how to compute the resistance of resistors arranged in series or in parallel. More complex networks can be broken into pieces in parallel or in series, and then these pieces can be combined. However, these calculations can be quite complicated. Here we'll investigate simplifications that can be used for resistors arranged in a Sierpinski gasket pattern.

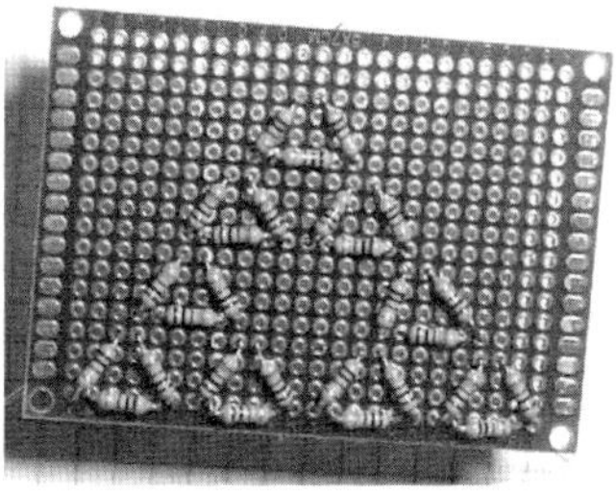

5.7.1 Purpose

We'll study how fractality in the construction of a resistor network can be reflected in a fractal aspect of the resistance calculation. We'll study scaling of the network resistance with network size, how resistance varies with the network node sampled, and how breaking some links in the network alters the resistance.

5.7.2 Materials

About a hundred resistors of the same resistance (we used 1 kΩ resistors), a breadboard, PCB circuit boards, and a volt-ohm meter.

5.7.3 Background

Recall some simple rules for combining resistances. See [181] for instance. For a series circuit of elements with resistance R_1 and R_2, the resistance R measured between points A and B is

$$R = R_1 + R_2 \tag{5.3}$$

Sufficiently old people who celebrated Christmas recall frustrating Christmas tree lights wired in series. If one light burned out, the entire string went dark. The burned out bulb could be found only by swapping the lights one at a time.

For Christmas lights wired in parallel, when one light burned out, it alone darkened. When parallel-wired lights first became available, they were understandably popular. Here the resistance formula is more complicated. The resistance R measured between points A and B is given by $1/R = 1/R_1 + 1/R_2$, so

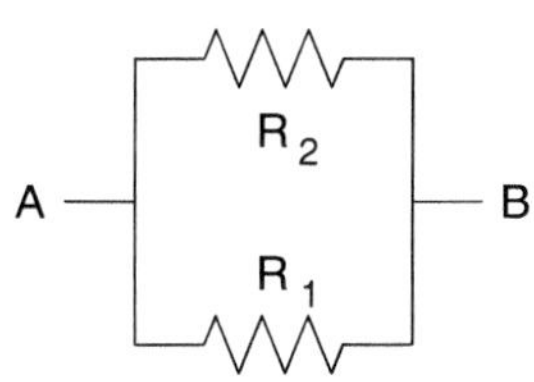

$$R = \frac{1}{\dfrac{1}{R_1} + \dfrac{1}{R_2}} \tag{5.4}$$

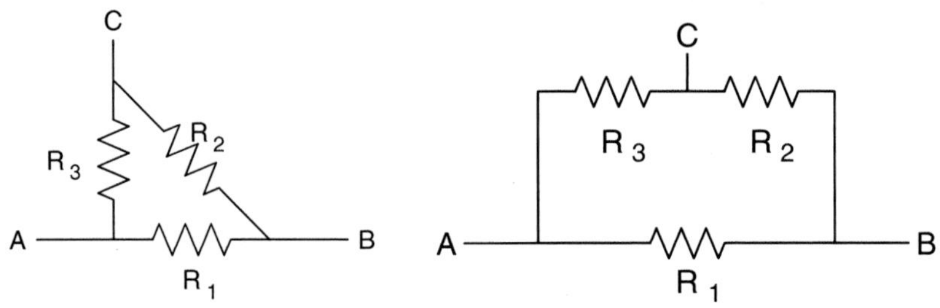

Figure 5.23: A triangular network and an equivalent representation.

Quick check of units: because we take the reciprocal of the sum of the reciprocals of the resistances, R, R_1, and R_2 all have units of ohms.

We'll be guided by the description in [182]. The first step in the construction of a Sierpinski gasket resistor network is a triangle network, called a Δ-network in some texts for obvious reasons, pictured on the left of Fig. 5.23. Suppose we measure the resistance between points A and B. At the moment, nothing happens at the point C. If we redraw the left image a bit, we get the right image. Along the upper path apply the series circuit rule (5.3) to find that the resistance along that path is $R_2 + R_3$. Then apply the parallel circuit rule (5.4) to deduce that the resistance between A and B is

$$R_{AB} = \frac{1}{\dfrac{1}{R_1} + \dfrac{1}{R_2 + R_3}} = \frac{R_1(R_2 + R_3)}{R_1 + R_2 + R_3} \tag{5.5}$$

Similar calculations show that

$$R_{AC} = \frac{R_3(R_1 + R_2)}{R_1 + R_2 + R_3} \quad \text{and} \quad R_{BC} = \frac{R_2(R_1 + R_3)}{R_1 + R_2 + R_3} \tag{5.6}$$

These direct applications of Eqs. (5.3) and (5.4) can be applied to more complicated circuits, though the calculations can become tedious. Happily, the calculation for the Sierpinski gasket network can be simplified significantly by a process called the *delta to star conversion* pictured on the right. Because we're talking about measuring resistances between any pair of points from A, B, and C, we say

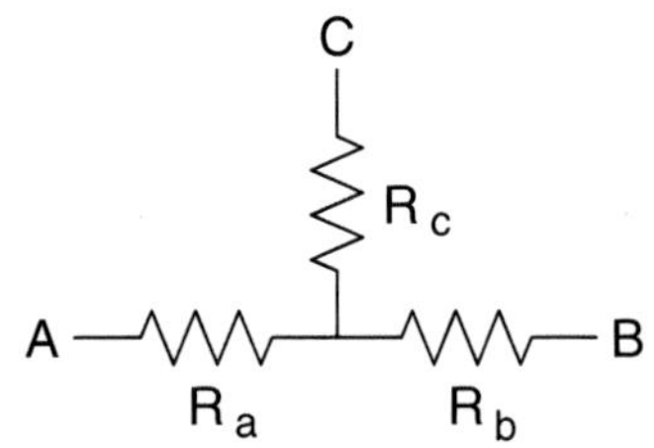

two networks are equivalent if both give the same values for R_{AB}, R_{AC}, and R_{BC}. In Appendix B.16 we'll show that the equivalent star network has

$$R_a = \frac{R_1 R_3}{R_1 + R_2 + R_3}, \quad R_b = \frac{R_1 R_2}{R_1 + R_2 + R_3}, \quad R_c = \frac{R_2 R_3}{R_1 + R_2 + R_3} \tag{5.7}$$

These star equivalents of triangle circuits are the basis for our calculation of the corner-to-corner resistances for Sierpinski gasket networks. We'll take all the resistors of these networks to have the same value, R. For the triangle network on the left side of Fig. 5.23 with $R_1 = R_2 = R_3 = R$, the equivalent star network has $R_a = R_b = R_c = R/3$ and so by the series rule (5.3)

$$R_{AB} = R_a + R_b = \frac{2R}{3} \tag{5.8}$$

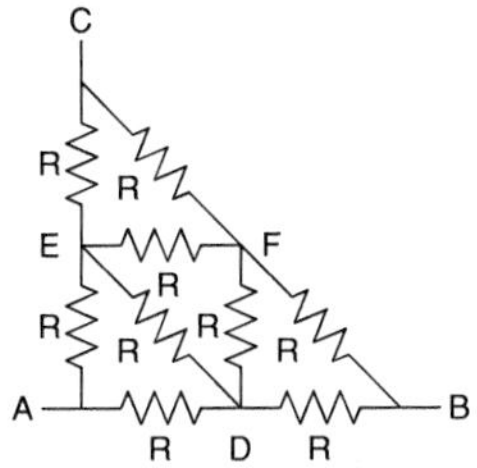

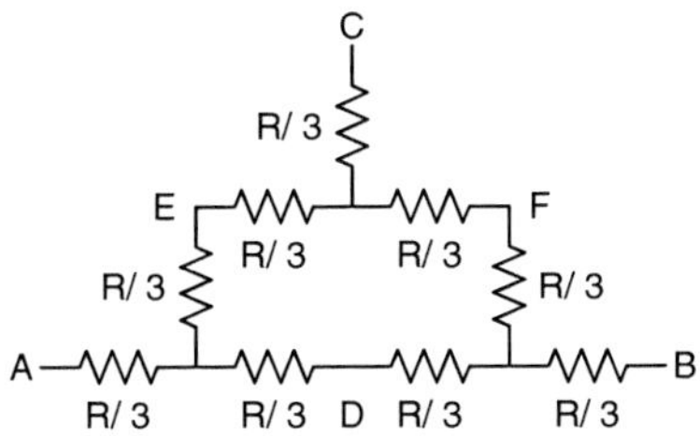

Figure 5.24: Left: the second level of the Sierpinski gasket resistor network. Right: star equivalents of three small network triangles.

On the left of Fig. 5.24 we see the second level in the construction of a Sierpinski gasket resistor network. On the right we see the star equivalents of the triangle with vertices A, D, E, the triangle with vertices D, B, F, and the triangle with vertices E, F, C.

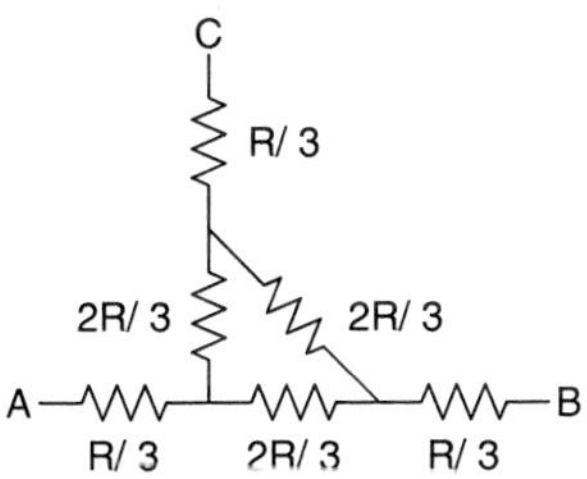

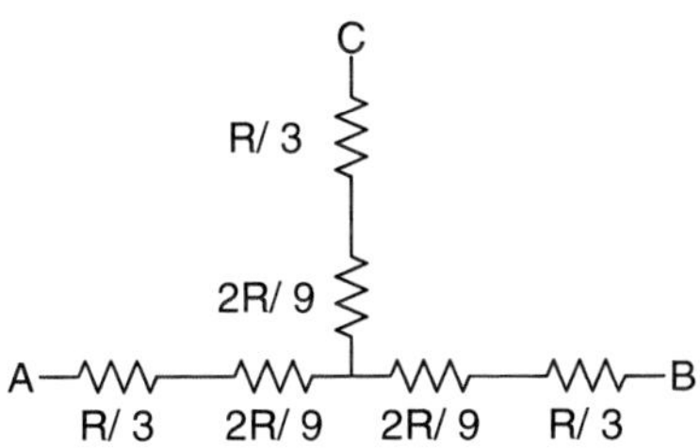

Figure 5.25: Left: applying the series rule. Right: star equivalents of the network triangle.

On the left of Fig. 5.25 we see the result of applying the series rule (5.3), in star equivalent network on the right side of Fig. 5.24, to the pair of resistors that share vertex D, to the pair that share vertex E, and to the pair that share vertex F. On the right we have found the star equivalent of the triangle on the left. The resistances of the equivalent star network are given by Eq. (5.7).

$$\frac{\frac{2R}{3} \cdot \frac{2R}{3}}{\frac{2R}{3} + \frac{2R}{3} + \frac{2R}{3}} = \frac{2R}{9}$$

Then to find R_{AB} for the second level in the Sierpinski gasket resistor network, apply the series rule to the four resistors between A and B.

$$R_{AB} = \frac{R}{3} + \frac{2R}{9} + \frac{2R}{9} + \frac{R}{3} = \frac{10R}{9} = \frac{5 \cdot 2R}{3^2} \tag{5.9}$$

By applying the delta to star conversion hierarchically in each level of the Sierpinski gasket resistor network construction, we find $R_{AB}(n)$, the resistance between vertices A and B for the nth level, is

$$R_{AB}(n) = \frac{5^{n-1} \cdot 2R}{3^n} \tag{5.10}$$

5.7.4 Procedure

We'll build several levels of a Sierpinski gasket resistor network and compare resistance measurements with the prediction of Eq. (5.10).

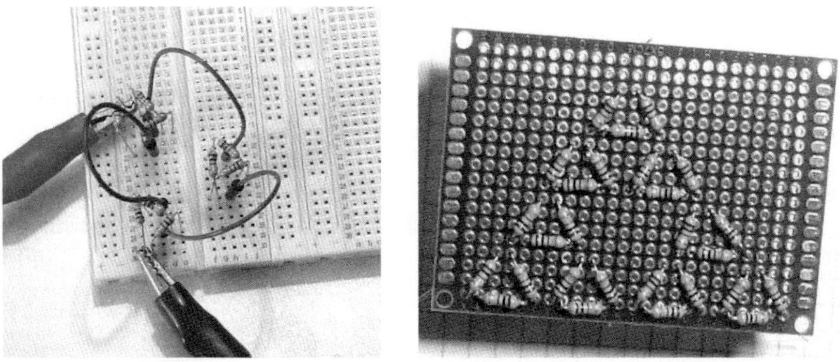

Figure 5.26: Physical realizations of Sierpinski gasket resistor networks, level 2 on the left, level 3 on the right.

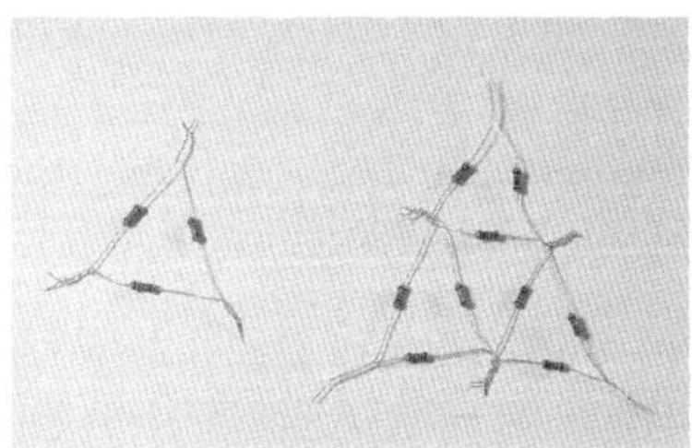

If the experiment involves removing some connections, the network is better assembled on a breadboard, as in the left image of Fig. 5.26, because elements and connectors can be added to and removed from the network with ease. If the experiment involves several measurements at different points, a better choice is a more stable arrangement such as the PVC circuit board pictured on the right of Fig. 5.26. As shown on the right here, these networks can be assembled with neither breadboards nor circuit boards. Just be careful to place the networks on an insulating surface. Details of several types of experiments are described in the samples.

5.7.5 Sample A. Measurements of $R_{AB}(n)$.

From Eq. (5.10) we expect these values of $R_{AB}(n)$ for $n = 1, 2, 3,$ and 4.

$$\frac{2}{3} \approx 0.67, \qquad \frac{10}{9} \approx 1.11, \qquad \frac{50}{27} \approx 1.85, \qquad \frac{250}{81} \approx 3.07$$

Construct the networks and measure

$$R_{AB}(1) \approx 0.65 \qquad R_{AB}(2) \approx 1.09$$
$$R_{AB}(3) \approx 1.84 \qquad R_{AB}(4) \approx 3.08$$

This is pretty good agreement with the prediction. Why isn't the match exact? One reason is that the wires have tiny resistances, but more significant is the fact that the resistors are not exactly 1 kΩ.

5.7.6 Sample B. The effect of breaking a connection.

For this we'll use a level 2 network built on a breadboard, shown in the left image of Fig. 5.26. We'll measure $R_{AB}(2)$ with various connections broken. We'll use the labels on the left image of Fig. 5.24 to describe which connections are broken.

First, though, we'll mention a result that may seem counterintuitive. When we remove the resistor between E and F, $R_{AB}(2)$ increases from 1.09 to 1.18. Shouldn't removing a resistor lower the network resistance? It would if we removed the resistor and replaced it with a wire. In fact, when we do that $R_{AB}(2)$ drops from 1.09 to 0.99. When we just remove a resistor, we've replaced a 1 kΩ resistance along that path with an infinite resistance. No surprise that $R_{AB}(2)$ increases.

If we break the level 2 gasket network at D, $R_{AB}(2)$ jumps from 1.09 to 1.96. If we break the level 2 gasket network at E or at F, $R_{AB}(2)$ jumps from 1.09 to 1.31. A qualitative understanding of this difference is straightforward. When we break the network at D, all paths between A and B pass through the triangle ECF. On the other hand, when we break the network at either E or F, no path between A and B passes through the triangle ECF; all paths between A and B pass through D and these broken networks have shorter paths between A and B than we find in the network with D broken.

5.7.7 Sample C. Measurement of the resistance between different points.

For this we'll use a level 3 network built on a PVC circuit board, shown in the right image of Fig. 5.26. We'll measure $R_{AB}(3)$, $R_{AJ}(3)$, $R_{AD}(3)$, and $R_{AG}(3)$.

$$R_{AB}(3) \approx 1.82 \qquad R_{AJ}(3) \approx 1.40$$
$$R_{AD}(3) \approx 1.01 \qquad R_{AG}(3) \approx 0.60$$

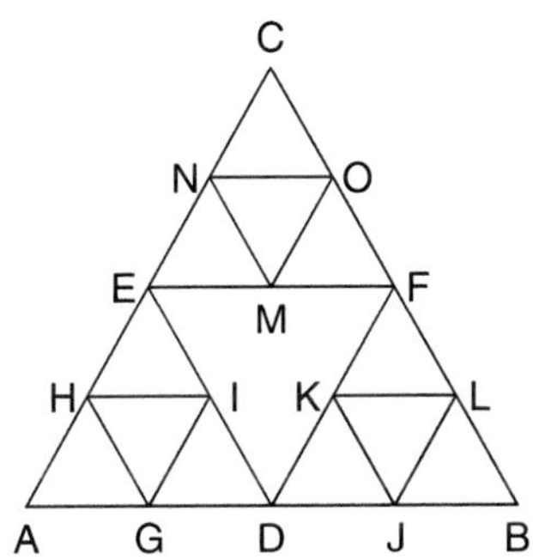

The resistance is approximately a linear function of the distance along AB between A and the other point between which the resistance is measured. Will this relation hold for other paths in the network? See Exercises 5.7.4–5.7.7.

5.7.8 Conclusion

The most commonly studied fractal resistor networks are random, proposed as models of conductivity in distorted media [183]. To study scaling, Sierpinski gasket networks are simpler to build and to analyze. While abstract calculations may be more complex for other fractals, experiments can lead to estimates of the power law exponents, if power laws are observed. Two natural questions are

• Do fractal resistor networks exhibit power law scalings and

• If there is a power law, is the power law exponent related to dimension of the fractal?

We expect this would be a very long-lived project, but each individual step could yield interesting results.

5.7.9 Exercises

Prob 5.7.1 Use the calculation of $R_{AB}(n)$ in Eq. (5.10) to find the power law exponent of how $R_{AB}(n)$ scales
(a) with the number of resistors on the edge AB if the gasket network, and
(b) with the total number of resistors in the network.
(c) Do either of these exhibit a simple relation to the dimension of the gasket?

Prob 5.7.2 Build your own Sierpinski gasket resistor networks of levels 1, 2, 3, and 4. Measure $R_{AB}(n)$ for $n = 1, 2, 3$, and 4. Do your results agree with the prediction of Eq. (5.10). Hint: to build a level 4 network, build three level 3 networks and connect them with wires between the appropriate corners. See Fig. 5.27.

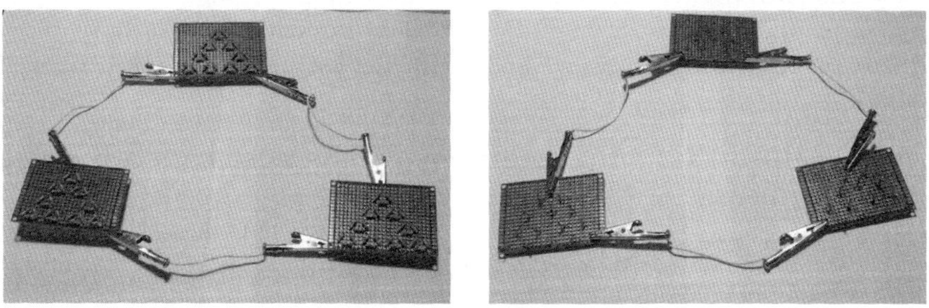

Figure 5.27: Physical realizations of a level 4 Sierpinski gasket resistor networks. Left: top view. Right: bottom view showing the connections of level 3 boards.

Prob 5.7.3 Build a level 3 network on a breadboard. Break connections one or two at a time. Can you find patterns in how the change in resistance is related to the positions of the removed resistors?

Prob 5.7.4 Refer to the labels in the diagram of Sample 5.7.7. In a level 3 network on a PVC circuit board, measure the resistances R_{DI}, R_{DE}, R_{DN},

R_{DM}, R_{DO}, and R_{DC}. Do you find a linear relation between the resistance and the length of the shortest path between the endpoints of the path? What does the symmetry of the gasket network suggest about R_{DN} and R_{DO}? Do your measurements agree with this expectation?

Prob 5.7.5 In a level 3 network on a PVC circuit board, find the resistance between A and various network nodes. Do you find linear trends similar to that seen in Sample 5.7.7.

Prob 5.7.6 In the level 4 network on PVC circuit boards shown in Fig. 5.27, find the resistance between A and various network nodes. Do you find linear trends similar to that seen in Sample 5.7.7 along some paths? Can you characterize those paths?

Prob 5.7.7 Repeat Exercise 5.7.5 when resistance is measured between two internal points, that is, points other than A, B, and C. That is, generalize the findings of Exercise 5.7.4.

Prob 5.7.8 Repeat Exercise 5.7.6 when resistance is measured between two internal points, that is, points other than A, B, and C.

5.8 Fractal networks of magnets

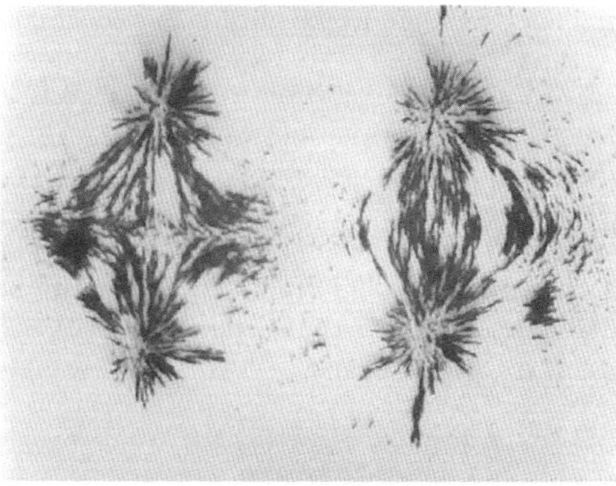

Here we'll study the magnetic field, revealed through patterns of iron filings, produced by magnets placed on the vertices of a fractal. We'll use stages of a Cantor set and of a Sierpinski gasket. On the right we see the patterns for two magnets. In the first image both magnets have the same pole pointing out of the page and the field lines appear to repel one another; in the second image the magnets have opposite poles pointing out of the page and the field lines appear to attract one another. We'll investigate whether a fractal arrangement of magnets produces similar field patterns on several scales.

So far as we know, each magnet has two poles, N and S. To date, an isolated N or an isolated S (these would be "magnetic monopoles") has not been discovered. We'll stick with magnetic dipoles. Every arrangement of magnets has many variations, many field patterns to explore.

5.8.1 Purpose

We'll study how the arrangement of magnets influences the magnetic field around those magnets, revealed through the pattern of iron filings on an acrylic sheet above the magnets. In particular, we'll investigate whether the placement of magnets on points that characterize a fractal (Cantor set interval endpoints or Sierpinski gasket triangle vertices, for example) can generate recognizably self-similar patterns in magnetic fields.

5.8.2 Materials

About 100 small cylindrical rare earth magnets (we used magnets with diameter 5 mm and height 1.5 mm), iron filings, a mortar and pestle, small flat-head tacks, glue, a push pin board, a tea strainer, a clear acrylic sheet, a piece of printer paper, and a digital camera or cellphone camera.

5.8.3 Background

The general idea behind this lab was inspired by a construction of Nathan Cohen [184]. His plan is to build an electromagnet whose core has a cross-section that is a Koch snowflake. While such a core could be produced by a 3-D printer, we decided on simpler fractals, a Cantor set and a Sierpinski gasket.

Cohen points out that fractal electromagnet cores can generate higher magnetic flux than electromagnets with Euclidean cores of the same size. Alternately, smaller fractal electromagnets can produce the same flux as larger Euclidean electromagnets. Specific applications are currently under development.

5.8.4 Procedure

In order to discern easily the orientation of the magnets, mark the same pole of each magnet with a spot of paint or permanent marker ink. Despite their small size, these rare earth magnets are quite strong. To immobilize them in the desired positions, count out the number of magnets you'll need for whatever pattern you choose and glue each of these magnets, in the same orientation, to a tack. To explore the effects of local field reversals, glue a few more magnets in the opposite orientation to tacks.

Even fine iron filings may be too coarse to detect subtle aspects of the magnetic field, so grind them further with the mortar and pestle.

The only delicate point of these experiments is assuring that the iron filings are spread sparsely enough to reveal the magnetic field lines. For us, this took some practice. Push the magnet pins into the push pin board in the desired pattern, cover with an acrylic sheet, and place a piece of unlined paper on the acrylic sheet. Sprinkle the iron filings sparingly over the paper above the magnets. Scoop a very small amount of iron filings from the mortar. With the strainer above the mortar drop the filings into the strainer. Some filings will fall back into the mortar. Move the strainer above the magnets and tap the side of the strainer while moving it across the paper. Repeat until the field pattern is clear. If some areas get too many filings, lift the paper, carefully pour the filings back into the mortar, and begin again.

5.8.5 Sample A. Cantor set patterns

Here we place eight magnets at the vertices of the second stage of the construction of a Cantor set. In the left image of Fig. 5.28 all the magnets have the same orientation; in the right image the orientations alternate.

Figure 5.28: Magnetic fields around the second stage of a Cantor set.

Certainly there are many other patterns of magnet orientations. Our goal is to account for the interactions of all the magnets, so with a higher iterate of the Cantor set construction, the interactions of the left-most and right-most magnets are too weak to detect. On the other hand, the fields of these rare earth magnets are so strong that if we move adjacent magnets closer together, if they have the same orientation one or both of the magnets fly out of the push pin boards and land on opposite sides of the table. If they have the opposite orientations, one or both of the magnets fly out of the push pin boards and stick together. Once this catalyzed other magnets leaving the board and sticking together, resulting in a spiky clump of magnets and tacks, a little magnetic hedgehog. We have seen all of these events.

Let's focus on the right configuration of Fig. 5.28 shown schematically in Fig. 5.29. North poles up are filled disks, south poles up are unfilled circles.

Figure 5.29: Cantor set schematic.

We begin the Cantor set construction by placing magnets A and H. This will produce weak field contours between these magnets. The next stage of the construction is to place magnets D and E. We'll see field contours between A and D. and between E and H, similar to those between A and H. This looks like we're set up to build a fractal pattern of field contours, but there are two complications. First is that the field between A and D is stronger than that between A and H. A smaller physical fraction of the Cantor set contributes a stronger component of the total magnetic field. The second issue is that magnet D does not interact only with A, and E does not interact only with H. The field from magnet E acts in opposition ot that of A, and D in opposition to that of H. And of course we find field contours between D and E.

So we see that at least in this geometry, a fractal arrangement of magnets produces an underlying fractal pattern of fields, but with complex modifications overlaid on this field. Do these modifications themselves follow a fractal pattern? How could we check?

5.8.6 Sample B. Sierpinski gasket patterns

In Fig. 5.30 we see magnets, all of the same orientation, placed on the vertices of the third stage of the construction of a Sierpinski gasket. Unlike the Cantor set of Sample 5.8.5, for the gasket no arrangement of magnets will result in all nearest neighbors of alternating orientations: any equilateral triangle must have all three vertices of the same orientation, or two with one orientation and the third of the opposite orientation. Sprinkle the filings on the acrylic sheet without the paper to more clearly guide where you sprinkle the filings.

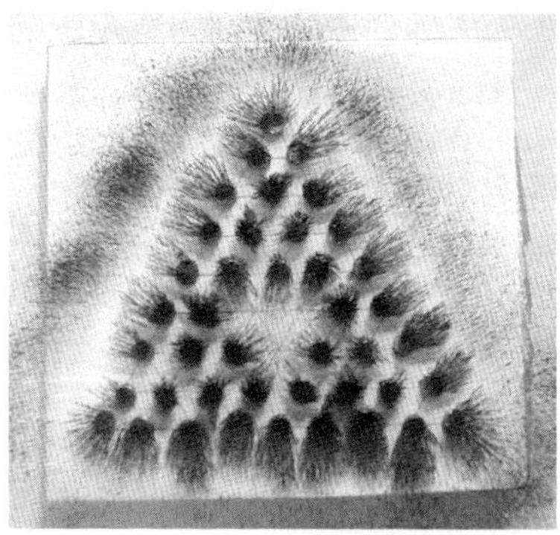

Figure 5.30: Magnetic fields around a stage of the Sierpinski gasket.

Here we have ample opportunity to experiment. Patterns within patterns. In your interpretations, keep in mind the two points from Sample 5.8.5: the interaction of nearby magnets is stronger than is that of distant magnets, and every magnet interacts with every other magnet. In Fig. 5.30, the nearest-neighbor interactions appear to swamp those with all other magnets. How might we detect the interactions of more distant magnets?

5.8.7 Conclusion

Fractal placements of magnets generate magnetic fields that have self-similar patterns but with variations caused by the interactions of the magnets. That magnets have two poles, and consequently can either attract or repel one another, adds an interesting complication to the notion of fractality.

5.8.8 Exercises

In all these problems, build the configuration of magnets described, investigate the magnetic fields with iron filings, and interpret the results. Assemble information to study the question of whether the variations exhibit a secondary pattern on top of the initial pattern.

Prob 5.8.1 Vary the orientations of the magnets in the Cantor set. For example, the pattern on the right.

Prob 5.8.2 Vary the orientations of the magnets in the Sierpinski gasket.

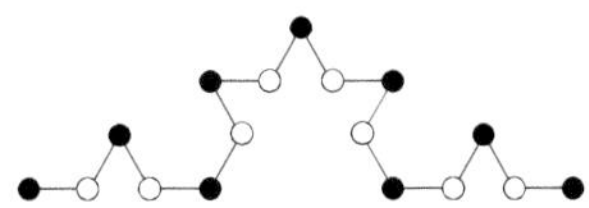

Prob 5.8.3 Place magnets on the vertices of the second iteration of a Koch curve.
(a) Place all the magnets in the same orientations.
(b) Place the magnets in alternating orientations, as shown above.
(c) Reverse the orientations of a few of the magnets.

Prob 5.8.4 Repeat Exercise 5.8.1 using a randomized Cantor set. At each stage of the Cantor set construction, independently pick the left and the right scaling factors to be $r = 1/2$ and $r = 1/3$ both with probability $1/2$.

Prob 5.8.5 Repeat Exercise 5.8.4 with a different sequence of random numbers.

Prob 5.8.6 Repeat Exercise 5.8.1 for a Cantor set that consists of 3 pieces each scaled by $1/5$. Place the pieces evenly in the unit interval. That is, the left endpoints of the three pieces are at $x = 0$, $x = 2/5$, and $x = 4/5$.

Prob 5.8.7 Another way to generate the pattern of iron filings is to sprinkle them across a region of the printer paper and then gently slide the acrylic sheet around on top of the magnets a bit. Does this reveal anything new?

Prob 5.8.8 Arrange the magnets at the vertices of an early stage of the fractal of Exercise (h) of Lab 1.1.

Prob 5.8.9 Arrange the magnets at the vertices of an early stage of the fractal of Exercise (l) of Lab 1.1.

Prob 5.8.10 Arrange the magnets at the vertices of an early stage of the fractal of Exercise (n) of Lab 1.1.

5.9 Synchronization in fractal networks of oscillators

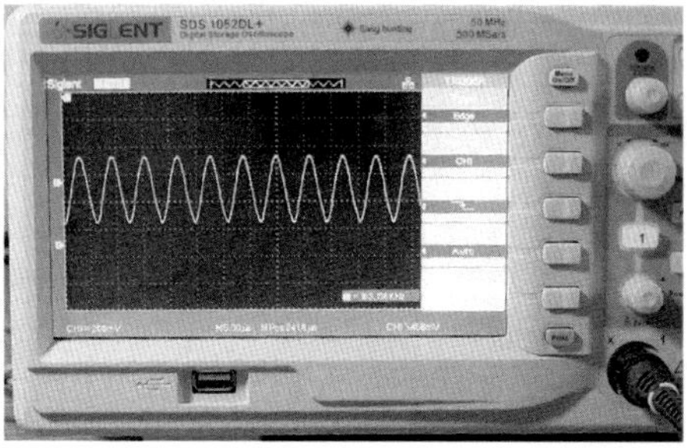

Here we'll investigate an analog version of Lab 3.2, with logistic maps replaced by oscillating circuits. An important variation is that while logistic maps with the same parameter have identical dynamics, electronic components with the same parameters need not be, and likely aren't, identical. An inductor listed as 220 microhenrys (μH) almost surely does not have exactly that inductance. So of necessity we will investigate synchronization in networks of oscillators with some noise added.

5.9.1 Purpose

We'll investigate how the coupling topology of a network of oscillators having about the same frequencies affects the synchronization of the oscillators. In addition, we'll see if evidence supports the idea that more oscillators coupled in a fractal pattern exhibit a richer variety of dynamical behaviors.

5.9.2 Materials

We'll use a variable DC power supply, a digital oscilloscope, six 830 hole breadboards, a 400 hole breadboard, jumper wires of several lengths, nine 2N3904 transistors, nine $220\mu H$ inductors, nine $0.001\mu F$ capacitors, eighteen $0.01\mu F$ capacitors, nine $1M\Omega$ resistors, nine $10k\Omega$ resistors, and 15 75Ω resistors. To vary the connection strengths we'll use a collection of other resistors, or some variable resistors.

5.9.3 Background

In Appendix B.17 we derive the formula for the frequency of an LC circuit, a circuit that consists of an inductor and a capacitor in series with a DC power source. How do we make this circuit oscillate? Charge the capacitor, then connect it to the inductor. Current begins to flow in an oscillating pattern between the capacitor and the inductor. Google can provide nice animations of how the energy stored in the circuit switches between an electric field in the capacitor and a magnetic field in the inductor. However, real circuit elements are not perfect conductors. As current flows through resistance it generates heat, so the current that oscillates through the LC circuit decreases. With real circuit elements, the LC circuit is a damped oscillator.

In order to detect synchronization, we need persistent oscillators. That is, we must find a way to maintain a constant energy in the circuit, replacing that lost to resistance. Solutions to this problem involve more complex circuits with transistors or op amps. These include the Hartley oscillator and the Colpitts oscillator. We'll build a simple version of the Colpitts oscillator with a 2N3904 transistor. We'll describe how to construct the oscillator in Sect. 5.9.4. A Colpitts oscillator compensates for the dampening of an LC circuit by feedback from a voltage divider that consists of two capacitors in series with an inductor. This signal is amplified by a transistor and fed back into the circuit. The degree of your curiosity about how the oscillator works and your familiarity with electronics determine the appropriate background about the Colpitts oscillator. To find the appropriate level of information, Google is your friend.

5.9.4 Procedure

Here we'll give a careful description of how to construct a Colpitts oscillator from the specific supplies listed in Appendix A. In Sample 5.9.5 we build and study the dynamics of a single oscillator. In Sample 5.9.6 we build a network of two coupled oscillators and begin to study synchronization. In Sample 5.9.7 we build a network of three oscillators and observe synchronization there. Finally, in Sample 5.9.8 we build a network of nine oscillators, coupled in a pattern determined by the second stage of the construction of the Sierpinsi gasket.

To build a Colpitts oscillator, first snap together two of the 830 hole breadboards on their long sides. For now ignore the − and + columns. We'll use them when we construct networks of oscillators. Each breadboard has 63 numbered rows of 10 holes. In each row, holes a-e are connected by a conducting strip, and holes f-j are connected by a conducting strip. To build an oscillator, first place the component leads in the specified holes, then attach the jumper wires between the specified holes. Component placement is much more difficult if some of the jumper wires are already in place. Because we've attached two breadboards, the array has a left set of holes a-e and f-j, and a right set of holes a-e and f-j. For example, we denote the hole in row 11 and column g of the right breadboard by R11g.
Attach a $220\mu H$ inductor between holes L63a and L53a.
Attach a $0.01\mu F$ capacitor between holes L57c and L55c.
Attach a $0.01\mu F$ capacitor between holes L55d and L53d.
Attach a $0.001\mu F$ capacitor between holes L51j and L49j.
Attach a $1M\Omega$ resistor between holes L53c and L40c.
Attach a $10k\Omega$ resistor between holes L51h and L40h.
Attach the emitter of a 2N3904 transistor to hole L55i, the base to hole L53g, and the collector to L51i. Note that the transistor is approximately a hemi-cylinder. The flat side of the transistor orients the three transistor leads, as indicted in the top view schematic on the right.

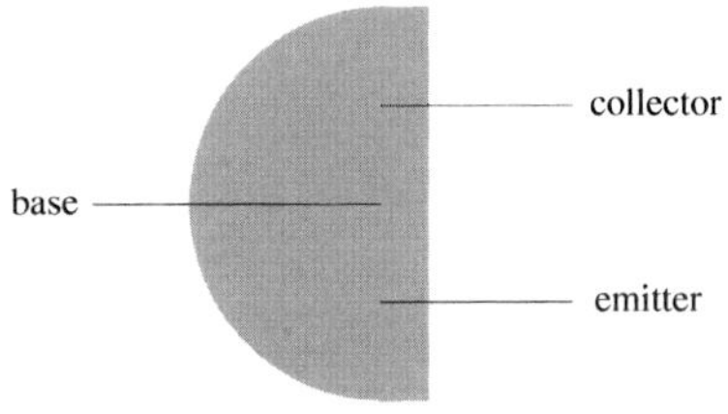

Now we add the jumper wires. We need five wires to connect the components, and three for DC power in, signal out, and common ground.

Attach a wire between L57b and L63b.
Attach a wire between L63c and L49f.
Attach a wire between L53e and L53f.
Attach a wire between L55e and L55h.
Attach a wire between L40d and L40i.
The DC power in wire is attached to L40j.
The signal out wire is attached to L49i.
The common ground wire is attached to L55j.

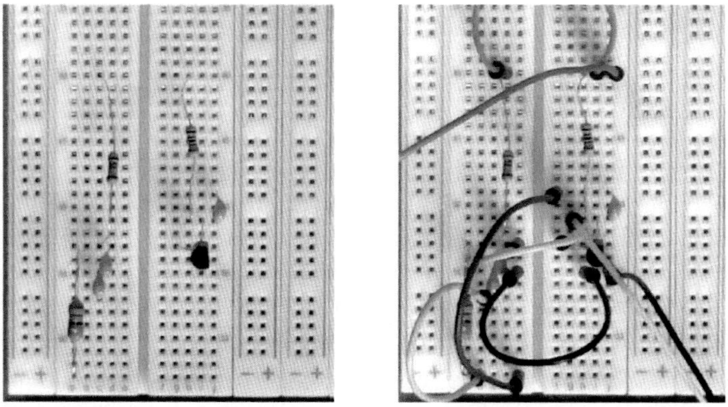

Figure 5.31: Left: the placement of oscillator components. Right: with the jumper wires added.

Applied to the left board and to the right breadboard (replace the Ls in the hole numbers by Rs), these give the lower left and lower right oscillators. To build the upper left oscillator, carry out the lower left oscillator construction but subtract 36 from all the row numbers. So for example, rather than

"Attach a $220\mu H$ inductor between holes L63a and L53a."

we write

"Attach a $220\mu H$ inductor between holes L27a and L17a."

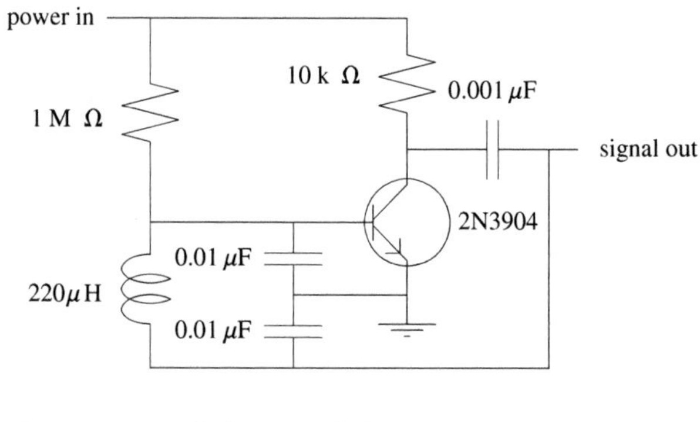

Figure 5.32: Colpitts oscillator schematic diagram.

On the right we see a schematic diagram of this oscillator. The two $0.01\mu F$ capacitors and the $220\mu H$ inductor on the lower left of the diagram constitute the voltage divider mentioned earlier. For networks, connect the red lead of the DC power supply to the + column on the side of the breadboard, and connect the power in of each oscillator to the + column. Similarly, connect the power supply black lead, the ground wire of each oscillator, and the ground of each oscilloscope probe to the − column.

In Fig. 5.33 we see a completed breadboard with three Colpitts oscillators, with power in, signal out, and ground wires not yet connected. To recognize the input and output wires, consider the upper left oscillator. The wire that passes out of the photo at the top is the power in wire. The wire from this oscillator that passes out of the photo at the left edge is the common ground. The wire that passes out of the photo at the upper right edge is the signal out wire.

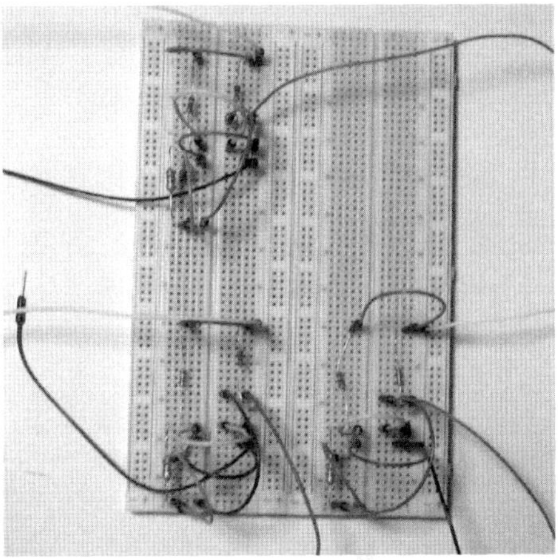

Figure 5.33: Three oscillators

When you've built these three oscillators, build two more groups of three oscillators. If you think that nine oscillators aren't enough to explore the effects of network fractality, that a network of 27, or better yet, 81 oscillators would allow for a more convincing study, of course you're right. But after you've built nine oscillators, decide if you want to build another 18, or another 72. Likely you'll agree that nine oscillators are enough for now.

We could construct a circuit simulation with 27, or 81, or 243 Colpitts oscillators, using for example LTSpice software downloaded from

https://ltspice.analog.com/software/LTspice.pkg

which requires MacOS 10.9 or newer. But then this would be a computer lab, not a physical lab. While computer labs are important, one of the points of this lab is that the idealized circumstance of simulation misses the messy complications of the physical world. Sometimes we need to see just how messy the physical world is.

For ideal systems we could test synchronization by comparing the oscilloscope traces of every pair of oscillators. In the YT display, synchronization would be revealed by the maxs and mins of both channels lining up. In the XY display, synchronization would be indicated by a simple closed curve. We'll illustrate both of these displays in Sample 5.9.5.

With imperfect systems—that is, with all real physical systems—we'll need other approaches, or other ways to interpret these approaches. We'll explore this in Samples 5.9.6, 5.9.7, and 5.9.8. .

5.9.5 Sample A. One oscillator.

Use the upper left oscillator on a breadboard to simplify the description of the location of the channel 2 lead. Connect the power supply red lead and the oscillator power in wire to the left + column of the breadboard. Connect the power supply black lead, the oscillator ground, and the grounds of both oscilloscope probes to the – column. Connect the channel 1 probe to the signal out wire. Connect the channel 2 lead to a wire attached to L15g. Set the power supply to 2.2 V. With the display set to YT we obtain the left image of Fig. 5.34. Both channels are set at 200 mV per (vertical) division; the time scale is 5μsec per (horizontal) division. With the display set to XY we obtain the right image of Fig. 5.34.

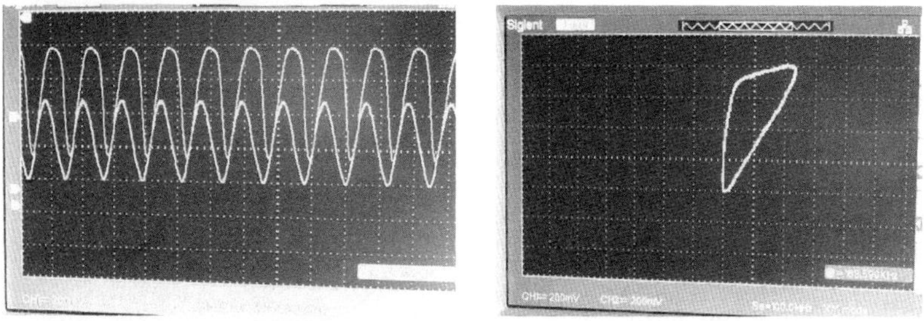

Figure 5.34: The YT display (left) and XY display (right) for two signals from a single oscillator.

In the YT display the channel 1 plot is approximately sinusoidal, while the channel 2 plot is a bit distorted: the max is wider than the min. We see that the channel 2 mins lead the channel 1 mins by the same amount. The signals are synchronized, though with a phase difference. (If there were no phase difference, the mins of both chanels would align.)

In the XY display, the channel 1 signal gives the X-value and the channel 2 signal gives the Y-value. If the signals in both channels are sine waves of the same frequency and phase, then the XY display is a diagonal line; if the phases differ—for example, $x = \sin(t)$ and $y = \sin(t+1)$—then the XY display is an ellipse. If one signal is a sine wave and the other is a more complicated periodic signal, then the XY display is a more complicated closed curve. This is what we see in the right image of Fig. 5.34.

5.9.6 Sample B. Two oscillators.

We begin with the three uncoupled oscillators of Fig. 5.33. We'll couple the two left oscillators and build the coupling circuits in the unused upper right quadrant of the breadboard. Both oscillators are driven by the DC power supply through the + column; both oscillators, along with the channel 1 and 2 oscilloscope probes, are grounded through the – column.

Coupling of the oscillators is achieved this way: the signal out from each oscillator goes to the corresponding probe and also through a resistor to the power in of the other oscillator. These resistors set the coupling strength between the oscillators.

In Fig. 5.35 we see the coupled network. Recall we denote the left and right halves of the breadboard by L and R. Attach wires between L13i and R5a, between L40f and R8a, between L4f and R18a, and between L49i and R15a. The upper left oscillator signal out lead is attached to R5b; the lower left oscillator signal out lead is attached to R15b. Attach a 75Ω resistor between R5e and R8e.

Our experiment will be to vary the resistances of these two couplings and search for synchronization. The uncoupled oscillators do not have quite the same frequency, so will not synchronize. If we couple the oscillators through very high resistances, we expect the network will be effectively uncoupled and the network will not synchronize. One question is: Can we find relatively small resistances where these oscillators synchronize? Another is: Can we estimate the range of resistances for which synchronization occurs?

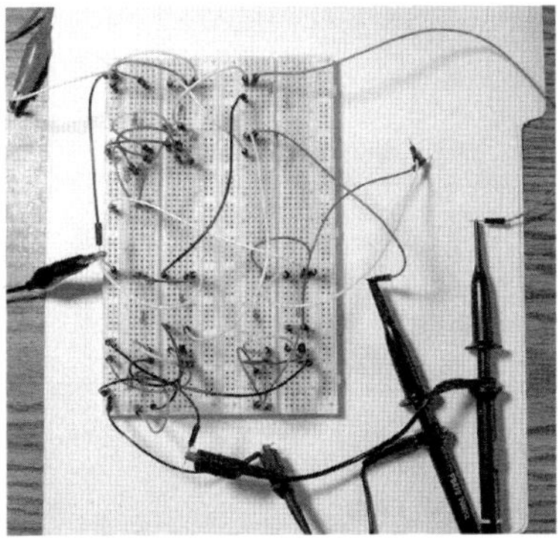

Figure 5.35: Two coupled oscillators.

Before we get to the results, an important point is that this is a fairly complicated circuit to build in a prototyping breadboard. Some of the component leads, particularly those of the resistors, are very thin and can come free of the board when jostled by the insertion of the jumper wires. If both vertical scales on the oscilloscope are set the same and one trace is a horizontal line while the other is waves, look for a component that has come disconnected. This has happened to us. With a bit of inspection, the initial "Oh, no" was replaced by "I got this."

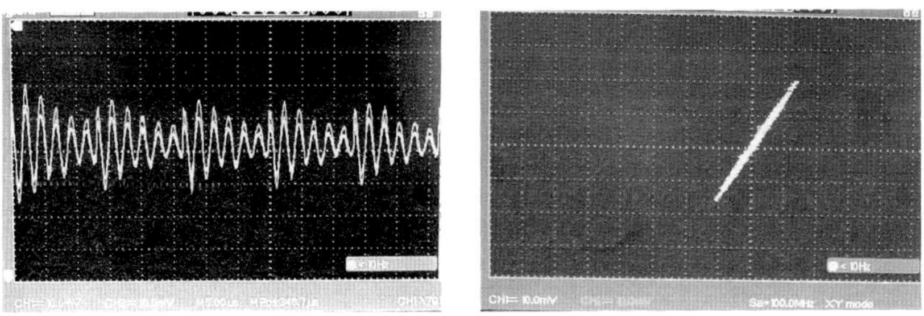

Figure 5.36: The YT display (left) and XY display (right) for two oscillators coupled through 75Ω resistors.

In Fig. 5.36 we see the YT display and the XY display of the two oscillators when coupled through 75Ω resistors. The DC power supply is set to 10.4 V. The spacing between horizontal lines is 10 mV; the spacing between vertical lines is 5μ sec. The most obvious feature of the YT display is that even though the two signals differ in amplitude, the traces line up peak for peak and valley for valley. The oscillators appear to have synchronized. This is reinforced by the XY display, whose points lie close to a diagonal line. On the oscilloscope, neither display is static; both change constantly. Pressing the Run/Stop button freezes the image, so it can be photographed.

Look again at the YT display. This differs from the YT display of Fig. 5.34 in that here the amplitudes vary. In Sample 5.9.5 the input voltage is constant, while in this sample the feedback of one oscillator to the other adds a variable voltage to the input. The varying amplitudes are not a surprise. But look more closely. After the first seven peaks, the maxes fall into a pattern that repeats every six maxes. The pattern of the mins still changes a bit. Would a longer run before sampling show repeating patterns of both maxes and mins?

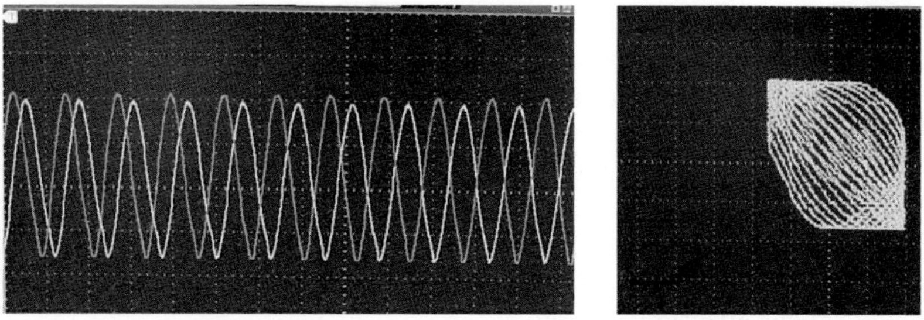

Figure 5.37: The YT display (left) and XY display (right) for two oscillators coupled with 51 kΩ resistors.

In Fig. 5.37 we see the YT display and the XY display when the oscillators are coupled through 51 kΩ resistors. Both plots show this network has not synchronized. On the left side of the YT plot the maxes of the lighter and darker signals are nearby, while on the right side the maxes have moved far apart. Reinforcing this, the XY display is far from a diagonal line or a simple closed curve. We know the uncoupled system cannot synchronize because the oscillators have slightly different frequencies, and the system lacks sufficient feedback to adjust frequency and phase. Coupling the oscillators through high enough resistances will simulate uncoupled oscillators. This experiment shows that 51 kΩ is close enough to uncoupled to destroy synchronization.

5.9.7 Sample C. Three oscillators.

In Fig. 5.38 we see three oscillators coupled through 75Ω resistors. We begin with the two coupled oscillators of Sample 5.9.6 and add connections to couple the third oscillator. Add wires between R8b and R40f, R18b and R40g, R25a and R49i, L40g and R28b, L4g and R28a. Attach the signal out lead to R25b. Add a 75Ω resistor between R25e and R28e.

In the figure we have attached the oscilloscope leads to the two lower oscillators. The symmetry of the circuit, so long as the three coupling resistors have the same resistance, suggests that the output of any two oscillators will produce similar displays. This is easily checked by switching the signal out leads to which the oscilloscope probes are attached.

Set the DC power supply to 9.8 V. The YT and XY displays with the oscilloscope probes connected to the two lower oscillators is shown in Fig. 5.39. Once again we see evidence that the oscillators, three in this case, have synchronized. Despite our symmetry argument, we checked the other two combinations. With all three coupling resistances at 75Ω, the three oscillators synchronize.

Based on our samples of networks of two and of three oscillators, we might surmise that all weakly coupled networks synchronize. But these networks are examples of *complete graphs*, every pair of distinct vertices is connected by an edge. Sometimes these graphs are given the more evocative name "everybody talks with everybody." In Sample 5.9.8 we'll investigate a network coupled along an early stage of a Sierpinski gasket. That is, we'll investigate synchronization on graphs that are not complete.

Next, we change the 75Ω resistor from the signal out of the top left oscillator to a $68K\Omega$ resistor. In Fig. 5.40 we see the YT displays, with oscilloscope leads from the two lower oscillators (left) and the two left oscillators (right). So we see that the oscillators of this network, no longer symmetrically connected, still synchronize.

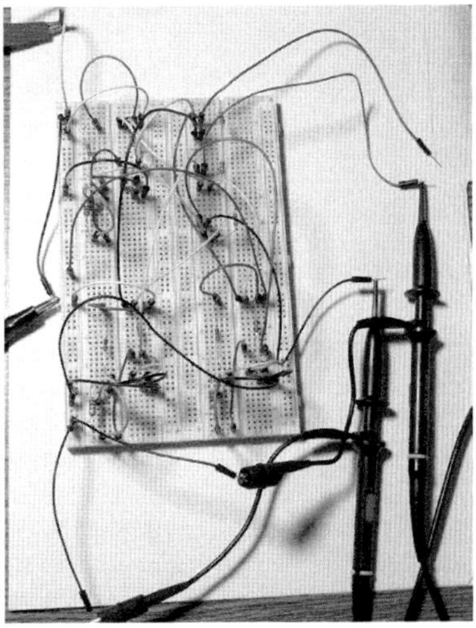

Figure 5.38: Three coupled oscillators.

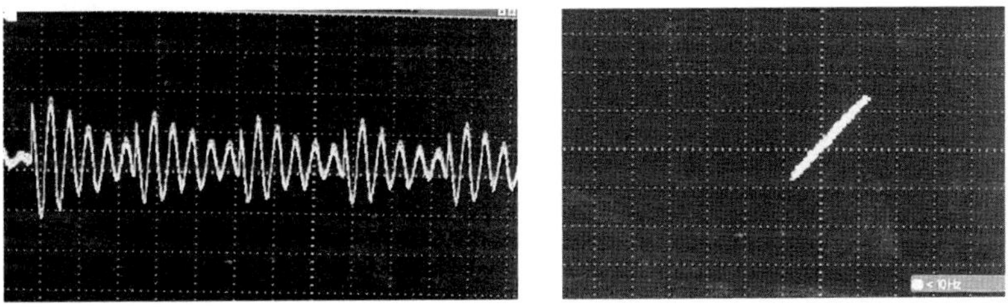

Figure 5.39: The YT display (left) and XY display (right) for three oscillators coupled through 75Ω resistors.

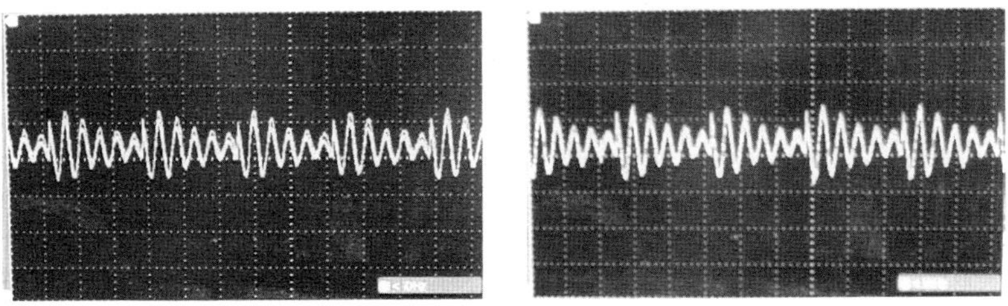

Figure 5.40: The YT display for three oscillators coupled through two 75Ω resistors and one $68K\Omega$ resistor.

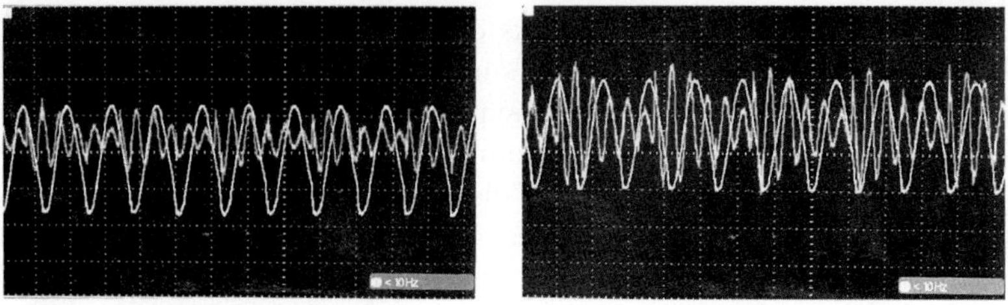

Figure 5.41: The YT display for three oscillators coupled through one 75Ω resistor and two 68KΩ resistors.

Finally, we change the 75Ω resistor from the signal out of the lower left oscillator to a 68KΩ resistor. In Fig. 5.41 we see the YT displays, with oscilloscope leads from the top left and lower right oscillators (left) and the two left oscillators (right). In neither case do we see synchronization. One channel appears to be approximately a sine wave, the other is more complicated. Comparing where the more complicated trace crosses the sine wave, we see no evidence of any kind of synchronization. To present more evidence of this, in Fig. 5.42 we see the XY display. Certainly this is not a diagonal line, and is much wilder than the relatively tame right plot of Fig. 5.37, itself an indicator of lack of synchronization. If we had plans to rank degrees of lack of synchronization—and there's no good reason for us to do this—we'd say the XY plot of this network is "more unsynchronized" than that of Fig. 5.37. This does sound silly, so we won't speak of it again.

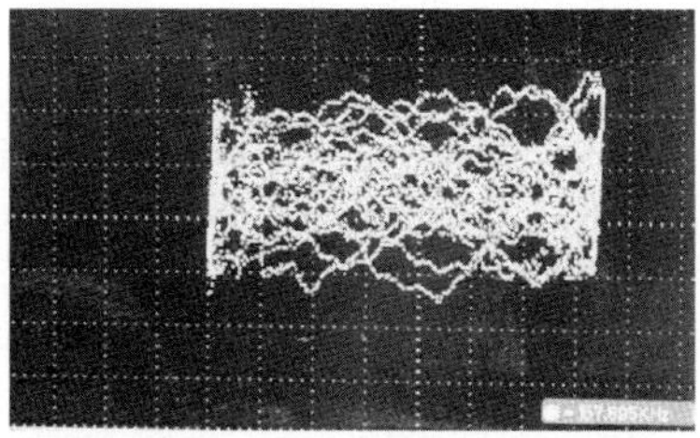

Figure 5.42: The XY display for the left YT display of Fig. 5.41.

5.9.8 Sample D. Nine oscillators.

Begin with three copies of our 3 oscillator network. The small breadboard is where we couple the groups together. We need 6 resistors and 18 wires to couple these networks together in the pattern shown in the schematic of Fig. 5.43. Also, we must move 6 wires from + columns to the power of 6 of the 9 oscillators. In a 400 hole breadboard, attach a 75Ω resistor between holes

1e and 1f, 3e and 3f, 10e and 10f,
12e and 12f, 20e and 20f, 22e and 22f.

Now to describe the attached wires, say breadboard 1 has oscillators 1, 2, and 3; breadboard 2 has oscillators 4, 5, and 6; and breadboard 3 has oscillators 7, 8, and 9. That is, in Fig. 5.43, 1 is the upper left breadboard, 2 is the lower left breadboard, and 3 is the lower right. Say the 400 hole breadboard is BB. For example, 1L36i is the hole in row 36 and column i of the left side of breadboard 1.

First, move these wires attached to + columns:

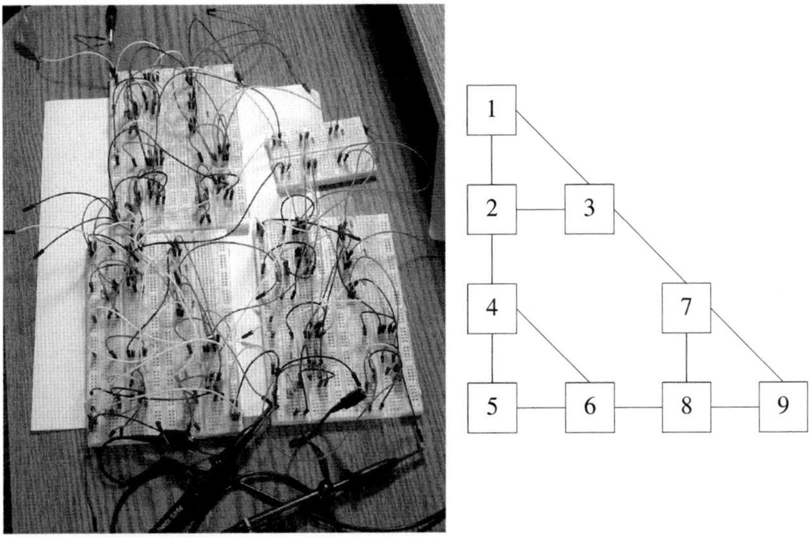

Figure 5.43: Our 9 oscillator network and schematic.

from 1L40j to 1L37j
from 1R40j to 1R37j
from 2L4j to 2L2j
from 2R40j to 2R37j
from 3L4j to 3L2j
from 3L40j to 3L37j

These are the new connections:
1L37i to BB1j, 1L40j to 1L37g, 1L49g to BB3j, 1R37i to BB22j, 1R40j to 1R37g, 1R49g to BB20j
2L2i to BB3a, 2L4j to 2L2g, 2L13g to BB1a, 2R49g to BB10a, 2R40j to 2R37g, 2R37i to BB12a
3L37i to BB10j, 3L40j to 3L37g, 3L4j to 3L2g, 3R49g to BB12j, 3L2i to BB20a, 3L4j to 3L2g, 3L13g to BB22a

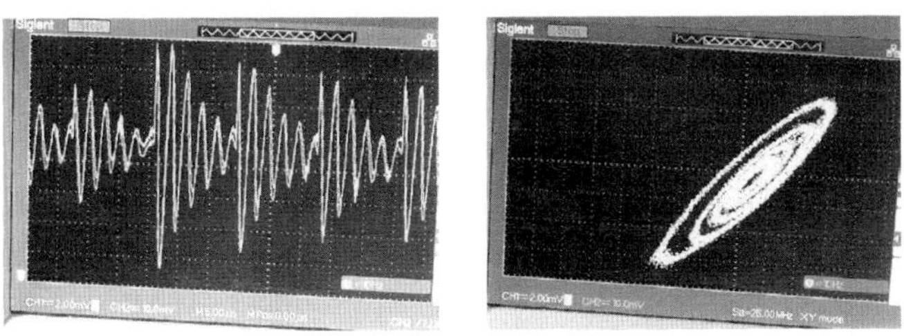

Figure 5.44: Left: the YT display for nine oscillators. Right: the XY display.

In Fig. 5.44 we see the YT and XY displays with the DC power supply at 3.9 V and the signals taken from oscillators 2 and 8. The YT display shows that these signals have synchronized; the XY display shows something new. This isn't a diagonal line, or even a fuzzy diagonal line. Nor is it a simple closed curve. Moreover, the static image of this XY display does not capture some of the system dynamics. Two later views are shown in Fig. 5.45. The transition between the right image of Fig. 5.44 and the left image of Fig. 5.45 *appeared* to involve a rotation in 3 dimensions. While this may sound like unsupported speculation, because this network consists of nine oscillators, the signal out leads generate a pattern in 9-dimensional space. By sampling the signals from two oscillators, we look at a cross-section of the attractor. Over time the orientation of this attractor can change; this change can be seen as an apparent rotation of the XY display.

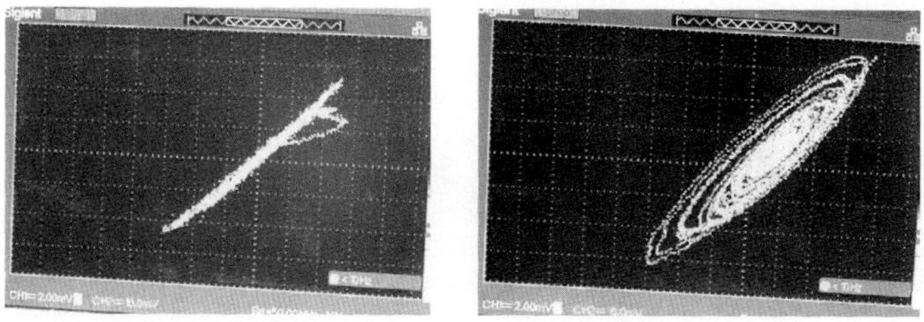

Figure 5.45: XY displays a few moments later.

These XY plots are reminiscent of plots of "strange attractors," the Lorenz attractor or the Rössler attractor are examples. If you've forgotten, or never knew, how these look, Google will show you. If this level of complexity still seems unlikely, consider this paper [185].

Andy Szymkowiak suggested that the slow growing and shrinking of these XY plots may be the results of beating between the lowest-frequency oscillations of the network. Recall that beats occur when periodic signals of nearby frequencies are added. For example, on the top of Fig. 5.46 we see graphs of $\sin(1.3x)$ and $\sin(1.4x)$ (the dashed curve); below this we see the graph of their sum.

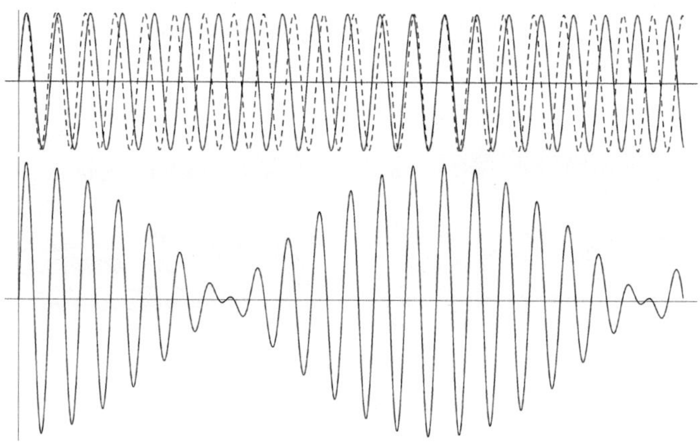

Figure 5.46: Two sine waves (top) and their sum (bottom)

Even this relatively simple network of nine oscillators (Okay, it doesn't seem so simple if you've wired it) has many paths connecting every pair of oscillators. This appears to give rise to a wealth of feedback behaviors that produce a complex signal. Does the hint of self-similarity of this arrangement of oscillators add another layer of complexity? How might we test this?

The nine oscillator network is the most complicated system we've built. It requires patience and an eye to detail unmatched by any other lab. So why did we stop the sample after only a simple experiment? This is our parting gift to you. The system is rich enough to offer dozens, maybe hundreds, of experiments. This sample shows that the network can offer surprises. Once we've established that, we don't want to prejudice your investigations. Bill Watterson gave the best advice in the very last *Calvin and Hobbes* comic:

> It's a magical world, Hobbes ol' buddy. Let's go exploring!

5.9.9 Conclusion

In Lab 3.2 we saw that networks of logistic maps can synchronize for some values of coupling strength. In this lab we investigate networks of Colpitts oscillators and find that the oscillators synchronize for some coupling strengths, where we've represented the coupling strength by resistors in the connection between the signal out of one oscillator and the power in of another. When the oscillators are coupled in a pattern based on an early stage of a Sierpinski gasket, the XY display resembles phase space plots of strange attractors. This opens the way to a whole new line of investigations, which we leave to you.

5.9.10 Exercises

We've mentioned already that constructing circuits on breadboards can be delicate, and may be aided by a fine set of tweezers. Now we'll say something about oscilloscope use. Suppose you've connected the probes and turned on the DC power supply, but you see nothing. Switch the display to XY and turn both vertical scales to about 1 V per division. Check the edges of the scope screen. If you see a yellow dot or line along one side, use the position controls to move that dot or blob near the center of the screen. Clockwise on the left control moves the dot right; counterclockwise moves it left. Clockwise on the right control moves the dot down; counterclockwise moves the dot up. You may need to turn the controls a lot, depending upon how far outside the scope window the dot lies. Once the dot is centered, reduce both vertical scales and recenter the dot. Until you get some practice, reducing these one at a time is useful. Continue until the blob or line occupies about a quarter of the screen. Then switch the display to YT. To photograph the screen, freeze the image by pressing the run/stop button. These descriptions work for the type of oscilloscope we used. Some variations may be necessary for a different scope. Consult the manual.

Prob 5.9.1 In the set-up of Sample 5.9.6,
(a) What happens if the resistances are 0? That is, replace the two resistors on the small breadboard with a jumper wire.

(b) Estimate the range of resistance values over which the network synchronizes.

Prob 5.9.2 In the set-up of Sample 5.9.6, can we achieve synchronization if the two coupling resistors have different resistances?
(a) Take both resistances different but in the synchronization range of Exercise 5.9.1.
(b) Take one resistance in the sybchronization range and one outside that range.

Prob 5.9.3 In the set-up of Sample 5.9.7,
(a) Use three coupling resistors of the same value and estimate the range of these values for which the network oscillators synchronize.
(b) Start with all three resistors near the highest value of this range. Decrease the resistance of one resistor and increase the resistance of the other. Can you achieve synchronization when one of the resistors is outside the range you determined in (a)?

Prob 5.9.4 Can you find coupling resistor values for which two of the oscillators synchronize, but the third doesn't?

Prob 5.9.5 In the set-up of Sample 5.9.7, each group of three oscillators uses 3 resistors to couple these oscillators, and 6 resistors to couple the signals between the groups. Call these *within-group* and *between-group* resistors.
(a) Use the same values for all 15 coupling resistors. Estimate roughly the upper value of resistance to observe synchronization between oscillators 1 and 9.
(b) With the within-group resistors at 75Ω, estimate roughly the upper value of the between-group resistances to observe synchronization between oscillator 1 and oscillator 9.
(c) With the between-group resistors at 75Ω, estimate roughly the upper value of the within-group resistances to observe synchronization between oscillators 1 and 9.

Prob 5.9.6 Return all 15 coupling resistors to 75Ω. Now change only the within-group resistances for breadboard 2. Can you observe synchronization between oscillators 1 and 9 even when the within-group resistances of breadboard 2 are increased above the maximum value obtained in Exercise 5.9.5 (c)?

Prob 5.9.7 To sample the output of all three oscillators of a breadboard, connect the signal out wire of each oscillator to holes 1a, 1b, and 1c of a 400 hole breadboard. Connect another wire to hole 1d and attach an oscilloscope probe to 1e. Attach the other oscilloscope probe to the signal out of an oscillator of another breadboard. Connect the probe grounds to the − column of a breadboard.
(a) For coupling resistor values that give synchronization, compare the oscilloscope traces.
(b) For coupling resistor values outside the range of synchronization, compare the oscilloscope traces.

Prob 5.9.8 To sample the outputs of all nine oscillators, start with the construction of Exercise 5.9.7. Then attach the signal out wire from each oscillator of the second 3-oscillator breadboard to 3a, 3b, and 3c of the first 400 hole breadboard, and attach the signal out wires from each oscillator of the third 3-oscillator breadboard to 5a, 5b, and 5c. Finally, attach 1d to 7a, 3d to 7b, and 5d to 7c. The attach an oscilloscope probe to 7d. Attach the other oscilloscope probe to the signal out of an oscillator of another breadboard. Connect the probe grounds to the − column of a breadboard.
(a) For coupling resistor values that give synchronization, compare the oscilloscope traces.
(b) For coupling resistor values outside the range of synchronization, compare the oscilloscope traces.

Prob 5.9.9 Rather than a Sierpinski gasket network, connect the oscillators in a circle network. That is, connect the signal out of oscillator 1 through a resistor to the power in of oscillator 2, the signal out of 2 through a resistor to the power in of 3, ..., the signal out of 9 to the power in of 1. Connect one oscilloscope probe to the signal out of oscillator 1, the other probe to the signal out of oscillator 5.
(a) Find values for the coupling resistors that give synchronization.
(b) Estimate roughly the range of resistor values that give synchronization.
(c) Do your answers to (a) and (b) change if you change the placement of the second oscilloscope probe?

Prob 5.9.10 Explore the 9 oscillator network. Be imaginative in your experimental design. Daydreaming a bit before you design your experiment may be useful.

Chapter 6

What Else?

Among the most commonly-given advice to young scientists is this, "If everything you try works, you aren't trying things that are crazy enough." The reason for this advice to that audience is to encourage a more aggressive deployment of imagination in research. When we planned these labs, we took this advice to heart. Some labs, IFS and box-counting for example, have a relatively long history. They introduce important topics in fractal geometry, and admit variations that allow each group to personalize its investigations. We had to include these old standards, but we wanted to explore also some less-familiar aspects of fractal geometry. So we designed other labs. Some worked well, others did not. Here we'll give brief sketches of some of the labs that didn't work so well. Each explores a topic that we think is natural and of interest, but our implementations did not help most people to grasp the main points.

If you think any of these has promise, and if you have some experience with the topic or a different idea of how to approach it, maybe you'll find an experiment more successful than ours.

6.1 Building block fractals

The point of this lab was to give another set of examples, physical rather than on a computer screen, of the relation between dimension and measure, and how dimension can signal the degree to which a fractal fills the space around it.

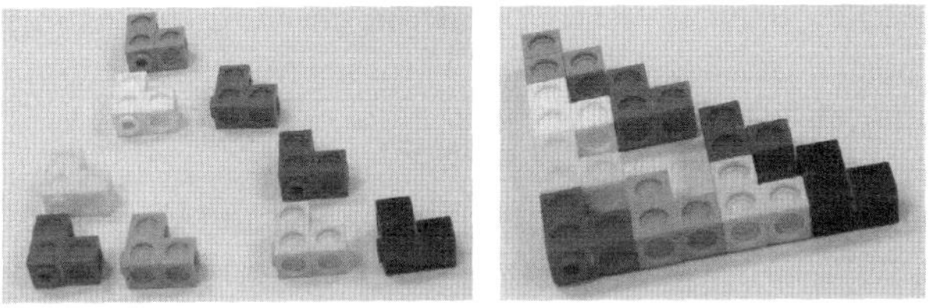

For the first model, we assemble groups of three blocks in an L-shape (left image). Think of these as first iterates in the formation of a Sierpinski gasket. Three of these are attached to form a second iterate, and three of those are attached to form a third iterate (right image). We build these models with *multilink cubes*; we need 216 cubes to build the third iterate of all three of these models. For the first model we need $27 = 3^3$ cubes for the third iterate. Each cube has an outward-pointing snap on one face and inward-pointing snaps on the other five faces, so some thought must go into how the blocks are assembled in order to attach each group to the the others.

In the second model, we assemble groups of four blocks, then four groups of these in the same pattern, and finally four groups of these. For this model we need $64 = 4^3$ cubes for the third iterate.

Observe the large central empty space. We saw this, for the equilateral case, in Lab 2.4.

In the third model, we assemble groups of five blocks, then five groups of these in the same pattern, and finally five groups of these. For this model we need $125 = 5^3$ cubes for the third iterate.

The left picture is of 22 five-block complexes; 25 are needed to build the third iterate.

If we visualize the limit to which these models would converge if we had a *lot* more blocks, the first model would produce a fractal that consists of $N = 3$ copies scaled by $r = 1/2$, the second $N = 4$ and $r = 1/2$, and the third $N = 5$ and $r = 1/2$. With the similarity dimension formula of Lab 2.3, we see these models are early steps in producing fractals of dimension

$$d_s = \frac{\log(N)}{\log(1/r)} = \frac{\log(3)}{\log(2)} \approx 1.585, \quad \frac{\log(4)}{\log(2)} = 2, \quad \frac{\log(5)}{\log(2)} \approx 2.322$$

Then we expect the first model will have infinite length and zero area, the second will have infinite length and zero volume, and the third will have infinite area and zero volume. These provide additional examples of our observation in Sample 2.1.6 of relations between dimension and measurement.

The only interesting math in these examples involved computation of the "perimeters" and "areas". Volumes are easy—number of cubes times the volume of each cube—because none of the cubes overlaps in a solid region. By perimeter we mean number of edges of each cube times the edge length, minus the lengths of those edges that are counted several times, that is, are edges common to several cubes. By area we mean the number of faces of each cube times the face area, minus the areas of those faces that are counted several times.

While these calculations were satisfying, some lab participants thought this was a lot of work for a result we'd seen (mostly) in Lab 2.4. The only new observation from the block models is that the base of the 5-block fractal, and consequently the base of each piece of the fractal, is 2-dimensional.

Perhaps other block models would provide some surprises, or point the way to other configurations students could build.

6.2 Non-Euclidean tilings

The motivation for this lab was M. C. Escher's *Circle Limit III* (fishes) and *Circle Limit IV* (angels and demons) woodcuts. A Google image search for "Escher circle limit" will return many copies of these pictures. Escher's *Circle limit* series was the culmination of his long search [186] to represent the infinite in a bounded region. While in the limit these tilings fill the entire Poincaré disk and so are not fractal, for some tilings the edges of the tiles form infinitely branched trees, the fractals that live in Escher's woodcuts. This offers the opportunity to learn how hyperbolic geometry differs from Euclidean.

A familiar formulation of the parallel axiom, though not Euclid's original version, is that for any line g and any point A not on g, exactly one line parallel to g passes through A. The alternatives are (i) no lines parallel to g pass through A (This is true in *spherical geometry*.) and (ii) infinitely many parallels to g pass through A (This is true in *hyperbolic geometry*.). Why are the only choices for the number of parallels 0, 1, or ∞? Why can't g have exactly 2 parallels that pass through A? Think about this a bit. What can you find?

For any geometry we need to say what are the lines and what are the angles between lines that intersect. In spherical geometry the lines are the *great circles*, the intersections of the sphere with planes through the center of the sphere. At an intersection point of two great circles, the angle is the angle between the (Euclidean) tangent lines at that point. All great circles intersect at two points, so spherical geometry has no parallel lines at all.

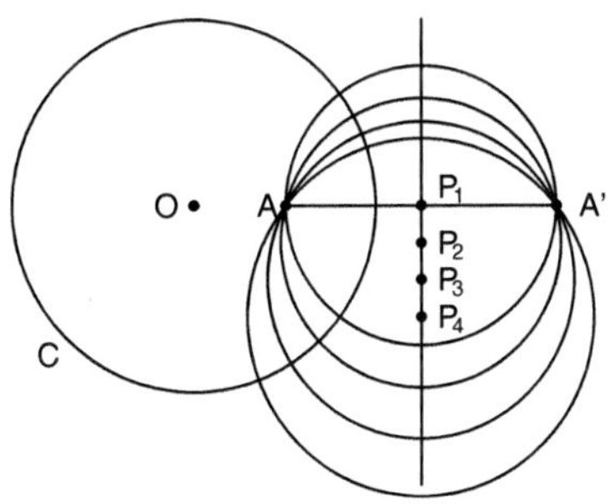

A common model (there are several) for hyperbolic geometry is the Poincaré disk, the interior of the unit disk, $\{(x, y) : x^2 + y^2 < 1\}$. The lines are diameters of the Poincaré disk, or arcs of (Euclidean) circles that intersect the boundary circle $\{(x, y) : x^2 + y^2 = 1\}$ at right angles between their tangent lines. The center of the circle on which a hyperbolic line lies is called the *pole* of that line. The pole always lies outside of the Poincaré disk. For hyperbolic lines that are diameters of the Poincaré disk, the pole is the point at ∞.

From the study of inversion in a circle (Lab 3.7)—specifically, that a circle S orthogonal to the inverting circle C inverts to itself, interchanging the arc of S inside C with the arc of S outside C—we know that if A and A' are inverses across C, then any circle that passes through both A and A' is orthogonal to

C. The centers of these circles, P_1, P_2, P_3, and P_4 in the figure above, are the poles of the hyperbolic lines determined by these circles.

The angles between hyperbolic lines that intersect are the angles between their tangent lines at the intersection points.

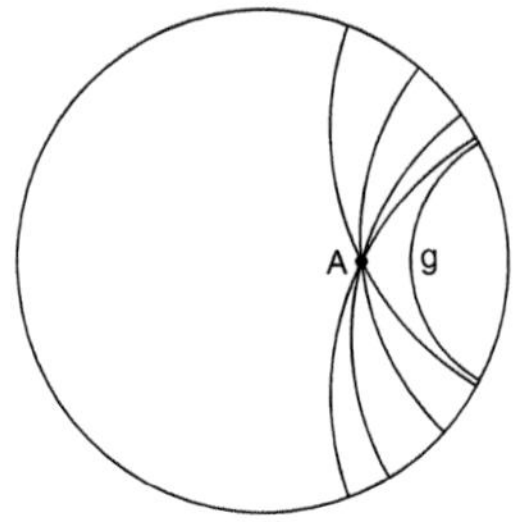

On the right we see four parallels to g that pass through the point A; these show how to find infinitely many parallels to g through A.

We'll use the Poincaré disk model. There are fairly complicated formulas to measure hyperbolic length and area, and we don't need these now. For a hyperbolic triangle T, that is, the triangle formed by the three hyperbolic lines that intersect in three points at angles α, β, and γ, the *angle defect* $\Delta(T)$ is

$$\Delta(T) = \pi - (\alpha + \beta + \gamma)$$

While the angles of a Euclidean triangle sum to π, those of a hyperbolic triangle sum to $< \pi$. With some work we can show that hyperbolic triangles with equal angle defects have equal areas, and so in hyperbolic geometry similar triangles are congruent. Methods of construction in the Poincaré disk model are described in [187].

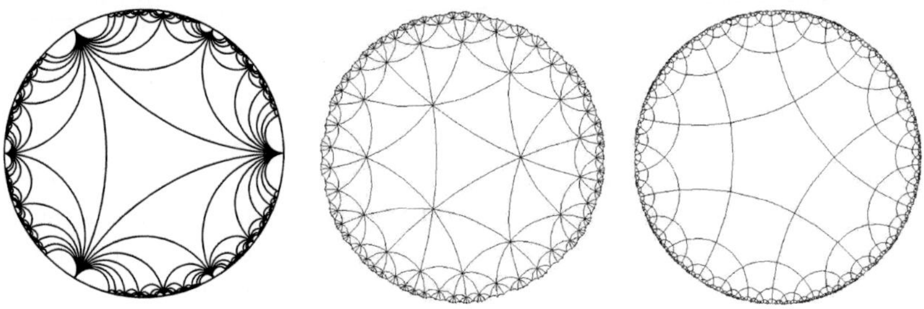

Figure 6.1: Three tilings of the Poincaré disk.

On the left of Fig. 6.1 we see a tiling by triangles that have all angles 0, so all have area π and consequently all are congruent in the Poincaré disk, despite the fact that those near C look smaller and smaller to our Euclidean eyes.

The center image of Fig. 6.1 is (the beginning of) a tiling of the Poincaré disk with triangles that have all angles $\pi/4$. Again, all triangles have the same angle defect and so are congruent.

The right image is a tiling by pentagons that have all vertex angles $\pi/2$. The angle defect formula can be extended to polygons with more than three sides by subdividing these polygons into triangles. The consequence is that because all these pentagons have equal angles, they are congruent in the Poincaré disk.

Some workshop participants were unfamiliar with the details of hyperbolic geometry, so the background was new and interesting for them. Many commented that their students would like a geometry without the raft of triangle congruence theorems. The ability to construct tilings with shapes other than triangles, squares, and hexagons intrigued many. Nevertheless, the tie to fractals through only the trees formed of tile edges seemed weak to many. In order for this lab to be effective, it must be developed in a direction with a more obvious fractal structure.

6.3 Fractal perimeters

The construction of this lab is a variation of the Koch snowflake, a shape with fractal perimeter. In Lab 6.2 we constructed tilings of the Poincaré disk; here the construction of the fractal perimeter tiles its interior with similar quadrilaterals. Their shape explains their name, kites.

That this is a compass and straightedge construction of a fractal was part of the appeal of the lab. Also, this fractal perimeter grows by accretion, as does the Koch snowflake, rather than by subtraction, as does the Sierpinski gasket. Finally, there is some interest in the fact that the steps are recursive in geometry, as were those of Labs 2.4, 2.5, and 2.6.

But this is a rigid construction. Everyone produced the same shape, exact to within the limits imposed by their facility with compass and straightedge. One variation is the assignment of colors to the interior tiles, but a "coloring book lab" didn't seem to be age appropriate. After we've constructed a few stages of the shape, we'll mention another possibility, inspired by work of M. C. Escher. Perhaps you can think of some geometric variations of the tile shape or general construction.

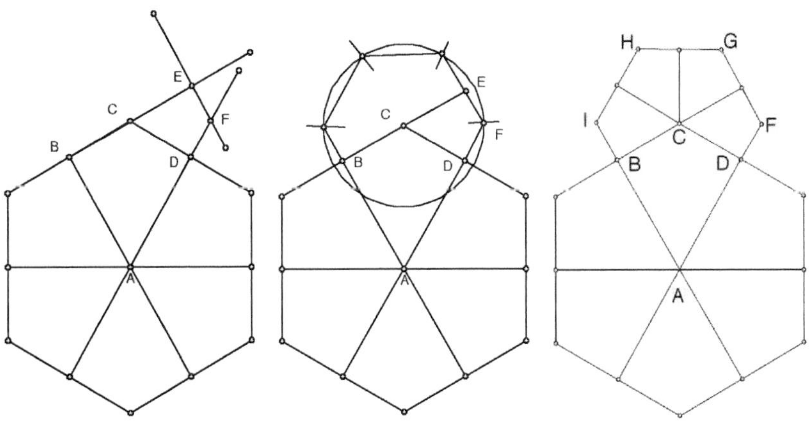

Figure 6.2: Left: Hexagon with an added external kite. Center: The circle to construct three more external kites. Right: Hexagon with four kites about vertex C.

Here are the first few steps of the construction. We begin with a regular hexagon. Draw segments that connect the midpoints of opposite sides of the hexagon. This divides the hexagon into six quadrilaterals, called *kites*, $ABCD$ in the right image of Fig. 6.2, for example. Here we see that to construct the first external kite, extend segments BC and AD. Mark the point E on the extension of BC so $BC = CE$. Draw the perpendicular to BE through E.

This intersects the extension of AD at a point F. Then $CEFD$ is similar to $ABCD$ (Do you see why?) so giving it the name "kite" is reasonable.

In the center image of Fig. 6.2 we draw a circle with center C and radius CF. Then start with F as the center and mark circles of radius CF around this circle. Connect the appropriate points to obtain three more small kites outside of the original hexagon. Erase the circle and extraneous lines to obtain the right image of Fig. 6.2, a hexagon with four small kites at one vertex.

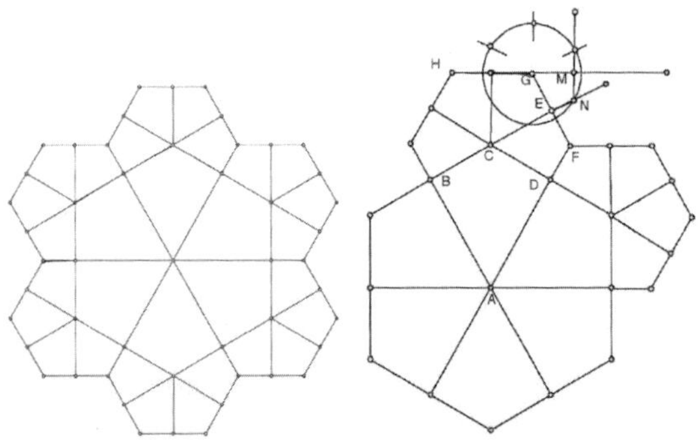

Figure 6.3: Left: Hexagon with second-generation kites. Center and right: construction of third-generation kites.

On the left image of Fig. 6.3 we see four small kites, which we'll call *second-generation* kites, around each vertex of the hexagon. In the center and right images we see steps in the construction of four third-generation kites around vertex G of a second-generation kite.

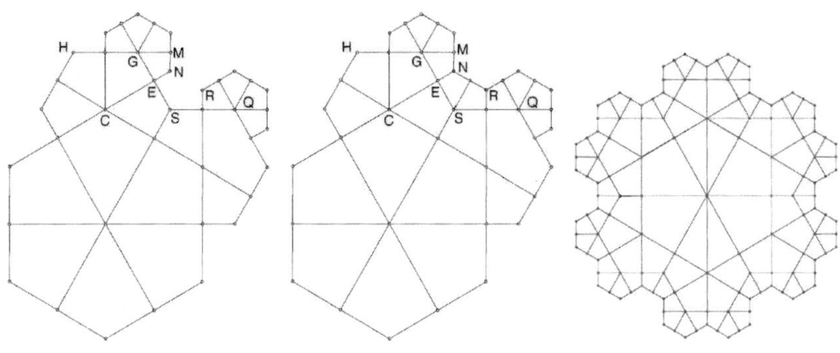

Figure 6.4: Left: The indentation between groups of third-generation kites. Center: Fill the indentation with kites. Right: Add the third-generation kites.

Now if you think we just continue this process until we are tired of it, there's a complication. The left image of Fig. 6.4 shows the problem. When we add the third-generation kites around vertex Q, we form an indentation between vertices N and R. As we add higher-generation kites along the sides of this indentation the kites might intersect. So we fill the indentation by adding the segment NR. Connect the midpoint of NR to the vertex S. This forms two additional third-generation kites that we see in the center image of Fig. 6.4.

In the right image of Fig. 6.4 we see the third-generation kites added around the perimeter of the shape.

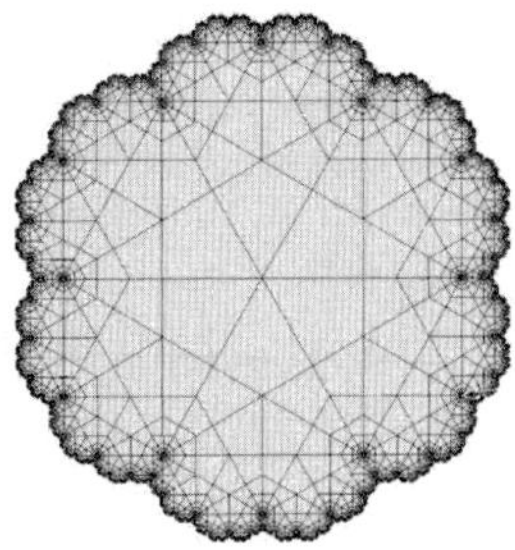

The addition of the third-generation kites is about the limit of the patience, and in some cases eyesight, of workshop participants. By this stage in the construction, the the geometric iteration was clear. What was not so clear, perhaps, was the roughness of the boundary of the shape that appears in the limit. This can be seen only in a later iterate. Tiled by kites, this shape invites artistic modification: convert each kite into a bird or a fish or something like that. Follow the examples of M. C. Escher. Peter Raedschelders and Robert Fathauer have made many examples of this sort of artwork. Google can find current collections of their work.

6.4 Multifractal finance

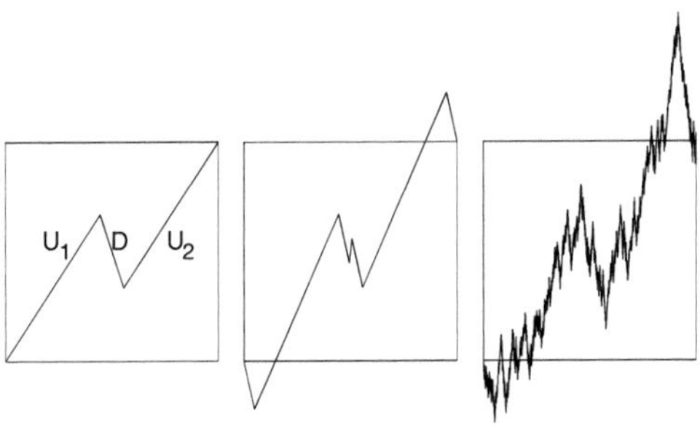

Figure 6.5: Building a multifractal finance cartoon.

One of the most highly visible components of Benoit's work is his multifractal approach to finance. (That anyone would find finance more interesting than geometry is a mystery I cannot understand. M.F.) The simplest examples are his multifractal cartoons. The first iteration are two segments, U_1 and U_2 that go up and one D that goes down. Benoit called this the *generator* of the cartoon. Each up segment is replaced by one of six arrangements U_1DU_2, U_1U_2D, DU_1U_2, DU_2U_1, U_2DU_1, and U_2U_1D, each starts at the lower left corner of the initial upward segment and ends at the upper right corner of that segment. Each down segment is replaced by one of six similarly scaled pieces, but now the two long segments point down and the short segment points up. The first and second images of Fig. 6.5 show the first iterate and one (out of 6^3) choices for the second iterate; the third is one instance of the ninth iterate.

One slightly tricky point is that we don't just graph the iterates, because we'll be comparing this graph to financial time series which are sampled at equal times (daily closing price, for instance). What we do plot is a high iterate sampled at equal time intervals.

First of all, Benoit called these "cartoons" rather than "models" because cartoons just reproduce appearance or behavior and make no claim to represent underlying mechanism. In Chapters 32 and 33 of [1], Benoit described what he called a "model" of galaxy formation. By this he meant a probabilistic process that produces a distribution statistically similar to the observed spread of galaxies across space. Despite this sentence, "The present method does not claim to account for the way galaxies had actually been generated," (pg. 294 of [1]) some astronomers criticized Benoit's model because it did not account for gravity, or the dynamics of galaxy motion, or the expansion of the universe, or really any physics. Most scientists think that a model should involve some aspects of mechanism, not just appearance. So to avoid similar criticisms by economists, Benoit called these finance *cartoons*.

Why are these multifractal? In the graphs of Fig. 6.5 the vertical axis represents money and the horizontal axis represents time. Say T_1, T_2, and T_3 are the time intervals that correspond to U_1, D, and U_2. Then the curve generated by iteration is a multifractal if not all three of the ratios

$$\frac{\log(U_1)}{\log(T_1)}, \; \frac{\log(|D|)}{\log(T_2)}, \text{ and } \frac{\log(U_2)}{\log(T_3)},$$

are equal. Each of these ratios is a measure of roughness; if the ratios are not all equal, then the roughness of the curve varies along the curve.

A final comment before we get to the lab: Benoit found a simple way to rescale time, called *trading time*, that converts a multifractal financial series into a fractal series (constant roughness). The rescaling is just an application of the Moran equation. Solve

$$|U_1|^h + |D|^h + |U_2|^h = 1$$

for h. Then the trading time intervals are

$$S_1 = |U_1|^h, \; S_2 = |D|^h, \text{ and } S_3 = |U_2|^h$$

If we take the first iterate to have the same vertical intervals, U_1, D, and U_2, and horizontal intervals S_1, S_2, and S_3, this generates a fractal, not a multifractal, curve.

Changing to trading time can be thought of this way: times of high volatility are stretched out, and times of low volatility are compressed. One imagines that estimating when a stock price will become more volatile would be of interest to those who understand these things. But we don't have the generator; we have just the time series. So how can we we go from time series to generator?

Applied to financial time series, that's the point of the lab. We won't look at time series, but rather at the differences of successive measured values, how much the price increases or decreases from one time period to the next.

We generated time series for ten different multifractal cartoons and for each time series, plotted the IFS driven by the differences with equal-weight bins. Then we took stock data from 18 months before the lab until 6 months before the lab and plotted the IFS driven by the differences of the time series values. Again we use equal-weight bins. Why equal-weight? Financial data can have infrequent very large positive or negative jumps which could result in very few points in bins 1 or 4, giving little structure to the driven IFS.

If you think driven IFS is too crude a tool to use with financial data, we point out that on page 264 of [66] Benoit shows a driven IFS plot generated by Joseph Thornton in 2002 as part of his project in M. F.'s fractal geometry course.

The length-2 address occupancies of the stock data are compared with those of the ten sample cartoons. The closest match is used as a surrogate for the stock data. When we generated the cartoon time series, we generated another six months of cartoon data. Local maxima of this data signaled times the lab participants would sell some of their stock, local minima signaled times to buy. Applying these buy and sell times to the next six months of real stock data, the goal of the lab was to find those strategies that made money at the end of the six months, and those that lost money. Compare the strategies. What do we learn?

The problem with implementation was that every group made money, so there was nothing to compare. After the lab one of the teachers pointed out that the market had strong growth for those six months, so most every strategy would make money.

Perhaps people with more financial knowledge can design a successful lab based on this idea. Much more information on these cartoons can be found in [188], [189], and [66].

6.5 Fractal music

Fractal aspects of music have been studied since the mid 1970s, before fractals were named. Richard Voss and John Clarke[190, 191] showed that both loudness fluctuations and pitch fluctuations exhibit a $1/f$ power law scaling (Eq. (1) with $b = -1$) over a variety of styles: classical, jazz, blues, rock, even talk radio if a long enough sample is taken. In the early 1990s Kenneth and Andrew Hsü [192] found a power law relation for the interval between successive notes and the frequency with which those intervals occur in a composition. In a different direction [193] the Hsüs began with a composition by Bach (they used Bach's *Invention No. 1 in C Major*) and constructed new "compositions" by removing every second note, then every second note of what remains, and finally every second note of what remains. The last piece is 1/8 as long as the original. The Hsüs believe all these "thinnings" sound Bach-like. Their motivation for this approach is based on the self-similarity of the composition.

> a musical composition could be represented by a music score of a different scale, using half, a quarter, or twice as many notes as were written by the composer.
>
> ...
>
> If a coastline has no definite length, could we state that Mozart's music has no definite number of notes or note intervals?

This lab, designed by Harlan Brothers, included a mini-tutorial in music theory which was of great interest to the workshop participants. To get participants to compose fractal music, Harlan took a different approach, suggested by the score shown at the start of this lab. Note the relation between the second line and each measure of the first line. This is called *structural scaling*, patterns of notes that repeat, in some fashion, on larger and larger scales.

Harlan illustrated structural scaling in the Bourree I from Bach's *Cello Suite No. 3*, and gave a very clear explanation of Michael Adelson's discovery of structural scaling in the second movement of Beethoven's *7th Symphony*. And he introduced *pulse trains* as another example of fractals in music, which he illustrated with Beethoven's *Ecossaise in* $E^\flat$.

Each group began with a motif, made copies of the motif translated in pitch and scaled in time to mimic the pattern of the original motif, then placed these copies one after other, with occasional modifications to make the composition sound better. The lab participants enjoyed this exercise, and some of the compositions sounded quite nice. Possibly this was a function of the previous experience with music by some of the teachers.

The difficulty with the lab was that the effort of only an hour or so was sufficient to produce no more than two levels of the pattern. From the beginning of our study of fractals we emphasized that in order to make a sensible claim that a shape is fractal, we must find copies of the same pattern on many different scales. The exercise sketched in this lab is better suited for a long-term project than for a single lab.

6.6 Other ideas

Finally, here are a few ideas that we discussed but didn't pursue to the level of physical implementation.

6.6.1 Lichtenberg figures

Lichtenberg figures are patterns formed when high-voltage electricity discharges rapidly into a medium that offers some resistance. A lightning strike on an isolated tree in a large open field can burn a branching pattern into the ground. A model, called dielectric breakdown, was proposed by L. Niemeyer, L. Pietronero, and H. Weismann [194]. The model points out a similarity between dielectric breakdown and DLA, which we observe in many examples.

Lichtenberg figures can be generated by discharging a high-voltage electrical pulse into a piece of plastic. We'd intended to include this as a lab, with a van de Graf generator as the power source. This produces a static charge so mistakes can be painful, but are not life-threatening. Our idea was to discharge into an overhead transparency, possibly dust graphite over the transparency to make the discharge pattern more visible. But we've been unable to do this without suffering an unpleasant electric shock. Consequently, we've decided not to propose this as a lab, but mention it for those who have more skill than we do with high-voltage electricity.

6.6.2 Fractality of a life

We thought of this as part narrative exercise, part discussion. About a week before the discussion, give the general background. The progression of a day is similar to the progression of a year, is similar to the progression of a life. Here's one way to read these similarities.

Morning: you wake up, organize your thoughts for the day, start to work. Afternoon: you work seriously, build on and rearrange the morning's ideas. Evening: the afternoon's ideas fall into place, you feel satisfied with what you've done. Night: you find some holes in the days work, become confused, think mostly of sleep.

Spring: as the animals and plants begin to stir, longer days and warmer weather energize you to start new projects. Summer: as animals and plants are so active, as are your thoughts. While a summer storm rolls in, you see new combinations of ideas. Autumn: your work for half a year organizes itself, relations previously unperceived now are clear. Winter: as animals and plants go into hiding for the harsh season, you find gaps, maybe errors, in the year's work. Shorter daylight and colder temperatures make you long to stay under the warm covers, sleep till spring.

Childhood: your mind wakes up, begins to make sense of self and of the world around you. Youth: you go to school, develop the skills you'll use throughout your life and you begin to use them. Adulthood: you use and refine the techniques you've learned, maybe develop new methods, do your major work, sculpt how you'll influence the world around you. Old age: you find gaps in what you've done, miscalculations about what's important and what isn't. Things mental and physical begin to fall apart.

The exercise is to find examples from your own life, from the lives of older friends to fill up the timeline, and try to support the claim that these patterns can describe self-similarity of life.

You can magnify smaller time scales, but unlike the seasons of a year, an individual life does not extend beyond an individual death. To expand the scale outward, replace a person with a group of people, a literary or musical style, a nation, or with something else. What can you find?

Comparisons of different approaches to this problem should be enlightening, both in their similarities and in their differences.

6.6.3 Fractals in literature

This is another long-term project. Fractals can appear both as subject and in the structure of a text. As guidance give an example or two of each kind.

In his novel *All the Names* [195] José Saramago describes the General Cemetery as

> an enormous felled tree, with a short, fat trunk, made up of the nucleus of the original graves, from which four stout branches reach out, all from the same growing point, but which, later, in successive bifurcations, extend as far as you can see, forming, in the words of an inspired poet, a leafy crown in which life and death are mingled, just as in real trees birds and foliage mingle.

About this description, in his diary *The Notebook* [196] Saramago wrote

> I found myself deeply involved in something as mysterious as fractal geometry [...] For those few days I was rubbing shoulders with no lesser a company than the best geometricians of the world. The point they had attained after so much hard effort, I realized I had reached through a sudden flash of scientific intuition, a realization from which, to tell you the truth, I still haven't recovered, despite the amount of time that has gone by.

If you don't think this is a clear enough mention, how about this from *Telegraph Avenue* [197] by Michael Chabon. Here Chabon describes the aftermath of a party:

> Bones and scattered beans, puddled sauces inter-oozing on discarded plates in Mandelbrot sets of grease and tomato.

For structural fractals, Lucy Pollard-Gott found Cantor set patterns of repetitions of words in some of the poetry of Wallace Stevens. Surely not placed by Stevens in order to construct a fractal, one possible explanation is that Stevens discovered this self-similarity unconsciously through rewrites, the repetition across scales setting in motion echoes that suggest infinite regress, strengthening the impression that a poem is a window on another world.

Then there's scene 7 of *Arcadia* [198] by Tom Stoppard. (Fractals occur as subject here, too.) Characters from two time periods, on the stage simultaneously, talk over and through one another. As her project in Yale's fractal geometry course, Josie Rodberg plotted a graph of when the time period of the speaker changes. The graph is a (random) Cantor set. We expect this was not engineered in, even though obviously Stoppard is familiar with fractals (fractals are represented correctly as a subject in *Arcadia*.), but rather arose, through rewrites, from the tension between a desired level of verbal complexity and the need to propel the narrative.

Let people pick favorite authors, or perhaps you provide a list of promising authors. For structural fractals, some discussion of the main points of John Yorke's *Into the Woods* [199] can be useful. Yorke writes

> Stories are built from acts, acts are built from scenes, and scenes are built from even smaller units called beats. All these units are constructed in three parts: fractal versions of the three-act whole. Just as a story will contain a set-up, an inciting incident, a crisis, a climax and a resolution, so will acts and so will scenes.

Be open-minded about suggested authors, provided you've emphasized how to distinguish a fractal from a non-fractal. It's a great joy to be shown a good example of a fractal that was previously unknown to you.

6.6.4 Fractal government

By this we don't mean "corruption at all levels" or "incompetence at all levels" or even "stupidity at all levels," though in early 2020 the case could be made for all three. Rather, we refer to the conjecture that city is to county, as county is to state, as state is to nation. Are the government structures similar? In what ways do they operate similarly, in what ways do they differ? Are the similarities sufficient to support the notion that government structures are fractal?

Groups could make use of local knowledge, familiarity with town and regional governments. If the case can be made for self-similar government, might this suggest some predictive or control options? Do subdivisions of government other than those of obvious geography reveal levels of similar design?

This is a "think about it for a week, then have an energetic discussion" kind of lab.

6.6.5 Found fractals: a photography assignment

Many people have cell phones with cameras, so the equipment for this lab already is distributed. Walk around and photograph things you think are fractal. Maybe divide into two groups, urban and rural (or suburban, or maybe parks). Or maybe manufactured and grown, though the manufactured can have natural erosion or plant growth on it. Look closely. Do you find more fractals on a small scale than on a large scale?

6.6.6 Fractal dance

This lab is based on a performance at the conference "Quantitative measures of dynamical complexity and chaos" held at Bryn Mawr College in June of 1989, and on a class project and performance in Yale's fractal geometry class.

The Bryn Mawr dance was choreographed by Linda Caruso-Haviland, and performed by Caruso-Haviland, Karen Anderson, and Renee Banson Shapiro. The performance, titled "Chaotic metamorphoses: a work in progress" [200] and subtitled "A fusion of poetry, dance and music incorporating chaotic dynamical structures and concepts," consisted of six acts.

"The second coming" presented three dancers in an allegorical composition accompanied by a reading of the poem "The second coming" W. B. Yeates.

"Chaotic vortex" involved a solo dancer whose movements were patterned after fluid vortices. The accompanying music included a tonal sequence generated by EEG data.

"The brain is wider than the sky" was a reading of Emily Dickenson's poem of that title, without musical accompaniment and with the auditorium dark.

"Music of the mind": three dancers' movements reflected the rhythms of an EEG signal. The fractal attractor of the EEG data was projected onto the stage so the dancers moved though the image. The accompanying music was "Baroque variations" by Lukas Foss.

"The wall between" was a reading of that poem by Tricia F. Harper on an empty stage with lighting in parallel with the rhythms of the poem.

"Logistic bifurcations" involved three dancers whose movements represent successive period-doublings of the logistic map through period 8, followed by chaotic motion and then reverse bifurcations back to the initial steady state.

Each period-doubling was signaled by increasingly complex movements by the dancers; in the chaotic regime the movements of the dancers illustrate sensitivity to initial conditions. The accompanying sounds and projected video display were generated by logistic map period-doublings.

The audience for the original performance included MF, who attended the Bryn Mawr conference with David Peak. Dave's understanding of modern dance is much more sophisticated than is Michael's; conversations with Dave after the performance provided valuable context for the appreciation of this event.

Now MF's memory of the Bryn Mawr performance is not so good, but the volume [201] recorded many of the details. Lacking such a record, including class notes from those years, memory of the Yale performance is much less filled-in. Nevertheless, I can say this: the core idea, swirls of swirls of swirls, is simple enough and does allow many variations, but implementation requires a group of people who are fairly athletic. While this may work for some classes of college-age students, it was not possible for a group of mostly oldish teachers, though we enjoyed the discussion of the likely missteps had we tried to implement such a lab.

As with the last act of the Bryn Mawr performance, this lab also can demonstrate sensitivity to initial conditions. Here's one way. Break the dancers into two groups. The movement of each dancer is guided by the movements of nearest neighbors. The precise rules are the point of the choreographic design. Both groups begin with the same motions. An outside source introduces a small perturbation in the movements of one group. Do the dances continue in about the same way, or do the small changes unfold in surprising and significant variations? What aspects of the rules determine if the dance is chaotic? This lab admits many variations.

Kimberley Cetron presents another approach in[202]. Her incorporation of fractals into dance is metaphoric, rather than geometrical. While this idea can open new avenues, the clarity of the fractal form can be more difficult to assess.

6.6.7 A fractal play

Based on the fractal structure of Scene 7 of Tom Stoppard's play *Arcadia*, a group of students wrote and performed a short play with a fractal structure. This offers the opportunity for students interested in playwriting to explore a novel aspect of the structure of plays. With this lab, students focused on the humanities will see a way math is relevant to their interests. Is there a self-similar structure different from the pattern of changes of speaker? This is another longer-term project, but can be ever so entertaining.

Our workshops ended in the first decade of the 21st century. By now the book of fractals has added many new chapters. With open eyes, a good imagination, and attention to the curiosity of your students, you may find possibilities undreamt by us.

Chapter 7

Why labs matter

When we were students, labs were not our favorite classes. We'll ignore the clumsiness of us and of our lab partners. The issue was that the labs weren't particularly interesting. All were more-or-less the same: follow a precise recipe and approximately reproduce the results of an experiment that established a fact we'd already learned in the theory part of the course. If this is experimental science, who would ever want to do it? Of course, much of experimental science isn't like this. For one thing, someone must do the experiments first to establish or validate the theory. For another, many experiments eliminate alternative theories by obtaining results that do not support the prediction.

Mostly, the problem was that the experiments weren't enjoyable, weren't interesting, had no potential for surprise. Nowadays experiments that involve satellites or particle accelerators or genome sequencers take much time, the work of many people, and buckets of money. They are very serious business.

Because kitchen science has very modest equipment needs, we have more freedom to follow for a while the inscrutable exhortations of our imaginations. We include some open-ended examinations, encourage you to make your own conjectures and design your own experiments to test them. While fractal geometry is applicable from the distribution of galaxies, to folding of our DNA, and maybe to the structure of spacetime at the Planck length (10^{-35} m), it offers surprises galore at our scale. Plenty of simple experiments can yield results whose details we cannot anticipate.

We want to help you see that experimental science can be fun. Models are wonderful and important, but unless you work in pure math, eventually all models must have some contact with experiments. Even then, Benoit often reminded his pure math friends about the story of the giant Antaeus who drew his strength from contact with the earth (his mother, Gaia) and was defeated when Hercules held Antaeus aloft, preventing contact with the earth, and crushed him. If math stays too long separate from the ground—from its interactions with the physical world—it will die.

This is easy to forget if you've spent years in MathWorld. Learn the rules of geometry or algebra or analysis and all that follows are deductions—sometimes quite clever—from combinations of these rules. Perfectly happy lives can be, and have been, spent this way. Certainly much of contemporary math is built layer upon layer of abstraction, so while physical roots can be hard to find, they are there if we look carefully. For example, one of the formative events of geometry was the need to survey the Nile basin after its annual flood. Experiments can help us remember where these abstractions began.

Nowadays there's another reason to do experiments. The first time you say this reason out loud, it may sound silly. The reason is to remind us that science is based on empirical evidence. There's a good chance that when you read this you thought, "Well, duhhh." This reminder is necessary because of an unanticipated, disturbing, and dangerous consequence of the internet. When your authors were young, the main source of complete nonsense was the rack of tabloids beside grocery store check-out lines. "I died and went to heaven and saw Elvis and brought back pictures to prove it" was an actual title MF saw. These could be laughed at and ignored.

But now this kind of drek, and worse, comes unchecked and unfiltered to our computer screens, sometimes in an apparently unrelated web search. But the most pernicious effect is not the superabundance of rubbish. It's the deterioration of our understanding of how we know what's true, or at least what's supported by current evidence. Smart people we respect have told us absurd conspiracy theories that they read on-line. Fiction, mythology, and to some extent philosophy, do not depend on empirical observations. But science does. Observations lead to models, models lead to predictions, then more observations show how to refine the models and make better predictions. This is how science works. This is how we learn to understand the natural world.

On-line you can find webpages that assert the earth is flat, or hollow, or 6,000 years old. All of these incorrect views are contradicted by empirical evidence, yet some people still believe this junk. Because we are writing this in the spring of 2021, we must mention the miasma of misinformation that has surrounded the covid-19 vaccines. On-line you could read (maybe still can read, though we hope not) that the vaccines cause infertility, turn people into zombies(!), and are a vehicle to implant geotrackers so the government can follow our every move. The first two are not supported by any data—none, not a bit—and as for the third, many people already can have their movements tracked by their GPS-enabled cellphones. How do people believe this dangerous nonsense about vaccines?

We need to remember that our understanding of the natural world is based on observations. These labs are a modest contribution to this goal.

And finally, we'd like to help you sharpen your curiosity. A good lab can ask more questions than it answers.

Appendix A

Specific Physical Supplies

Lab 1.1. Any ruler with a mm scale, any overhead transparencies, and tracing paper. We used HP transparency films and Strathmore 370-9 300 series Tracing paper.
Lab 1.2. Any mm scale ruler, protractor, and calculator.
Lab 1.3. Any mm scale ruler, graph paper with grid size between 5 and 10 mm. We used Mr. Pen graph paper with 4 squares per inch.
Lab 1.4. Any mm scale ruler, protractor, and overhead transparency. We used HP transparency films and Strathmore 370-9 300 series Tracing paper.
Lab 1.5. Any scissors, glue sticks, clear tape, and white paper. Printer paper works well.
Lab 1.6. Any mm scale ruler, die, clear adhesive tape, overhead transparencies, and permanent marking pen. We used HP transparency films and Sharpie ultra fine point permanent marking pens. We wonder how many schools still have overhead projectors. If you must use a videocamera to project the images on the stacked transparencies, you'll need to fid a way to backlight the transparencies. The backlighting box described in Lab 4.7 can be adapted here.
Lab 1.8. Any paper, pencil, and rulers with mm scale and inch scales.
Lab 2.1. Photocopy the grids onto overhead transparencies, graph paper, and an erasable marker. We used HP transparency films, Mr. Pen graph paper with 4 squares per inch, and Frixion erasable markers.
Lab 2.2. Any mm scale ruler, compass (for drawing circles, not for finding north) or dividers, plastic wrap, dried garbanzo or navy beans, construction paper, printer paper, and tracing paper, a sharp serrated knife. For more accuracy, replace counting the beans with weighing them using a laboratory scale. We used Crayola construction paper, Strathmore 370-9 300 series Tracing paper, and a Vinmax high-precision digital laboratory scale.
Lab 2.4. Any ruler with a mm scale, pencil, scissors, pencil, clear tape, and 256 letter-size ($3\frac{5}{8} \times 6\frac{1}{2}$ inch) envelopes.
Lab 2.5. Any straightedge (a ruler will work), scissors, clear tape, and at least 24 lb printer paper. We used HP BrightWhite 24.
Lab 2.6. Scissors and about 800 pipe cleaners. We used EpiqueOne 1200 pipe cleaners.

Lab 3.1. Any ruler and pencil will do.
Lab 3.3. Any straightedge or ruler and pencil, and graph paper. We used Mr. Pen graph paper with 4 squares per inch.
Lab 3.4. Any pencil and eraser. Photocopy the Pascal triangle template to printer paper.
Lab 3.5. Any pencil and eraser. Photocopy the Pascal triangle template to printer paper.
Lab 3.7. Any ruler with a mm scale, pencil, printer paper, and a compass (for drawing circles).
Lab 3.8. Any ruler with a mm scale, protractor, printer paper, pencil, and erasers.
Lab 4.1. Any ruler with a mm scale, dividers (a compass is an acceptable substitute), watercolor paper, a fine point marking pen, a magnifying glass, and acrylic paints. We used Canson 140 lb. watercolot paper, a Sharpie ultra fine point permanent marker, and Artist's Loft acrylic paint, metallic lemon yellow and phthalo blue. (These were our color choices. The pigments may have a mild influence on paint flow, so we encourage you to experiment with paints of different colors.) Artist's Loft watercolor paints, and Crayloa washable finger paints can be substituted for acrylic paints.
Lab 4.2. Benjamin Moore Color Samples, 14.5 fluid oz. latex in Bright Yellow, Big Country Blue, and Bonfire (red), combined these to make purple, green, and orange.
Lab 4.3. Glazed ceramic tiles that are 4 in. × 4 in. or larger, Craftsmart white acrylic paint, Floetrol latex paint additive, a container for the paint mixture, isopropyl alcohol (at least 70%), a dropper to add alcohol, Dr. Ph Martin's Bombay India Ink (your choice of color), a comb, a ball stylus, a plastic paint tray palette, a razor blade scraper, verhead transparencies. See Fig. A.1.

Figure A.1: Left to right: left to right: Floetrol, Craftsman white acrylic paint, Dr. Ph Martin's Bombay India Ink, plastic paint tray palette, comb, razorblade scraper, 4 in. × 4 in. glazed tile, ball stylus.

Lab 4.4. We used a RAM universal double ball mount with two round plates made by Acr Electronics, Artist's Loft Acrylic paint and Pouring Medium, and Canson watercolor paper. We used the lid of a cat litter bucket for the top of the tilt table, and rather than a drop cloth, we put the tilt table in a (clean) cat litter box. Use the resources that surround you. For me (MF) many of these resources involve cats.
Lab 4.5. Craftsmart Navy Matte acrylic, the canvas was 18″ × 24″ stretched on a frame bought at Michael's craft store, and a Plugable USB 2.0 Digital Microscope.

Lab 4.6. Origami paper, sharp scissors or an exacto knife, and a mm scale ruler. We used BUBU 6 in. square origami paper.
Lab 4.7. Two 8″ × 10″ sheets of clear glass or Plexiglass, two 2″ × 4″ blocks, and a light source (we used a press and stick LED light) to place under the light box, a fine point marking pen, and a digital camera. And of course a selection of leaves.
Lab 4.8. A sample from your garden or the produce section of your local grocery.
Lab 4.9. No particular components here, just implements and ingredients in any moderately well-stocked kitchen. We used a nonstick frying pan, egg beater, spatula, flour, salt, eggs, butter, and milk.
Lab 5.1. We used Anpro 18mm diameter ceramic disk magnets, but any similar magnets will work. Be sure that you use magnets of (about) the same strength for the fixed points of the basins of attraction. Suction cup, overhead transparency, thread, tape, balsa wood rod, fishing swivel, glue, thumb tack, and carpenter's clamp are generic. As in Lab 1.6, the overhead projector can be replaced by a digital videocamera and projector.
Lab 5.2. Plastic silvered ornament balls, about 6″ diameter, with smooth shiny surfaces, not faceted or covered with glitter. Garden gazing globes would work well, but they are relatively expensive. Other components— supports for the silvered ornaments, colored folders, light source, laser pointers—are generic.
Lab 5.3. We used a SONY videocamera with yellow vider-out (the red and white out cables are for the audio channels, unneeded for these experiments) and an analog tv with yellow-white-red input. Nowadays digital videocameras with output consistent with digital tv input may be more readily avaialble. You'll need a tripod to support the video camera. To assemble the mirror mount, use about 10 ft of 1″ × 2″ pine boards, a saw, wood screws, and L brackets and flat L brackets. Rather than mirrors, we used Gardner 12″ × 12″ mirror tiles, available in building supply stores.
Lab 5.4. We used zinc sulfate monohydrate from Alpha Chemicals, a mortar and pestle, UPlasma 120mm ×20mm Petri dishes, TekPower TP3003E 30V 3A DC adjustable switching power supply, Rotometals 0.02″ × 8″ × 11″ zinc sheets, tin snips, RElectronix Express helping hand alligator clampswith magnifier, ThreeBulls metal alligator clamps, Intllab magnetic stirrer, Vinmax high precision digital laboratory scale, Roynes Lifescience 100 ml graduated cylinder, Karter Scientific 1000 ml Erlenmyer flask and 1000 ml beaker, and 31/2″ diameter Brew Rite circular coffee fileter.
Lab 5.5. We used Glycerine Supplier vegetable glycerine, Liquitex acrylic ink, two USAMade 12″ × 12″ × 0.12″ clear acrylic sheets, ATP clear PVC tubing with 1/8″ inside diameter and 1/4″ outside diameter, an electric drill with 1/4″ bit, epoxy or a similar glue, Karlling 10 ml syringe with tip outside diameter 4 mm, Pure guar gum powder, an Intllab magnetic stirrer, a Vinmax digital laboratory scale, a Karter Scientific 1000 ml beaker, and four 11/4″ binder clamps.
Lab 5.6. We used Hygloss mylar sheets, very sharp scissors, Small Parts polycarbonate tubing with inside diameter 11/8″ and outside diameter 11/4″, a 1/4″ dowel, a plastic disk of diameter slightly less than 11/8″, a small tack, tape, a Sharpie ultra fine point permanent marking pen, a mm scale ruler, and

a hands-free magnifying glass. For this we used Helping Hand alligator clamps with a magnifying glass.

Lab 5.7. About 100 1 KΩ resistors (E-Projects 100EP5141K00), Bonefu 6″ precision side cutter pliers, Smraza double-sided PCB board kit, REXQualis 830 point solderless breadboards, Elegoo breadboard jumper wires, a TekPower TP3003E 30V 3A DC adjustable switching power supply, and a Micronta 22-185A digital multimeter. Thanks to my brother-in-law Roger Maatta for recommending this multimeter.

Lab 5.8. Deryun 200 pack of size 502 (diameter about 5 mm, height about 1.5 mm) rare earth magnets, Eisco Scientific fine iron filings, a mortar and pestle, small flat-head tacks (4.5 mm diameter, 10 mm length, type unknown—they were leftovers from MF's father's workshop), Gorilla glue, a push pin board, two clear acrylic sheets used in Lab 5.5, a sheet of unlined printer paper, a digital camera to record the iron filing patterns.

Lab 5.9. Two SunFounder 5 piece breadboard kits, a Siglent SDS 1052DL digital oscilloscope, a TekPower TP3003E 30V 3A DC adjustable switching power supply, Z&T solderless breadboard jumper wires (you'll need two 100 pc sets), SIQUK 1480 pc resistor set, 3PDT Electronics 200 pc inductor assortment, Walfront 500 pc ceramic capacitor set, HiLetgo 100 pc transistor assortment.

Appendix B

Technical Notes

B.1 Notes for finding IFS, Lab 1.1

Here we sketch some of the main points of the argument that the deterministic algorithm for applying IFS rules generates a sequence of sets that converges to a limiting shape, the fractal produced by the IFS. More details can be found in Sect. 9.1 of [16] and in Sect. 2.2 of [20]. An early criticism of fractal geometry is that it was "just pretty pictures," so many supporters of fractal geometry embraced the opportunity to demonstrate some real math that underlies the pictures.

B.1.1 Compact sets

The theorem that an IFS generates a fractal and only one fractal—the "existence and uniqueness" theorem—holds in the collection of all compact subsets of the plane. In general topology the definition of compact set is a bit involved, but in Euclidean space the Heine-Borel theorem (Theorem 2.41 of [36], for example) gives a simpler characterization:

> A subset of Euclidean space is compact if and only if it is closed and bounded.

A set is *closed* if its complement is open. So what's an open set? As with compact sets, in general topology the definition is a bit involved, but in Euclidean space openness has a simpler characterization.

> A subset A of Euclidean space is *open* if for every point p of A there is a disk D, centered at p and that lies entirely in A.

For example, in the plane $A = \{(x, y) : x^2 + y^2 < 1\}$ is open, while $B = \{(x, y) : x^2 + y^2 \leq 1\}$ is not open. In fact, B is closed.

A subset of the plane is *bounded* if it lies entirely inside some sufficiently large disk. This extends in the obvious way to subsets of Euclidean space of any dimension. The intuition is that a bounded set doesn't run away to infinity in any direction.

B.1.2 Hausdorff distance

To understand how a sequence of images $B_0, B_1, B_2, \ldots$ can converge, we need a way to measure a distance between images. That is, a way to measure how close one shape is to another. Is there a way to measure how much we must squint at a picture to make one blurry image look like another? There is: the Hausdorff distance, based on the notion of an ϵ-neighborhood, A_ϵ, of a set A. This is the set of all points lying within a distance ϵ of some point in the set A:

Figure B.1: Some ϵ-neighborhoods.

$$A_\epsilon = \{(w, x) : d((w, x), (y, z)) \leq \epsilon \text{ for some point}(y, z) \in A\}$$

where $d((w, x), (y, z))$ is the Euclidean distance $\sqrt{(w-y)^2 + (x-z)^2}$. In Fig. B.1 we see A_ϵ for A a point, a circle, and a line segment.

Given two sets A and B, the *Hausdorff distance* $h(A, B)$ is

$$\min\{\epsilon : A \subseteq B_\epsilon \text{ and } B \subseteq A_\epsilon\}$$

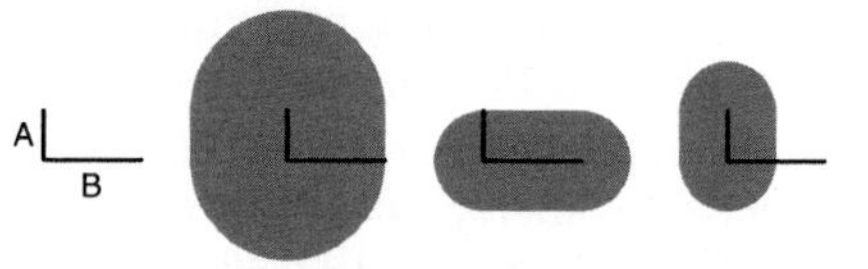

Figure B.2: Left to right: A and B, $B \subseteq A_1$, $A \subseteq B_{1/2}$, and $B \not\subseteq A_{1/2}$.

For example, suppose A is the vertical segment from $(0, 0)$ to $(0, 1/2)$ and B is the horizontal segment from $(0, 0)$ to $(1, 0)$. See Fig. B.2. We observe that $B \subseteq A_1$ and $A \subseteq B_{1/2}$, so a careless reading of the min in the definition of $h(A, B)$ might suggest we deduce $h(A, B) = 1/2$. However, on the right side of Fig. B.2 we see $B \not\subseteq A_{1/2}$. The correct deduction from these observations is that $h(A, B) = 1$. The point to note is that the *same* ϵ must work for both containments $A \subseteq B_\epsilon$ and $B \subseteq A_\epsilon$.

B.1.3 Convergence of the deterministic algorithm

At two places in this section we employ more advanced terminology to give signposts to readers familiar with basic real analysis. These points are not necessary in order to understand the example calculation, but they are essential to understand the general proof that the deterministic algorithm converges. Readers who trust the insight provided by the example and by the software explorations can skip these details. These comments are enclosed in square brackets [].

The deterministic algorithm converges because the Hausdorff distance between successive iterates B_i and B_{i+1} decreases at a constant rate. Specifically, suppose the T_i are similarities, that is, $r_i = s_i$ so r_i is the contraction factor of T_i. Then

$$\begin{aligned} h(\mathcal{T}(A), \mathcal{T}(B)) &= h(T_1(A) \cup \cdots \cup T_N(A), T_1(B) \cup \cdots \cup T_N(B)) \\ &\leq \max\{r_i\} \cdot h(A, B) \end{aligned}$$

That is, the function $\mathcal{T} = T_1 \cup \cdots \cup T_N$ acting on the space of all images [technically, the space of all compact subsets of the plane] is a contraction in Hausdorff distance. It follows that the sequence

$$B, \mathcal{T}(B), \mathcal{T}^2(B), \mathcal{T}^3(B), \mathcal{T}^4(B), \cdots$$

converges to a unique set A in the sense that

$$h(\mathcal{T}^k(B), A) \to 0 \quad \text{as } k \to \infty \tag{B.1}$$

regardless of the initial set B. Moreover, this set A is the unique set that satisfies Eq. (1.2). [All this is a consequence of the observation that the space of compact subsets of the plane, equipped with the Hausdorff distance, is complete. This guarantees that any Cauchy sequence of compact subsets converges to a compact subset.]

Example B.1.1 Suppose B is the filled-in unit square and T_1, T_2, and T_3 are the gasket transformations

$$T_1(x,y) = (x/2, y/2),\ T_2(x,y) = (x/2, y/2) + (1/2, 0),$$
$$T_3(x,y) = (x/2, y/2) + (0, 1/2)$$

Then $\mathcal{T}(B)$ is image (a) of Fig. 1.11, $\mathcal{T}^2(B)$ is image (b), $\mathcal{T}^3(B)$ is image (c), and so on. We see that

$$\mathcal{T}(B) \subseteq B,\ B \subseteq (\mathcal{T}(B))_{1/2} \quad \text{so} \quad h(B, \mathcal{T}(B)) \le \frac{1}{2}$$
$$\mathcal{T}^2(B) \subseteq \mathcal{T}B,\ \mathcal{T}(B) \subseteq (\mathcal{T}^2(B))_{1/4} \quad \text{so} \quad h(\mathcal{T}(B), \mathcal{T}^2(B)) \le \frac{1}{4}$$
$$\mathcal{T}^3(B) \subseteq \mathcal{T}^2B,\ \mathcal{T}^2(B) \subseteq (\mathcal{T}^3(B))_{1/8} \quad \text{so} \quad h(\mathcal{T}^2(B), \mathcal{T}^3(B)) \le \frac{1}{8}$$

and in general, $h(\mathcal{T}^{i-1}(B), \mathcal{T}^i(B)) \le 1/2^i$. So the distance between successive iterates of the deterministic algorithm goes to 0 and we understand why iterating this algorithm gives convergence to a limit.

B.2 Notes for spiral fractals, Lab 1.2

Here we present two more complicated arguments from the background of the Spiral Fractals Lab. First we show that the fixed point of each T_i is a point of the fractal generated by every IFS that includes the transformation T_i. Second, we sketch the proof that for a given IFS, the deterministic and the random algorithms generate the same fractal.

B.2.1 Fixed points and IFS

First, how do we know that each of the T_i has a fixed point? This is a consequence of a powerful result called the Contraction Mapping Principle: every contraction map on a complete metric space (and the plane is complete metric space) has a unique fixed point. So each T_i has a unique fixed point.

Suppose the IFS that consists of $T_1, \ldots, T_n$, and suppose p is the fixed point of T_1. Then

$$\mathcal{T}(p) = T_1(p) \cup T_2(p) \cup \cdots \cup T_n(p) = p \cup T_2(p) \cup \cdots \cup T_n(p)$$

and so p belongs to $\mathcal{T}^k(p)$ for all k.

Now recall Eq. (B.1), $h(\mathcal{T}^k(B), A) \to 0$ as $k \to \infty$, holds for all B, so take $B = p$. We want to show that $p \in A$, so suppose $p \notin A$. Then for some $\epsilon > 0$, $h(p, A) = \epsilon$. This ϵ is just the distance from p to the point of A closest to p. Then because $p \in \mathcal{T}^k(p)$ for all k,

$$h(\mathcal{T}^k(p), A) \geq h(p, A) = \epsilon > 0$$

But this contradicts Eq. (B.1). So our assumption that $p \notin A$ is wrong and we see that the fixed point of each T_i belongs to A.

B.2.2 Comparing the random and deterministic algorithms

Central to understanding why the deterministic and random algorithms produce the same picture is the notion of the *address* of a part of a fractal. This is the topic of Lab 1.7. Here we review the relevant concepts, illustrating them with this spiral IFS.

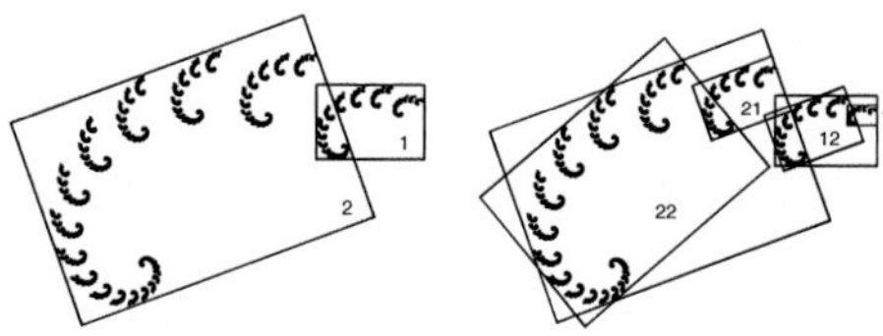

Figure B.3: Length-1 and -2 address regions of the spiral.

Denote the spiral fractal by A. Then from Eq. (1.2) we have

$$A = T_1(A) \cup T_2(A) = A_1 \cup A_2$$

These pieces of the fractal—the right-most subspiral and everything else—are enclosed by boxes labeled 1 and 2 in the left side of Fig. B.3. We call 1 and 2 the length-1 addresses of these two pieces of A.

This decomposition can be continued:

$$\begin{aligned} A = T_1(A) \cup T_2(A) &= T_1(T_1(A) \cup T_2(A)) \cup T_2(T_1(A) \cup T_2(A)) \\ &= T_1(T_1(A)) \cup T_1(T_2(A)) \cup T_2(T_1(A)) \cup T_2(T_2(A)) \\ &= A_{11} \cup A_{12} \cup A_{21} \cup A_{22} \end{aligned}$$

We call 11, 12, 21, and 22 the length-2 addresses of these four pieces of A, indicated on the right side of Fig. B.3. Note that *the order of the address digits of a region is identical to the order of the composition of transformations giving rise to that region.* A simple consequence of this is the effect on the address of a region of applying a transformation to that region:

$$T_a(A_{b\ldots c}) = T_a(T_b(\cdots T_c(A)\cdots)) = A_{ab\ldots c}$$

That is, shift the current address digits one place to the right and insert the new address digit on the left.

Because the T_i are contractions, longer addresses specify smaller subregions of the fractal; infinitely long addresses correspond to points. For example, the fixed point of T_2 has address $222\cdots = 2^\infty$. To see this, denote by P the point with address 2^∞. Then $T(P)$ has address $2(2^\infty) = 2^\infty$. That is, P and $T(P)$ have the same (infinite) address, and so they are the same point.

To see that the random algorithm and the deterministic algorithm generate the same images, it is enough to show that the sequence of points produced by the random algorithm include points that lie in all finite length address regions of the fractal. (The set of limit points of this sequence then will be the whole fractal. We do not pursue this issue now, but mention it to be honest: the full argument involves a bit more than we present here.)

Take the initial point (x_0, y_0) to be the fixed point of T_2, hence with address 2^∞. The points $(x_1, y_1), (x_2, y_2), \ldots$ of the sequence are generated by applying the T_i in random order:

$$\begin{aligned}
(x_1, y_1) &= T_{i_1}(x_0, y_0)\\
(x_2, y_2) &= T_{i_2}(x_1, y_1) = T_{i_2}(T_{i_1}(x_0, y_0))\\
(x_3, y_3) &= T_{i_3}(T_{i_2}(T_{i_1}(x_0, y_0)))
\end{aligned}$$

and so on. The addresses of these points are

$$(x_0, y_0) = 2^\infty,\ (x_1, y_1) = i_1(2^\infty),\ (x_2, y_2) = i_2 i_1(2^\infty),\ (x_3, y_3) = i_3 i_2 i_1(2^\infty)$$

To finish the argument we need this property of infinite random sequences: with probability 1 a random sequence $i_1, i_2, i_3, \ldots$ contains all finite sequences. Consequently, every finite address occurs as an initial segment of the address of some point of the sequence. The result follows by noting the inclusions

$$A_a \supset A_{ab} \supset A_{abc} \supset \cdots$$

because then we see the point with addess $i_1 i_2 i_3 \ldots$ lies in A_{i_1}, in $A_{i_1 i_2}$, in $A_{i_1 i_2 i_3}$, and so on.

B.3 Notes for cumulative gasket pictures, Lab 1.6

Here we use a coin-toss example to illustrate a method for estimating the likelihood of certain patterns in the chaos game. The most straightforward application of this method, explored in Exercise 1.6.6, is to computing the probability of a march of points, of a specified length, toward one of the vertices. This approach can be adapted to other patterns as well.

To illustrate the method, we compute the probability of obtaining five consecutive H in 100 tosses of a fair coin. We do this by defining six states of the experiment.

S_1: five consecutive Hs have not occurred, and the current toss is T.
S_2: five consecutive Hs have not occurred, the current toss is H and the previous toss (if there was one) is T.

S_3: five consecutive Hs have not occurred, the current toss is H, before that is H, and before that (if there was a toss before that) is T.
S_4: five consecutive Hs have not occurred, the current toss is H, before that are two Hs, and before that (if there was a toss before that) is T.
S_5: five consecutive Hs have not occurred, the current toss is H, before that are three Hs, and before that (if there was a toss before that) is T.
S_6: five consecutive Hs have occurred.

Each successive coin toss moves the experiment from one state to another:

Next toss is T	Next toss is H
$S_1 \to S_1$	$S_1 \to S_2$
$S_2 \to S_1$	$S_2 \to S_3$
$S_3 \to S_1$	$S_3 \to S_4$
$S_4 \to S_1$	$S_4 \to S_5$
$S_5 \to S_1$	$S_5 \to S_6$
$S_6 \to S_6$	$S_6 \to S_6$

We represent the probabilities of the transitions in a transition matrix M, where M_{ij}, the entry in the ith row and the jth column, is the probability of going from S_j to S_i. And we take the vector $\vec{v}_n$ to represent the probability of being in each of the six states after n coin tosses. The first coin toss gives T or H with equal probability, that is, the system is in state S_1 or S_2, both with probability 0.5. Then

$$M = \begin{bmatrix} 0.5 & 0.5 & 0.5 & 0.5 & 0.5 & 0 \\ 0.5 & 0 & 0 & 0 & 0 & 0 \\ 0 & 0.5 & 0 & 0 & 0 & 0 \\ 0 & 0 & 0.5 & 0 & 0 & 0 \\ 0 & 0 & 0 & 0.5 & 0 & 0 \\ 0 & 0 & 0 & 0 & 0.5 & 1 \end{bmatrix} \quad \text{and} \quad \vec{v}_1 = \begin{bmatrix} 0.5 \\ 0.5 \\ 0 \\ 0 \\ 0 \\ 0 \end{bmatrix}$$

The first two coin tosses have four possible outcomes: TT (the system winds up in state S_1), TH (that is, first T, then H; the system winds up in S_2), HT (in S_1), and HH (in S_3). Each of these four pairs has probability $0.5 \cdot 0.5 = 0.25$, so after two tosses, the system has probability 0.5 of being in S_1, 0.25 of being in S_2, and 0.25 of being in S_3. We can obtain this by

$$M\vec{v}_1 = \begin{bmatrix} 0.5 & 0.5 & 0.5 & 0.5 & 0.5 & 0 \\ 0.5 & 0 & 0 & 0 & 0 & 0 \\ 0 & 0.5 & 0 & 0 & 0 & 0 \\ 0 & 0 & 0.5 & 0 & 0 & 0 \\ 0 & 0 & 0 & 0.5 & 0 & 0 \\ 0 & 0 & 0 & 0 & 0.5 & 1 \end{bmatrix} \begin{bmatrix} 0.5 \\ 0.5 \\ 0 \\ 0 \\ 0 \\ 0 \end{bmatrix} = \begin{bmatrix} 0.5 \\ 0.25 \\ 0.25 \\ 0 \\ 0 \\ 0 \end{bmatrix} = \vec{v}_2$$

and it is easy to see that $\vec{v}_{n+1} = M^n \vec{v}_1$, for example, $\vec{v}_3 = M\vec{v}_2 = M^2\vec{v}_1$. If five consecutive Hs have occurred, the system will be in S_6. The last entry of $\vec{v}_{100} = M^{99}\vec{v}_1$ is 0.8011, the probability of observing at least five consecutive Hs in 100 tosses of a fair coin.

This approach—the definition of various states of the system and transition probabilities between these states—is called a *Markov chain*. In this example,

S_6 is an absorbing state: once the system enters that state, it never leaves. This may sound odd, but this Markov chain is designed to detect the occurrence of five consecutive Hs in a sequence of coin tosses. The system enters state S_6 when five consecutive heads have occurred, and once that has happened, later coin tosses have no influence over the fact that five consecutive Hs already gave occurred.

In case you think calculating M^{99} is a daunting amount of work, here's a short cut:

$$M^2 = M \cdot M,\ M^4 = M^2 \cdot M^2,\ M^8 = M^4 \cdot M^4,\ M^{16} = M^8 \cdot M^8,$$
$$M^{32} = M^{16} \cdot M^{16},\ M^{64} = M^{32} \cdot M^{32}$$

and so $M^{99} = M^{64} \cdot M^{32} \cdot M^2 \cdot M$. Not too many multiplications at all. If we continue this calculation, we find that in a sequence of 400 coin tosses the probability of obtaining five consecutive Hs is 0.9989.

B.4 Notes for IFS with more memory, Lab 1.10

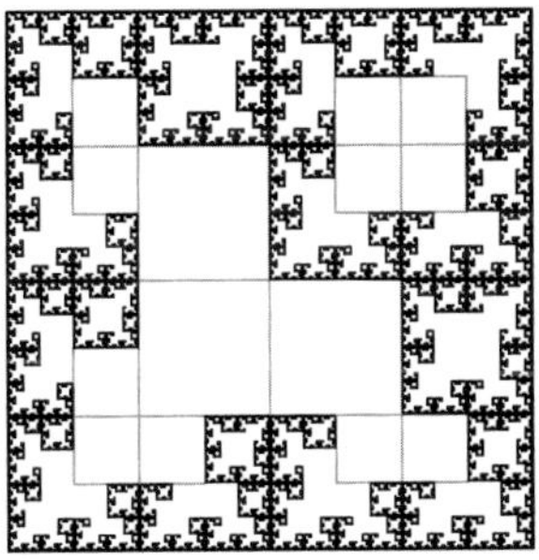

Here we'll show that the attractor of a 2-step memory IFS with transformations (1.7) always can be reduced to a 1-step memory IFS possibly with a larger set of transformations. The construction can be extended to more steps of memory, but the number of transformations can grow considerably. Immediate references are [40, 41]. For higher block shifts you can't do better than [42]. We'll use this example from Sect. 1.10.3.

The transformations are the allowed pairs of the transformations (1.7). This is the "block" of higher block shift. From the attractor image we see that this IFS forbids only three pairs: $T_1(T_4)$, $T_2(T_3)$, and $T_3(T_2)$. The equivalent 1-step memory IFS has 13 transformations. If possible, names should remind us of the object named. (In the topology seminar at the University of Chicago in 1981, at one point the board contained 6 different manifolds, all denoted M, and 10 different functions, all denoted f. The seminar topic was interesting, so this notational superposition was frustrating. If you blinked, you were lost for the rest of the talk.) We'll write T_{ij} for $T_i(T_j)$. The transformations of this IFS are T_{11}, T_{12}, T_{13}, T_{21}, T_{22}, T_{24}, T_{31}, T_{33}, T_{34}, T_{41}, T_{42}, T_{43}, and T_{44}.

When can T_{kl} follow T_{ij}? Two conditions must be satisfied:
(1) $i = l$, and
(2) kij is an allowed triple. Label the transition matrix rows and columns in the order presented above. Then the transition matrix for this 1-step memory

IFS is

$$\begin{bmatrix}
1 & 1 & 1 & 0 & 0 & 0 & 0 & 0 & 0 & 0 & 0 & 0 & 0 \\
0 & 0 & 0 & 1 & 1 & 1 & 0 & 0 & 0 & 0 & 0 & 0 & 0 \\
0 & 0 & 0 & 0 & 0 & 0 & 1 & 1 & 1 & 0 & 0 & 0 & 0 \\
1 & 1 & 1 & 0 & 0 & 0 & 0 & 0 & 0 & 0 & 0 & 0 & 0 \\
0 & 0 & 0 & 1 & 1 & 1 & 0 & 0 & 0 & 0 & 0 & 0 & 0 \\
0 & 0 & 0 & 0 & 0 & 0 & 0 & 0 & 0 & 1 & 1 & 1 & 1 \\
1 & 1 & 1 & 0 & 0 & 0 & 0 & 0 & 0 & 0 & 0 & 0 & 0 \\
0 & 0 & 0 & 0 & 0 & 0 & 1 & 1 & 1 & 0 & 0 & 0 & 0 \\
0 & 0 & 0 & 0 & 0 & 0 & 0 & 0 & 0 & 1 & 1 & 1 & 1 \\
1 & 1 & 1 & 0 & 0 & 0 & 0 & 0 & 0 & 0 & 0 & 0 & 0 \\
0 & 0 & 0 & 1 & 1 & 1 & 0 & 0 & 0 & 0 & 0 & 0 & 0 \\
0 & 0 & 0 & 0 & 0 & 0 & 1 & 1 & 1 & 0 & 0 & 0 & 0 \\
0 & 0 & 0 & 0 & 0 & 0 & 0 & 0 & 0 & 0 & 1 & 1 & 1
\end{bmatrix}$$

Many 0s in this matrix result from condition (1), but the 0 in row 13 and column 10 is a consequence of condition (2) because the triple 441 is forbidden.

To render the image, we must be careful with what the compositions mean. For instance, T_{12} can follow T_{24} because the triple 124 is allowed. This transition corresponds to the composition $T_1(T_2(T_4))$, not to $T_1(T_2(T_2(T_4)))$. Details are presented in [40].

B.5 Notes on entropy and partitions, Lab 1.11.

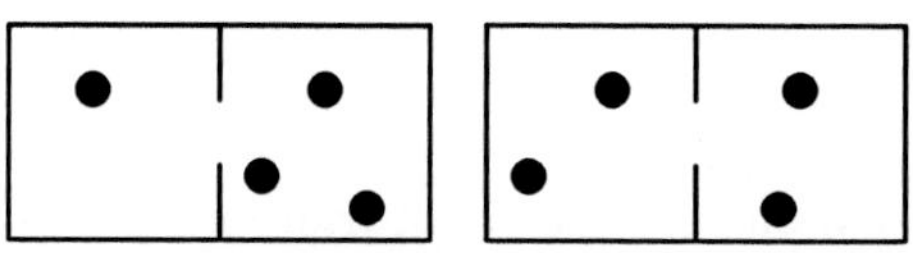

For equal-weight bins the bin boundaries b_1, b_2, and b_3 are chosen so the four bins have about the same number of points x_i. This also is called the *maximum entropy partition* because of the similarity with a standard statistical mechanics illustration of entropy. Imagine a rectangular box divided into two chambers by a wall with a hole. Suppose we are interested in only whether a particle is in the left chamber or in the right chamber. In the first picture we see one particle on the left and three on the right, so this system is determined by which of the four particles is in the left chamber. There are 4 "microstates," arrangements of particles into the chambers. In the second picture we see two particles in the left chamber, two in the right. This is realized by $4!/(2!(4-2)!) = 6$ microstates. Random movement of the four particles will visit the right configuration more often than any of the others, each of which has some order, more points in one chamber or the other. The configuration with points evenly distributed, the one visited most often, has maximum entropy. This is why equal-weight bins are called the maximum entropy partition.

B.6 Notes on linear regression, Lab 2.1

Here we'll derive the linear regression formula and discuss the correlation coefficient r, a measure of how well the data fit the line determined by the regression formula. For this we'll use the two-variable equivalent of the one-variable local max test: a point x_0 is a local maximum for the function f if $f'(x_0) = 0$ and $f''(x_0) < 0$. The condition $f'(x_0) = 0$ guarantees that the graph of $y = f(x)$ has a horizontal tangent line at the point $(x_0, f(x_0))$, and the condition $f''(x_0) < 0$ means the graph is concave down at $(x_0, f(x_0))$, so the point is a local maximum.

For a function $f(x, y)$ of two variables, the corresponding conditions are

$$\frac{\partial f}{\partial x}(x_0, y_0) = \frac{\partial f}{\partial y}(x_0, y_0) = 0$$

and evaluated at (x_0, y_0),

$$\frac{\partial^2 f}{\partial x^2}\frac{\partial^2 f}{\partial y^2} - \left(\frac{\partial^2 f}{\partial x \partial y}\right)^2 > 0 \text{ and } \frac{\partial^2 f}{\partial x^2} > 0 \tag{B.2}$$

This is a common result in third semester calculus courses, see the second derivative test in section 14.7 of [48], for example. Here the symbol ∂—called "curly d", really—denotes the *partial derivative*. To compute, say $\partial f/\partial x$, use the familiar derivative rules while treating x as the variable and y as a constant. For example, with $f(x, y) = x^3y^2 + x^2 + y^4$ we have

$$\frac{\partial f}{\partial x} = 3x^2y^2 + 2x \quad \text{and} \quad \frac{\partial f}{\partial y} = x^3 2y + 4y^3$$

Suppose we have data points $(x_1, y_1), \dots, (x_n, y_n)$. We want to find the line $y = mx + b$ that comes closest to passing through the data points. How can we do this? For each x_i the line predicts a y-value, namely $mx_i + b$. We'd like to choose m and b to minimize the sum of the differences $y_i - (mx_i + b)$. But it's a bit trickier than this: some differences can be positive, others negative, and so they can subtract out. We could add the absolute values, but differentiating absolute values can be a bit messy. So instead, we'll find m and b to minimize the sum of $(y_i - (mx_i + b))^2$. So define a function

$$f(m, b) = (y_1 - (mx_1 + b))^2 + \cdots + (y_n - (mx_n + b))^2$$

Why are the variables m and b? For one thing, $(x_1, y_1), \dots, (x_n, y_n)$ aren't variables, they are the fixed data points. For another thing, we're trying to find the best-fitting line, that is, to find m and b. So we want to find m and b that make

$$\frac{\partial f}{\partial m} = 0 \quad \text{and} \quad \frac{\partial f}{\partial b} = 0 \tag{B.3}$$

To hold in check the lengths of the expressions in the derivation, we'll use shorthand notation: for example, $\sum x_i y_i$ to represent $x_1y_1 + \cdots + x_ny_n$. Then the two equations of Eq. (B.3) become

$$\sum 2(y_i - (mx_i + b))(-x_i) = 0 \quad \text{and} \quad \sum 2(y_i - (mx_i + b))(-1) = 0 \tag{B.4}$$

We can solve the second equation for b:

$$b = \frac{\sum y_i - m \sum x_i}{n} \tag{B.5}$$

Here the n in the denominator comes from the fact that the sum in the second equation of (B.4) produces n copies of b.

Now rearranging the first equation of (B.4) gives

$$m\sum x_i^2 = \sum x_i y_i - b\sum x_i$$

Substitute in the expression for b from (B.5) and solve for m:

$$m = \frac{n\sum x_i y_i - (\sum x_i)(\sum y_i)}{n\sum x_i^2 - (\sum x_i)^2} \tag{B.6}$$

To find the best-fitting line, first use Eq. (B.6) to find m and substitute that value into Eq. (B.5) to find b.

In order to be sure this is the minimum value of f, we apply the second derivative test (B.2). So we compute the second partial derivatives.

$$\frac{\partial^2 f}{\partial m^2} = 2\sum x_i^2, \quad \frac{\partial^2 f}{\partial m \partial b} = 2\sum x_i, \quad \frac{\partial^2 f}{\partial b^2} = 2n$$

Then

$$\frac{\partial^2 f}{\partial x^2}\frac{\partial^2 f}{\partial y^2} - \left(\frac{\partial^2 f}{\partial x \partial y}\right)^2 = 4\left(n\sum x_i^2 - (\sum x_i)^2\right)$$

The right-hand side is ≥ 0. The $n = 3$ case is enough to see why this is true.

$$\begin{aligned}
&3(x_1^2 + x_2^2 + x_3^2) - (x_1 + x_2 + x_3)^2 \\
&= 3x_1^2 + 3x_2^2 + 3x_3^2 - \left(x_1^2 + x_2^2 + x_3^2 + 2x_1x_2 + 2x_1x_3 + 2x_2x_3\right) \\
&= 2x_1^2 + 2x_2^2 + 2x_3^2 - 2x_1x_2 - 2x_1x_3 - 2x_2x_3 \\
&= \left(x_1^2 - 2x_1x_2 + x_2^2\right) + \left(x_1^2 - 2x_1x_3 + x_3^2\right) + \left(x_2^2 - 2x_2x_3 + x_3^2\right) \\
&= (x_1 - x_2)^2 + (x_1 - x_3)^2 + (x_2 - x_3)^2
\end{aligned}$$

In addition, the right hand side $= 0$ if and only if $x_1 = \cdots = x_n$. This is a case we can ignore, because it corresponds to all the data points lying on a vertical line above the common value of x_i.

For the other part of condition (B.2) we see that

$$\frac{\partial^2 f}{\partial m^2} = 2\sum x_i^2$$

which is positive, unless all the $x_i = 0$, a case we can ignore because it implies that all the data points would lie on the y-axis.

So the slope and intercept given by Eqs. (B.5) and (B.6) do determine the best-fitting line through the data points $(x_1, y_1), \dots, (x_n, y_n)$. But how well does this line fit those points?

This is measured by the *correlation coefficient* r. Explaining what this is and how to calculate it will take four steps.

First, the *expected value* of a collection of values $x_1, \dots, x_n$ is $\mathbb{E}(X) = (x_1 + \cdots + x_n)/n$, this is just the average of the x_i, also called the *mean* μ. The name "expected value" is common in probability and statistics, but it can lead to some confusion: the expected value needn't be a value that can occur at all. For example, the expected value when a fair die (singular of dice) is tossed is $(1+2+3+4+5+6)/6 = 3.5$, and yet we never see 3.5 dots when the die is tossed. The expected value has two nice properties. For any constants a and b,

$$\mathbb{E}(aX) = a\mathbb{E}(X) \quad \text{and} \quad \mathbb{E}(X+b) = \mathbb{E}(X) + b \tag{B.7}$$

We say the expected value is a linear function. These two properties are easy to prove. By the way, if you're curious about why the x is capitalized in $\mathbb{E}(X)$, common practice in probability is to denote the outcomes of individual measurements by lower-case letters, $x_1, x_2, dots$, and to use an upper-case letter to denote the variable, called a random variable, that stands for all possible measurements.

Second, the *variance* is a measure of how widely the x_i are spread about the mean. A first guess might be $\mathbb{E}(X - \mathbb{E}(X))$, the expected value of how far the measurements x_i depart from their average value. A small value should indicate measurements tightly clustered about the average. The problem with the formulation $\mathbb{E}(X - \mathbb{E}(X))$ is that positive and negative values can subtract out, giving a small expected value when the measurements are widely spread. The solution is to take the expected value of the squares of the differences. That is, the variance $\sigma^2(X)$ and the standard deviation $\sigma(X)$ are

$$\sigma^2(X) = \mathbb{E}((X - \mathbb{E}(X))^2) \quad \text{and} \quad \sigma(X) = \sqrt{\mathbb{E}((X - \mathbb{E}(X))^2)}$$

The definition of variance has a useful equivalent.

$$\begin{aligned} \sigma^2(X) &= \mathbb{E}((X - \mathbb{E}(X))^2) = \mathbb{E}(X^2 - 2\mathbb{E}(X)X + (\mathbb{E}(X))^2) \\ &= \mathbb{E}(X^2) - 2\mathbb{E}(X)\mathbb{E}(X) + (\mathbb{E}(X))^2 \\ &= \mathbb{E}(X^2) - (\mathbb{E}(X))^2 \end{aligned} \tag{B.8}$$

where the second line is the result of the linearity of the expected value. Unlike the expected value, the variance is not a linear function. How is $\sigma^2(2X)$ related to $\sigma^2(X)$?

Third, the *covariance* of two random variables X and Y is

$$\begin{aligned} \text{cov}(X, Y) &= \mathbb{E}((X - \mathbb{E}(X)) \cdot (Y - \mathbb{E}(Y))) && \text{(B.9)} \\ &= \mathbb{E}(XY) - \mathbb{E}(X)\mathbb{E}(Y) - \mathbb{E}(X)\mathbb{E}(Y) + \mathbb{E}(X)\mathbb{E}(Y) && \text{(B.10)} \\ &= \mathbb{E}(XY) - \mathbb{E}(X)\mathbb{E}(Y) && \text{(B.11)} \end{aligned}$$

where Eq. (B.10) follows by applying Eq. (B.7) to Eq. (B.9). Suppose we measure pairs of values $(x_1, y_1), \dots (x_n, y_n)$. (Now you may begin to see how this long excursion is tied back to our original scattering of data points.) If both x_i and y_i tend to be greater than their average values, or both tend to be smaller than their average values, then usually their covariance is positive. If x_i and y_i tend to be on opposite sides of their means, the covariance is negative.

Fourth, the numerical value of the covariance isn't exactly what we want. we really want the size of the covariance compared with the spread of the x_i and the spread of the y_i. So the *correlation coefficient* $r(X, Y)$ is defined by

$$r(X, Y) = \frac{\text{cov}(X, Y)}{\sigma(X)\sigma(Y)} = \frac{\mathbb{E}(XY) - \mathbb{E}(X)\mathbb{E}(Y)}{\sigma(X)\sigma(Y)} \tag{B.12}$$

where the second equality is an application of Eq. (B.11). To derive the expression we'll use for $r = r(X, Y)$, first we must work a bit with Eq. (B.8).

$$\begin{aligned}\sigma^2(X) &= \mathbb{E}(X^2) - (\mathbb{E}(X))^2 = \frac{\sum x_i^2}{n} - (\frac{\sum x_i}{n})^2 \\ &= \frac{1}{n^2}\left(n\sum x_i^2 - (\sum x_i)^2\right)\end{aligned}$$

and so

$$\sigma(X) = \frac{1}{n}\sqrt{n\sum x_i^2 - (\sum x_i)^2}$$

Then we find the expression we'll use to compute the correlation coefficient.

$$\begin{aligned}r &= \frac{\mathbb{E}(XY) - \mathbb{E}(X)\mathbb{E}(Y)}{\sigma(X)\sigma(Y)} \\ &= \frac{\frac{\sum x_i y_i}{n} - \frac{\sum x_i}{n}\frac{\sum y_i}{n}}{\frac{1}{n^2}\sqrt{n\sum x_i^2 - (\sum x_i)^2}\sqrt{n\sum y_i^2 - (\sum y_i)^2}} \\ &= \frac{n\sum x_i y_i - (\sum x_i)(\sum y_i)}{\sqrt{n\sum x_i^2 - (\sum x_i)^2}\sqrt{n\sum y_i^2 - (\sum y_i)^2}} \qquad \text{(B.13)}\end{aligned}$$

One final point: if in general the y_i increase as the x_i increase, r is positive. On the other hand, if the y_i decrease as the x_i increase, r is negative. Even if the data points lie very close to the best-fitting line, r still is negative. Because we are interested in how well the data points are modeled by a straight line, and not so much whether the slope of the line is positive or negative. we'll measure correlation by r^2. With our definition of r, we have $0 \le r^2 \le 1$. Then r^2 near 0 signals that the best-fitting line has little predictive power, while r^2 near 1 implies that $mx_i + b$ is close to y_i.

At last, here are two examples.

data 1 = $(2, 2.96)$, $(3, 4.73)$, $(4, 7.61)$, $(5, 9.80)$, $(6, 7.25)$, $(7, 9.17)$, $(8, 14.20)$, $(9, 13.96)$, $(10, 11.75)$, $(11, 17.16)$, $(12, 20.41)$, $(13, 16.74)$, $(14, 14.14)$, $(15, 18.92)$, $(16, 23.76)$, $(17, 18.34)$, $(18, 23.23)$, $(19, 29.65)$, $(20, 24.35)$

data 2 = $(2, 3.7)$, $(3, 5.33)$, $(4, 6.09)$, $(5, 8.29)$, $(6, 9.74)$, $(7, 10.06)$, $(8, 11.58)$, $(9, 11.20)$, $(10, 12.04)$, $(11, 14.46)$, $(12, 17.21)$, $(13, 16.50)$, $(14, 18.81)$, $(15, 22.09)$, $(16, 23.78)$, $(17, 22.94)$, $(18, 25.43)$, $(19, 26.02)$, $(20, 27.53)$

On the left we see the plot of the points of data 1, together with the best-fitting line. The y-intercept b and slope m are found by Eqs. (B.5) and (B.6). With this data we find $b = 1.89$ and $m = 1.21$. While the points generally move in the direction of the line, they are spread fairly widely about that line. Then with Eq. (B.13) we find $r^2 = 0.86$.

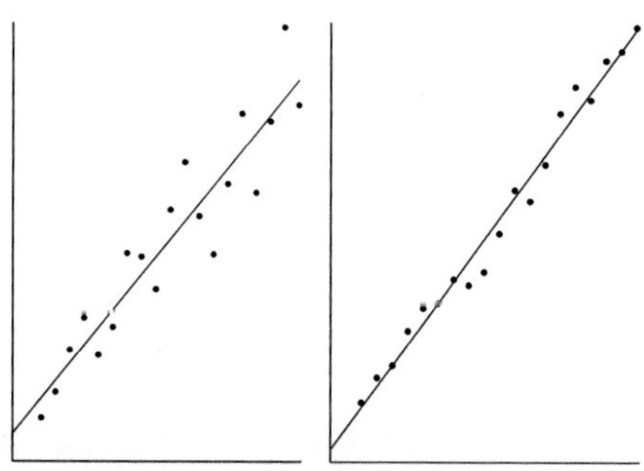

On the right we see the plot of the points of data 2, and the best-fitting line through those points. Here we have $b = 0.77$ and $m = 1.33$. The correlation coefficient is $r^2 = 0.98$. We see that the points of data 2 lie more closely to the best-fitting line that do those of data 1.

B.7 Notes on the algebra of dimensions, Labs 2.1 and 2.2.

We described the box-counting dimension d_b in Lab 2.1 and the similarity dimension d_s in Lab 2.3. A powerful generalization of the similarity dimensionis the Hausdorff dimension d_h. The definition is subtle, but the idea is an elaboration of Sample 2.1.6 of Lab 2.1. Denote by $\mathcal{H}^d(A)$ the d-dimensional Hausdorff measure of a set A. Then the Hausdorff dimension $d_h(A)$ is value if d where $\mathcal{H}^d(A)$ jumps from ∞ to 0. That is,

$$\mathcal{H}^d(A) = \begin{cases} \infty & \text{if } d < d_h(A) \\ 0 & \text{if } d > d_h(A) \end{cases}$$

Often $d_s = d_h$.

Here are the rules for the algebra of dimensions for d_b and d_h.

Monotonicity rule If $A \subseteq B$ then $d_b(A) \leq d_b(B)$ and $d_h(A) \leq d_h(B)$.

Union rule Both the box-counting and the Hausdorff dimensions satisfy the finite union rule,

$$d_b(A_1 \cup \cdots \cup A_n) = \max\{d_b(A_1), \ldots, d_b(A_n)\}$$

and

$$d_h(A_1 \cup \cdots \cup A_n) = \max\{d_h(A_1), \ldots, d_h(A_n)\}.$$

The Hausdorff dimension satisfies the countable union rule

$$d_h(A_1 \cup A_2 \cup \cdots) = \sup\{d_h(A_1), d_h(A_2), \ldots\},$$

the box-counting dimension does not.

Invariance rule For every similarity transformation f, $d_b(f(A)) = d_b(A)$ and $d_h(f(A)) = d_h(A)$. The invariance rule holds for a much more general class of functions. For every pair of points p and q of A, if the distance between $f(p)$ and $f(q)$ is bounded above and below by fixed multiples of the distance between p and q—that is, if there are constants m and M with

$$m \cdot \text{dist}(p, q) \leq \text{dist}(f(p), f(q)) \leq M \cdot \text{dist}(p, q)$$

for all points p and q in A—then the invariance rule holds.

Product rule For spaces well-behaved enough that $d_h(A) = d_b(A)$,

$$d_b(A \times B) = d_b(A) + d_b(B) \text{ and } d_h(A \times B) = d_h(A) + d_h(B)$$

In general,

$$d_b(A \times B) \leq d_b(A) + d_b(B) \text{ and } d_h(A \times B) \geq d_h(A) + d_h(B)$$

Intersection rule For sets A and B in n-dimensional space, and for all translations f,

$$d_h(A \cap f(B)) \leq \max\{0, d_h(A) + d_h(B) - n\}$$

and for a "large" group of translations f,

$$d_h(A \cap f(B)) \geq d_h(A) + d_h(B) - n$$

For sets that are reasonably well-behaved, a vague condition we can't sharpen up without a lot more math (Chapter 8 of [16], for example), we have

$$d(A \cap f(B)) = d(A) + d(B) - n$$

B.8 Notes on eigenvalues and the Moran equation, Lab 2.3.

Here we'll define the eigenvalues and eigenvectors of a matrix, and sketch the proof that the Moran equation has a unique solution.

B.8.1 Eigenvalues and eigenvectors

First we'll show how to multiply a matrix by a vector shown on the left below, and a vector by a number. We'll use a 2×2 matrix for illustration; the generalization to $n \times n$ matrices is straightforward.

$$M\vec{v} = \begin{bmatrix} a & b \\ c & d \end{bmatrix} \begin{bmatrix} x \\ y \end{bmatrix} = \begin{bmatrix} ax + by \\ cx + dt \end{bmatrix} \quad \text{and} \quad \lambda \begin{bmatrix} x \\ y \end{bmatrix} = \begin{bmatrix} \lambda x \\ \lambda y \end{bmatrix}$$

A vector $\vec{v}$ is an *eigenvector* of the matrix M with *eigenvalue* λ if

$$M\vec{v} = \lambda\vec{v} \tag{B.14}$$

Certainly, Eq. (B.14) is true for all λ if $\vec{v}$ is the 0-vector $\vec{0}$, the vector with all entries equal to 0. So, we're interested in solutions only for $\vec{v} \neq \vec{0}$. This requires several bits of matrix algebra: how to add and multiply matrices, and how to tell if a matrix has a multiplicative inverse. Again, we'll stick to 2×2 matrices to simplify the illustrations.
Matrix addition and multiplication. Two matrices A and B can be added if both have the same number of rows and the same number of columns. Then to add matrices, just add the corresponding entries. To multiply AB multiply the entries of the rows of A by the entries of the columns of B. So to form the product AB the number of entries in the rows of A must equal the number of entries in the columns of B. That is, the number of columns of A must equal the number of rows of B.

$$A + B = \begin{bmatrix} a & b \\ c & d \end{bmatrix} + \begin{bmatrix} e & f \\ g & h \end{bmatrix} = \begin{bmatrix} a+e & b+f \\ c+g & d+h \end{bmatrix}$$

and

$$AB = \begin{bmatrix} a & b \\ c & d \end{bmatrix} \begin{bmatrix} e & f \\ g & h \end{bmatrix} = \begin{bmatrix} ae + bg & af + bh \\ ce + dg & cf + dh \end{bmatrix}$$

That is, we go across the rows of the left matrix and down the columns of the right matrix. While matrix addition is commutative, $A + B = B + A$, in general, matrix multiplication is not.

There is a special matrix, the *identity matrix* I, that acts like the number 1 in the multiplication of numbers:

$$AI = \begin{bmatrix} a & b \\ c & d \end{bmatrix} \begin{bmatrix} 1 & 0 \\ 0 & 1 \end{bmatrix} = \begin{bmatrix} a & b \\ c & d \end{bmatrix} \quad \text{and} \quad IA = \begin{bmatrix} 1 & 0 \\ 0 & 1 \end{bmatrix} \begin{bmatrix} a & b \\ c & d \end{bmatrix} = \begin{bmatrix} a & b \\ c & d \end{bmatrix}$$

If the matrix I acts like the number 1 for multiplication, we say the matrix B is the *inverse* of the matrix A if $AB = BA = I$. In order for a number x to have an inverse (reciprocal), x must satisfy the condition $x \neq 0$. The corresponding condition for matrices is

$$A \text{ has an inverse if and only if } \det(A) \neq 0 \tag{B.15}$$

where for 2×2 matrices the *determinant* is defined by

$$\det(A) = \det\left(\begin{bmatrix} a & b \\ c & d \end{bmatrix}\right) = ad - bc$$

In a moment we'll show ways to compute the determinant for $n \times n$ matrices.

We can show one direction of (B.15) pretty easily by using

$$\det(AB) = \det(A)\det(B) \tag{B.16}$$

For 2×2 matrices this is a simple calculation; for larger matrices the proof is trickier. Google "Determinant of a product" if you're interested.

Suppose A is invertible, so there is a matrix B with $AB = I$. Then

$$\det(A)\det(B) = \det(AB) = \det(I) = 1$$

and so $\det(A) \neq 0$.

Back to Eq. (B.14). Move all the terms to one side,

$$\vec{0} = M\vec{v} - \lambda\vec{v}$$

We'd like to factor out $\vec{v}$, but that would leave us with a matrix minus a number, not possible. However, $\vec{v} = I\vec{v}$. Then,

$$\vec{0} = M\vec{v} - \lambda I\vec{v} = (M - \lambda I)\vec{v}$$

If $M - \lambda I$ is invertible, then multiply both sides of this equation by $(M - \lambda I)^{-1}$. This gives $\vec{v} = \vec{0}$, but we want non-zero eigenvectors and so $M - \lambda I$ must not be invertible. But $\det(M - \lambda I) = 0$ implies that $M - \lambda I$ is not invertible. Consequently, the eigenvalues are the solutions of the *eigenvalue equation*,

$$\det(M - \lambda I) = 0 \tag{B.17}$$

To find eigenvalues of an $n \times n$ matrix for $n > 2$, we need to compute the determinant of an $n \times n$ matrix, and to solve a polynomial equation of degree n. For the polynomial equation, there are cubic and quartic equations, but nothing higher. Sometimes factoring can work, but don't count on it. Numerical solutions may be the only option.

As for the determinant, we can reduce the computation of the determinant of an $n \times n$ matrix to the computation of the determinants of n matrices of size $(n-1) \times (n-1)$. We'll illustrate this with two examples.

$$\det \begin{bmatrix} a & b & c \\ d & e & f \\ g & h & i \end{bmatrix} = a \det \begin{bmatrix} e & f \\ h & i \end{bmatrix} - d \det \begin{bmatrix} b & c \\ h & i \end{bmatrix} + g \det \begin{bmatrix} b & c \\ e & f \end{bmatrix}$$
$$= -d \det \begin{bmatrix} b & c \\ h & i \end{bmatrix} + e \det \begin{bmatrix} a & c \\ g & i \end{bmatrix} - f \det \begin{bmatrix} a & b \\ g & h \end{bmatrix}$$

The first equality is called expanding down the first column; the second equality is called expanding along the second row. Multiply each element in the selected row or column by the determinant of the matrix formed by deleting the row and column that contain that entry. In these expansions, the sign of the entry in row i and column j is $(-1)^{i+j}$. Certainly, this extends to $n > 3$, though the calculations can become tedious. Note in particular that if a row or column is mostly 0, the corresponding expansion can be shorter.

B.8.2 Moran equation solutions

Here we'll show that the Moran equation $r_1^d + \cdots + r_n^d = 1$ has a unique solution, so long as $n \geq 2$ and for each i, $0 < r_i < 1$. To do this we'll see how the left-hand side of the Moran equation varies with d. So define

$$f(d) = r_1^d + \cdots + r_n^d$$

First we see that $f(0) = r_1^0 + \cdots + r_n^0 = n$.

Next, as $d \to \infty$, each $r_i^d \to 0$, so $\lim_{d\to\infty} f(d) = 0$.

Third, $f'(d) = r_1^d \ln(r_1) + \cdots + r_n^d \ln(r_n)$. Because each r_i lies between 0 and 1, each $\ln(r_i) < 0$ and so $f'(d) < 0$. That is, f is a decreasing function of d.

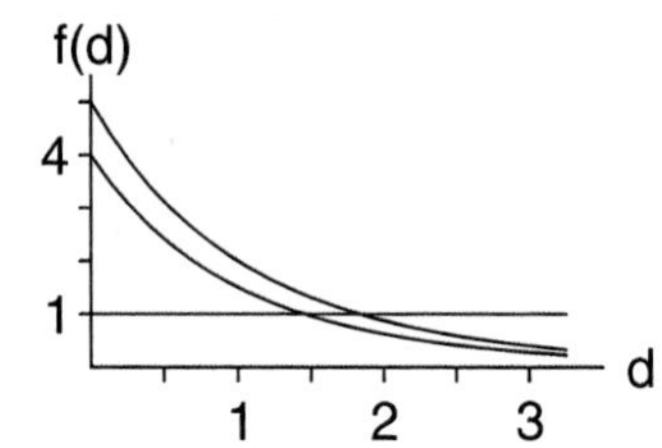

So we see that the graph of f starts at $n > 1$ for $d = 0$. Then the graph of f decreases steadily toward 0, so crosses the line $f = 1$ at exactly one value of d, the unique solution of the Moran equation.

The two curves on the right are graphs of f for $r_1 = r_2 = r_3 = 1/2$ and $r_4 = r_5 = 1/4$ (the top curve) and $r_1 = r_2 = 1/2$ and $r_3 = r_4 = 1/4$ (bottom curve).

B.9 Notes on multifractal analysis, Lab 2.7

First, we'll show that for each real q, Eq. (2.13) has a unique solution $\beta(q)$. We'll try to mimic the proof that the Moran equation has a unique solution, but we'll need to make some minor adjustments. Define $g(\beta) = \sum_i p_i^q r_i^\beta$. We want to show that for each q, $g(\beta) = 1$ has a unique solution β, which we'll call $\beta(q)$ because the solution β can be different for different q.

Because for all r_i, $0 < r_i < 1$, we see that

$$\lim_{\beta\to\infty} p_i^q r_i^\beta = p_i^q \lim_{\beta\to\infty} r_i^\beta = 0 \quad \text{so} \quad \lim_{\beta\to\infty} g(\beta) = 0$$

and

$$\lim_{\beta\to-\infty} p_i^q r_i^\beta = p_i^q \lim_{\beta\to-\infty} r_i^\beta = \infty \quad \text{so} \quad \lim_{\beta\to-\infty} g(\beta) = \infty$$

The last piece we need is that g is a decreasing function of β.

$$\frac{dg}{d\beta} = \sum_i p_i^q r_i^\beta \ln(r_i) < 0 \quad \text{because each term is negative}$$

Put all this together: $g(\beta)$ is large and positive for large negative β, $g(\beta)$ approaches 0 for large positive β, and g is a continuous, decreasing function. Then by the Intermediate Value Theorem, for some value $\beta(q)$ we have $g(\beta(q)) = 1$.

B.9.1 Hölder exponents

Now we'll define the Hölder exponent and sketch how it is a measure of roughness.

Denote by r_i the scaling factor of a similarity transformation T_i, by p_i the probability of applying T_i in the random IFS algorithm, and by A the attractor of the IFS $\{T_1, \dots, T_n\}$. Then the part of the attractor with address $i_1 \dots i_k$ is

$$A_{i_1 \dots i_k} = T_{i_1} \circ \cdots \circ T_{i_k}(A)$$

In the random IFS algorithm, each application of a transformation is independent of which transformation was applied before it. Recall that for independent events E and F,

$$\text{Prob}(E \text{ and } F) = \text{Prob}(E) \cdot \text{Prob}(F)$$

Consequently,

$$\text{Prob}(A_{i_1 \cdots i_k}) = p_{i_1} \cdots p_{i_k} \tag{B.18}$$

As a surrogate for the roughness of physical fractal, for this mathematical construction we'll use a measure based on the probability that a point will land in that region. This may seem to be an odd choice, but one way to understand it is this: imagine above every region of the attractor A we build a histogram whose height is the fraction of the total number of random IFS points that have landed in that region. Then the variation of the heights is related to the roughness of this histogram surface.

We've mentioned that dimension is a measure of roughness[61], so roughness is an exponent in a power law scaling, if there is such a scaling. Then we define the Hölder exponent $\alpha(i_1 \dots i_k)$ of the region $A_{i_1 \dots i_k}$ by

$$\text{Prob}(A_{i_1 \dots i_k}) = (r_{i_1} \cdots r_{i_k})^{\alpha(i_1 \dots i_k)} \tag{B.19}$$

Combining Eqs. (B.18) and (B.19) we find a simple way to calculate $\alpha(i_1 \dots i_k)$,

$$\alpha(i_1 \dots i_k) = \frac{\log(p_{i_1} \cdots p_{i_k})}{\log(r_{i_1} \cdots r_{i_k})} \tag{B.20}$$

B.9.2 Concavity of the $f(\alpha)$ curve

Now we'll show that so long as the $\ln(p_i)/\ln(r_i)$ are not all equal, the $f(\alpha)$ curve is concave down. Recall Eq. (2.20) $f(\alpha) = q\alpha + \beta(q)$. Then

$$\frac{df}{d\alpha} = q \quad \text{and so} \quad \frac{d^2 f}{d\alpha^2} = \frac{dq}{d\alpha} \tag{B.21}$$

Because $\alpha = -d\beta/dq$ (Eq. (2.17)),

$$\frac{d\alpha}{dq} = -\frac{d^2\beta}{dq^2} \tag{B.22}$$

To show that the $f(\alpha)$ curve is concave down, we'll show that $d^2 f/d\alpha^2 < 0$. We've seen that $d^2 f/d\alpha^2 = dq/d\alpha$, so it's enough to show that $dq/d\alpha < 0$. With Eq. (B.22) we'll be able to show that $d\alpha/dq < 0$. How are $dq/d\alpha$ and $d\alpha/dq$ related? If $d\alpha/dq \neq 0$, then by the inverse function theorem, $dq/d\alpha = 1/(d\alpha/dq)$, so $d\alpha/dq < 0$ implies $dq/d\alpha < 0$, the result we want.

If the inverse function theorem is unfamiliar, pages 193–195, and especially Eq. (31), of [36] is a good source. So is Google. Here's a simple example. Suppose $y = \sin(x)$, so $x = \sin^{-1}(y)$. Now $\sin^{-1}(y)$ is defined for $-\pi/2 < y < \pi/2$. The inverse function theorem is an example of what's called a local result: any argument that depends on derivatives can be applied only near the points in question. For this example we'll work only for x near 0. Here's what the inverse function theorem says about derivatives

$$\frac{dy}{dx} = \frac{d\sin(x)}{dx} = \cos(x)$$
$$\frac{dx}{dy} = \frac{d\sin^{-1}(y)}{dy} = \frac{1}{\sqrt{1-y^2}} = \frac{1}{\sqrt{1-\sin^2(x)}} = \frac{1}{\sqrt{\cos^2(x)}} = \frac{1}{\cos(x)} = \frac{1}{dy/dx}$$

where we've used $\sqrt{\cos^2(x)} = \cos(x)$, which is true as long as $\cos(x)$ is positive. (If $\cos(x) < 0$ then $\sqrt{\cos^2(x)} = -\cos(x)$.) Because x is close to 0, $\cos(x)$ is positive.

So by Eq. (B.22) we need to compute $d^2\beta/dq^2$ and show it is positive. To ease the burden of keeping track of the terms, we'll rewrite Eq. (2.14) in a more compact notation.

$$\frac{d\beta}{dq} = -\frac{\sum_i p_i^q r_i^{\beta(q)} \ln(p_i)}{\sum_i p_i^q r_i^{\beta(q)} \ln(r_i)} \tag{B.23}$$

Now compute the second derivative $d^2\beta/dq^2 =$

$$-\frac{(\sum_i p_i^q r_i^\beta \ln(r_i))(\sum_i p_i^q r_i^\beta \ln(p_i))' - (\sum_i p_i^q r_i^\beta \ln(p_i))(\sum_i p_i^q r_i^\beta \ln(r_i))'}{(\sum_i p_i^q r_i^\beta \ln(r_i))^2}$$

where $'$ denotes d/dq and we've written β rather than $\beta(q)$. The denominator certainly is positive. Expand the derivatives in the numerator. Half subtract out and the remaining terms are

$$\left(\sum_{i,j} p_i^q r_i^\beta \ln(r_i) p_j^q r_j^\beta \ln(r_j) \frac{d\beta}{dq}\right) - \left(\sum_{i,j} p_j^q r_j^\beta \ln(p_j) p_i^q r_i^\beta \ln(r_i) \frac{d\beta}{dq} \ln(r_j)\right)$$

Because $\ln(p_i)$ and $\ln(r_i)$ are negative, from Eq. (B.23) we see that $d\beta/dq$ is negative. Then we see that $d^2\beta/dq^2 > 0$ and consequently the $f(\alpha)$ curve is concave down.

But what about the condition that not all $\ln(p_i)/\ln(r_i)$ are equal? To see what goes wrong, suppose $\ln(p_i)/\ln(r_i) = K$, a constant, for $i = 1, \dots, n$. Now look at Eq. (B.23). Multiply each term of the numerator by $\ln(r_i)/\ln(r_i)$,

$$\begin{aligned} \frac{d\beta}{dq} &= -\frac{\sum_i p_i^q r_i^{\beta(q)} \ln(r_i)\ln(p_i)/\ln(r_i)}{\sum_i p_i^q r_i^{\beta(q)} \ln(r_i)} \\ &= -\frac{\sum_i p_i^q r_i^{\beta(q)} \ln(r_i) K}{\sum_i p_i^q r_i^{\beta(q)} \ln(r_i)} = -K\frac{\sum_i p_i^q r_i^{\beta(q)} \ln(r_i)}{\sum_i p_i^q r_i^{\beta(q)} \ln(r_i)} = -K \end{aligned}$$

Consequently, $d^2\beta/dq^2 = 0$ and we cannot conclude that the $f(\alpha)$ curve is concave down.

If you're interested, look back over our derivation of $d^2 f/d\alpha^2 < 0$ and see where we used the condition not all $\ln(p_i)/\ln(r_i)$ are equal.

B.9.3 Attractor dimension and the $f(\alpha)$ curve

This argument is much simpler than the previous one. From Eq. (B.21) we see $df/d\alpha = q$, so the only critical point occurs at $q = 0$. Because the $f(\alpha)$ curve is concave down, at the critical point the graph has a maximum.

Then for $q = 0$ the generalized Moran equation (2.13) becomes $p_1^0 r_1^{\beta(0)} + \cdots + p_0^q r_n^{\beta(0)} = 1$. That is, $r_1^{\beta(0)} + \cdots + r_n^{\beta(0)} = 1$. This is just the familiar Moran equation for the dimension d of the IFS attractor, with $d = \beta(0)$. Finally, recall Eq. (2.20), $f(\alpha) = q\alpha + \beta(q)$. At $q = 0$ this becomes $f(\alpha) = \beta(0)$. That is, the dimension of the attractor is the maximum height of the $f(\alpha)$ curve.

B.9.4 A formula for the $f(\alpha)$ curve

The full proof that $f(\alpha) = q\alpha + \beta$ is the dimension of the IFS attractor uses more advanced math than we'll use here. Full developments are in Sect. 17.3 of [16] and Sect. 11.2 of [20]. Here we'll sketch just one part of the proof.

The dimension used for this proof is the Hausdorff dimension, a subtle mathematical construction that underlies our notion of similarity dimension. Hausdorff dimension is based on the Hausdorff measure, so we'll start with how to find the Hausdorff measure of a set A. For every $\delta > 0$, a *δ-cover* of A is a collection of sets $\{B_i\}$ that satisfies two properties:

- this collection is a cover of A, that is, $A \subseteq \cup_i B_i$, and
- the diameter $|B_i|$ of each B_i satisfies $|B_i| \leq \delta$. For every $\delta > 0$ and every dimension $d > 0$, define

$$\mathcal{H}_\delta^d(A) = \inf\left\{\sum_i |B_i|^d : \text{where } \{B_i\} \text{ is a } \delta\text{-cover of } A\right\}$$

The computation of this quantity can be tricky, because we must consider *every* δ-cover of A.

Now if $\delta_1 < \delta_2$, every δ_1-cover of A also is a δ_2-cover of A. Consequently, $\delta_1 < \delta_2$ implies $\mathcal{H}^d_{\delta_1}(A) \geq \mathcal{H}^d_{\delta_2}(A)$. That is, as $\delta \to 0$, $\mathcal{H}^d_\delta(A)$ cannot wobble around and cannot get smaller, so $\lim_{\delta\to 0} \mathcal{H}^d_\delta(A)$ exists, though the limit may be infinite. Then the d-dimensional Hausdorff measure of A is

$$\mathcal{H}^d(A) = \lim_{\delta\to 0} \mathcal{H}^d_\delta(A)$$

We've seen, in Sample 2.1.6 for instance, that when measured in a dimension lower than that of the shape gives ∞ (remember, this side of the argument is challenging), and when measured in a dimension higher than that of the shape gives 0. These examples are precursors of this definition: the *Hausdorff dimension* $d_h(A)$ is defined by

$$d < d_h(A) \text{implies } \mathcal{H}^d(A) = \infty \text{ and } d > d_h(A) \text{ implies } \mathcal{H}^d(A) = 0$$

Usually an upper bound for $d_h(A)$ is easier to establish than is a lower bound. This is because for an upper bound a clever choice of δ-covers can establish an upper bound, but for a lower bound we must identify a behavior common to all δ-covers. Here we'll establish that $\alpha q + \beta$ is an upper bound for $d_h(A)$; we'll leave the lower bound for other sources.

The argument is a single calculation. For this we'll need some notation. Recall the setting for our construction of $f(\alpha)$ curves. The set A is the attractor of an IFS $\{T_1, \ldots, T_n\}$, where the T_i are similarity transformations, r_i is the contraction factor of T_i and p_i is the probability that T_i is applied.
- For any $R \subseteq A$, $\mu(R)$ is the probability that a point generated by the random IFS algorithm falls in the region R.
- $A_{i_1\ldots i_k} = T_{i_1} \circ \cdots \circ T_{i_k}(A)$.
- For any point $x \in A$, $A_{i_1\ldots i_k}(x)$ is the address length-k region that contains x.
- $|A_{i_1\ldots i_k}| = r_1 \ldots r_k$, the diameter of $A_{i_1\ldots i_k}$.
- I_k is the list of all length-k addresses $i_1 \ldots i_k$. There are n^k of these addresses.

For each α between $\alpha_{\min}$ and $\alpha_{\max}$,

$$A(\alpha) = \left\{ x \in A : \lim_{k\to\infty} \frac{\log(\mu(A_{i_1\ldots i_k}(x)))}{\log(|A_{i_1\ldots i_k}(x)|)} = \alpha \right\}$$

Recall that we are looking for power law scalings where the probability of falling in a region is approximately proportional to the a power of the diameter of the region. We'll formalize this with the definition

$$Q_k(\alpha) = \{(i_1 \ldots i_k) : |A_{i_1\ldots i_k}|^\alpha \approx \mu(A_{i_1\ldots i_k})\}$$

This is an approximation of the region on A with Hölder exponent α.We want to show that the Hausdorff dimension of this region is $\alpha q + \beta$. For this we'll estimate the $(\alpha q+\beta)$-dimensional measure of this region. That is,

$$\begin{aligned}
\sum_{(i_1\dots i_k)\in Q_k(\alpha)} |A_{i_1\dots i_k}|^{\alpha q+\beta} &= \sum_{(i_1\dots i_k)\in Q_k(\alpha)} (|A_{i_1\dots i_k}|^{\alpha})^q |A_{i_1\dots i_k}|^{\beta} \\
&\approx \sum_{(i_1\dots i_k)\in Q_k(\alpha)} \mu(A_{i_1\dots i_k})^q |A_{i_1\dots i_k}|^{\beta} \\
&\leq \sum_{(i_1\dots i_k)\in I_k} \mu(A_{i_1\dots i_k})^q |A_{i_1\dots i_k}|^{\beta} \\
&= \sum_{i_1,\dots,i_k=1}^{n} (p_{i_1}\cdots p_{i_k})^q (r_{i_1}\cdots r_{i_k})^{\beta} \qquad \text{(i)} \\
&= \sum_{i_1,\dots,i_k=1}^{n} (p_{i_1}^q r_{i_1}^{\beta})\cdots(p_{i_k}^q r_{i_k}^{\beta}) \\
&= \left(\sum_{i=1}^{n} p_i^q r_i^{\beta}\right)^k = 1^k = 1 \qquad \text{(ii)}
\end{aligned}$$

Here (i) is a consequence of the observations $\mu(A_{i_1\dots i_k}) = p_{i_1}\cdots p_{i_k}$ and $|A_{i_1\dots i_k}| = r_{i_1}\cdots r_{i_k}$. The first equality of (ii) follows from the multinomial expansion, for example, $(x+y+z)^2 = x^2+xy+xz+yx+y^2+yz+zx+zy_z^2$ and the second equality of (ii) is the Moran equation.

We've shown that $\sum_{(i_1\dots i_k)\in Q_k(\alpha)} |A_{i_1\dots i_k}|^{\alpha q+\beta} \leq 1$. To see how this is related to Hausdorff measure, observe that for every $\delta > 0$ there is a k large enough that $(\max\{r_i\})^k < \delta$. Then $\{A_{i_1\dots i_k} : (i_1\dots i_k)\in Q_k(\alpha)\}$ is a δ-cover of $A(\alpha)$. So because $\sum_{(i_1\dots i_k)\in Q_k(\alpha)} |A_{i_1\dots i_k}|^{\alpha q+\beta} \leq 1$, we see that

$$\inf\left\{\sum_i |B_i|^d : \text{where } \{B_i\} \text{ is a } \delta\text{-cover of } A(\alpha)\right\} \leq 1$$

That is, $\mathcal{H}_\delta^d(A(\alpha)) \leq 1$. This argument can be applied for any $\delta > 0$, so $\mathcal{H}^d(A(\alpha)) = \lim_{\delta\to 0}\mathcal{H}_\delta^d(A(\alpha)) \leq 1$.

Recall that $f(\alpha) = d_h(A(\alpha))$ is the number where $\mathcal{H}^d(A(\alpha))$ jumps from ∞ to 0. Then $\mathcal{H}_\delta^d(A(\alpha)) \leq 1$ means $d_h(A(\alpha)) \leq \alpha q+\beta$. That's as far as we'll go with the proof.

B.10 Notes on the Mandelbrot set and Julia sets, Lab 3.6

Here we'll establish the escape criterion for $z^n + c$; describe some geometry of the Mandelbrot set including the main unsolved problem, MLC, the Mandelbrot set local connectivity conjecture (MLC); and the linear change of variables to remove the next-to-the-highest order term in a polynomial. This is a tiny bit of the substantial and significant math that underlies the Mandelbrot set. Once again we'll point out that the claim that the study of the Mandelbrot set is "just pretty pictures" is so very far from either right or relevant.

B.10.1 Escape to infinity criteria

Here we'll find an escape criterion for $z^n + c$, then take $n = 2$ to deduce the condition we presented in Sect. 3.6.3.

First, we suppose that $\|z_k\| > \max\{2^{1/(n-1)}, \|c\|\}$. Then

$$\|z_k\| > 2^{1/(n-1)}$$

Now raise both sides to the $(n-1)$st power,

$$\|z_k\|^{n-1} > (2^{1/(n-1)})^{n-1} = 2$$

so

$$\|z_k\|^{n-1} = 2 + \epsilon \tag{B.24}$$

for some $\epsilon > 0$.

The triangle inequality $|a + b| \leq |a| + |b|$ for absolute values is valid for complex number moduli. We'll find a lower bound for $\|z_k^n + c\|$ with the triangle inequality.

$$\|z_k^n\| = \|z_k^n + c - c\| \leq \|z_k^n + c\| + \| - c\| = \|z_k^n + c\| + \|c\|$$

where the $\leq$ in this string is an application of the triangle inequality.

Now subtract $\|c\|$ from both sides and rearrange a bit,

$$\begin{aligned}\|z_k^n + c\| &\geq \|z_k^n\| - \|c\| \overset{1}{\geq} \|z_k^n\| - \|z_k\| \overset{2}{=} \|z_k\|^n - \|z_k\| \\ &= (\|z_k\|^{n-1} - 1)\|z_k\| \overset{3}{=} (1+\epsilon)\|z_k\|\end{aligned}$$

Here inequality 1 holds because $\|z_k\| \geq \|c\|$, equality 2 because $\|(x+iy)^n\| = \|x+iy\|^n$ (check $n = 2$ to see the idea), and equality 3 follows from Eq. (B.24).

Since $z_{k+1} = z_k^n + c$, we have shown

$$\|z_{k+1}\| \geq (1+\epsilon)\|z_k\|$$

Iterating, we obtain

$$\|z_{k+m}\| \geq (1+\epsilon)^m \|z_k\|$$

Consequently, if $\|z_k\| > \max\{2^{1/(n-1)}, \|c\|\}$, then $\|z_{k+m}\| \to \infty$ as $m \to \infty$.

Now we can state the general escape criterion for $z_{k+1} = z_k^n + c$:

If for some k we have $\|z_k\| > 2^{1/(n-1)}$, then later iterates escape to ∞.

What we've shown so far is that if $\|z_k\| > \max\{2^{1/(n-1)}, \|c\|\}$ then later iterates escape to ∞. To replace the condition with $\|z_k\| > 2^{1/(n-1)}$ we must consider two cases, (i) $\|c\| \leq 2^{1/(n-1)}$ and (ii) $\|c\| > 2^{1/(n-1)}$.

In case (i), $\|z_k\| > 2^{1/(n-1)}$ implies $\|z_k\| > \max\{2^{1/(n-1)}, \|c\|\}$, so $\|z_k\| > 2^{1/(n-1)}$ shows that later iterates run to ∞.

In case (ii), $\|c\| > 2^{1/(n-1)}$ implies $\|c\|^{n-1} > 2$. Then when $z_0 = 0$, we have $z_1 = c$ and $z_2 = c^n + c$, so

$$\|z_2\| = \|c^n + c\| = \|c\|\|c^{n-1} - 1\| \geq \|c\|(\|c^{n-1}\| - 1) > \|c\| = \max\{2^{1/(n-1)}, \|c\|\}$$

The first inequality is another application of the triangle inequality

$$\|c^{n-1}\| = \|c^{n-1} - 1 + 1\| \le \|c^{n-1} - 1\| + 1$$

together with $\|c^{n-1}\| = \|c\|^{n-1}$. So again, the argument we gave above shows that later iterates run to ∞.

Note that if $n = 2$ the escape criterion is $\|z_k\| > 2$, as stated in Sect. 3.6.3.

B.10.2 Some Mandelbrot set geometry

Since the discovery of the Mandelbrot set around 1980, many of its properties have been deduced. The Mandelbrot set is the set of all complex numbers c for which the sequence (3.3) doesn't escape to ∞. One way, but not the only way, the sequence can not escape to ∞ is if it converges to a repeating pattern, that is, to a stable cycle. For any positive integer n, an *n-cycle component* of the Mandelbrot set is a maximal connected region of the Mandelbrot set on which the iterates of $z_0 = 0$ converge to an n-cycle. By "connected" we mean that if c_1 and c_2 belong to an n-cycle component, then there is a path between c_1 and c_2, and for every point c on this path, the sequence (3.3) converges to an n-cycle. By "maximal" we mean that for all c immediately outside this component, the sequence (3.3) converges to some other cycle, runs away to ∞, or wanders around chaotically but doesn't run away to ∞.

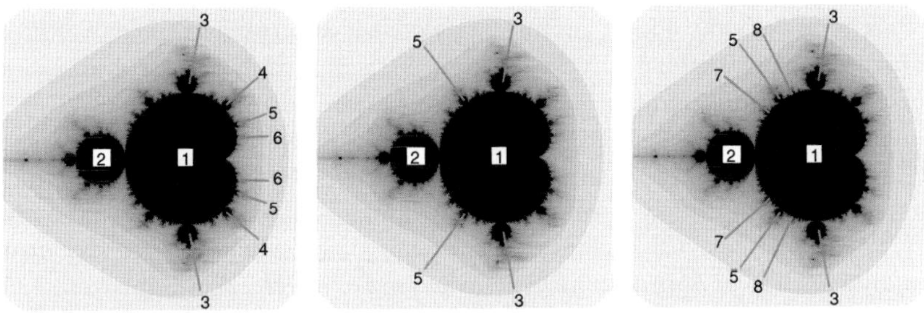

Figure B.4: Left: part of the principal sequence. Middle and right: the start of the Farey sequence.

So far as we have seen, every Mandelbrot set component is either approximately a cardioid or approximately a disk. Of these, only one, the *main cardioid* (labeled 1 in Fig. B.4) is a true cardioid, and only one (labeled 2 in Fig. B.4) is a true disk. For every c in the main cardioid, the sequence (3.3) converges to a fixed point, hence the label 1. For every c in the true disk, the sequence (3.3) converges to a 2-cycle. The arrangement of components in the Mandelbrot set is well understood.

First we'll describe the arrangement of the n-cycle disks attached to the main cardioid. This is illustrated in Fig. B.4. On the left we see the start of the *principal sequences.* From the 2-cycle disk, to the right along both the top and the bottom of the main cardioid, the largest disk we encounter is a 3-cycle disk. Continuing to the right, the next largest disk is a 4-cycle disk, to the right beyond that the next largest is a 5-cycle disk, and so on.

Many other disks are attached to the main cardioid. One more sequence fills in all of these. In the middle picture of Fig. B.4 we see the start of the *Farey sequence*, named for the British geologist John Farey. Farey's observation was about a way to order the rational numbers; at that time (early 19th century) it had nothing to do with the Mandelbrot set, still over 150 years in the future. Along the main cardioid between the 2-cycle disk and a 3-cycle disk, the largest disk is a $2 + 3 = 5$-cycle disk. The same rule applies between any consecutive disks in the principal sequence: between an n-cycle disk and an $(n + 1)$-cycle disk attached to the main cardioid, the largest disk is an $n+(n+1) = (2n+1)$-cycle disk.

The right picture of Fig. B.4 shows how to continue the Farey sequence. Between the 2-cycle disk and the 5-cycle disk, the largest disk is a $2 + 5 = 7$-cycle disk; between the 5-cycle disk and the 3-cycle disk, the largest disk is a $5 + 3 = 8$-cycle disk. This pattern continues forever. A consequence is that for all integers $n > 1$, attached to the main cardioid we find (at least one, but more often, many) n-cycle disks.

The relation between the disk size and its cycle number is the $1/n^2$-*rule* [85, 86]: for an n-cycle disk attached to the main cardioid, the radius of the disk is about $1/n^2$. The longer the cycle, the smaller the disk. If you compare the 5-cycle disks of the principal sequence (left picture of Fig. B.4) with the 5-cycle disks of the Farey sequence (middle picture of Fig. B.4), you'll notice the Faery sequence disk is larger than the principal sequence disk. The $1/n^2$ rule has a correction factor, roughly how far from the cusp of the main cardioid the disk is attached. Think of the 2-cycle disk as attached $m/n = 1/2$ way around the cardioid, the 3-cycle disks as attached $m/n = 1/3$ and $2/3$ way around the cardioid, and the 4-cycle disks as attached $m/n = 1/4$ and $3/4$ way around the cardioid. Then the principal 5-cycle disks are attached $m/n = 1/5$ and $4/5$ way around the cardioid, while the Farey 5-cycle disks are attached $m/n = 2/5$ and $3/5$ way around the cardioid. A more precise statement of the $1/n^2$ rule is this: the radius of an n-cycle disk attached m/n way around the main cardioid is approximately $\sin(m\pi/n)/n^2$. Even this improvement has some variations, which are themselves arranged in a fractal pattern. This was discovered by Kerry Mitchell. His graphs appear on pgs. 96–97 of [88]. See Appendix A.75 of [17], for example.

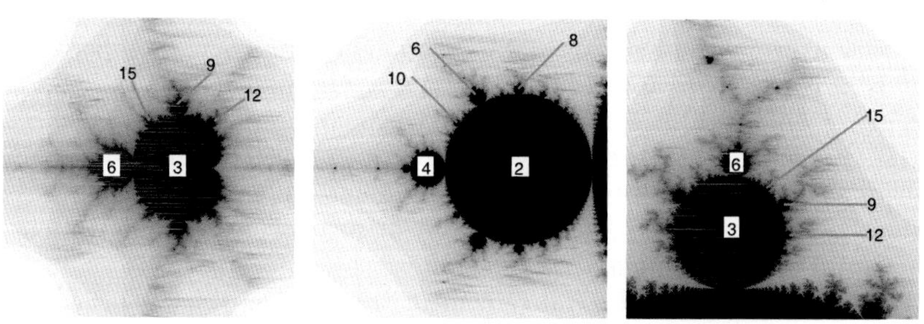

Figure B.5: The multiplier rule for a cardioid and for two discs.

This gives the cycle number (and approximate radius) of all the discs attached to the main cardioid. What about the disks attached to other cardioids, or the disks attached to disks? Those cycle numbers are determined by the *multiplier rule*. The left picture of Fig. B.5 shows the cycle numbers of the disks attached to the 3-cycle cardioid, centered on the real axis to the left of the main cardioid. The cycle numbers of each of these disks are 3 times the cycle numbers of the corresponding disks attached to the main cardioid. The middle and right pictures of Fig. B.5 show the multiplier rule applied to the 2-cycle and upper 3-cycle disk on the main cardioid. To anchor the orientation of these disks, the cusp of the cardioid corresponds to the point of tangency of the disk to a larger disk or to a cardioid.

This is enough for now, but we'll mention several other properties of the Mandelbrot set. The number of branches of an antenna above a disk is related to the cycle number of the disk: note the three branches above the 3-cycle disk in the right picture of Fig. B.5. For disks attached to disks, the multiplier rule is expressed for different branching numbers in the antennas.

The cycle numbers of cardioids, and their location relative to other features, can be found by *Lavaurs' algorithm* [89]. See Sect. 5.4 of [17], for example.

And there is a set of c-values, dense in the boundary of the Mandelbrot set, for which a small bit of the Mandelbrot set near that c looks like a small bit of the Julia set for that c. Recall that a set A is *dense* in a set B if arbitrarily close to every point of B we can find a point of A. So the boundary of the Mandelbrot set is filled with little bits of Julia sets. This is a theorem of Tan Lei [90].

If we look closer and closer to the boundary of the Mandelbrot set, we see ever more branches. In fact, the branches are so dense that the Mandelbrot set boundary has Hausdorff dimension 2. This is a theorem of Mitsuhiro Shishikura [91, 92, 94].

Although computer-generated pictures suggest that the small cardioids around the Mandelbrot set are islands, in fact they are connected to the main cardioid. That the Mandelbrot set is connected is a theorem of John Hubbard and Adrien Douady [95].

Douady and Hubbard proved another result. In 1983 James Curry, Lucy Garnett, and Dennis Sullivan [96] observed that the Mandelbrot set appears in a space associated with finding roots of cubic polynomials by Newton's method. The function iterated here is nothing like z^2+c. In fact, for these examples the Newton function is a ratio of polynomials. This appearance of the Mandelbrot set was a surprise, but then in 1985 Douady and Hubbard [97] proved that the Mandelbrot set is universal, in the sense that it appears naturally in a large class of functions. See also [98]. The Mandelbrot set is remarkable, but so also is the fact that it's all over the place in math, if you know where and how to look. Benoit was the first person who looked there.

The big Mandelbrot set question that remains open is whether the Mandelbrot set is locally connected. This is called the *MLC conjecture* [95]. A set A is *locally connected* if for every point p of A, the portion of A in every sufficiently small disk centered at p is

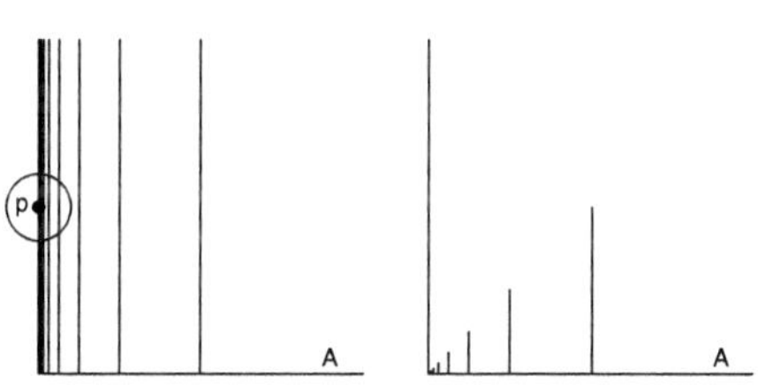

connected. In the left picture, the set A is not locally connected. Do you see why the set A in the right picture is locally connected?

Brilliant mathematicians have worked on this problem, and although some progress has been made, the general result so far has not given up its secrets. It is known [100] that the Mandelbrot set for the family of cubic polynomials $f_{a,b}(z) = z^3 + az + b$ is not locally connected. See also [101].

B.10.3 Simplified polynomials

First, we'll show that any quadratic polynomial $a_2z^2 + a_1z + a_0$ can be converted by a linear change of variable into a polynomial $z^2 + c$. The change of variables is $z = w - a_1/(2a_2)$. This gives

$$\begin{aligned} a_2z^2 + a_1z + a_0 &= a_2(w - a_1/(2a_2))^2 + a_1(w - a_1/(2a_2)) + a_0 \\ &= a_2w^2 + a_0 - a_1^2/(4a_2) \end{aligned}$$

To translate the iteration formula

$$z_{k+1} = a_2z_k^2 + a_1z_k + a_0$$

in terms of w_k and w_{k+1}, substitute again with $z_k = w_k - a_1/(2a_2)$ and $z_{k+1} = w_{k+1} - a_1/(2a_2)$. Now the iteration becomes

$$\begin{aligned} w_{k+1} - a_1/(2a_2) &= a_2(w_k - a_1/(2a_2))^2 + a_1(w_k - a_1/(2a_2)) + a_0 \\ &= a_2w_k^2 + a_0 - a_1^2/(4a_2) \end{aligned}$$

We see the w_k term has disappeared from the right side of the equation. Isolating w_{k+1} on the left side, we obtain

$$w_{k+1} = a_2w_k^2 + a_0 + (2a_1 - a_1^2)/(2a_2)$$

This is almost what we want. The only problem is that we want the coefficient of w_k^2 to be 1, but it is a_2. So: one more pair of substitutions, $u_k = a_2w_k$ and $u_{k+1} = a_2w_{k+1}$. Then the iteration becomes

$$\begin{aligned} u_{k+1} &= u_k^2 + a_2(a_0 + (2a_1 - a_1^2)/(2a_2)) \\ &= u_k^2 + (a_2a_0 + a_1 - a_1^2/2) = u_k^2 + c \end{aligned}$$

The constant c is pretty messy, but it is still a constant.

This argument can be generalized to polynomials with any exponent n by the substitution

$$z_{k+1} = w_{k+1} - \frac{a_{n-1}}{na_n} \quad \text{and} \quad z_k = w_k - \frac{a_{n-1}}{na_n}$$

Then the iteration

$$z_{k+1} = a_nz_k^n + a_{n-1}z_k^{n-1} + a_{n-2}z_k^{n-2} + \cdots + a_1z_k + a_0$$

is transformed into

$$w_{k+1} = b_nw_k^n + b_{n-2}w_k^{n-2} + \cdots + b_1w_k + b_0$$

where the coefficients $b_n, b_{n-2}, \ldots, b_1$, and b_0 are determined by the coefficients $a_n, a_{n-1}, \ldots, a_1$, and a_0. One more step changes the coefficient b_n into 1. The general expressions for these coefficients are *not* worth the effort to write down, unless you try to code an iteration of the general polynomial, which we won't do.

The point here is that for Lab 3.6 to study the general cubic Mandelbrot set we'll use the polynomial $f_{c_1,c_1}(z) = z^3 + c_1 z + c_2$.

B.11 Notes on circle inversion fractals, Lab 3.7

Here we'll derive some basic properties of circle inversion and present some examples of limit sets.

B.11.1 Circle inversion properties

From Sect. 3.7.3.2 recall the list of properties of inversion I_C in a circle C with center O and radius r.

1. Every point of C is left fixed by I_C.
2. I_C interchanges the inside and the outside of C.
3. I_C^2 is the identity transformation.
4. I_C is a contraction for sets on the outside of C, and an expansion for sets inside C.
5. I_C takes to itself every circle that intersects C orthogonally.
6. I_C takes every circle that does not pass through O to a circle that does not pass through O.
7. I_C takes every circle passing through O to a straight line.

The first three we left as exercises, so we'll start with property 4, I_C is a contraction on sets outside of the circle C and an expansion on sets inside of C.

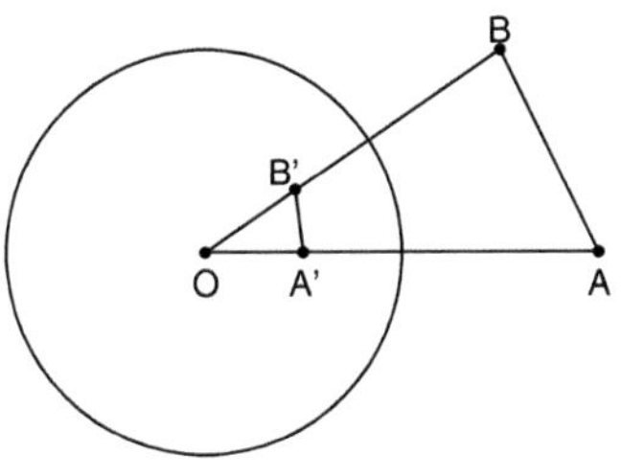

Inversion and contraction

Suppose A and B are points that lie outside the inverting circle C. Denote by $|AB|$ the distance between A and B, that is, the length of the line segment between A and B. Because A' is the inverse of A and B' is the inverse of B, we have

$$|OA'| \cdot |OA| = r^2 = |OB'| \cdot |OB| \qquad \text{so } \frac{|OA'|}{|OB'|} = \frac{|OB|}{|OA|}$$

Certainly $\angle A'OB' = \angle BOA$ so $\Delta OAB \sim \Delta OB'A'$. (If your geometry is a bit rusty, triangles are similar if two sides of one triangle have the same ratio as the corresponding sides of the other triangle, and if the angles determined by these sides are equal.) Now if you think, "Wait a minute: how can these triangles be similar? The sides AB and $A'B'$ aren't parallel," notice the orientations of the triangles: $\Delta OAB \sim \Delta OB'A'$. In particular, $\angle OA'B' = \angle OBA$.

Now the similarity of ΔOAB and $\Delta OB'A'$ guarantees that all pairs of corresponding sides have the same ratio, so in particular

$$\frac{|A'B'|}{|AB|} = \frac{|OA'|}{|OB|} = \frac{|OA| \cdot |OA'|}{|OA| \cdot |OB|} = \frac{r^2}{|OA| \cdot |OB|}$$

From this we see that if both A and B lie outside the inverting circle C, then $|A'B'| < |AB|$. That is, inversion is a contraction for points outside of C.

Together with property 3, that I_C is a contraction on sets outside of C implies that I_C is an expansion on sets inside of C.

Inversion and stereographic projection

Before we can continue to derive formulas for inversion of a line and of a circle, we should clear up a detail. As we have defined it, inversion is not a transformation of the whole plane, because the center of the inverting circle does not invert to any point on the plane. This can be resolved by adding a point at infinity to the plane. Call this extended plane $\hat{\mathbb{R}}^2 = \mathbb{R}^2 \cup \{\infty\}$; by the method of *stereographic projection* the extended plane is equivalent to a sphere. Place a sphere on the plane, the south pole of the sphere resting on the plane. To each point b of the plane we associate a point a, the intersection of the sphere and the line determined by b and N, the north pole of the sphere. As the point b moves farther from the south pole of the sphere, the point a moves closer to N. In this sense, the point at infinity corresponds to the north pole of the sphere.

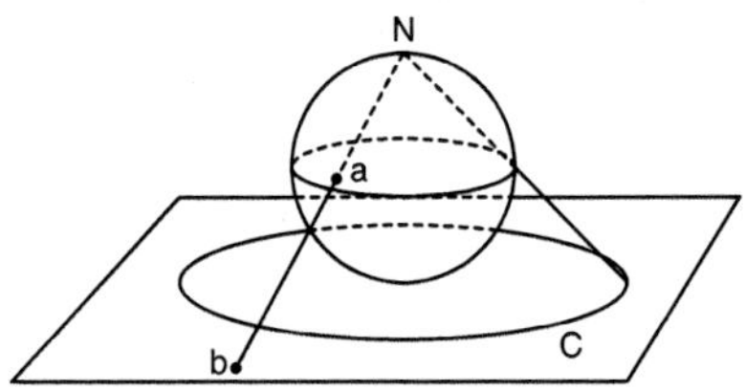

Note an interesting connection between inversion and stereographic projection. The equator of the sphere projects to a circle C with center the south pole of the sphere and radius the diameter d of the sphere. We show that inversion in C can be achieved by projecting to the sphere, reflecting across the equator, and projecting back to the plane.

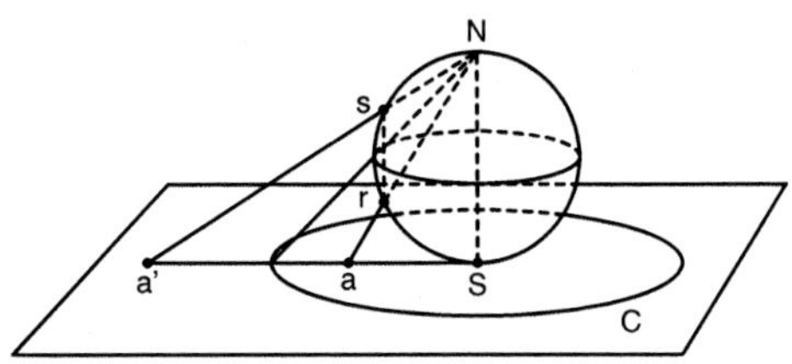

Suppose r is the projection of the point a to the sphere, s the reflection across the equator, and a' the projection of s to the plane. We show a' is the inverse of a in C. Note that $|NS|$ is the diameter of the circle C.

First, note $\tan(\angle aNS) = |Sa|/|NS|$ and $\tan(\angle a'NS) = |Sa'|/|NS|$. Because r and s are reflections across the equator of the sphere, $\tan(\angle rNS) = 1/\tan(\angle sNS)$. (This is a straightforward calculation in analytic geometry. Can you find a synthetic proof?) Then we see $|Sa|/|NS| = |NS|/|Sa'|$ and it follows that a and a' are inverses in C.

Orthogonal circles

Now we'll prove inversion property 5: a circle S that intersects a circle C orthogonally (at both points of intersection, the tangent of C is orthogonal, that is perpendicular, to the tangent of S) is invariant under inversion across C.

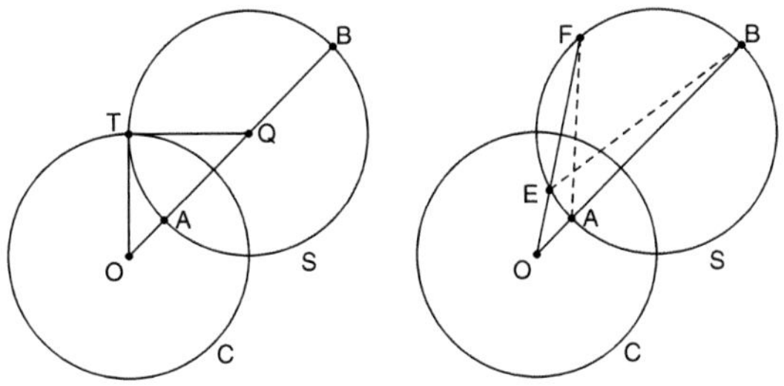

In the left image above, S intersects C orthogonally means that the line through the points O and T is tangent to S at T, and so is perpendicular to the radius TQ. That is, the triangle ΔOTQ is a right triangle. Apply the Pythagorean theorem to see

$$\begin{aligned}|OT|^2 &= |OQ|^2 - |QT|^2\\ &= (|OQ| - |QT) \cdot (|OQ| + |QT|)\\ &= (|OQ| - |QA|) \cdot (|OQ| + |QB|)\\ &= |OA| \cdot |OB|\end{aligned}$$

where the penultimate equality follows because $|QT| = |QA| = |QB|$: all are radii of the circle S. We have shown that A and B lie on the same ray from O, and $|OA| \cdot |OB| = |OT|^2$. Because $|OT|$ is the radius of the circle C, we have shown that A and B are inverses in the circle C.

Now consider any other ray from O that intersects the circle S in two points, E and F. We'll show that $|OE| \cdot |OF| = |OA| \cdot |OB|$ and so E and F are inverses in the circle C. Draw two auxiliary lines, one from F to A, the other from B to E. Because $\angle EFA$ and $\angle EBA$ are inscribed in the circle S and subtend the same arc, $\overset{\frown}{EA}$, we see $\angle EFA = \angle EBA$. Then the triangles ΔOEB and ΔOAF are similar and we see $|OE|/|OA| = |OB|/|OF|$. Cross-multiplication gives $|OE| \cdot |OF| = |OA| \cdot |OB|$. That is, the points E and F are inverses in the circle C.

A product of the form $|OE| \cdot |OF|$, where E and E are the points of intersection with the circle S of a line through a point O outside of S, is called the *power* of the point O with respect to S. The preceding paragraph is a proof that the power of O and S does not depend on the line chosen.

We have shown that inversion in the circle C preserves every circle S that intersects C orthogonally, although inversion interchanges the part of S inside C and the part of S outside C.

Lines invert to circles

Now we'll show a line L inverts to a circle passing through the center O of the inverting circle. The left image of Fig. B.6 shows a typical case of a line L outside of C. A similar argument works for a line that intersects C in one or two points, so long as the line L does not pass through the center of C. If L does pass through the center of C, inversion takes that line to itself, though it interchanges the portion of L that lies inside C with the portion that lies outside of C.

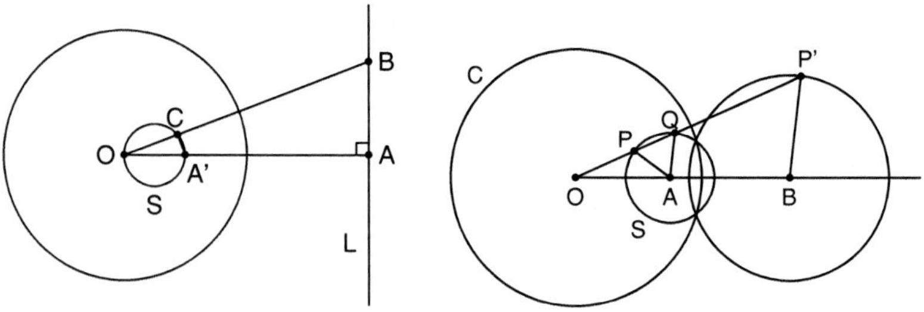

Figure B.6: Line inverts to circle, circle inverts to circle.

Take the point A on L closest to O, so OA is perpendicular to L. Take A' to be the inverse of A and draw the circle S with diameter OA'. Pick any point B on L and let C denote the point of intersection of S and the line OB. The triangles $\Delta OCA'$ and ΔOAB are similar (angle $\angle OCA'$ is right: a triangle inscribed in a circle, with one triangle side a circle diameter, is a right triangle), so $OB/OA = OA'/OC$. Then we see that $OB \cdot OC = OA \cdot OA' = r^2$, where r is the radius of the inverting circle. That is, C is the inverse of B, making the circle S the inverse of the line L.

By basic property 3 of inversion, this implies that a circle that passes through the center of the inverting circle inverts to a line.

(Most) Circles invert to circles

What is the inverse of a circle not passing through the center of the inverting circle? Suppose C is the inverting circle and O its center. Suppose S is the circle to be inverted, A is the center of S and r the radius of S. Pick any point P on S and let P' denote its inverse in C. The line OP' intersects S in another point Q, or is tangent to S. We consider only the first case, the second (tangent intersection) is left as an exercise. Through P' draw a line parallel to QA, and let B denote its intersection with OA. To show the inverse of S is a circle, we show (i) the distance $|OB|$ is independent of P, and (ii) the distance $|P'B|$ is independent of P. Then as P traces out points on the circle S, P' traces out points on the circle with center B and radius BP'. See the right image of Fig. B.6.

Because P and P' are inverses, the product $|OP| \cdot |OP'| = R^2$, where R is the radius of C.

Next, we'll show that $|OP| \cdot |OQ|$ remains constant as P and Q move around S. One way to see this is analytic. This involves just the circle S, the point O outside of S, and a line from O that intersects S. To simplify the calculations, we'll assume $O = (0,0)$, and S has radius r and center $A = (a,0)$. (Note that $a > r$.) The line of slope m from O intersects S at points P and Q with x-coordinates $(a \pm \sqrt{r^2(m^2+1) - m^2a^2})/(m^2+1)$. The distance from these points to O is $(a \pm \sqrt{r^2(m^2+1) - m^2a^2})/(\sqrt{m^2+1})$, and the product of the distances $|OP| \cdot |OQ|$ is $a^2 - r^2$.

Then
$$\frac{|OP'|}{|OQ|} = \frac{|OP| \cdot |OP'|}{|OP| \cdot |OQ|} = \frac{r^2}{a^2 - r^2} = k$$
The number k depends on only the radius and center of the circle S, and so is a constant: it does not change as different points P on S are selected. Because AQ is parallel to BP', the triangles ΔOAQ and $\Delta OBP'$ are similar, and so we see
$$\frac{|OB|}{|OA|} = \frac{|OP'|}{|OQ|} = k$$
and consequently $|OB| = |OA|k$. That is, all points P on the circle S determine the same point B.

Also from $\Delta OAQ \sim \Delta OBP'$ we see
$$\frac{|BP'|}{|AQ|} = \frac{|OB|}{|OA|} = k$$
That is, $|BP'| = |AQ|k'$, another constant because $|AQ|$ is the radius of S. So we see that as P traces out S, P' traces out points a fixed distance from a fixed point B. The inverse of S is a circle.

Inverted circle formulas

Now we'll derive a formula for the center and radius of a circle inverted in another circle.

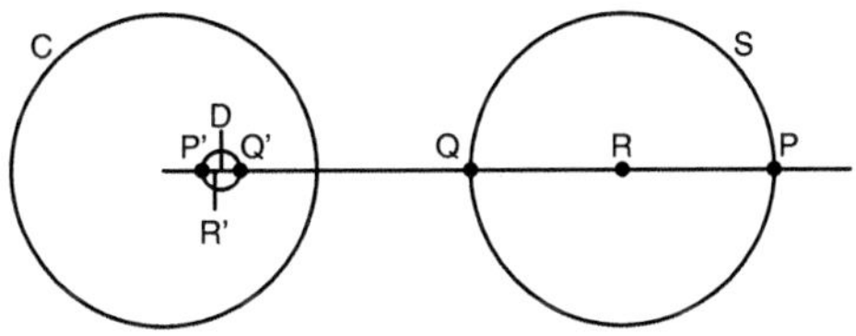

Suppose C is the circle with center (a, b) and radius r, and S is the circle with center $R = (c, d)$ and radius s. To invert S in C, first draw the ray from (a, b) through R. This ray intersects S in two points, Q and P. The inverses of these are P' and Q', endpoints of a diameter of the inverse S'. To find Q' and P', we must have expressions for Q and P. To find these, take the unit vector
$$\frac{1}{\sqrt{(a-c)^2 + (b-d)^2}}(a - c, b - d)$$
Then
$$P = (c, d) + \frac{s}{\sqrt{(a-c)^2 + (b-d)^2}}(a - c, b - d)$$
and
$$Q = (c, d) - \frac{s}{\sqrt{(a-c)^2 + (b-d)^2}}(a - c, b - d)$$
Apply the point inverse formula (3.6) to P and Q and we find
$$P' = (a, b) + \frac{r^2}{\left(\sqrt{(a-c)^2 + (b-d)^2} - s\right)\sqrt{(a-c)^2 + (b-d)^2}}(c - a, d - b)$$
and
$$Q' = (a, b) + \frac{r^2}{\left(\sqrt{(a-c)^2 + (b-d)^2} + s\right)\sqrt{(a-c)^2 + (b-d)^2}}(c - a, d - b)$$

The center of the inverted circle is the midpoint of P' and Q', that is

$$(a, b) + \frac{r^2}{(a-c)^2 + (b-d)^2 - s^2}(c-a, d-b)$$

Note that the center of S does *not* invert to the center of S'.

The diameter of the inverted circle is the distance between P' and Q', so the radius is

$$\frac{r^2 s}{(a-c)^2 + (b-d)^2 - s^2}$$

B.11.2 Limit set examples

Here we'll work through a few examples of limit sets.

Example B.11.1 *If X is a finite set of points, then Λ is empty.*

To see this, first note that because X is finite, there is a minimum distance, $\delta_0 > 0$ between the elements of X. If there were a limit point q of X, then take the distance $\delta < \delta_0/3$. Now any point $w \in X$ satisfying $0 < \text{dist}(q, w) < \delta$ lies in the disc D of radius δ and center q. Because the diameter of D is $2\delta < \delta_0$, no point of X, other than w, lies in D. Let $\delta_1 = \text{dist}(q, w)$. Then no point $v \in X$ can satisfy $0 < \text{dist}(q, v) < \delta_1/2$, because such $v \neq w$ and w is the only point of X lying in D.

Example B.11.2 *If X is the set of all points (x, y) with both x and y rational, then Λ is the whole plane.*

To see this, we must show that for any point (a, b) in the plane, and for any distance $\delta > 0$, there is a point $(w, v) \in X$ with $0 < \text{dist}((a, b), (w, v)) < \delta$. All points of a square S with one corner at (a, b) and diagonal of length δ lie within a distance δ of (a, b). Because the rational numbers are dense in the real numbers (every real number z can be approximated arbitrarily closely by a rational—think of increasingly long decimal expansions of z), on a horizontal side of the square S lies a point with x-coordinate w, a rational number. Similarly, on a vertical side of the square S lies a point with y-coordinate v, a rational number. Then the point $(w, v) \in X$ lies in S, and so $\text{dist}((a, b), (w, v)) < \delta$. Taking w and v near the middle of the sides of the square S, we guarantee $0 < \text{dist}((a, b), (w, v))$.

Example B.11.3 *If X is a Cantor set, $\Lambda = X$.*

This is a bit more work. To keep the notation manageable, we consider only the Cantor middle thirds set $\mathcal{C}$ as a subset of the line. We can construct $\mathcal{C}$ as the attractor of the IFS $T_1(x) = x/3$ and $T_2(x) = x/3 + 2/3$. Now follow the successive iterations of the deterministic IFS algorithm with initial shape C_0 the interval $[0, 1]$. The first two iterates are

$$\begin{aligned}
C_1 &= T_1(C_0) \cup T_2(C_0) \\
C_2 &= T_1(C_1) \cup T_2(C_1) = T_1(T_1(C_0)) \cup T_1(T_2(C_0)) \cup T_2(T_1(C_0)) \cup T_2(T_2(C_0))
\end{aligned}$$

The other iterates are similar, but form longer lists.

Certainly, $C_0 \supset C_1 \supset C_2 \supset \cdots$, and $\mathcal{C} = \cap_{i=0}^{\infty} C_i$. Note C_i consists of 2^i intervals, each of length $1/3^i$. The endpoints of every interval in every C_i belong to all C_i, and so belong to the intersection $\mathcal{C}$. We show that every point of $\mathcal{C}$ is a limit point of $\mathcal{C}$, and every limit point of $\mathcal{C}$ is a point of $\mathcal{C}$.

First, take any point $q \in \mathcal{C}$. Note $q \in C_i$ for all $i \geq 0$. For any distance $\delta > 0$ we must find a point $w \in \mathcal{C}$ with $0 < \text{dist}(q, w) < \delta$. We consider two cases: q is an endpoint of an interval in one of the C_i, and q is not an endpoint of an interval in any of the C_i.

If q is an endpoint of an interval in one of the C_i, then q is an endpoint of an interval in all C_j for $j \geq i$. Take j large enough that $1/3^j < \delta$ and $j \geq i$. Then q is an endpoint of some interval I in C_j. Let w be the other endpoint of I. Then $w \in \mathcal{C}$ and $0 < \text{dist}(q, w) < 1/3^j < \delta$. This shows q is a limit point of $\mathcal{C}$.

Second, suppose q is not an endpoint of an interval in any of the C_i. Given $\delta > 0$, take j large enough that $1/3^j < \delta$. Because q belongs to $\mathcal{C} = \cap_{i=0}^{\infty} C_i$, $q \in C_i$ for all i, so $q \in C_j$ and consequently q belongs to some interval I in C_j. Take w to be one of the endpoints of I. Then $w \in \mathcal{C}$ and $0 < \text{dist}(q, w) < 1/3^j < \delta$. This argument is valid for all $\delta > 0$, so q is a limit point of $\mathcal{C}$.

That is, we have shown every point of $\mathcal{C}$ is a limit point of $\mathcal{C}$.

Now suppose q is a limit point of $\mathcal{C}$, and for the purpose of contradiction, suppose $q \notin \mathcal{C}$. Then q belongs to $(-\infty, 0)$, to $(1, \infty)$, or to some open interval removed when passing from some C_{i-1} to C_i. That is, q belongs to some open interval (a, b) disjoint from $\mathcal{C}$. Let δ be the smaller distance $\text{dist}(a, q)$ and $\text{dist}(q, b)$. Then no point of $\mathcal{C}$ lies within δ of q, so q is not a limit point of $\mathcal{C}$.

B.12 Notes on fractal painting: dripping, Lab 4.5

In [128], Richard Taylor, Adam Micolich, and David Jonas applied the box-counting dimension (Lab 2.1) to quantify the complexity of Jackson Pollock's drip paintings. The various layers were digitally separated, box-counting was applied to each layer, then the layers were combined to find the box-counting dimension of the entire painting.

They found two scaling ranges, 1mm to 5cm and 5cm to 2.5m, both cover a range of lengths that varies by a factor of 50, less than the usual 2 orders of magnitude preferred to signal fractality, but pretty close. Films of Pollock at work suggest that the smaller-size range is due to the behavior of paint drips, while the larger-size range is a signature of his movement around the canvas.

For a large collection of Pollock's drip paintings, they compared the box-counting dimension on the smaller scale, which they called the *drip dimension.* In the preliminary period (1943–45) they measured the drip dimension as slightly larger than 1. As Pollock refined his technique, the drip dimension increased throughout the transitional period (1945–47), to achieve maximum values of about 1.7 for paintings in the classical period (1948–52).

In his 2002 *Scientific American* article [129], Taylor recounts his experiences in 1994 at the Manchester School of Art, where an experiment suggested to him that Pollock "must have adopted nature's rhythms when he painted." Here

Taylor gives some more details of the calculation and discusses connections between fractal dimension and emotional response to a painting. This last topic is explored in more detail in [137] and in [130], which includes a striking series of magnifications in Fig. 5 that illustrates scale ambiguity of Pollock's paintings.

Katherine Jones-Smith and Harsh Mathur [132] criticized the results of Taylor's group. For example, they drew a collection of free-hand five-pointed stars and established power law scalings over two ranges, with a total box size range covering three orders of magnitude.

Taylor's group responded [133] with other free-hand five pointed star sketches for which they found no linear behavior in the log-log plot, and raised other criticisms of Jones-Smith and Mathur's methods.

In [134] Taylor and eight coauthors explored the possibility of using fractal dimension computations as a component of the authentication process of drip paintings that might be previously unknown Pollocks. As part of their argument to support this approach, 37 University of Oregon undergraduate students attempted to replicate drip paintings in Pollock's style. None of these examples exhibited fractal characteristics similar to Pollock's work.

J. R. Minkel's "Pollock or Not? Can Fractals Spot a Fake Masterpiece?" [135] reports on the use of Taylor's technique to authenticate Pollocks. Next, on an analysis by Jones-Smith, Mathur, and Lawrence Krauss [136] of three Pollocks not studied by Taylor's group. They did not find similar fractal characteristics. Finally, Minkel points out that Taylor has criticisms for these criticisms. The situation remains fluid.

One more example, related but not identical. Google "Norman Rockwell" and "The Connoisseur". You'll find what looks a bit like a Pollock in a museum or gallery, and a well-dressed older gentleman inspecting the painting. Rockwell had seen a film of Pollock making drip paintings, and he did his drip painting mimicking Pollock's technique. Then he added the art lover image after. Rockwell took two sections of his drip painting, signed them with pseudonyms, and entered them in exhibitions, one at the Cooperstown Art Association, the other at the Berkshire Museum. He won first place at Cooperstown and honorable mention at the Berkshire. This was long before Taylor's fractal analysis.

By the way, if you think Norman Rockwell produced only illustrations of mid-century American nostalgia, Google his "Rosie the Riveter," "Four Freedoms," "Murder in Mississippi," and "The Problem We All Live With." You'll find that the artist responsible for almost five decades of *Saturday Evening Post* covers also explored considerably deeper emotional and social themes.

B.13 Notes on power law measurements, Lab 4.9

We'll stick to scalings of collections of circles, either cooking marks on crepes or craters on the moon. Suppose only a finite collection or radii, $r_1 >$

$r_2 > \cdots > r_n$ are observed. Let

$$N(r) = \text{the number of circles with radius } r$$

and

$$N_+(r) = \text{the number of circles with radius} \geq r$$

If $N(r) = k(1/r)^d$, must $N_+(r) = k_+(1/r)^{d_+}$, and if it does, is $d_+ = d$?

Near the end of the calculation we'll need to make an assumption about how quickly the r_i decrease. For now, let's see how far we can get before we need this.

First, for $1 \leq i \leq n$, we have $N(r_i) = k(1/r_i)^d$ and

$$\begin{aligned} N_+(r_i) &= k\left(\left(\frac{1}{r_1}\right)^d + \cdots + \left(\frac{1}{r_{i-1}}\right)^d\right) + k\left(\frac{1}{r_i}\right)^d \\ &< k(i-1) \max_{j=1,\ldots,i-1}\left\{\left(\frac{1}{r_j}\right)^d\right\} + k\left(\frac{1}{r_i}\right)^d \end{aligned}$$

For x and d positive, x^d is an increasing function of x $((x^d)' = dx^{d-1} > 0)$. Then $r_1 > \cdots > r_{i-1}$ implies $1/r_1 < \cdots < 1/r_{i-1}$ and so $(1/r_1)^d < \cdots < (1/r_{i-1})^d$. That is, for $j = 1, \ldots, i-1$ we have $\max\{(1/r_j)^d\} = (1/r_{i-1})^d$. This gives us a useful upper bound on $N_+(r)$,

$$\begin{aligned} N_+(r) &< k(i-1)\left(\frac{1}{r_{i-1}}\right)^d + k\left(\frac{1}{r_i}\right)^d = k\left(\frac{1}{r_i}\right)^d\left((i-1)\left(\frac{r_i}{r_{i-1}}\right)^d + 1\right) \\ &< N(r_i)((i-1)+1) = N(r_i)i \end{aligned}$$

where the last inequality comes from the observation that $r_i/r_{i-1} < 1$ so $(r_i/r_{i-1})^d < 1^d = 1$.

Combining what we have so far, $N(r_i) < N_+(r_i) < N(r_i)i$. This gives

$$\log(N(r_i)) < \log(N_+(r_i)) < \log(N(r_i)i) = \log(N(r_i)) + \log(i) \tag{B.25}$$

So long as all $r_i < 1$, by taking the units of length to be $2r_1$ for example, $\log(1/r_i) > 0$ and Eq. (B.25) gives

$$\frac{\log(N(r_i))}{\log(1/r_i)} < \frac{\log(N_+(r_i))}{\log(1/r_i)} < \frac{\log(N(r_i))}{\log(1/r_i)} + \frac{\log(i)}{\log(1/r_i)} \tag{B.26}$$

Finally, we're at the place where we need to say something about how quickly the radii shrink. We need to choose the radii so that $\log(i)/\log(1/r_i)$ gets very small for large i. For example, $r_i = 1/2^i$ works fine because as $i \to \infty$ l'Hôpital's rule gives $\log(i)/\log(2^i) \to 0$. Although note that this convergence is slow, so for even reasonable length data sets the power law exponent computed with $N(r_i)$ can be a bit less than that computed with $N_+(r_i)$.

Then by Eq. (B.26), in the small r_i limit we have

$$\frac{\log(N(r_i))}{\log(1/r_i)} \to d \quad \text{implies} \quad \frac{\log(N_+(r_i))}{\log(1/r_i)} \to d$$

This last means that $N_+(r_i) \approx k_+(1/r_i)^d$.

The advantage of computing the power law exponent with $N_+(r_i)$ is that we needn't worry about how to bin the measured radii. Just pick a radius r_1 and count the number of circles with radii $\geq r_1$. Repeat this for a sequence of of radii $r_2 > r_3 > \cdots > r_n$ that decrease at least geometrically, $r_i \leq \epsilon r_{i-1}$ for some constant $\epsilon < 1$. If the points of the log-log plot are close to a straight line, that establishes a power law relationship over at least the range of sizes where the points lie near the line. The slope of the line is the power law exponent.

B.14 Notes on magnetic pendulum differential equations, Lab 5.1

Here we'll derive the differential equations for a simplified model of a pendulum executing low-amplitude swings in a gravitational field and above magnets that attract the pendulum. As long as the amplitude of the pendulum swing is low, the part of the sphere through which it swings is close to a portion of the xy-plane. So we'll suppose the magnets lie a fixed distance d below the xy-plane. The pendulum is suspended above the point $(0,0)$ of the xy-plane, so in this simplification the effect of gravity is to force the pendulum back toward the origin with a force proportional to the distance from the pendulum location (x,y) to the origin. In order to direct the force toward the origin, write the vector $\langle x,y\rangle$ as

$$\langle x,y\rangle = \left\langle \frac{x}{\sqrt{x^2+y^2}}, \frac{y}{\sqrt{x^2+y^2}} \right\rangle \sqrt{x^2+y^2} = \vec{u}\sqrt{x^2+y^2}$$

where $\vec{u}$ is the unit vector in the direction of $\langle x,y\rangle$. Suppose c is the proportionality constant of this gravitational force. To direct the gravitational force toward the origin, we'll use

$$\text{gravitational force} = -c\vec{u}\sqrt{x^2+y^2} = -c\langle x,y\rangle$$

Next, friction slows the motion. We'll take the frictional force to be proportional to the pendulum speed and in the direction opposite the motion. That is,

$$\text{frictional force} = -r\langle x',y'\rangle$$

(If this isn't clear, modify the gravitational force argument to use the unit vector in the direction of the velocity $\langle x,y\rangle$.)

Finally, the magnetic force exerted by the magnet at the point $(p_i, q_i, -d)$ on the pendulum at $(x,y,0)$ is in the direction from $\langle x,y\rangle$ to $\langle p_i,q_i\rangle$ (recall that motion is restricted to the xy-plane), and proportional to

$$\frac{1}{(\sqrt{(p_i-x)^2+(q_i-y)^2+d^2}\ )^2}$$

the reciprocal of the square of the distance between (p_i, q_i, d) and $(x,y,0)$. The unit vector from $\langle x,y\rangle$ to $\langle p_i,q_i\rangle$ is

$$\frac{\langle x-p_i, y-q_i\rangle}{\sqrt{(p_i-x)^2+(q_i-y)^2}}$$

We'll use

$$\frac{\langle x - p_i, y - q_i \rangle}{\sqrt{(p_i - x)^2 + (q_i - y)^2 + d^2}}$$

because the magnets are located a distance d below the xy-plane. Write the magnetic force proportionality constant as f. Then the total magnetic force of the three magnets on the pendulum at (x, y) is

$$f \sum_{i=1}^{3} \frac{\langle p_i - x, q_i - y \rangle}{((p_i - x)^2 + (q_i - y)^2 + d^2)^{3/2}}$$

where we've reversed the sign of the vector because the magnetic force acts to decrease the distance between the pendulum and the magnet.

When we apply these forces to Newton's second law (force = mass × acceleration) we obtain

$$\langle x'', y'' \rangle = f \sum_{i=1}^{3} \frac{\langle p_i - x, q_i - y \rangle}{((p_i - x)^2 + (q_i - y)^2 + d^2)^{3/2}} - c\langle x, y \rangle - r\langle x', y' \rangle$$

We'll make one modification to prepare for the computer code: we'll turn this pair of equations that involve second derivatives (second-order equations) into four first-order equations. Define new variables u and v by $u = x'$ and $v = y'$. Then the four equations for the magnetic pendulum are

$$\begin{aligned}
x' &= u \\
y' &= v \\
u' &= f \sum_{i=1}^{3} \frac{p_i - x}{((p_i - x)^2 + (q_i - y)^2 + d^2)^{3/2}} - cx - ru \\
v' &= f \sum_{i=1}^{3} \frac{q_i - y}{((p_i - x)^2 + (q_i - y)^2 + d^2)^{3/2}} - cy - rv
\end{aligned}$$

B.15 Notes on molarity calculations, Lab 5.4

The concentration of a solution is measured by *molarity*, the number of moles of the solute (the substance to be dissolved in the solution) per liter of solution. A *mole* of solute is Avogadro's number, 6.022×10^{23}, of solute molecules. Certainly we won't try to count out an Avogadro's number of molecules. For one thing, the number is way too large, and for another, a molecule is way too small. (This was a joke in freshman chemistry at RPI in the fall semester of 1969. Typically, jokes in chemistry class are not nearly as funny as jokes in math class. Chemistry teachers may have a different opinion, but they're wrong.) Rather, we compute the weight of a mole of solute, dissolve that in a bit less than a liter of water, and after the solute dissolves, add a bit of water to bring the solution volume up to a liter. We start with less than a liter of water and add some after the solute is dissolved because adding the solute might increase the volume of the solution.

The *molarity* is the number of moles of solute dissolved in a liter of solution.

30
Zn
Zinc
65.39

The weight of one mole of the solute is the sum of the atomic weights of the solute atoms. The atomic weight (nowadays it's often called atomic mass) of an atom is the number at the bottom of the box of that element in a periodic table. If you don't have a periodic table, Google can find copies for you. Roughly, it's the number of protons and neutrons in an atom. More precisely, the atomic weight is the mass of the nucleus compared with that of carbon 12, defined to be 12.0 atomic mass units. The number at the bottom of the periodic table box is calculated from the atomic weight of each isotope (same number of protons, different number of neutrons) combined proportionally to their relative abundances. The number at the top of the box is the atomic number, the number of protons.

To make a 1 molar solution of zinc sulfate, $ZnSO_4$, first note that the atomic weights of zinc, sulfur, and oxygen are 65.39, 32.066, and 15.999. Then the atomic weight of zinc sulfate is

$$65.39 + 32.066 + 4 \cdot 15.999 = 161.452$$

So, dissolve 161.452 g of zinc sulfate in a bit less than 1 L of water, then add water to bring the volume of the solution up to 1 L. This produces a 1 molar solution of zinc sulfate.

This scales linearly. If we don't want a L of 1 molar zinc sulfate solution, dissolve 80.726 g in 500 ml of water, or 40.363 g in 250 ml of water. To make a 0.5 molar solution in 250 ml, dissolve 20.182 g in 250 ml of water.

This is all we need to do if we have pure zinc sulfate. We used zinc sulfate monohydrate, $ZnSO_4 \cdot H_2O$, so must adjust the molarity calculation to account for the water already in the solute. Here's the adjustment.

$$\begin{aligned}\frac{ZnSO_4}{\text{liter}} &= \frac{ZnSO_4 \cdot H_2O}{\text{liter}} \times \frac{ZnSO_4}{ZnSO_4 \cdot H_2O}\\ &= \frac{ZnSO_4 \cdot H_2O}{\text{liter}} \times \frac{161.452}{161.452 + 2 \cdot 1.008 + 15.999}\\ &= \frac{ZnSO_4 \cdot H_2O}{\text{liter}} \times 0.900 = \frac{ZnSO_4 \cdot H_2O}{1.11 \text{ liter}}\end{aligned}$$

B.16 Notes on fractal resistor networks Lab 5.7

For the star network on the right of Fig. B.7 to be equivalent to the triangle network on the left, we must have

$$R_a + R_b = R_{AB} \tag{B.27}$$

by the series rule (5.3), because R_c contributes nothing to the resistance between A and B. Similarly,

$$R_a + R_c = R_{AC} \quad \text{and} \quad R_b + R_c = R_{BC} \tag{B.28}$$

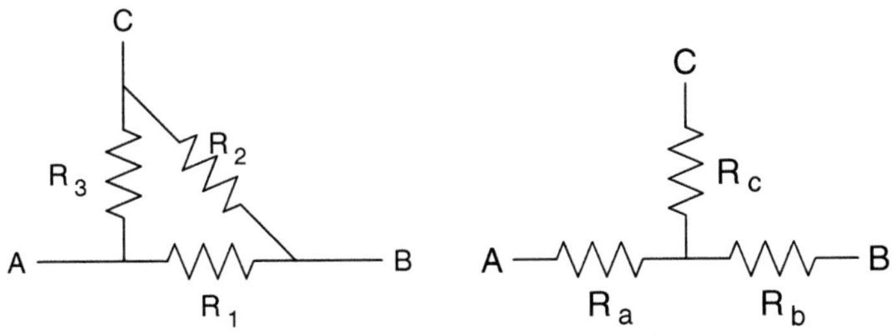

Figure B.7: A triangular network and equivalent star network.

Recall the expressions for R_{AB}, R_{AC}, and R_{BC} from Eqs. (5.5) and (5.6),

$$R_{AB} = \frac{R_1(R_2+R_3)}{R_1+R_2+R_3}, \quad R_{AC} = \frac{R_3(R_1+R_2)}{R_1+R_2+R_3},$$
$$R_{BC} = \frac{R_2(R_1+R_3)}{R_1+R_2+R_3} \tag{B.29}$$

We can solve equations (B.27) and (B.28) for R_a, R_b, and R_c in terms of R_1, R_2, and R_3. For example, adding Eqs. (B.27) and (B.28), and using Eqs. (5.5) and (5.6), we find

$$R_a + R_b + R_c = \frac{R_1R_2 + R_1R_3 + R_2R_3}{R_1+R_2+R_3} \tag{B.30}$$

Then by Eqs. (5.6), (B.28) and (B.30)

$$R_a = (R_a + R_b + R_c) - (R_b + R_c) = \frac{R_1R_3}{R_1+R_2+R_3} \tag{B.31}$$

and similar calculations give

$$R_b = \frac{R_1R_2}{R_1+R_2+R_3} \quad \text{and} \quad R_c = \frac{R_2R_3}{R_1+R_2+R_3} \tag{B.32}$$

and we have calculated the resistances of Eq. (5.7).

B.17 Notes on synchronization in fractal networks of oscillators, Lab 5.9

Here we'll derive the formula for the resonant frequency of an LC circuit, a circuit that consists of an inductor and a capacitor. The voltage across a capacitor is $V = q/C$, where q is the charge and C is the capacitance, measured in farads. The voltage across an inductor is given by $V = LdI/dt$, where I is the current and L is the inductance, measured in henrys. If the capacitor and inductor are arranged in a series with a constant applied voltage V, then Kirchhoff's voltage law gives

$$V = L\frac{dI}{dt} + \frac{q}{C}$$

Differentiating, and recalling that V is constant so $dV/dt = 0$, we find

$$0 = L\frac{d^2 I}{dt^2} + \frac{1}{C}\frac{dq}{dt} = L\frac{d^2 I}{dt^2} + \frac{1}{C}I$$

where the last equality follows because current is the rate of change of charge, that is, $I = dq/dt$. This gives a second-order differential equation I,

$$\frac{d^2 I}{dt^2} = -\frac{1}{LC}I$$

What functions do we know whose second derivative is proportional to the negative of the function? Sine and cosine, of course. We'll take the solution

$$I = I_0 \sin(\omega t + \phi)$$

where ϕ is the phase, a measure of the departure from synchronization of two oscillators, and $\omega = 1/\sqrt{LC}$ is the angular resonant frequency of the oscillator. In units of Hz the resonant frequency is $f = 1/(2\pi\sqrt{LC})$. For example, an LC circuit that consists of a 1 μF (microfarad) capacitor and a $2mH$ (millihenry) inductor has a resonant frequency of $3.559kHz$.

References

Prologue

[1] B. Mandelbrot, *The Fractal Geometry of Nature*, Freeman, San Francisco, 1982.

Note that some sources, for example [1], are cited in several labs. Bibliographical details are given in the first lab that uses that source.

[2] B. Mandelbrot, "How long is the coast of Britain? Statistical self-similarity and fractional dimension," *Science* **156** (1967), 636–638.

[3] B. Mandelbrot, *The Fractalist: Memoirs of a Scientific Maverick*, Pantheon, New York, 2012.

[4] M. Kleiber, "Body size and metabolism," *Hilgardia* **6** (1932), 315–353,

[5] M. Rubner, "Über den Einfluss der Körpergrösse auf Stof- und Kraftwechsel," *Zeitschrift für Biologie* **19** (1993), 535–562.

[6] S. Lovejoy, B. Mandelbrot, "Fractal properties of rain, and a fractal model," *Tellus* **A 37** (1985), 209–232.

[7] B. Mandelbrot, J. Wallis, "Noah, Joseph, and operational hydrology," *Water Resources Research* **4** (1968), 909–918.

[8] B. Mandelbrot, "The Pareto-Lévy law and the distribution of incomes," *Int. Economic Rev.* **1** (1960), 79–106.

[9] B. Mandelbrot, "The variation of certain speculative prices," *J. Business Univ. Chicago* **36** (1963), 294–419.

[10] B. Mandelbrot, "Intermittent turbulence in self-similar cascades; divergence of high moments and dimension of the carrier," *J. Fluid Mech.* **62** (1974), 331–358.

[11] B. Mandelbrot, "A population birth and mutation process, I: Explicit distributions for the number of mutants in an old culture of bacteria," *J. Appl. Propb.* **11** (1974), 437–444.

[12] S. Campbell, R. Campbell, *Mysterious Patterns: Finding Fractals in Nature*, Boyds Mills, Honesdale, PA, 2014.

[13] K. Falconer, *Fractals: A Very Short Introduction*, Oxford, Univ. Pr., Oxford, 2013.

[14] D. Peak, M. Frame, *Chaos Under Control: The Art and Science of Complexity*, Freeman, New York, 1994.

[15] M. Frame, B. Mandelbrot, *Fractals, Graphics, and Mathematics Education*, Math. Assoc. Amer., Washington D. C., 2002.

[16] K. Falconer, *Fractal Geometry: Mathematical Foundations and Applications*, 3rd ed., Wiley, Chichester, 2014.

[17] M. Frame, A. Urry, *Fractal Worlds: Grown, Built, and Imagined*, Yale Univ. Pr., New Haven, 2016.

[18] H-O. Peitgen, H. Jürgens, D. Saupe, *Chaos and Fractals: New Frontiers in Science*, 2nd ed., Springer, New York, 2004.

[19] Y. Pesin, V. Climenhaga, *Lectures on Fractal Geometry and Dynamical Systems*, Amer. Math. Soc., Providence, 2009.

[20] K. Falconer, *Techniques in Fractal Geometry*, Wiley, Chichester, 1997.

[21] R. Devaney, *Chaos, Fractals, and Dynamics: Computer Experiments in Mathematics*, Addison-Wesley, Menlo Park, 1990.

[22] H.-O. Peitgen, H. Jürgens, D. Saupe, E. Maletsky, T. Perciante, and L. Yunker, *Fractals for the Classroom: Strategic Activities*, Springer, New York, 1991 (volume 1), 1992 (volume 2), and 1999 (volume 3).

[23] J. Choate, R. Devaney, and A. Foster, *Iteration: A Tool Kit of Dynamics Activities*, Key Curriculum Press, Emeryville, 1999.

[24] J. Choate, R. Devaney, and A. Foster, *Fractals: A Tool Kit of Dynamics Activities*, Key Curriculum Press, Emeryville, 1999.

[25] R. Devaney and J. Choate, *Chaos: A Tool Kit of Dynamics Activities*, Key Curriculum Press, Emeryville, 2000.

1 IFS labs

[26] J. Hutchinson, "Fractals and self-similarity," *Indiana Univ. J. Math.* **30** (1981), 713–747.

[27] M. Barnsley and S. Demko, "Iterated function systems and the global construction od fractals," *Proc. Roy. Soc. London A* **399** (1985), 243–275.

[28] M. Barnsley and A. Sloan, "A better way to compress images," *Byte* **13** (1988), 215–223.

[29] M. Barnsley, "Lecture notes on iterated function systems," pgs. 127–144 of [30].

[30] R. Devaney and L. Keen, eds., *Chaos and Fractals: The Mathematics behind the Computer Graphics*, Amer. Math. Soc., Providence, 1989.

[31] M. Barnsley, *Fractals Everywhere* 2nd ed., Academic Press, Boston, 1993.

[32] M. Barnsley, *Superfractals: Patterns of Nature*, Cambridge Univ. Pr., Cambridge, 2006.

[33] T. Taylor, "Connectivity properties of Sierpiński relatives," *Fractals* **19** (2011), 481–506.

[34] T. Taylor, C. Hudson, and A. Anderson, "Examples of using binary Cantor sets to study the connectivity of Sierpiński relatives," *Fractals* **20** (2012), 61–75.

1.1 Finding IFS for fractal images

[35] M. Barnsley, V. Ervin, D. Hardin, and J. Lancaster, "Solution of an inverse problem for fractals and other sets," *Proc. Nat. Acad. Sci. USA* **83** (1986), 1975–1977.

[36] W. Rudin, *Principles of Mathematical Analysis*, 2nd ed., McGraw-Hill, New York, 1964.

1.5 Fractal wallpaper

[37] V. Fegers and M. B. Johnson, "Fractals—energizing the mathematics classroom," pgs. 69–103 of [15].

1.9 IFS with memory

[38] M. Frame and J. Lanski, "When is a recurrent IFS attractor a standard IFS attractor?" *Fractals* **7** (1999), 257–266.

[39] J. Lyman and T. Womack, "Linear Markov iterated function systems," *Comp. Graph.* **14** (1990), 343–353.

1.10 IFS with more memory

[40] R. Bedient, M. Frame, K. Gross, J. Lanski, and B. Sullivan, "Higher block IFS 1: Memory reduction and dimension computations," *Fractals*, **18** (2010), 145–155.

[41] R. Bedient, M. Frame, K. Gross, J. Lanski, and B. Sullivan, "Higher block IFS 2: Relations between IFS with different levels of memory," *Fractals*, **18** (2010), 399–408.

[42] D. Lind and B. Marcus, *An Introduction to Symbolic Dynamics and Coding*, Cambridge U. Pr., Cambridge, 1995.

1.11 Data analysis by driven IFS

[43] F. Medivil, "Fractals, graphs, and fields," *Amer. Math. Monthly* **110** (2003), 503–515.

[44] I. Stewart, "Order within the chaos game," *Dynamics Newsletter* **3** (1989), 4–9.

[45] H. Jeffrey, "Chaos game representation of gene structure," *Nucl. Acid Res.* **18** (1990), 2163–2170.

[46] H. Jeffrey, "Chaos game visualization of sequences," *Comp. & Graphics* **16** (1992), 25–33.

2.1 Dimension by box-counting

[47] K. Falconer, *The Geometry of Fractal sets*, Cambridge Univ. Pr., Cambridge, 1985.

[48] J. Stewart, *Multivariable Calculus*, 7th ed., Brooks/Cole, Belmont, 2012.

2.2 Paper ball and bean bag dimensions

[49] B. Mandelbrot and M. Frame, "A primer of negative test dimensions and degrees of emptiness for latent sets," *Fractals* **17** (2009), 1–14.

2.3 Calculating similarity dimension

[50] R. Mauldin and S. Williams, "Hausdorff dimension in graph-directed constructions," *Trans. Amer. Math. Soc.* **309** (1988), 811–829.

[51] T. Bedford and M. Urbanski, "On box and Hausdorff dimension of self-affine sets," *Ergod. Th. Dyn. Syst.* **10** (1990), 627–644.

[52] G. Edgar, "Fractal dimension of self-affine sets: some examples," *Measure Theory Oberwolfach 1990* **28** (1992), 341–358.

[53] K. Falconer, "The Hausdorff dimension of self-affine fractals," *Proc. Camb. Phil. Soc.* **111** (1988), 330–350.

[54] D. Gatzouras and S. Lalley, "Statistically self-affine sets: Hausdorff and box dimensions," *J. Theoret. Prob.* **7** (1994), 437–468.

[55] I. Hueter and Y. Peres, "Self-affine carpets on the square lattice," *Comb. Prob. Comput.* **6** (1997), 197–204.

[56] C. McMullen, "The Hausdorff dimension of general Sierpinski carpets," *Nagoya Math. J.*, **96** (1984), 1–9.

2.5 Koch tetrahedron

[57] H. von Koch, "Sur une courbe continue sans tangente obtenue par une construction géométrique élémentaire," *Arkiv för Matematik, Astronomi och Fysik* **1** (1904), 681–702.

[58] G. Edgar, *Classics on Fractals*, Addison-Wesley, Reading, 1993.

[59] S. Wagon, *Mathematica in Action*, Freeman, New York, 1991.

2.6 Sierpinski hypertetrahedron

[60] T. Banchoff, *Beyond the Third Dimension: Geometry, Computer Graphics, and Higher Dimensions*, Freeman, New York, 1990.

2.7 Basic multifractals

[61] B. Mandelbrot, D. Passoja, and A. Paullay, "The fractal character of fracture surfaces of metal," *Nature* **308** (1984), 721–722.

[62] P. Ivanov, et al., "Multifractality in human heartbeat dynamics," *Nature*, **399** (1999), 461–465.

[63] S. Lovejoy and D. Schertzer, *The Weather and Climate: Emergent Laws and Multifractal Cascades*, Cambridge Univ. Pr. Cambridge, 2013.

[64] J. McCauley, "Introduction to multifractals in dynamical systems theory and fully developed turbulence," *Phys. Reports* **185** (1990), 225–266.

[65] T. Hirabayashi, K. Ito, T. Yoshi, "Multifractal analysis of earthquakes," *Pure Appl. Geophysics* **138** (1992), 591–610.

[66] B. Mandelbrot and R. Hudson, *The (Mis)Behavior of Markets: A Fractal View of Risk, Ruin, and Reward*, Basic Books, New York, 2004.

[67] M. Keßeböhmer and B. Stratmann, "A multifractal formalism for growth rates and applications to geometrically finite Kleinian groups," *Ergodic Th. Dynamical Syst.* **24** (2004), 141–170.

[68] A. Feldman, A. Gilbert, and W. Willinger, "Data networks as cascades: investigating the multifractal nature of Internet WAN traffic," *Computer Communications Review* **28** (1998), 42–55.

[69] S. Drożdż, et al., "Quantifying origin and character of long-range correlations in narrative text," *Information Sciences* **331** (2016), 32–44.

[70] R. Lopes and N. Betrouni, "Fractal and multifractal analysis: a review," *Medical Image Analysis* **13** (2009), 634–649.

[71] Y. Pesin and H. Weiss, "The multifractal analysis of Birkhoff averages of large deviations," *Global Analysis of Dyn. Syst.* (2001), 419–431.

[72] Z. Yu, V. Anh, R. Eastes, D. Wang, "Multifractal analysis of solar flare indices and their horizontal visibility spectrum," *Nonlin. Processes Geophys.* **19** (2012) 657–665.

[73] M. Hassan, "Scale-free network topology and multifractality in a weighted planar stochastic lattice," *New J. Phys.* **12** (2010), 093045.

[74] C. Meneveau and K. Sreenivasan, "The multifractal nature of turbulent energy dissipation," *J. Fluid Mech.* **224** (1991), 429–484.

3.1 Visualzing iteration patterns

[75] R. Devaney, *An Introduction to Chaotic Dynamical Systems* 2nd ed., Addison-Wesley, Redwood City, 1989.

[76] R. Bedient and M. Frame, "Carrying surfaces for return maps of averaged logistic maps," *Computers & Graphics* **31** (2007), 887–895.

3.2 Synchronized chaos

[77] S. Strogatz, *Sync: How Order Emerges from Chaos in the Universe, Nature, and Daily Life*, Hyperion, New York, 2003.

[78] S. Strogatz, *Sync: The Emerging Science of Spontaneous Order*, Hachette, New York, 2015.

[79] J. Gleick, *Chaos: Making a New Science*, Viking, New York, 1987.

3.3 Domains of compositions

[80] N. Neger and M. Frame, "Clarifying compositions with cobwebs," *College Math. J.* **34** (2003), 196–204.

3.4 Fractals and Pascal's triangle

[81] M. Frame and N. Neger, "Dimensions and the probability of finding odd numbers in Pascal's triangle and its relatives," *Computers & Graphics*, **34** (2010), 158–166.

[82] M. Bardzell and K. Shannon, "The PascGalois project: visualizing abstract algebra," *Focus* **22** (2002), 4–5.

[83] M. Bardzell, K. Shannon, D. Spickler, faculty.salisbury.edu/~despickler/pascgalois/

[84] T. Osler, "Variations on a theme from Pascal's triangle," *College Math. J.* **32** (2003), 226–233.

3.6 Mandelbrot sets and Julia sets

[85] J. Guckenheimer, R. McGehee, "A proof of the Mandelbrot n^2 conjecture," Institute Mittag-Leffler, Report 15, 1984.

[86] B. Mandelbrot, "On the dynamics of iterated maps. iii: The individual molecules of the M-set self-similarity properties, the empirical n^2 rule, and the n^2 conjecture," pgs. 213–224 of [87].

[87] P. Fischer, W. Smith, eds., *Chaos, Fractals, and Dynamics*, Dekker, New York, 1985.

[88] B. Mandelbrot, *Fractals and Chaos: the Mandelbrot Set and Beyond*, Springer, New York, 2004.

[89] P. Lavaurs, "Une description combinatoire de l'involution définie par M sur les rationnels à dénominateur impair," *C. R. Acad. Sci. Paris* **303** (1986), 143–146.

[90] L. Tan, "Similarity between the Mandelbrot set and Julia sets," *Commun. Math. Phys.* **134** (1990), 587–617.

[91] B. Mandelbrot, "On the dynamics of iterated maps. V: conjecture that the boundary of the M-set has a fractal dimension equal to 2," pgs. 235–238 of [87].

[92] J. Milnor, "Self-similarity and the hairiness of the Mandelbrot set," pgs. 211–257 of [93].

[93] M. Tangora, ed., *Computers in Geometry and Topology*, Dekker, New York, 1989.

[94] M. Shishikura, "The Hausdorff dimension of the boundary of the Mandelbrot set and Julia sets," *Ann. of Math.* **147** (1998), 225–267.

[95] A. Douady and J. Hubbard, "Iteration des pôlynomes quadratiques complexes," *C. R. Acad. Sci. Paris* **294** (1982), 123–126.

[96] J. Curry, L. Garnett, and D. Sullivan, "On the iteration of a rational function: computer experiments with Newton's method," *Commun. Math. Phys.* **91** (19830, 267–277.

[97] A. Douady and J. Hubbard, "On the dynamics of polynomial-like maps," *Ann. Scient. Ec. Norm. Sup.* **18** (1985), 287–343.

[98] C. McMullen, "The Mandelbrot set is universal," pgs. 1–17 of [99].

[99] L. Tan, *The Mandelbrot Set, Theme and Variations*, Cambr. Univ. Pr. Cambridge, 2000.

[100] P. Lavaurs, "Systèmes dynamiques holomorphes: Explosion de points périodiques," thesis, Univ. Paris-Sud, Orsay, 1989.

[101] J. Milnor, "Remarks on iterated cubic maps," *Experimental Math.* **1** (1992), 5–24.

3.7 Circle inversion fractals

[102] B. Mandelbrot, "Self-inverse fractals osculated by sigma-discs and the limit sets of inversion groups," *Math. Intelligencer* **5** (1983), 9–17.

[103] D. Mumford, C. Series, and D. Wright, *Indra's Pearls: The Vision of Felix Klein*, Cambr. Univ. Pr. Cambridge, 2002.

3.8 Fractal tiles

[104] D. Schattschneider, "Will it tile? Try the Conway criterion!" *Math. Magazine* **53** (1980), 224–233.

[105] C. Bandt, "Self-similar sets in $\mathbb{R}^n$" *Proc. Amer. Math. Soc.* **112** (1991), 549–562.

[106] R. Darst, J. Palagallo, and T. Price,"Fractal tilings in the plane," **71** (1998), 12–23.

[107] S. Hagey and J. Papagallo, "Fractal tilings derived form complex bases," *Mathematical Gazette* **85** (2001), 194–201.

[108] B. Grünbaum and G. Shepard, *Tilings and Patterns*, Freeman, New York, 1987.

[109] G. Gelbrich, "Crystallographic reptiles," *Geom. Dedicata* **51** (1994), 235–256.

[110] G. Gelbrich and K. Giesche, "Fractal Escher salamanders and other animals," *Math. Intelligencer* **20** (1998), 31–35.

[111] R. Fathauer, "Fractal tilings based on kite- and dart-shaped prototiles," *Computers & Graphics* **25** (2001), 323–331.

[112] R. Fathauer, "Fractal tilings based on v-shaped prototiles," *Computers & Graphics* **26** (2002), 635–643.

4.1 Fractal painting: decalcomania 1

[113] H. Janson, *History of Art*, 4th ed, Abrams, New York, 1991.

[114] M. Feigenbaum, "Quantitative universality for a class of nonlinear transformations," *J. Stat. Phys.* **19** (1978), 25–52.

4.2 Fractal painting: decalcomania 2

[115] atomicshrimp.com/post/2014/08/16/Cohesive-Dendritic-Painting

4.3 Fractal painting: bleeds

[116] https://www.youtube.com/watch?v=hZy4kGqoJq8

[117] https://www.youtube.com/watch?v=u2SgW9n6fV4

[118] B. Mandelbrot and M. Frame, "The canopy and shortest path in a self-contacting fractal tree," *Math. Intell.*, **21** (1999), 18–27.

[119] E. Ben-Jacob, O. Shochet, I. Cohen, A. Tenenbaum, A. Czorók, and T. Vicsek, "Cooperative strategies in Formation of complex bacterial patterns," *Fractals* **3** (1995), 849–868.

4.4 Fractal painting: mixing

[120] J. Ottino, *The Kinematics of Mixing: Stretching, Chaos, and Transport*, Cambridge Univ. Pr., Cambridge, 1989.

4.5 Fractal painting: dripping

[121] L. Emmerling, *Pollock (Taschen Basic Art)*, Taschen, Cologne, 2003.

[122] I. Engelmann, *Jackson Pollock and Lee Krasner*, Prestel, Munich, 2007.

[123] B. Friedman, *Jackson Pollock: Energy Made Visible*, Da Capo Press, New York, 1972, 1995.

[124] R. Goodnough, "Pollock paints a picture," *Art News* **50** (May 1951), 38–41, 60–61.

[125] J. Pollock, *Autumn Rhyhm (Number 30)*, 1950.

[126] R. Taylor, "Francis O'Connor and Jackson Pollock's fractals," https://cpb-us-e1.wpmucdn.com/blogs.uoregon.edu/dist/e/12535/files/2016/02/OConnorandPollockFractals.pdf

[127] J. Potter, *To a Violent Grave: An Oral Biography of Jackson Pollock*, Putnam, New York, 1985.

[128] R. Taylor, A. Micolich, and D. Jonas, "Fractal analysis of Pollock's drip paintings," *Nature* **399** (1999), 422.

[129] R. Taylor, "Order in Pollock's chaos," *Sci. Am.* **287** (Dec. 2002), 116–121.

[130] R. Taylor, B. Newell, B. Spehar, C. Clifford, "Fractals: A resonance between art and nature," pgs. 53–63 of [131].

[131] M. Emmer, *Mathematics and Culture II. Visual Perfection: Mathematics and Creativity*, Springer, Berlin, 2005.

[132] K. Jones-Smith and H. Mathur, "Revisiting Pollock's drip paintings," *Nature* **444** (2006), E9–E10.

[133] R. Taylor, A. Micolich, and D. Jonas, "Taylor et al. reply," *Nature* **444** (2006), E10–E11.

[134] R. Taylor, R. Guzman, T. Martin, G. Hall, A. Micolich, D. Jonas, B. Scannell, M. Fairbanks, and C. Marlow, "Authenticating Pollock paintings using fractal geometry," *Pattern Recognition Lett.* **28** (2007), 695–702.

[135] R. Minkel, "Pollock or Not? Can Fractals Spot a Fake Masterpiece?" *Scientific American* (2007), https://www.scientificamerican.com/article/can-fractals-spot-genuine/

[136] K. Jones-Smith, H. Mathur, and L. Krauss, "Drip paintings and fractal analysis," *Phys. Rev. E* **79** (2009), 046111-1–12.

[137] R. Taylor, B. Spehar, P. van Donkelaar, and C. Hagerhall, "Perceptual and physiological responses to Jackson Pollock's fractals," *Frontiers in Human Neuroscience* **5** (2011), DOI=10.3389/fnhum.2011.00060.

[138] R. Taylor, "Personal reflections on Jackson Pollock's fractal paintings," *Historía, Ciências, Saúde-Manguinhos,* **13** (2006), supplement 0, Rio de Janeiro, 108–123.

4.6 Fractal folds

[139] D. Uribe, *Fractal Cuts: Exploring the Magic of Fractals with Pop-Up Designs*, Parkwest Publications, Miami, 1994.

[140] E. Simmt, B. Davis, "Fractal cards: a space for exploration in geometry and discrete mathematics," *Math. Teacher* **91** (1998), 102–108.

[141] E. Demaine, M. Demaine, and A. Lubiw, "Folding and one straight cut suffice," *Proc. Tenth Annual ACM-SIAM Symp. Discrete Algorithms* (1999), 891–892.

[142] M. Gardner, pgs. 58–69 of *New Mathematical Diversions from Scientific American*, Simon & Schuster, New York, 1966.

[143] J. O'Rourke, *How to Fold It: The Mathematics of Linkages, Origami, and Polyhedra*, Cambridge Univ. Pr. Cambridge, 2013.

[144] H. Houdini, *Houdini's Paper Magic; the Whole Art of Performing with Paper, Including Paper Tearing, Paper Folding and Paper Puzzles*, Dutton, New York, 1922.

[145] B. Gallican, "How to fold paper in half twelve times: an impossible challenge soved and explained," *Historical Society of Pomona Valley*, 2002.

4.7 A closer look at leaves

[146] L. Sack and C. scoffoni, "Leaf venation: structure, function, development, evolution, ecology, and applications in the past, present and future," *New Physiologist* **198** (2013), 983–1000.

[147] https://science.jrank.org/pages/3869/Leaf-Venation.html#ixzz6Yzpk5jHb

4.9 Cooking fractals

[148] D. Turcotte, "Fractals in geomogy and geophysics," pgs. 3822–3826 of [149], vol. 4.

[149] R. Meyers, *Encyclopedia of Complexity and Systems Science*, Springer, New York, 2009.

[150] A. Thomas, *The Vegetarian Epicure*, Random House, New York, 1972.

[151] I. Newton, *Memoirs of the Life Writings, and Discoveries of Sir Isaac Newton*, D. Brewster, ed., Edmonston and Douglas, Edinburgh, 1860.

5.1 Magnetic pendulum

[152] A. Cayley, "The Newton-Fourier imaginary problem," *Amer. J. Math.* **1** (1879), 97.

[153] K. Yoneyama, "Theory of continuous sets of points," *Tohoku Math./ J./* **11-12** (1917), 43–158.

5.2 Optical gasket

[154] D. Camp, "Reflecting on Wada basins: some fractals with a twist" pgs. 49–53 of [15].

5.3 video feedback fractals

[155] D. Hofstadter, *Gödel, Escher, Bach: An Eternal Golden Braid*, Basic Books, New York, 1979.

[156] S. Wolfram, "Universality and complexity in cellular automata," *Physica D* **10** (1984), 1–35.

[157] J. Crutchfield, "Space-time dynamics of video feedback," *Physica D* **10** (1984), 220–245.

[158] J. Courtial, J. Leach, and M. Padgett, "Fractals in pixellated video feedback," *Nature* **414** (2001), 864.

[159] J. Leach, M. Padgett and J. Courtial, "Fractals in pixellated video feedback," *Contemporary Physics* **44** (2003), 137–143.

[160] C. Epstein, W. Sendewicz, and D. Brown, "Structure of the path length set in asymmetric trees," *Fractals*, **13** (2005), 293 - 297.

[161] T. Taylor, "Homeomorphism classes of self-contacting symmetric binary fractal trees," *Fractals*, **15** (2007), 9–25.

[162] M. Frame and N. Neger, "Fractal video feedback as analog iterated function systems," *Fractals* **16** (2008), 275–285.

5.4 Electrodeposition

[163] R. Brady and R. Ball, "Fractal growth of copper electrodeposits," *Nature*, **309** (1984), 225–229.

[164] M. Matsushita, M. Sano, Y. Hayakawa, H. Honjo, and Y. Sawada, "Fractal structures of zinc metal leaves grown by electrodeposition," *Phys. Rev. Lett.* **53**, 286–289.

[165] B. Mandelbrot, *Fractals: Form, Chance, and Dimension*, Freeman, San Francisco, 1977.

[166] H. Fujikawa and M. Matsushita, "Fractal growth of *Bacillus subtilis* on agar plates," *J. Phys. Soc. Japan* **58** (1989), 3875–3878.

[167] M. Matsushita and H. Fujikawa, "Diffusion-limited growth in bacterial colony formation," *Physica A* **168** (1990), 498–506.

[168] T. Matsuyama, R. Harshey, and M. Matsushita, "Self-similar colony morphogenesis by bacteria as the experimental model of fractal growth by a cell population," *Fractals* **1** (1993), 302–311.

5.5 Viscous fingering

[169] J. Feder, *Fractals*, Plenum, New York, 1988.

[170] G. Homsy, "Viscous fingering in porous media," *Ann. Rev. Fluid Mech.* **19** (1987), 271–311.

[171] P. Saffman and G. Taylor, "The penetration of a fluid into a medium or Hele-Shaw cell containing a more viscous liquid," *Proc. Roy. Soc.* **245** (1958), 312–329.

[172] J.-D. Chen, "Growth of radial viscous fingers in a Hele-Shaw cell," *J. Fluid Mech.* **201** (1989), 223–242.

[173] J.-D. Chen, "Radial viscous fingering patterns in a Hele-Shaw cell," *Experiments in Fluids* **5** (1987), 363–71.

5.6 Crumpled paper patterns

[174] G. Zipf, *Human Behavior and the Principle of Least Effort*, Addison-Wesley, Cambridge, 1949.

[175] D. Blair and A. Kudrolli, "The geometry of crumpled paper," *Phys. Rev. Lett.* **94** (2005), 166107.

[176] O. Gottesman, J. Andrejević, C. Rycroft, and S. Rubinstein, "A state variable for crumpled thin sheets," *Commnications Physics* **1** (2018) DOI:10.1038/s42005-018-0072-x.

[177] J. Andrejević, L. Lee, S. Rubinstein, and C. Rycroft, "A model for the fragmentation kinetics of crumpled thin sheets," *Nature Communications* **12** (2021)
DOI:10.1038/s41467-021-21625-2.

[178] David Peak, personal communication. In addition to his own experiments with cabosil powder, including one experiment flown in the space shuttle to test compaction in microgravity environments, Peak mentioned that Mormons observed this logarithmic compaction effect when preparing the ground for their Salt Lake City temple. With a crane they repeatedly raised and dropped a cannon (two consecutive ns, so a piece of field artillery, not a clergyman) to the ground. They observed "diminishing returns," each additional drop compacted the soil less than the previous drop. The data fit a logarithmic model reasonably well.

[179] B. Daviss, "Dust demon: Physicist David Peak has big plans for small dust balls," *Discover* March, 1992.

[180] P. Meakin and B. Donn, "Aerodynamic properties of fractal grains: implications for the primordial solar nebula," *Astrophys. J.* **329** (1988), L39–L41.

5.7 Fractal resistor networks

[181] http://physics.bu.edu/py106/notes/Circuits.html

[182] W. Ching, M. Erockson, P. Garik, P. Hickman, J. Jordan, S. Schwarzer, and L. Shore, "Overcoming resistance with fractals: A new way to teach elementary circuits," *Pys. Teacher* **32** (1994), 546–551.

[183] K. Bahr, "Electrical anisotropy and conductivity distribution functions of fractal random networks and of the crust: The scale effect of connectivity," *Geophys. Int. J.* **130** (1997), 649–660.

5.8 Fractal networks of magnets

[184] N. Cohen, "Fractal switching systems and related electromechanical devices," U. S. patent application US 2018/0366251 A1.

5.9 Fractal network synchronization

[185] L. Minati, et al., "High-dimensional dynamics in a single-transistor oscillator containing Feynman-Sierpiński resonators: Effects of fractal depth and irregularity," *Chaos* **28** 093112 (2018).

6.2 Non-Euclidean tessellations

[186] M. Escher, *Escher on Escher*, Abrams, New York, 1989.

[187] C. Goodman-Strauss, "Compass and straightedge in the Poincaré disk," *Math. Assoc. Amer. Monthly* **108** (2001), 38–49.

6.4 Multifractal finance

[188] B. Mandelbrot, *Fractals and Scaling in Finance: Discontinuity, Concentration, Risk*, Springer, New York, 1997.

[189] B. Mandelbrot, "A multifractal walk down Wall street," *Sci. Am.* Feb. 1999, 50–53.

6.5 Fractal music

[190] R. Voss and J. Clarke, "$1/f$ noise in music and speech," *Nature* **258** (1975), 317–318.

[191] R. Voss and J. Clarke, " '$1/f$ noise' in music: music from $1/f$ noise," *J. Acoust. Soc. Am.* **63** (1978), 258–263.

[192] K. Hsü and A. Hsü, "Fractal geometry of music," *Proc. Nat. Acad. Sci. USA* **87** (1990), 938–941.

[193] K. Hsü and A. Hsü, "Self-similarity of the '$1/f$ noise' called music," *Proc. Nat. Acad. Sci. USA* **88** (1991), 3507–3509.

6.6.1 Lichtenberg figures

[194] L. Niemeyer, L. Pietronero, and H. Wiesmann, "Fractal dimension of dielectric breakdown," *Phys. Rev. Lett.* **52** (1984), 1033–1036.

6.6.3 Fractals in literature

[195] J. Saramago, *All the Names*, Mariner, New York, 2001.

[196] J. Saramago, *The Notebook*, Verso, New York, 2011.

[197] M. Chabon, *Telegraph Avenue*, Harper, New York, 2012.

[198] T. Stoppard, *Arcadia*, Farrar, Straus, and Giroux, New York, 1993.

[199] J. Yorke, *Into the Woods: How Stories Work and Why We Tell Them*, Penguin, New York, 2013.

6.6.6 Fractal dance

[200] L. Caruso-Haviland and P. Rapp, "Chaotic metamorphoses: a work in progress," pgs. 29–31 of [201].

[201] N. Abraham, A. Albano, A. Passamante, and P. Rapp, *Measures of Complexity and Chaos*Plenum, New York, 1989.

[202] K. Cetron, *Fractals: The Invisible WOrls of Fractals Made Visible through Theater and Dance*, Lucid Publishing, Atlanta, 2021.

B.17 Acknowledgments

[203] N. Lesmoir-Gordon, *Mandelbrot's World of Fractals*, Key Curriculum Press, New York, 2005.

Figure Credits

In Fig. 1 the first, second, and eighth images are from NASA, the third is a photo by Donna Laine, the fourth from a ClipArt file, fifth and sixth from GoogleMaps, and seventh is a photo by one author (MF) of a lung cast made by Dr. Robert Henry DMV. The image of Fig. 4.56 is from NASA. All line drawings and other photographs were prepared by us.

Acknowledgements

First we must thank Benoit Mandelbrot for his efforts to move scaling and self-similarity from occasional appearances in art and mathematics to an essential tool for analysis and modeling in natural and social sciences and in the humanities. For many years this was a lonely journey, but when people began to understand what Benoit had done, science experienced a tectonic shift. Many people began to see the world through the lens of fractals.

Personally, we both enjoyed, and learned a great deal from, our work with Benoit. He was a good friend. Aliette still is a good friend. We are very lucky to have known Benoit and to know Aliette.

These notes grew out of lab exercises we wrote for the teacher training workshops we ran with Benoit from 2000 through 2005. These workshops were supported by NSF grant DMS-0203203. The Yale Mathematics Department provided facilities, support, and encouragement.

The idea to use chaos and fractals as an introduction to quantitative thinking for non-science students was developed by David Peak, then in the Union College Physics Department. Dave and MF developed and taught a course, OCAM (Order and Chaos, Art and Magic) at Union. This course was the reason Benoit arranged for Yale to offer MF a position in its Math Department. That led to his undergraduate fractal geometry course at Yale, and that led to the summer workshops, the source of this book. So: thanks, Dave. Also, thanks for the enjoyable years when we wrote [14].

Many people assisted us with these labs. Engaged and curious students are the most important component of any classroom. Regardless of your own enthusiasm for the day's topic, the response of your students determines the success of the class. Bored students (or nowadays, students tapping out melodies on their cellphones—we are *so* lucky we retired before this plague spread) can turn a lesson about your favorite topic into a slog. On the other hand, to watch the light come on behind students' eyes when the idea fits into their heads, when they see how it works and why, elevates any topic to a day you'll remember fondly for years. We were very fortunate with our students—college, high school, and some middle school teachers—who populated our workshops. A great deal of the fun we had came from enthusiastic interactions with Kristen Amon, Cathy Aucoin, Kathleen Bavelas, the late Robert Bolger, Cecila Boys, Melkana Brakalova, Anita and Gwendolyn Bright, Alice Burstein, Barbara Camp, Maryanne Cavanaugh, Allen Cook, Emeka Dan-Udekwa, Katherine Demers, Gene Des Jarlais, Carol-Ann and Frank Dobek, Carrie Embleton, Michael Fisher, Janet Flament, Michael and Trisha Fraboni, the late Miguel Garcia, Larry Gould, Evelyn Grovesteen, Susan Hart, Kevin Hart, Casey Hawthorne, Anthony Hellmann, Bonnie Hole, Joe Iwanski, the late Alan Johnson, Virginia Jones, Charles Joscelyne, Annmarie Kennedy, Roger Loiseau, Patricia and Thomas McGrath, Bonita Messman, Trisha Moller, Sandra Naughton, the late Edward O'Neill, Mari O'Rourke, Barbara Paskov, Barbara Pfannkuch, Phyllis Pruzinsky, Dennis Riordan, the late Ann Robertson, the late Natalia Romalis, Dan Ryan, Gene Scheck, Marilyn Scheck, Svyatoslav Sharapov, Victor Sharapov, Alan Sherman, Geoffrey Smith, Stephen Smith, Rita Smith, Ralph Soden, Kathleen Stankewicz, Frances Stern, Ruth Sullo, Gail Tibbals, Gayle Town, James Vance, Janice Vulo, Charlie Waveris, Paul Welch, and William F. Widulski. Particularly helpful were the contributions of Harlan Brothers and of Richard Bedient.

Emeka Dan-Udekwa suggested bisecting the crumpled paper balls of Lab 2.2. In particular, his idea was the basis for Exercise 2.2.6, an important step in visualizing the complex folds-within-folds of crumpled paper. Thanks, Emeka.

Mary Laine shared hours of fascinating conversations about the structures of plants and pointed out the fractality of celosia blossoms. Donna Laine took photos of the sutures of a deer skull in her garden and shared interesting observations about patterns found in nature. Thanks, Mary and Donna.

Jane Cerilli introduced us to the work of Jovana Andrejević on quantifying the creases of crumpled paper. Jane's suggestion is the motivation of Lab 5.6, an important addition to our investigation complex patterns in physical objects. Thanks, Jane.

A conversation with Roger Howe suggested the ideas of Lab 1.8. Colleen Clancy introduced one of us (MF) to the construction of fractal folds of Lab 4.6. Many of our early experiments with decalcomania and with paint bleed were conducted with Brianna Murratti. Thanks to Patti and John Reid who introduced MF to the Monster paintings of their son Driscoll. These were the inspiration for the last image of Lab 4.1.

Part of the NSF grant funded the production of the dvd *Mandelbrot's World of Fractals* [203] by the accomplished filmmaker Nigel Lesmoir-Gordon. We had a great time working with Nigel on the script and with Nigel and his crew on filming the dvd.

Andy Szymkowiak was an excellent guide and tutor in Python. In fact, he convinced us to provide Python as as well as Mathematica programs to accompany the computer labs. In addition, he ran every Python program on his computer, an independent check that they work and do what we think they do. Any transgressions against the Python religion land squarely on the shoulders of MF.

Also, Andy provided invaluable guidance in electronics for Lab 5.9. MF's circuit theory course was in the fall semester of 1970. Andy's experience with oscilloscopes, transistors, resistors, and inductors is considerably more recent, and more detailed, than mine.

Finally, we thank our wives Jean Maatta and Liz Neger. Their support and patience through the workshop years and beyond was essential. Teachers' spouses are an integral part of teaching. Jean's many years of work in clinical laboratories made her a wonderful lab assistant for the experiments of Chapter 5. We hope Jean and Liz know how lucky we are.

Index